LATTICE DEFECTS IN SEMICONDUCTORS

LATTICE DEFECTS IN SEMICONDUCTORS

Edited by

Ryukiti R. Hasiguti

Professor of Applied Solid State Physics
Department of Metallurgy
University of Tokyo

UNIVERSITY OF TOKYO PRESS
Tokyo
THE PENNSYLVANIA STATE UNIVERSITY PRESS
University Park and London

UTP No. 67202
Printed in JAPAN

Published jointly by

UNIVERSITY OF TOKYO PRESS
Tokyo

and

THE PENNSYLVANIA STATE UNIVERSITY PRESS
University Park and London

Library of Congress Catalog Card Number 67-27364

PREFACE

Over the years, the study of lattice defects in semiconductor crystals has developed into a unique branch of lattice defects science, and many new developments have been introduced. For example, the chemical purity of covalent elemental semiconductors such as silicon and germanium is extremely high, and therefore the exact behavior of very low density lattice defects could effectively be examined; in addition, characteristic semiconductive properties have been advantageously used in lattice defects research.

Radiation damage is also an important problem in the study of lattice defects in semiconductor crystals, and in fact has been more intensively investigated than lattice defect properties in unirradiated crystals, as the number of papers dealing with this question in the present book will show.

The book is divided into several parts. Part I covers various fundamental properties and energies of defect states. Part II deals with the formation, migration and binding energies of various single and complex point defects, and with the annealing behavior relating to these point defect energies. Lattice defects in compound semiconductors are discussed in Part III. Part IV covers various aspects of radiation damage. The crystal surface, one of the particular problems in semiconductor materials, is discussed in Part V. And, finally, papers on dislocations and plastic behavior are presented in Part VI. All of the papers presented at the International Symposium on Lattice Defects in Semiconductors held on September 5th and 6th, 1966 at the Hotel New Otani in Tokyo have been collected in this volume.

RYUKITI R. HASIGUTI

CONTENTS

Part V. Surface

Part VI. Dislocations

Part 1

Defect States

Localized Modes due to Boron and Lithium in Silicon

M. Balkanski,* R. J. Elliott,† W. Nazarewicz‡ and P. Pfeuty*

* *Laboratorie de Physique de l'Ecole Normale Supérieure, Paris*

‡ *Clarendon Laboratory, Oxford, England*

and

† *Instutut Fizyki Doswiadczalnej Warszawa uf Hoza Nr. 69 Poliogne*

ABSTRACT

The position and the intensity of the absorption bands due to localized modes in boron doped lithium compensated silicon have been obtained at different temperatures from transmission measurements. Isotopic effects have been studied to determine the impurity origin of the different peaks. With isotopic impurities one line is mainly associated with lithium and the two others with boron vibrations. Thus we deduce from symmetry considerations that boron occupies a substitutional site and lithium an interstitial site. With this model for the pair defect B-Li and from the calculated eigenelements of the lattice dynamic in pure silicon the localized mode frequencies and absorption intensities have been claculated using a Green's function technique. The conclusions depend closely on the degeneracy of the different lines deduced from intensity measurements which could be confirmed by uniaxial stress experiments.

I. INTRODUCTION

In the last five years, the direct observation of localized modes in the infrared spectra,[1–4, 13] together with the detection of resonance modes from the temperature dependence of the thermal conductivity of alkali halides containing impurities, have contributed to a revival of interest in the study of the effects of lattice defects on vibrational properties of crystals.

Considerable recent refinement of the theory[5] and the use in calculations of realistic models for the host crystal and for the defect has led to a mutual stimulation between theory and experiments, giving a good detailed understanding of the phenomena.

It is now possible, using the experimental determination of the localized mode frequencies, to undertake exact calculations involving the force constant variations and the symmetry properties of the defect. The physical models used give a detailed description of the nature of the defect and its precise situation in the lattice. It is then possible to use the lattice dynamical properties to study the detailed structure of defects in crystals.

For a point defect with only changed mass there are two main effects depending on the mass of the impurity. For heavy impurities with mass M' larger than the host lattice atom mass M, the frequencies of the normal modes are slightly shifted to lower frequencies, but the shift is smaller than the distance to the next normal mode frequency, and it is therefore negligible for all practical purposes. The heavy impurity vibrations propagate through the lattice by the normal modes but the amplitude near the defect may have considerable enhancement by a resonance in a particular part of the spectrum.

The light impurity with $M'<M$ induces a negligible shift in the normal mode frequencies toward the high frequencies and ejects one frequency out of the range of allowed normal mode frequencies. This so-called localized mode corresponds to a sharp resonance band with spatial attenuation of the vibrational amplitude which falls off exponentially with distance from the defect.

If the mass difference $M'-M$ is small it is possible to observe by infrared spectroscopy the singularities of the dispersion curve practically unchanged by the presence of the defects.[2] Infrared spectroscopy has thus been used for the determination of the normal mode frequencies, when activated by the presence of defects in homopolar crystals for which the first order photon-phonon interaction is normally forbidden.

Experimental information concerning the localized vibrations may be obtained from the study of neutron scattering,[7] Mossbauer effect[8] and infrared absorption.[1–3]

When studied by infrared spectroscopy the boron doped lithium compensated silicon shows eight absorption bands[1] for impurities in natural abundance. This complicated spectrum has to be studied on the basis of a boron lithium complex which constitutes a sort of molecular impurity in the lattice. The study of the effects of such defects on the vibrations of crystal lattices[5] shows that the number and the degeneracy of the observed localized absorption peaks depends essentially on the symmetry of the defect.

Introduced in the silicon lattice, which has the diamond structure with the space group O_h,[7] lithium is supposed from experimental and theoretical considerations, to occupy an interstitial site[9,12] and boron a substitutional site.[2]

The substitutional boron when isolated in the perfect silicon lattice forms covalent bands with the four nearest neighbors at the corners of a tetrahedron. The local site symmetry is given by the point group T_d. Selection rules for optical absorption indicate an allowed dipole transition from the ground state with the representation Γ_1 to the triply degenerate excited state with representation Γ_{15}. Only one absorption band should be observed in the infrared spectrum for the local mode due to isolated substitutional boron[2] as expected by the calculations developed by Dawber and Elliott.[6]

Two interstitial positions in the diamond lattice are available:

a) One tetrahedral site with four nearest neighbors at the distance 0.433 a_0 and six next nearest neighbors at the distance 0.5 a_0, a_0 being the unit cell edge length. Such interstitials occupy the five positions along the body diagonals, i.e. the $\langle 111 \rangle$ direction in the unit cell as shown in Fig. 4.

b) One hexagonal site with six nearest neighbors at the distance 0.415 a_0 and eight next nearest neighbors at the distance 0.649 a_0. Such sites occur at the saddle points separating the two tetrahedral sites.

Recent investigations[9] of the electronic spectrum of lithium in silicon under uniaxial stress indicates that the site symmetry of isolated interstitial lithium is tetrahedral and has the T_d point group site symmetry. Isolated lithium-like boron will then give rise to only one absorption band due to its localized mode of vibration.

The simultaneous presence of lithium and boron in silicon leads to pair formation.[10] Lithium in an interstitial position approaches a negatively charged B^- ion, and since neither boron nor silicon have electrons available for the formation of covalent bands with lithium present in the lattice in the form of Li^+ ion, the interaction of boron with lithium may have a Coulomb character. The interaction of lithium with the host silicon atoms arises from ion induced dipole attractive forces and the repulsive forces between the electron cores.

In this paper the whole infrared absorption spectrum is interpreted on the basis of this model for the lithium-boron complex. A specific calculation of the position and the intensity of the lines using variable effective force constants for the defect and the normal mode spectrum of silicon calculated by Johnson was developed for the different pairs of isotopic impurities and compared with the experimental results. A detailed account of the theory will be given elsewhere.[11]

II. THEORY

1. *Method of derivation of the localized modes frequencies and the corresponding absorption coeffients*

The theory of the localized mode spectrum of different impurities in silicon has been worked out in detail by Dawber and Elliott[6] assuming a substitutional defect with isotopic mass changes but maintaining the force constant parameters of the host lattice unchanged. The formation of large complex defects leads to considerable variation of the bonding forces in the immediate vicinity of the defect pairs. These variations of the force constants have been explicitly taken into account in calculating the localized mode frequencies and absorption spectra of the different defect pairs formed by the isotopes of boron and lithium.

The dynamical behavior of the perturbed crystal is described by the equations of motion:

$$M_{\alpha\nu}(l)\frac{d^2}{dt^2}u_{\alpha\nu}(l,t)+\sum_{\beta\nu' l'}[\Phi_{\alpha\nu,\beta\nu'}(l,l')-C_{\alpha\nu,\beta\nu'}(l,l')]u_{\beta\nu'}(l',t)=0 \qquad (1)$$

in terms of the displacements $u_{\alpha\nu}(l, t)$ of a given atom at time t in the unit cell specified by 1 centered at $\mathbf{x}(l)$; α defines the three cartesian coordinates and ν the atoms in the cell. $M_{\alpha\nu}(l)$ is the atomic mass and $\Phi_{\alpha\nu,\beta\nu'}(l, l')$ the elements of the force constant matrix of the perfect crystal. $C_{\alpha\nu,\beta\nu'}(l, l')$ is the change in the force constant due to the defect. In a perfect crystal the translation symmetry gives normal modes with the wave-vector $\mathbf{q}$ taking all values in the first Brillouin zone. $\sigma_{\alpha\nu}{}^{j'}(\mathbf{q})$ and $\omega_j^2(\mathbf{q})$ are the eigenvectors and eigenvalues of the dynamical matrix with elements

$$\Psi_{\alpha\nu,\beta\nu'}=\sum_{l}\frac{\Phi_{\alpha\nu,\beta\nu'}(l,l')\exp\{i\mathbf{q}[\mathbf{x}(l)-\mathbf{x}(l')]\}}{M_{\alpha\nu 1}{}^{1/2}M_{\beta\nu' l'}{}^{1/2}}$$

The frequencies of the modes are given by

$$-M_{\alpha\nu}(l)\omega^2u_{\alpha\nu}(l)+\sum_{\beta\nu' l'}[\Phi_{\alpha\nu,\beta\nu'}(l,l')-C_{\alpha\nu,\beta\nu'}(l,l')]u_{\beta\nu'}(l')=0 \qquad (2)$$

which can be put into the matrix form

$$(-M\omega^2I+\Phi-C)U=0 \qquad (3)$$

where I is a unit matrix.

C is a finite perturbation matrix with the dimension of the defect subspace; we neglect any long range perturbation due to the introduction of the defect in the lattice. The inverse matrix known as the non-perturbed lattice Green's function matrix $g=(-M\omega^2I+\Phi)^{-1}$. Then Eq. (3) becomes

$$(I-gC)U=0 \tag{4}$$

where g is defined using the eigenelements of the dynamical matrix

$$g_{\alpha\beta}(\nu l, \nu' l', \omega^2) = \sum_{j\mathbf{q}} \frac{\sigma_{\alpha\nu}{}^{j*}(\mathbf{q})\,\sigma_{\beta\nu'}{}^{j}(\mathbf{q}) \exp\{i\mathbf{q}[\mathbf{x}(l)-\mathbf{x}(l')]\}}{M_{\alpha\nu l}{}^{1/2} M_{\beta\nu' l'}{}^{1/2}}\, g_{jj',\mathbf{q}\mathbf{q}'} \tag{5}$$

where

$$g_{jj',\mathbf{q}\mathbf{q}'} = \frac{\delta_{jj'}\,\delta(\mathbf{q}-\mathbf{q}')}{\omega_j{}^2(\mathbf{q})-\omega^2}$$

We have only to consider the g matrix related to the defect subspace. If n in the dimension of this subspace the matrices g and C are $n\times n$ matrices and the secular equation

$$\mathrm{Det}\,|I-gC| = 0 \tag{6}$$

gives the frequency of the localized modes.

Symmetry considerations can be used to solve Eq. (6) and find the number, the degeneracy and the frequency of the different localized modes associated with the defect. We suppose that the symmetry at the defect site belongs to a point group G. The displacement vectors of the defect atoms transform as a representation R which can be reduced according to the symmetry point group G into a certain number n_s of irreducible representations s of degeneracy γ_s. The matrices g, C, and gC remain unchanged by the operations of the group G.

Using correct symmetry coordinates, the matrix gC is reduced into block matrices corresponding to the irreducible representations. The dimension of the s block matrix is equal to n_s. Equation (6) can now be written:

$$\prod_s \mathrm{Det}\,[I-(gC)_s] = 0 \tag{7}$$

which gives the solutions ω_s corresponding to localized modes which transform like the representations s.

The optical absorption depends on the electric susceptibility which is related [Kubo (1958)] to the Green's function of the perturbed crystal. The absorption coefficient $\alpha=\sum_s \alpha_s$ where α_s is the absorption coefficient

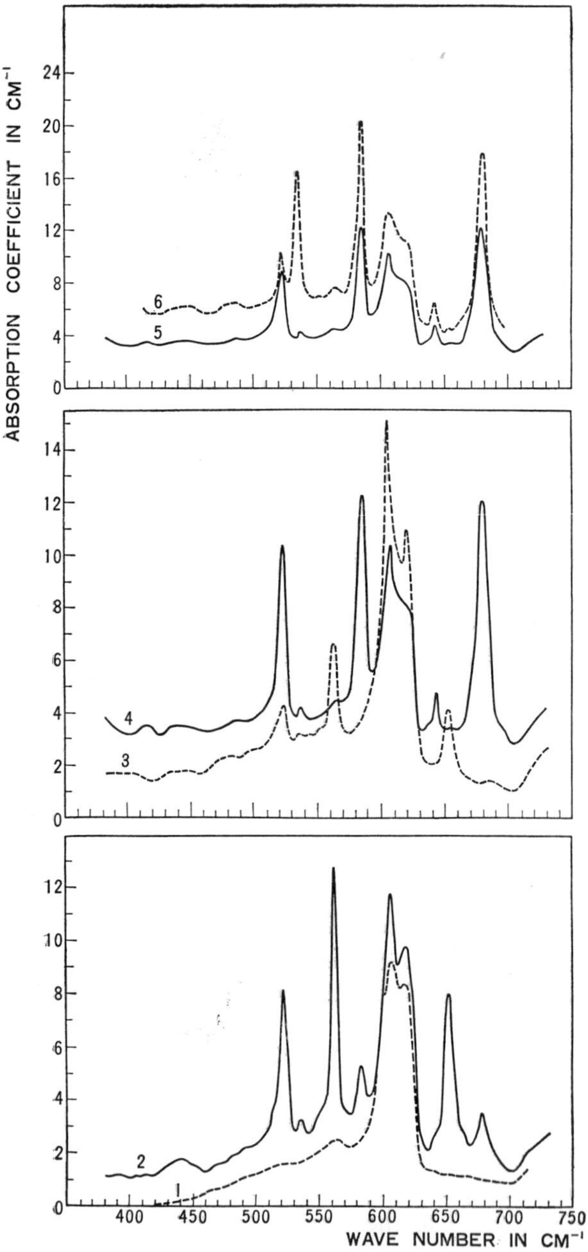
ABSORPTION COEFFICIENT IN CM^{-1}
24
20
16
12
8
4
0
6
5
14
12
10
8
6
4
2
0
4
3
12
10
8
6
4
2
0
2
1
400
450
500
550
600
650
700
750
WAVE NUMBER IN CM^{-1}

Fig. 1.

related to the defect modes with the representation *s*,

$$\alpha_s = \frac{\gamma_s N}{3} \frac{4\pi\lambda\omega}{ncV} \mathrm{Im} \sum G_s(\omega+i\varphi), \qquad \varphi\to 0 \tag{8}$$

where λ is the local field correction,
n the refractive index,
c the light velocity,
N the number of defects,
and V the volume.

$\sum G_s(\omega)$ is a linear combination of Green's functions of the perturbed lattice according to the position of the effective charges in the defect,

$$G_s = [I-(gC)_s]^{-1} g_s \tag{9}$$

2. *Calculation of the frequencies and the absorption intensity due to the localized modes of the boron-lithium complex*

a) *Symmetry considerations*

In the defect as shown in Fig. 5, the boron occupies a silicon site in the cubic diamond lattice and is surrounded by four first neighbor silicon atoms at the corners of a tetrahedron and by six second neighbors. We will only consider three of these which together with boron form a tetrahedron with lithium at the center. We consider first a simpler model consisting only of this latter tetrahedron of five atoms.

The defect subspace has the dimension $n=15$ and belongs to the point group C_{3V}. In this group the representation R has the following decomposition

$$R = 4\Gamma_1^{(1)} + \Gamma_2^{(2)} + 5\Gamma_3^{(2)}$$

(The notation of Koster *et al.*[16] will be used throughout.)

Only modes corresponding to the representations Γ_1 and Γ_3 are infrared active. They are associated with vibrations along and perpendicular to the pair axis respectively.

Fig. 1. Isotopic effects. Infrared spectrum of boron doped lithium compensated silicon at room temperature. Different curves correspond to the following isotopic composition of impurities.

Curve 1: Pure silicon.
Curve 2: B and Li in natural isotopic abundances.
Curve 3: ^{11}B and natural Li.
Curves 4 and 5: ^{10}B and natural Li.
Curve 6: ^{10}B and ^{6}Li.

b) *Green's function and perturbation matrices*

1° *The Green's function matrix.* The Green's function matrix g is a finite matrix 15×15 whose elements depend on the eigenelements of the dynamical matrix of the perfect crystal. By symmetry considerations the number of different elements is reduced to four:

$$g_1(\omega) = g_{xx}(00, 00, \omega^2)$$
$$g_4(\omega) = g_{xx}(00, 01, \omega^2)$$
$$g_5(\omega) = g_{zz}(00, 01, \omega^2)$$
$$g_6(\omega) = g_{xy}(00, 01, \omega^2)$$

where

$$\mathbf{x}(l) = \mathbf{i}+\mathbf{j}$$
$$g(\omega+i\varphi)_{\varphi\to 0} = g(\omega)+i\nu(\omega)$$

The real part $g(\omega)$ and the imaginary part $\nu(\omega)$ of these four functions are plotted in Figs. 6 and 7.

This matrix is reduced into four block matrices: a single (1×1) corresponding to Γ_1, a (4×4) from Γ_2 and two identical (5×5) matrices corresponding to the Γ_3 representation.

2° *The perturbation matrix.* The perturbation matrix depends on the changes of force constants due to the introduction of the impurities in the lattice and on the mass of the impurities.

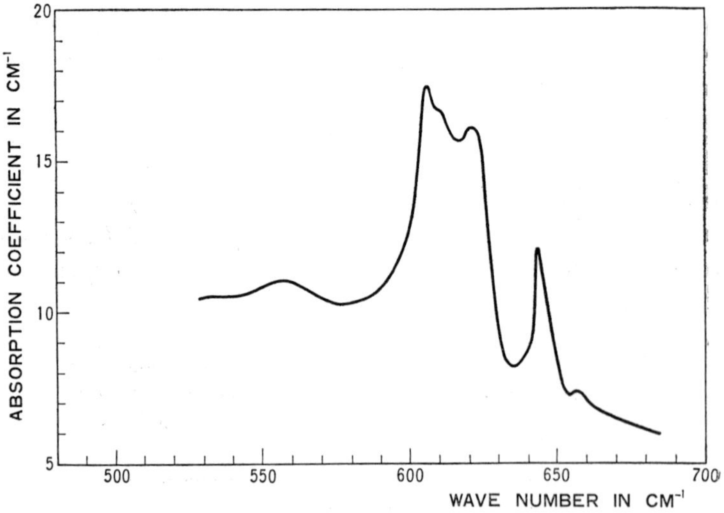

Fig. 2. Absorption coefficient of silicon crystal doped with enriched boron (^{10}B 92%) and natural antimony at room temperature.

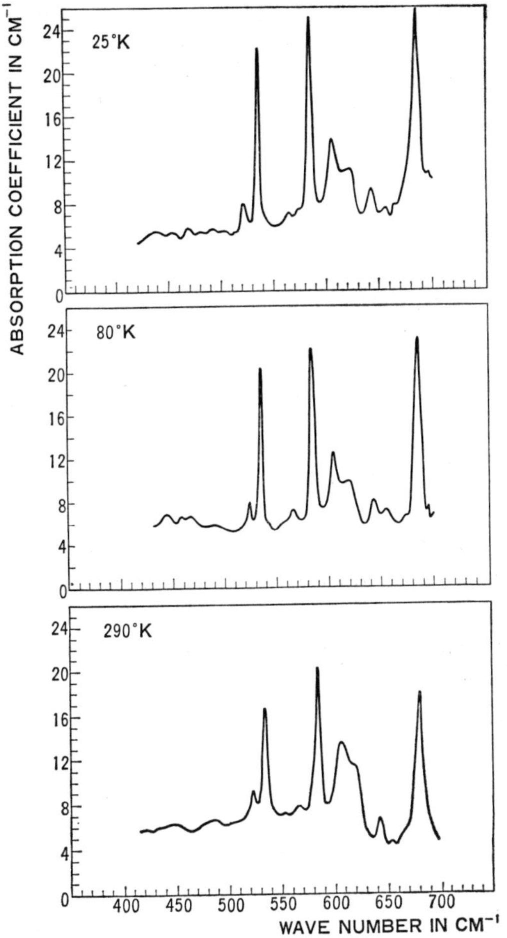

Fig. 3. Temperature effects. Infrared spectrum of silicon crystal doped with ^{10}B and ^{6}Li at three different temperatures 290°K, 80°K, and 25°K.

We use actually the following parameters:

$\nu = M_i/M$ where M_i is the interstitial mass,

$\varepsilon = (M - M_s)/M$ where M_s is the substitutional atom mass,

β represents the isotropic central force constant between the interstitial and the surrounding atoms,

α represents the central change of force constant along the pair axis between boron and lithium.

The perturbation matrix is (15×15) and can be reduced in the same way as g. We are only concerned with the (5×5) matrix C_{Γ_3} and the (4×4) matrix C_{Γ_1}. These two matrices are shown in Appendix. We obtain easily the two matrices $(gC)_{\Gamma_1}$ and $(gC)_{\Gamma_3}$ and we solve separately the two Eqs. Det $|I-(gC)_{\Gamma_1}|=0$ and Det $[I-(gC)_{\Gamma_3}]=0$ whose solutions give the frequencies of the localized modes. The intensities of the local modes are calculated in the way presented in the Eqs. (8) and (9).

III. EXPERIMENTAL RESULTS

1. *Results*

The absorption spectrum has been obtained from transmission measurements at room temperature, liquid nitrogen, and liquid hydrogen temperatures, between 500 and 700 cm^{-1}. Absorption spectra for samples containing different combinations of boron and lithium isotopes with concentrations of about 10^{-19} atoms per cm^3 are shown in Fig. 1. The different observed peaks are listed in Table I. Table III gives the integrated intensities of different localized absorption peaks for two different

TABLE I. List of frequencies of prominent absorption bands arising from changes in isotopic composition.

Isotopic composition of impurities	Observed peaks (cm^{-1})							
B+Li	681	653	644	620	584	564	534	522
$^{10}B+^{7}Li$	681	—	644	—	584	—	—	522
$^{10}B+^{6}Li$	681	—	644	—	584	—	534	—
$^{11}B+^{7}Li$	—	653	—	620	—	564	—	522
$^{11}B+^{6}Li$	—	653	—	620	—	564	534	—

TABLE II. Summary of the interpretation of absorption bands in silicon doped with boron and lithium.

Position of peaks (cm^{-1})		Isotopic attribution	Local site symmetry	Degeneracy
1	681	^{10}B	C_{3V}	2
2	653	^{11}B	C_{3V}	2
3	644	^{10}B	T_D	3
4	620	^{11}B	T_D	3
5	584	^{10}B	C_{3V}	1
6	654	^{11}B	C_{3V}	1
7	534	^{6}Li	C_{3V} or T_d	2 or 3
8	522	^{7}Li	C_{3V} or T_d	2 or 3

TABLE III. Intensity of absorption peaks.

Isotopic composition of impurities	Peak (cm^{-1})	Integrated intensity		Relative intensity (584 & 681 cm^{-1} peaks)	
		300°K	80°K	300°K	80°K
B+Li	522	33	45		
	584	40	44	1	1
	681	77	75	1.9	1.7
^{10}B+Li	522	65	53		
	584	67	72	1	1
	681	113	139	1.7	1.9

TABLE IV. Effect of temperature on the position and width of the strongest peaks in the localized mode spectrum.

Peak position (cm^{-1})			Line width			Relative line width in units of 10^{-2}		
290°K	80°K	25°K	290°K	80°K	25°K	290°K	80°K	25°K
522	525	525	8	7	5	1.53	1.33	0.95
535	537	537	6	4	4	1.12	0.75	0.75
564	566	566	7	5	5	1.24	0.88	0.88
584	587	587	7	6	6	1.20	1.02	1.02
653	656	656	10	9	9	1.53	1.37	1.37
681	683	683	10	9	9	1.47	1.32	1.32

samples. The absorption spectra at low temperature are given in Fig. 3. The positions of the strongest peaks for the different temperatures and the corresponding relative linewidths are listed in Table IV.

2. *Discussion*

a) *Position and intensity of the lines*

On the basis of the previous measurements, the impurity and isotopic origin is obtained for all observed peaks and the interpretation is summarized in Table II. The two peaks at 620 cm^{-1} and 644 cm^{-1} correspond to the two boron isotopes vibrating alone; the same lines have been observed with boron-antimony as shown in Fig. 2, or boron-phosphorus doped silicon samples.[2] Isolated borons have tetrahedral symmetry and only one threefold degenerate local mode corresponds to this defect.

The spectrum associated with the pair defects consists of two pairs of peaks associated respectively with ^{10}B and ^{11}B and two single peaks associated with ^{6}Li and ^{7}Li. If the lithium is alone one threefold degenerate local mode peak is expected.

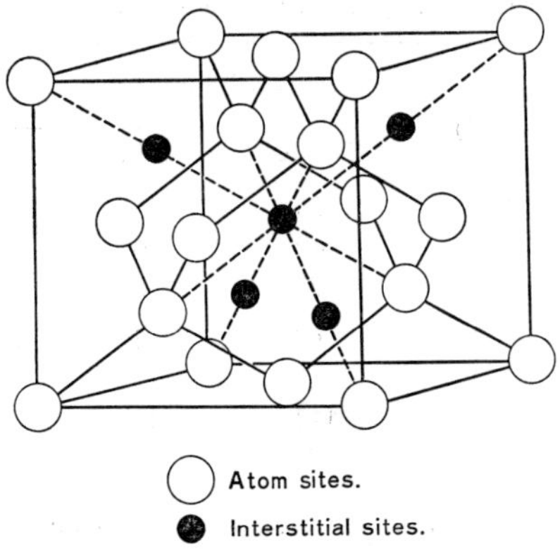

Fig. 4. Diamond type lattice showing the five tetrahedral interstitial positions entirely within one cubic unit cell. The relative coordinates of these interstitial positions are: 1/2, 1/2, 1/2; 1/4, 1/4, 3/4; 1/4, 3/4, 1/4; 3/3, 1/4, 1/4; and 3/4, 3/4, 3/4.

By putting the lithium impurity near the boron, the symmetry of the defect site is changed and the previous two triplets associated with lithium alone and boron alone are split, the number of lines depending on the symmetry of the new defect site. We know that the boron occupies a substitutional site.

Lithium can occupy three different places:

1° a substitutional site in the same unit cell as boron,
2° a tetrahedral interstitial site,
3° an hexagonal interstitial site.

The tetrahedral interstitial sites are shown in Fig. 4. We shall discuss successively the three possible configurations.

1° It is difficult to believe that lithium is substitutional because of its small size and lack of bonding electrons. For a substitutional lithium in the immediate vicinity of boron the two triplets are split into two doublets and two singlets.[11] Because of the light mass the lithium lines would be higher in frequency than the boron lines, which is contrary to the experimental results as shown in Table IV, unless the force constants of the surrounding bonds were drastically reduced.

2° In this case the defect subspace belongs to the point group C_{3v}; two lines associated with boron and two lines associated with lithium are expected. Experimentally we obtain one lithium line, but it is possible that one localized mode has not appeared out of the band because the force constants are too weak.

3° In this case, the defect subspace belongs to the point group C_{2v} and in the boron-lithium complex, three lines should appear for lithium and for each of the boron isotopes. This prediction is in contradiction with the experimental observed spectrum.

From these considerations we deduce that lithium is found in a tetrahedral interstitial site and in all the calculations and discussions this model has been used.

Let us now analyze the previous results more carefully and consider a ^{11}B-^{7}Li pair. Three lines at 522, 564 and 653 cm^{-1} are observed. The isotopic effects have shown that the 522 cm^{-1} line is mainly associated with lithium and the two other with boron vibrations. In order to determine the frequency of the modes associated respectively with vibrations along and perpendicular to the pair axis, one has to consider the fact that the first one is non-degenerate and the second is twofold degenerate. The measurement of the integrated intensity of the absorption peaks allowed the determination of the respective degeneracy. This is correct if we suppose that the two impurities are vibrating independently. In reality the observed ratio of the intensities as shown in Table III is not equal to the ratio of the degeneracies. The two impurities cannot be considered as vibrating strictly independently. In the calculations the vibration amplitudes of the two impurities for the different local modes have been found.

b) *Residual line width and temperature effects*

The main contribution to the residual line width when $T \to 0$ arises from spontaneous decay of the excited localized modes into several band phonons. The energy conservation requirement $\sum \omega_k = \omega$ must be obeyed in this process. In the case of localized modes in silicon due to boron, $\omega < 2\omega_m$. Hence decay into two band modes through the third order anharmonic potential is possible and will be the dominant process at normal temperatures.[13,14] The estimate of this broadening for localized modes of boron in silicon gives the value $\Delta\omega/\omega \sim 10^{-2}$ in good agreement with experimental results as shown in Table IV. The experimental results summarized in Table IV show a weak temperature dependence of position and width of localized mode peaks. These effects are also connected with the anharmonic interaction between the local vibrations and other vibrational modes of the crystal.

IV. COMPARISON BETWEEN THE EXPERIMENTS AND THE CALCULATIONS

To better understand the pairing effect on the localized mode spectrum of each impurity we shall first mention the results for boron alone and for lithium alone. If boron is isolated in the crystal as a substitutional impurity one localized triplet is predicted[6] and has been observed.[2] If lithium is isolated in the crystal as an interstitial impurity one localizedt triplet is predicted.[11] In Fig. 8 $[(\omega/\omega_M)-1]$ is plotted versus $\varepsilon=1-\nu$ for different values of β, where ω_M is the limit frequency of the normal mode spectrum (the notation was introduced in II-2-b-2). The frequencies of the absorption bands corresponding to lithium observed in the boron-lithium localized spectrum can be calculated from such a model by taking a force constant

TABLE V. Comparison between the observed and calculated values.

Observed values			Calculated values		
Isotopic attribution	Frequency (cm^{-1})	Degeneracy	Li isolated $\beta = 22$ kergs/cm^2	B-Li 9 atoms model $\beta = 23$ kergs/cm^2 $\alpha = -60$ kergs/cm^2	B-Li 9 atoms model $\beta = 23$ kergs/cm^2 $\alpha+\varphi = -58$ kergs/cm^2 $\gamma = -7$ kergs/cm^2
^{10}B	681	2		707	678
^{11}B	653	2		679 (^7Li)	651 (^7Li)
				683 (^6Li)	655 (^6Li)
^{10}B	584	1		609	584
^{11}B	564	1		590	565.5
^6Li	534	2 or 3	533.5	533	533
^7Li	522	2 or 3	522	522	522

TABLE VI. Integrated intensities (cm^{-2}).

Composition of the pair	Concentration	Peaks (cm^{-1})	Calculated	Observed
		522+534	105	50
B+Li	5×10^{18}	564+584	56	68
		653+681	72	99
		522+534	168	65
^{10}B+Li	8×10^{18}	584	85	67
		681	131	113

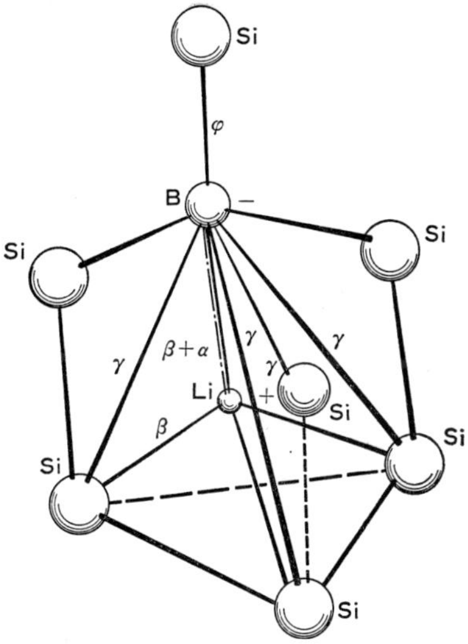

Fig. 5. Configuration of the B-Li defect with the changes in different force constant parameters used in the different models.

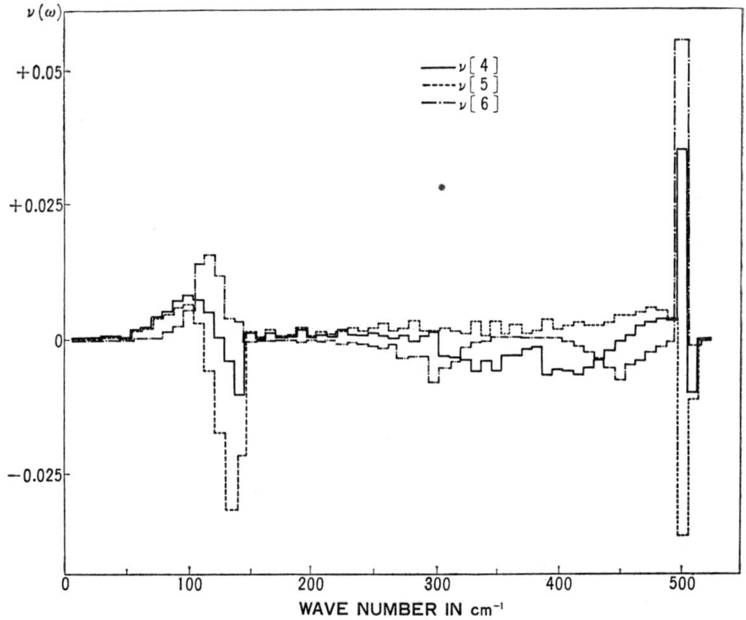

Fig. 6. Imaginary part $\nu(\omega)$ of three of the Green's functions g_4, g_5, and $g_6(\omega)$.

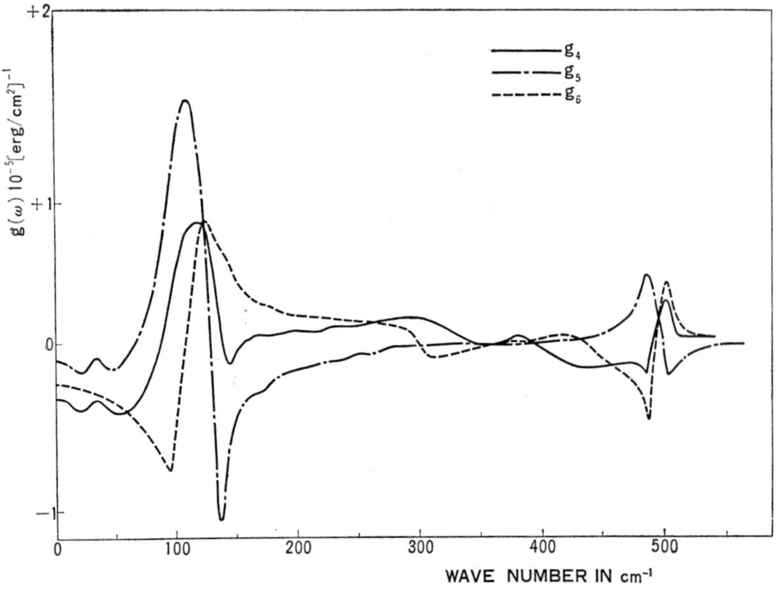

Fig. 7. Real part $g(\omega)$ of the functions g_4, g_5 and g_6.

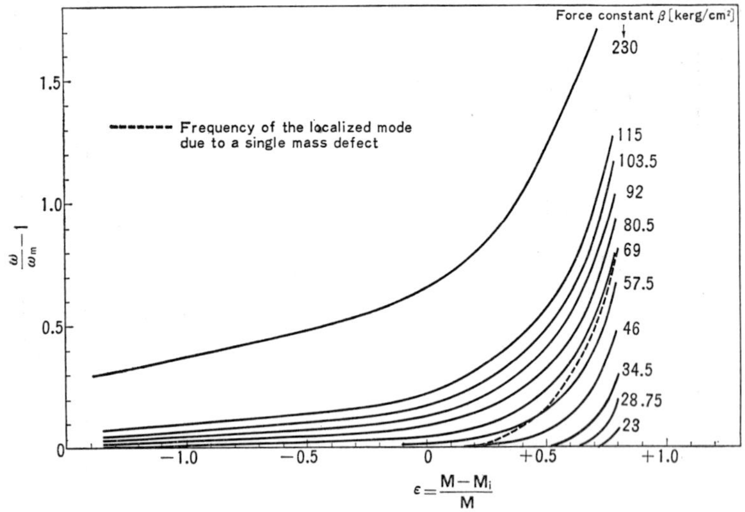

Fig. 8. Frequency of the localized mode due to an interstitial impurity in a tetrahedral site in silicon as a function of $\varepsilon=(M-M_i)/M$ for different values of the coupling force constant F.

$\beta=22$ kergs/cm^2 which is about half of the force constant for the nearest neighbor pure silicon ($\beta_{Si}=55$ kergs/cm^2).[15] The two isotopes ^{6}Li and ^{7}Li give rise to two separate absorption bands, the corresponding isotopic shift is correctly obtained as shown in Table V. If independent harmonic oscillators are considered for the two lithium isotopes vibrating independently of the lattice one obtains an isotopic shift of 42 cm^{-1} which is far from the experimental value 12 cm^{-1} and that calculated from the present model 11.5 cm^{-1}. The reduction of the isotopic shift is due to the fact that the localized mode frequency is close to the limit of the observed normal mode frequencies, and the mode extends in space.

For the boron-lithium pair calculations using a reduced model containing the lithium and the four surrounding atoms as shown in Fig. 5 give the following results.

As in the previous model the absorption lines corresponding to the different isotopes remain triply degenerate if the force constant β remains unchanged and is the same to both Li-Si and Li-B. The frequencies corresponding to the boron and the lithium triplets move as the force constant F is changed as shown in Fig. 9. Without lithium in the lattice or for a force constant equal to zero one obtains the two boron isotope lines at 654 cm^{-1} and 679 cm^{-1}. Introduction of lithium in the model with a force constant above 20 kergs/cm^2 leads to the appearance of lithium local-

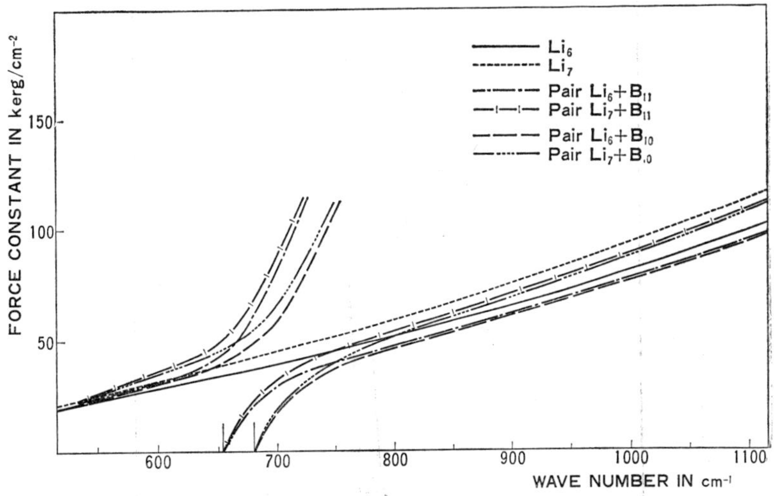

Fig. 9. Calculated frequencies of the localized triplets due to B-Li pairs when we only consider the force constant β coupling Li with the surrounding atoms.

ized mode frequencies above 518 cm^{-1}. This simplified model does not give the correct frequencies for the boron localized lines in the presence of lithium.

In reality the coupling of lithium with the surrounding atoms is anisotropic and we take this into account by changing the force constant along the pair axis by α. This may be regarded as a consequence of the ionic interaction between the ionized impurities. In the isotropic model with $\alpha=0$ the localized modes belonged to the representations Γ_1 and Γ_3 but the two matrices $(gC)_{\Gamma_1}$ and $(gC)_{\Gamma_3}$ led to the same values giving an accidental degeneracy into triplets.

Introducing the change of force constant α along the pair axis B-Li leads to solutions for the Γ_3 representation which remains identical to the previous ones but the solutions for Γ_1 depend on the value of α. The degeneracy is therefore lifted and the frequency corresponding to the Γ_1 representation is shifted.

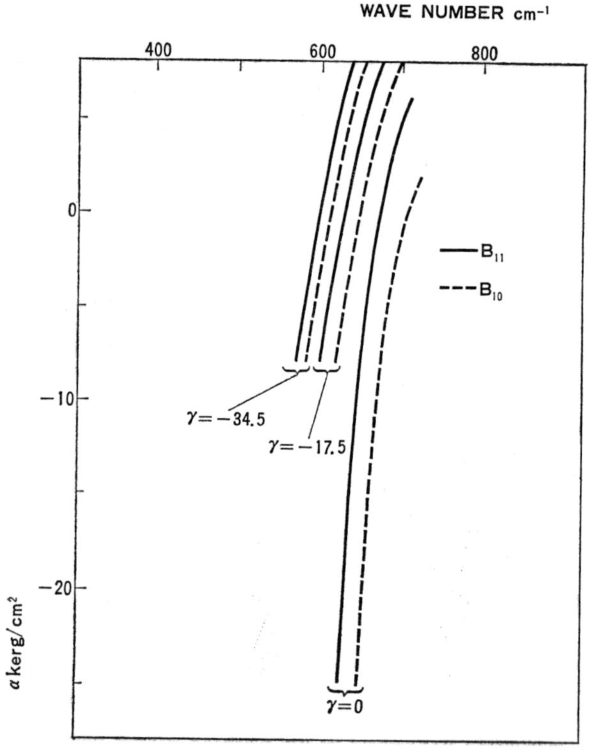

Fig. 10. Effect of the force constant parameter α on the Γ_1 local made frequency.

If we admit that the integrated absorption for a doubly degenerate state is bigger than for a single state, the experimental results indicate a peak corresponding to Γ_3 at higher frequency than that corresponding to Γ_1. This would then mean that the vibrational energy for the mode perpendicular to the B-Li axis is higher than for the mode along this axis. Considering that boron and lithium are ionized one might have expected the contrary. To fit the experimental frequencies of the two modes one calculates a force constant parameter α of the order of -60 kergs/cm^2. This gives a negative total force constant for the B-Li bond and is clearly unsatisfactory. However we shall pursue the details of this model and consider its justification later.

The variation of the Γ_1 boron frequency with α is shown in Fig. 10 with β fixed at 23 kergs/cm^2. When α is negative the lithium localized line disappears rapidly. The frequencies vary only slowly with α; therefore the values obtained would depend strongly on the precision with which the localized mode frequency is determined. Experimentally we see (Table I) that the change of mass of lithium has no effect on the lines associated with boron and the change of boron has no effect on the lines associated with lithium. The calculations (Fig. 9) do give lithium lines independent of the boron mass, but there is some variation in the boron lines with the lithium mass. The results are summarized in Table V.

The integrated intensities of the different localized peaks have been calculated in this model from the Eqs. (8) and (9). We supposed that a positive effective charge was centered on the lithium atom and a negative charge on the boron atom. Calculated and observed values are listed in Table VI. We deduce from the expression of the Green's functions used in the calculations that the two impurities are not vibrating independently to give rise to the different localized absorption peaks. Because of this calculated intensities of the boron lines are not in the ratio of the respective degeneracies. From the comparison between the calculated and experimental results, we find the best fit if the effective charge on boron is nearly equal to -1 and the effective charge on lithium to $+0.7$.

It is clear that a proper description of the defect should also include a modification of the force constants between boron and neighboring silicon atoms. As a first step we have introduced a parameter φ representing the change in the coupling force constants between boron and silicon atoms along the Li-B pair axis (the 111 axis). The effect of this force constant on the localized mode frequencies can be extimated using a two substitutional pair defect model[11] and we find that φ has the same effect as α: only the Γ_1 frequency is modified and by a similar amount.

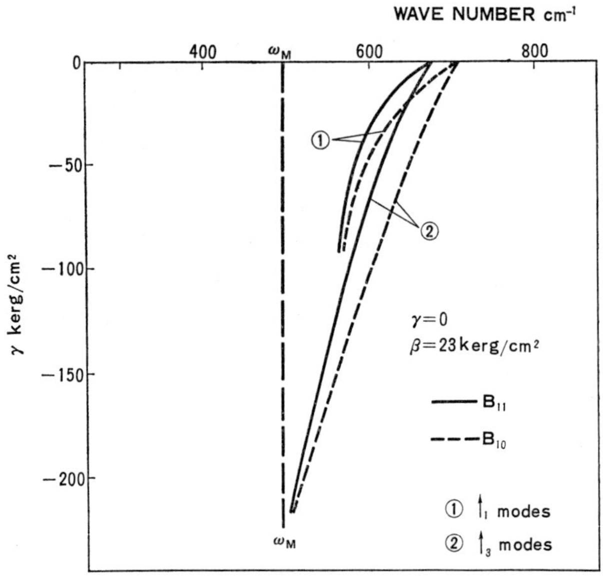

Fig. 11. Effect of the change of force constant γ on the Γ_1 and Γ_3 local mode frequencies.

Along the other directions the change of force constants with both the first and second neighbor is represented by another force constant parameter γ. The effect of this parameter γ on the localized mode frequencies occurs for both Γ_1 and Γ_3 modes. The localized mode frequencies are shifted to low frequencies when the change of force constant γ is negative as shown in Fig. 11. The comparison between the experimental and calculated values are summarized in Table V for the different models. Suitable agreement can now be obtained without making the B-Li force constant negative, if we assume appreciable weakening of the B-Si n.n. force constant on the opposite side of the B to the Li. If this is the case it suggests a strong modification of the electron bonds by the introduction of Li, and in particular a weakening of the B-Si bond opposite the Li, possibly by partly removing one of its electrons.

It can be clearly seen that this interpretation comes from the fact that the experimental results indicate that the doubly degenerate mode is the high frequency one. Before pursuing the model further some verification of this degeneracy is still needed. In order to check either the 681 cm^{-1} and 653 cm^{-1} modes belong to the representation Γ_3, we are attempting studies of the effect on the defect of uniaxial stress. The defect pairs are oriented along the four main diagonals of the cube and a uniaxial stress would have different effects on the different pairs. The observed expected effect is the sum of the effects on each pair. The only direction for which the effect is identical for all the pairs is the [100] axis. The symmetry of the defect is reduced from C_{3}v to C_{2}v and the doublet+singlet structure is transformed into three singlets. The difficulties in observing such a splitting arise from the finite width of the lines and from the high stress we have to apply.

Acknowledgement

The theoretical aspects of this work were made possible by the partial support and sponsorship of the U.S. Department of the Army through its European Research Office.

The experimental work has been sponsored in part by the D.R.M.E. and by the Air Force Cambridge Research Laboratories OAR through the European Office of Aerospace Research.

REFERENCES

1. M. Balkanski, W. Nazarewicz, *J. Phys. Chem. Solids*, **27**, 671 (1966); R. M. Chrenko, R. S. McDonald, E. M. Pell, *Phys. Rev.*, **138**, A 1775 (1965); M. Waldner, M. A. Hiller and W. G. Spitzer, *Phys. Rev.*, **140**, A 172 (1965).
2. J. F. Angress, A. R. Goodwin, S. D. Smith, *Proc. Roy. Soc.*, **A 287**, 64 (1965).
3. W. G. Spitzer and M. Waldner, *J. Appl. Phys.*, **36**, 2450 (1965); W. Hayes, *Phys. Rev.*, **138**, A 1227 (1965); A. R. Goodwin and S. D. Smith, *Physics Letters*, **17**, 203 (1965).
4. A. J. Sievers, A. A. Maradudin and S. S. Jaxwal, *Phys. Rev.*, **138**, A 272 (1965); K. F. Renk, *Phys. Letters*, **14**, 281 (1965).
5. E. M. Lifshitz, *Nuovo Cimento, Suppl.*, **3**, 716 (1956); A. A. Maradudin, *Reports on Progress in Physics*, **XXVIII**, 331 (1965); R. J. Elliott, *Aberdeen Summer School*, Oliver and Boyd (1966).
6. P. G. Dawber and R. J. Elliott, *Proc. Roy. Soc.*, **A 273**, 222 (1963); *Proc. Roy. Soc.*, **81**, 453 (1963).
7. B. Moser, K. Otnes and O. W. Myers, *Phys. Rev. Letters*, **8**, 278 (1962).
8. A. J. F. Boyle and H. E. Hall, *Reports on Progress in Physics*, **XXV**, 441 (1962).
9. R. L. Aggarwal, P. Fisher, W. Mourzine and A. K. Ramdas, *Phys. Rev.*, **138**, A 882 (1965).
10. H. Reiss, C. S. Fuller, and F. J. Morin, *Bell Syst. Tech. J.*, **35**, 535 (1956).
11. P. Pfeuty, to be published.
12. K. Weisser, *Phys. Rev.*, **126**, 1427 (1962).
13. R. J. Elliott, W. Hayes, G. D. Jones, H. F. McDonald and C. T. Sennett, *Proc. Roy. Soc.*, **A 289**, 1 (1965).
14. A. A. Maradudin, *Ann. Phys.*, **30**, 371 (1964).
15. Helen Smith, *Phil. Trans. Roy. Soc. London*, **A 241**, 45 (1948).

16. G. F. Koster, J. O. Dimmock, R. G. Wheeler and H. Statz, *Properties of 32 Point Groups*, M.I.T. Press.

APPENDIX

Form of the Green's function and perturbation submatrices used in the calculation of the localized mode frequencies. We use the notation $j = \exp(2\pi i/3)$.

$$g_{\Gamma_3} = \begin{bmatrix} -\frac{1}{M\omega^2} & 0 & 0 & 0 & 0 \\ 0 & g_1 & g_5 & jg_4+j^2g_6 & j^2g_4+jg_6 \\ 0 & g_5 & g_1+g_6 & jg_4 & j^2g_4 \\ 0 & j^2g_4+jg_6 & j^2g_4 & g_1 & jg_5-j^2g_6 \\ 0 & jg_4+j^2g_6 & jg_4 & j^2g_5-jg_6 & g_1 \end{bmatrix}$$

$$g_{\Gamma_1} = \begin{bmatrix} -\frac{1}{M\omega^2} & 0 & 0 & 0 \\ 0 & g_1 & g_5 & (g_4+g_6)\sqrt{2} \\ 0 & g_5 & g_1+2g_6 & \sqrt{2}\,(g_4) \\ 0 & (g_4+g_6)\sqrt{2} & g_4\sqrt{2} & (g_1+g_5-g_6) \end{bmatrix}$$

$$C_{\Gamma_3} = \begin{bmatrix} -4F-(1-\nu)M\omega^2 & F & F & jF & j^2F \\ F & -F-\varepsilon M\omega^2 & 0 & 0 & 0 \\ F & 0 & -F & 0 & 0 \\ j^2F & 0 & 0 & -F & 0 \\ jF & 0 & 0 & 0 & -F \end{bmatrix}$$

$$C_{\Gamma_1} = \begin{bmatrix} -4F-\varphi-(1-\nu)M\omega^2 & F+\varphi & F & F\sqrt{2} \\ F+\varphi & -F-\varphi-\varepsilon M\omega^2 & 0 & 0 \\ F & 0 & -F & 0 \\ F\sqrt{2} & 0 & 0 & -F \end{bmatrix}$$

Alteration of Crystal Wave Functions by Lattice Defects*

DAVID REDFIELD

School of Engineering and Applied Science, Columbia University, New York, N.Y. 10027, U.S.A.

ABSTRACT

Most types of lattice defects are shown to be capable of altering the wave functions of the intrinsic electronic states of solids. This alteration is caused by the strong electric fields arising at charged defects of all types. The effect is similar to that by which a uniform field changes Bloch functions to Houston functions. A number of consequences follow from the changes caused by the modified wave functions in the matrix elements for electronic transitions. Changes in the fundamental optical absorption edge and in the exciton spectrum are shown to be among these consequences. The application of these results to scattering of current carriers in heavily doped semiconductors is suggested.

I. INTRODUCTION

Although lattice defects have many well-known effects on the properties of solids, there is an additional effect which has received very little attention but may have considerable importance in certain cases. The purpose of this paper is to point out this effect and some of its consequences. In brief, the effect referred to is the alteration of the wave functions of the band states of a solid caused by the presence of charged defects.

Defects of all the various classes—points, lines and planes—frequently exist in conditions in which they carry a net charge. Thus, besides any other effects they produce, charged defects in non-metallic solids are always surrounded by regions of strong electric fields. Although the detailed magnitudes and distributions of these fields depend on both the nature of the defects and the screening of the charges, typical field strengths are 10^4–10^5 V/cm.[1] If one generalizes the definition of defects to include

* Supported in part by the U.S. Office of Naval Research.

lattice vibrations, then even stronger fields can occur in polar crystals, but these will not be discussed here.

II. RESULTS AND DISCUSSION

The central point of this discussion is that the electron wave functions of the crystal itself, which are normally Bloch functions, are changed by the presence of the electric fields originating at the defects. To illustrate this change, consider the time-dependent Bloch function in an ideal field-free crystal

$$\Psi_k(\mathbf{r}, t) = u_k(\mathbf{r}) \exp(i\mathbf{k}\cdot\mathbf{r}) \exp(-iE_k t/\hbar) \tag{1}$$

where $\mathbf{k}$ is the wave vector, $u_k(\mathbf{r})$ has the periodicity of the lattice and E_k is the energy of the state $\mathbf{k}$. The effect of the electric field is illustrated most simply by considering the case of a uniform field $\mathbf{F}$ in the manner given by Houston.[2] He showed that the time-dependence of the wave vector caused by such a field could be accounted for by using functions of the form

$$\Psi(\mathbf{r}, t) = u_{k-yt}(\mathbf{r}) \exp[i(\mathbf{k}-\mathbf{y}t)\cdot\mathbf{r}] \exp\left[-(i/\hbar)\int^t E_{k-y\tau}d\tau\right] \tag{2}$$

where $y \equiv eF/\hbar$. As is well-known, this is equivalent to a uniformly increasing wave vector.

When the electric field is non-uniform, but slowly varying with position, similar alterations in the wave functions will occur. For example, the monotonically varying field in the space charge region near a charged plane (a surface or junction) does not change appreciably over a distance comparable with the lattice constant. Thus any layer parallel to the charged plane and thin enough to be approximated by a constant field value may be characterized by Houston functions with the local value of $\mathbf{F}$. In this Thomas-Fermi spirit we may then describe the wave functions as a weighted distribution of Houston functions.

In the vicinity of an unscreened point charge (impurity, interstitial, etc.) the spatial portion of the wave function would take on some of the characteristics of the Coulomb wave functions. These are too complicated for convenient use, but in most solids they, too, would require modification anyway since the Coulomb field of the point charge would be altered by two effects. One is the screening effect of possible free charges; the second is interactions with neighboring defects which distort the potential distribution. Thus there is no analytical way of correctly ex-

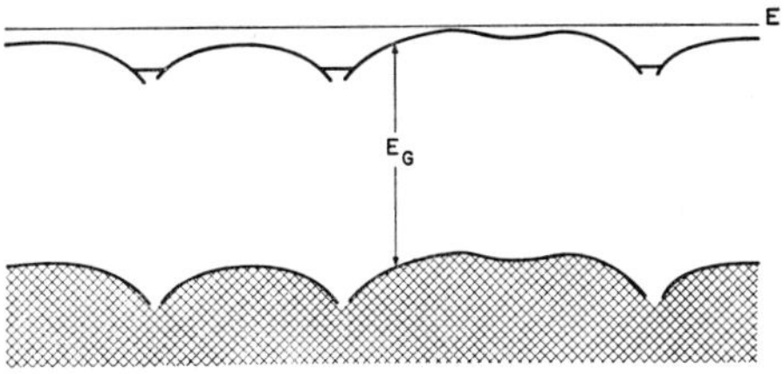

Fig. 1.

pressing the wave functions of the band states in the presence of normal concentrations of charged impurities in a semiconductor. The situation may best be described by a diagram such as Fig. 1 in which positive charges can be seen to vary the levels of the band edges. This produces a position-dependence of the kinetic energy of a conduction band state having a fixed total energy E.

Another way of viewing these effects is to say that the wave number **k** is no longer a good quantum number when strong electric fields are present. This is obvious from Fig. 1 when we use the effective mass relation for kinetic energy $\hbar^2k^2/2m^*$. The same conclusions evidently apply to cases other than point charges. We may say in general that the correct wave function is a composite of field-altered Bloch functions and is spatially non-uniform.

In the absence of analytical expressions for such wave functions, it is necessary to resort to a statistical form. This requires that the probability distribution of the potential or the electric fields be known in any specific case. From this distribution, one might use a Thomas-Fermi approach to evaluate the probability distribution of wave function-alterations. A related but somewhat different approach was used in the one area where these considerations have already been applied. This will be mentioned in the following discussion of some of the consequences of these changes in wave functions.

Because of the central role of the crystal wave functions in many physical problems, it is possible for these effects to have a number of consequences. The present discussion is aimed at stressing the processes in which electronic transitions are important. It is obvious that, in principle, any change in the electronic wave functions may cause a change in the matrix elements which determine the transition probabilities. Consider, for example, the effect of a uniform, externally applied electric field on the fundamental optical absorption edge of a solid (the Franz-Keldysh effect). The direct, interband transition probability in the field-free case contains the well-known threshhold for optical absorption at the photon energy which equals the energy gap. But in the presence of the electric field, the change in the wave functions removes this sharp threshhold and an absorption " tail " is allowed at photon energies less than the gap energy.

A similar effect on the optical absorption edge caused by the fields of charged defects has been calculated for the cases of points[3] and planes.[4] In these calculations the actual wave functions were not evaluated. Instead, the approximation of slowly-varying fields permitted the local application of the Franz-Keldysh formulas to obtain local values for the field-dependent absorption coefficient. The total absorption coefficient was then found by performing a weighted integration of these local values. In both of these cases the results of such calculations indicated that significant changes of the band-to-band absorption edge could be caused by the lattice defects.

It is apparent from the work of Nikitine and Grosmann[5] that exciton states are also affected by defects in a similar way. They observed the exciton absorption spectrum of Cu_2O as a function of the dosage level of neutron bombardment. It was found that the radiation damage caused the exciton spectrum to change in just the same way that is caused by applying an electric field to pure, undamaged crystals. From this they were able to infer the average internal field strength and thus the concentration of charged damage centers.

There is another area in which transition probabilities may be altered by lattice defects but which has apparently not been treated. This concerns the intraband transitions, or scattering, of current carriers and the consequent effect on conductivity. The present discussion does not refer to the scattering by charged impurities which has been long known. That conventional impurity scattering considers the transitions among ordinary Bloch states caused by a charged center. The point of the present discussion is that Bloch functions may not be adequate descriptions of the states of an impure crystal, particularly when there are high concentrations of charged impurities. Thus for the case of impurity scattering this may be stated as an argument for a second order perturbation calculation—one which treats the perturbation of the wave function.

This is clearly not a simple problem. But in view of the strong current interest in the properties of heavily doped semiconductors, it may be worth the effort. Furthermore, the success of the statistical approach in the optical transition problem suggests that it may be useful in scattering problems also.

REFERENCES

1. David Redfield, *Phys. Rev.*, **130**, 914 (1963).
2. W. V. Houston, *Phys. Rev.*, **57**, 184 (1940).
3. David Redfield, *Phys. Rev.*, **130**, 916 (1963).
4. David Redfield, *Phys. Rev.*, **140**, A2056 (1965).
5. S. Nikitine and M. Grosmann, *J. de Phys.*, **24**, 525 (1963).

T-matrix Theory of Localized Electronic States due to a Vacancy with Application to Diamond

K. H. BENNEMANN

Institute for the Study of Metals, The University of Chicago, Chicago 37, Illinois, U.S.A.

I. INTRODUCTION

The problem of determining the electronic state of a vacancy in semiconductors, particularly in diamond, has met with some interest in the past and, due to the remarkably sensitive derivative techniques used with great success at the present time in studying optical properties of solids, there is good reason to believe that this problem will soon gain some new actuality. First, I would like to describe briefly the work which has been done in the past on the vacancy problem. Then I would like to propose a new method for determining the electronic structure of a vacancy and to apply the theory to diamond. Finally, I would like to introduce a new aspect on charged vacancies.

In the past essentially Coulson and Kearsley,[1] and Yamaguchi,[2] and more recently Stoneham[3] made a detailed study of localized electronic states at neutral and (positively or negatively) charged vacancies in diamond. In all these studies the electronic structure of the vacancy was determined by applying a molecular-orbital technique using one-electron wave functions centered at the vacancy or using linear combinations of atomic orbitals (L.C.A.O.). The molecular orbitals were constructed from the tetrahedrally hybridized 2s and 2p atomic orbitals. Then from such " vacancy "-electronic orbitals Slater determinantal wave functions were formed representing a special electronic vacancy configuration taking into account properly lattice symmetry of the vacancy and electron spins. Finally several Slater wave functions with the same spin multiplicity and space symmetry were linearly combined. The spectrum of energies for the localized electronic states were then determined from the appropriate secular determinant including configuration interaction and interaction among the localized electrons and among the localized electrons and lattice atoms. The results obtained differed mainly because of uncertainties in determining the interatomic electronic Coulomb interaction matrix elements.

Instead of improving on these previous methods which are all very much in the spirit of "dangling bonds," we would like to propose a very different method for treating vacancies in semiconductors. Recently the pseudo-potential theory as developed by Kleinman, Phillips, Cohen, Heine *et al.* which replaces the crystal potential by an effective potential which is in many cases much weaker and smoother, has been successfully applied to semiconductors such as C, Si, Ge, etc. Using such an effective crystal potential permits one to treat the valence electrons approximately as quasi-free-electrons. This suggested determining the electronic structure of defects in semiconductors using scattering theory, which is in a way an extension of Friedel's theory of defects in metals.

Using t-matrices therefore a theory of the electronic structure of vacancies in semiconductors was developed and applied to diamond. This theory unifies the effective mass approximation, which is adequate for treating shallow localized states (Kohn, Luttinger *et al.*), and the localized perturbation theory (Lifshitz, Clogston-Slater *et al.*) which is adequate for treating deeply bound states.

II. THEORY

The most remarkable feature of the diamond type lattice is that each atom sits at the center of a regular tetrahedron formed by its four nearest neighbors to which it is covalently bound. Therefore we have four broken covalent bands around the vacancy. This gives rise to a strong change in the valence electron distribution in the vicinity of the vacancy and leads to a lattice distortion around the vacancy. Therefore, it is not reasonable to determine the redistribution of the valence electrons around the vacancy by perturbation theory using the perfect crystal as the unperturbed state. Instead, we determine self-consistently the electronic distribution and crystal potential in the vicinity of the vacancy by scattering theory. By including multiple electron scattering involving different lattice atoms we take explicitly into account the atomic configuration of the lattice. A vacancy means that an atomic scatterer is missing in the lattice. In detail the crystal potential is obtained as follows: after replacing the crystal potential V by the appropriate pseudo-potential V_S, resulting from using orthogonalized wave functions, we introduce the one-electron Green's function G and the t-matrix T by

$$G = G_0 + G_0 T G_0 \tag{1}$$

and

$$T = V + V G_0 T \tag{2}$$

The scattering matrix T describes the electron scattering by the self-consistent crystal potential V. The atomic configuration of the crystal comes in through T as is seen by the following expansion:

$$T(k, q; E) = \sum_l T_l(k, q; E) \exp [i(\mathbf{q}-\mathbf{k})\cdot\mathbf{r}_l] \tag{3}$$

with

$$T_l(k, q; E) = t_l(k, q; E) + \sum_{l' \neq l} t_l(k, p; E) G_0(p, E) \times$$
$$\times T_{l'}(p, q; E) \exp [i(\mathbf{p}-\mathbf{k})\cdot\mathbf{r}_{l'l}] d^3p + \ldots\ldots \tag{4}$$

The higher terms result from multiple scattering due to 3, 4, and more atoms. T_l gives the electron scattering associated with the atom l. If a vacancy is present, then $l \neq V$, where V denotes the vacant lattice site. t_l results from V_l neglecting the crystalline environment of the ion l, and T_l results if the multiple electron scattering involving the surrounding lattice atoms of the atom l is taken into account.

From T the electron distribution in the vicinity of the vacancy is determined. Figure 1 shows the valence electron charge density along a (111)-axis and Fig. 2 shows the charge density at the vacancy along a (1, −1, 0)-axis. The lattice relaxation has been neglected (which may lead at most to about 10% correction). These calculations shed some light on the validity of the often used concepts of dangling covalent bonds, of stretching

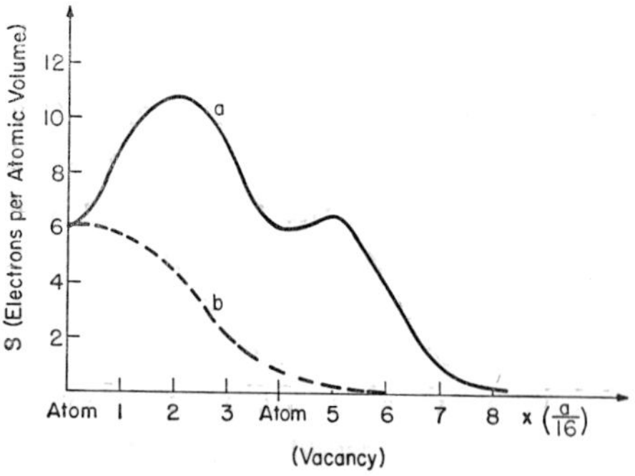

Fig. 1. Curve " a " refers to the perfect crystal, and curve " b " refers to a crystal with a vacancy present.

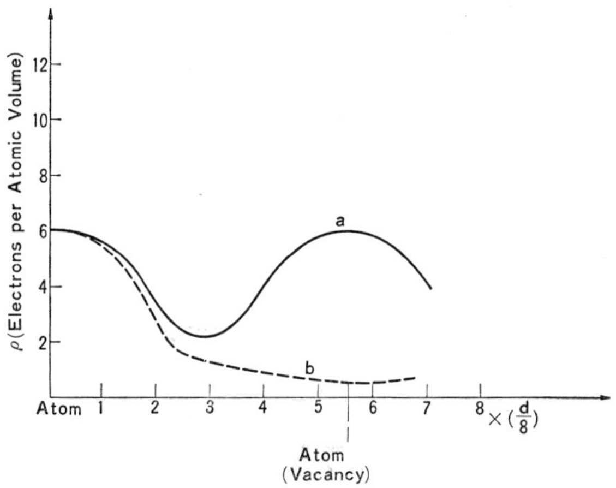

Fig. 2. The curves " a " and " b " refer to the perfect crystal and to a crystal with a vacancy present, respectively.

covalent bonds, and of reforming covalent bonds from the four electrons in the broken covalent bonds. The results indicate the strong deviation from spherical symmetry of the potentials at the nearest neighbor lattice sites of the vacancy and suggest that a point ion approximation and a dielectric screening treatment, neglecting *local field corrections*, are bound to be poor approximations for such strong lattice perturbations as given by a vacancy.

After having determined the crystal potential the localized states in the resulting crystal potential at the vacancy are given by the poles of the *t*-matrix. To obtain the localized states within the forbidden gap we look for the poles of

$$T(E) = VG_0(E)T(E) \tag{5}$$

In determining the poles of $T_{l_0}(E)$ it is necessary to take into account the poles associated with the other three $T_{l_0}(E)$ which appear for the same energies E for symmetry reasons. Such a self-consistent determination of the poles is probably equivalent to configuration interaction. However, contrary to configuration interaction such a self-consistent procedure does not lift up the degeneracy of the localized states at the four nearest neighbors. This might happen via Jahn-Teller lattice distortion. If the wave functions associated with the localized states mentioned earlier do not overlap, then it is also possible that this degeneracy just persists (see degeneracy of 1s core electrons, ε_{1s} same for each atom). It follows that

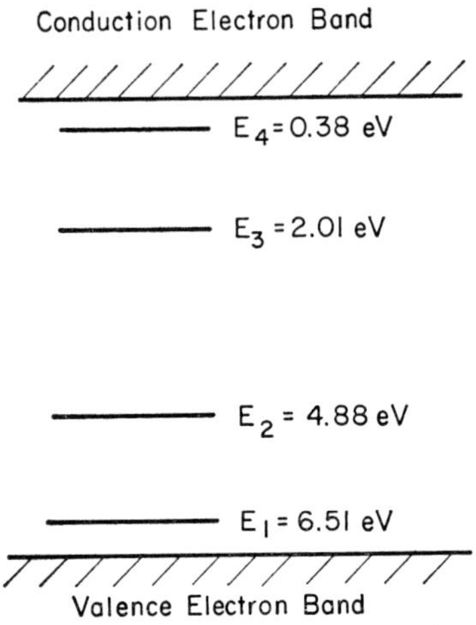

Fig. 3.

the spectrum of bound states is approximately given by

$$1-\frac{1}{\Omega}\sum_{p} V_{l_0}(k, p)G_0(p, E)$$

$$-\frac{1}{\Omega}\sum_{p,l'} \exp\left[i(\mathbf{p}-\mathbf{k})\cdot\mathbf{r}_{l'l}\right]V_{l'}(k, p)G_0(p, E) = 0 \tag{6}$$

if (a) the potential V is highly localized in real space,
and *if* (b) the Fourier transformed potential is sharply peaked, which case is adequate for shallow bound states described by the effective mass approximation.

Using Eq. (6) the spectrum of bound states is determined. The results are given in Fig. 3. Note that the top of the valence band refers to the crystal symmetry point Γ_{25} and the bottom of the conduction electron band to Γ_{15}. They symmetry of the bound states is: $E_1(l=0)$; $E_2(l=1)$; E_3 $(l=2)$; $E_4(l=3)$.

III. CHARGED VACANCY

The vacancy might trap electrons in the localized states and also might release electrons from the broken covalent bonds to the conduction band and become charged in this way. It would be nice to get these charged vacancy states out of the theory, without introducing explicitly wave functions for the electrons in the localized states. I do not know right now how this can be done, but perhaps the theory yields these charged vacancy states if one performs a self-consistent (bootstrap-like) calculation taking carefully into account the localized states in determining the crystal potential which just gives rise to the bound states.

Charged vacancies and also neutral vacancies might be in a paramagnetic state. They act then like paramagnetic impurities having an exchange interaction with the conduction electrons. Therefore, paramagnetic vacancies should give rise to anomalous conduction electron scattering (Kondo Effect) possibly exhibited in resistivity, magneto-resistivity, magnetic nuclear relaxation rate, etc. Possibly the observed anomalies in the conductance of semiconductor-oxyde-metal junctions arise due to anomalous electron exchange scattering from vacancy like paramagnetic defects in the interface of semiconductors and oxide. As will be shown in detail in another paper these paramagnetic defects in semiconductors and their interaction with conduction electrons can be treated very elegantly by using Anderson's Hamiltonian used in connection with localized magnetic moments in metals. With respect to the optical properties of the vacancy it should be noted that the exchange scattering of the electrons involves spin flips and consequently the selection rule $\Delta S=0$ for optical transitions is no longer valid for paramagnetic vacancies.

It would be of considerable interest to include electron-phonon interaction in the theory. This would be important, for example, in determining the lifetime of the higher excited bound states at defects in semiconductors.

It is hoped that the contrast between previous theories and the one proposed here will help in studying the shortcomings and limitations of the existing theories.

REFERENCES

1. C. A. Coulson and H. J. Kearsley, *Proc. Roy. Soc.*, **A241**, 433 (1957).
2. T. Yamaguchi, *Journal of the Physical Society of Japan*, **17**, 1359 (1962).
3. Stoneham, *Preprint* (Harwell) (1966).

Energy Levels Associated with Isolated Vacancies in Silicon*

JOSEPH CALLAWAY† AND A. JAMES HUGHES‡

† *Department of Physics, University of California, Riverside, California, U.S.A.*
and
‡ *Applied Research Laboratory, Aeronutronic Division of Philco Corporation Newport Beach, California, U.S.A.*

We have studied the energy levels associated with the neutral vacancy in silicon, using a method based on solid state scattering theory.[1] The following is a preliminary acount of this investigation. A more complete description will be published elsewhere.

In a formal sense, the method is quite simple, and is applicable to other localized defects as well. Let V be the potential representing the defect, and G be the Green's function for the perfect crystal defined by:

$$G = (E^{+} - H_0)^{-1} \tag{1}$$

where H_0 is the Hamiltonian for the perfect crystal (including the periodic potential). A bound state of energy E will exist if the equation

$$D(E) = 0 \tag{2}$$

is satisfied. The function D is a determinant:

$$D(E) = \det (I - GV) \tag{3}$$

Equation (2) will possess solutions for real energies only if the energy E does not lie within the continuous spectrum of H_0. In the case of a semiconductor, this means that we are interested in energies E within the fundamental gap.

We will express the operators G and V on the basis of Wannier functions. These localized functions $a_n(\mathbf{r}-\mathbf{R}_s)$ are related to the Bloch functions $\Psi_n(\mathbf{k}, \mathbf{r})$ which are the eigenfunctions of H_0 by

$$a_n(\mathbf{r}-\mathbf{R}_s) = \frac{\Omega^{1/2}}{(2\pi)^{3/2}} \int \exp(-i\mathbf{k}\cdot\mathbf{R}_s)\Psi_n(\mathbf{k}, \mathbf{r})d^3k \tag{4}$$

Ω is the volume of the unit cell.

* Supported by the Air Force Cambridge Research Laboratories under Contract AF-ESD/AF19 (628)–5660.

Although Wannier functions have been used in formal discussions of defects in crystals for many years, to our knowledge this is the first work in which they have been used in numerical calculations of the properties of defect states.

The matrix elements of G on the Wannier function basis are given by

$$(\mathrm{sn}\,|G|\,\mathrm{tm}) = \int a_n^*\,(\mathbf{r}-\mathbf{R}_s)\frac{1}{E^+-H_0}a_m(\mathbf{r}-\mathbf{R}_t)d^3r$$
$$= \delta_{nm}\frac{\Omega}{8\pi^3}\int d^3k\frac{\exp[-i\mathbf{k}\cdot(\mathbf{R}_s-\mathbf{R}_t)]}{E^+-E_n(\mathbf{k})} \qquad (5)$$

These integrals can be evaluated in a straightforward way if the band structure is known. In order for the present method to be a useful one, it is necessary that the defect potential V can be approximated on the Wannier basis by a matrix which has only a finite number of non-zero elements.

We have calculated the matrix elements of G and V numerically. The lowest eight bands (four valence bands and four conduction bands) have been considered. Let us suppose that the defect is located in the cell at the origin. The missing atom has coordinates $a/8(1, 1, 1)$ where a is the lattice parameter. One of its nearest neighbors is in the same cell, and the other three neighbors are in cells whose lattice vectors are $\frac{1}{2}a(1, 1, 0)$, $\frac{1}{2}a(1, 0, 1)$, and $\frac{1}{2}a(0, 1, 1)$. These four cells are included in our calculation, so that we have altogether 32 basis functions to consider. Some simplifications are afforded by crystal symmetry, as will be discussed below.

In order to avoid complications associated with core electron states, we have adopted a pseudo-potential approach. We assume that the periodic potential in the perfect crystal can be represented as a sum of atomic pseudo-potentials, and therefore that a vacancy can be represented as one missing pseudo-potential. On a plane wave basis, let the pseudo-potential be described by a function $V(\mathbf{k})$. When $\mathbf{k}=\mathbf{K}_n$, where $\mathbf{K}_n$ is any reciprocal lattice vector, the function V can be taken to be determined from the semi-empirical calculations of energy bands, which use experimental information concerning the band structure to determine pseudo-potential parameters. We have used values for these parameters determined by Brust.[2] The pseudo-potential is not known unambigously for other values of $\mathbf{k}$ and we have simply fitted a polynomial to the known values. We took $V(0)=-0.61ry$, in accord with Harrison.[3]

In order to compute the matrix elements of potential, we require the Bloch functions $\Psi_n(\mathbf{k}, \mathbf{r})$. Let these be expanded in plane waves as follows:

$$\Psi_n(\mathbf{k}, \mathbf{r}) = \frac{1}{\Omega^{1/2}} \sum_j b_n(\mathbf{k}, \mathbf{K}_j) \exp [i(\mathbf{k}+\mathbf{k}_j)\cdot\mathbf{r}] \tag{6}$$

Then the matrix elements of the perturbation can be expressed as

$$(\text{sn}\,|V|\,\text{tm}) = \sum_j \int\int d^3k d^3p \exp [-i(\mathbf{p}+\mathbf{K}_j)\cdot(\mathbf{R}_t-\mathbf{d})] \exp [i\mathbf{k}\cdot(\mathbf{R}_s-\mathbf{d})]$$

$$\times [\sum_l b_n{}^*(\mathbf{k}, \mathbf{K}_l) b_m(\mathbf{p}, \mathbf{K}_l+\mathbf{K}_j)] U(\mathbf{p}+\mathbf{K}_i-\mathbf{k}) \tag{7}$$

In order to keep the numerical work within reasonable bounds, we have employed only the smallest 15 reciprocal lattice vectors $\mathbf{K}_j$ in Eq. (6). The defect potential is assumed to be centered about $\mathbf{r}=\mathbf{d}$.

There are some ambiguities connected with the question of the phase of the Bloch function.[4] We have resolved this problem as follows. First, we note that the coefficients $b_n(\mathbf{k}, \mathbf{K}_j)$ in Eq. (6) can be chosen to be real. When this is done, the question of phase reduces to one of sign. The coefficients b_n are actually calculated as the eigenvectors of the matrix H_0, for certain points within 1/48′th of the Brillouin zone. The wave function for some particular $\mathbf{k}$ within a given 1/48′th of the zone is related to that for any other $\mathbf{k}$ in the star by a symmetry operation. We will ignore questions connected with the degeneracies of the energy bands (and only general points were included in our evaluation of the matrix element of Eq. (7)). Consider an operation $\{\alpha|\mathbf{W}_\alpha\}$ of the space group. Thc cocfficients b_n obey the relation

$$b_n(\alpha\mathbf{k}, \mathbf{K}_j) = \chi(n, \alpha) \exp [-i\mathbf{K}_j\cdot\mathbf{W}_\alpha]\, b_n(\mathbf{k}, \alpha^{-1}\mathbf{K}_j) \tag{8}$$

in which $\chi(n, \alpha)$ is the character for one of the one-dimensional irreducible representations of the point group. The character will be the same for all $\mathbf{k}$. The phase factor in Eq. (8) is real. We now have two things to do. Within the 1/48′th of the zone under consideration, the signs of the coefficients b_n are chosen to insure continuity. There are certain interesting problems connected with accidental degeneracies and near degeneracies which we will not describe here. The characters $\chi(n, \alpha)$ are determined from the behavior of the coefficients on the symmetry planes bounding the 1/48′th of the zone. When this is done, Eq. (8) gives b_n for every $\mathbf{k}$ in the star.

The double sums over reciprocal lattice vectors in Eq. (7) lead to a very lengthy calculation, and with the computing facilities at our disposal, we were limited to a selection of 10 points within 1/48′th of the zone. Even with this small number of points, each potential matrix element required about 5 minutes to evaluate on a high speed computer. The evaluation of the Green's functions was much simpler, since there is only a single integral over the Brillouin zone, and since the Green's function is invariant under rotations in the point group.

One valence band and one conduction band turn out to have charcters χ corresponding to representations which are odd under interchange of k_x and k_y, etc. These bands have small matrix elements of potential, and do not couple strongly to other bands. It appears to be sufficient therefore to consider six bands only. Since we have chosen the origin of coordinates at the midpoint of the line joining two atoms, our potential is invariant under the group C_{3V}. Symmetrized combinations of functions transforming according to the representations of C_{3V} were constructed, and used to factor the determinant $D(E)$ as described in ref. 1. From examination of the matrix elements involved, it is apparent that a localized state, if any, must belong to the symmetric representation. When the restriction to six bands is imposed, we are left with a 12×12 matrix $I - GV$. The determinant of this matrix was evaluated numerically for energies within the band gap.

The results are as follows: For the original pseudo-potential, no solutions of Eq. (2) were found. The assumed potential was apparently too weak to produce a localized state. However, when all matrix elements of the potential were multiplied by 1.15, a root appeared close to the valence band, in agreement with experiment.[5] This factor may be crudely interpreted as a measure of the error in our calculation: that is, we may say roughly that we are in error by 15%. In view of the approximations made in the course of this work, this should be regarded as quite reasonable agreement. As the strength of the potential is increased, the energy of the localized state rises, and ultimately (if the multiplying factor is increased to 1.6) disappears into the conduction band. The wave function of the defect state, expressed in terms of coefficients of Wannier functions can be determined, and appears to be dominated by the contributions from the lowest two valence bands. Compared to the valence bands, the conduction bands seem to play a relatively minor role in the formation of the defect state. However, interband couplings among the valence bands are quite important.

We hope to discuss other characteristics of the vacancy in future work.

REFERENCES

1. J. Callaway, *J. Math. Phys.*, **5**, 783 (1964).
2. D. Brust, *Phys. Rev.*, **124**, A1337 (1964).
3. W. A. Harrison, *Physica.*, **31**, 1692 (1965).
4. R. E. Turner and D. A. Goodings, *Proc. Phys. Soc.* (London), **86**, 87 (1965).
5. G. D. Watkins, in *Radiation Damage in Semiconductors*, p. 97, Academic Press, New York, 1965.

Piezo-spectroscopic Effects of Impurity States in Semiconductors*

P. Fisher, A. K. Ramdas and J. H. Reuszer

Department of Physics, Purdue University, Lafayette, Indiana, U.S.A.

ABSTRACT

Electronic spectra associated with imperfections in silicon and germanium, when studied under an applied uniaxial stress and examined with polarized light, yield valuable information about both the imperfection and the host lattice. The piezo-spectroscopic effects observed for the shallow Group V donors in silicon and germanium have been interpreted within the framework of the effective mass theory and deformation potential theory. The symmetry assignments of the ground states and the excited states of the impurity have been verified from the observations. The piezo-spectroscopic effects for isolated lithium donors and for the lithium-oxygen complex in silicon have also been examined. This study has shown that isolated lithium has a unique ground state structure, one which is " inverted " relative to the Group V impurities in silicon. The excitation spectra of substitutional thallium and zinc acceptors in germanium, and Group III acceptors in silicon under uniaxial stress have been interpreted using group theoretical arguments. The degeneracies of the ground states and the excited states have been deduced for many of the transitions on the basis of the number of stress induced components, their polarization features and their intensity variation with stress.

I. INTRODUCTION

A number of interesting phenomena arise from the presence of impurities in semiconductors. In the present article, we will concern ourselves with those impurities in silicon and germanium which exhibit line spectra associated with the localized states of the bound electron or hole. The Group V impurities, phosphorus, arsenic, antimony and bismuth are

* Work supported in part by an Office of Naval Research Contract, the U.S. Army Research Office, Durham, and the Advanced Research Projects Agency.

the classic examples. The line spectra, besides being of interest per se, provide significant information about the host material, since the localized states have been described very successfully in terms of the band structure of the crystal.

As in atomic spectroscopy, the effects of external perturbing fields are particularly useful in arriving at symmetry assignments for the energy states. Further, the response of the energy bands of the crystal to the perturbation is reflected by the localized impurity states, thus providing additional information about the host crystal. The object of the present review is to discuss the effects of uniaxial stress on the excitation spectra (piezo-spectroscopic effects[1]) of various donor and acceptor impurity systems in silicon and germanium.

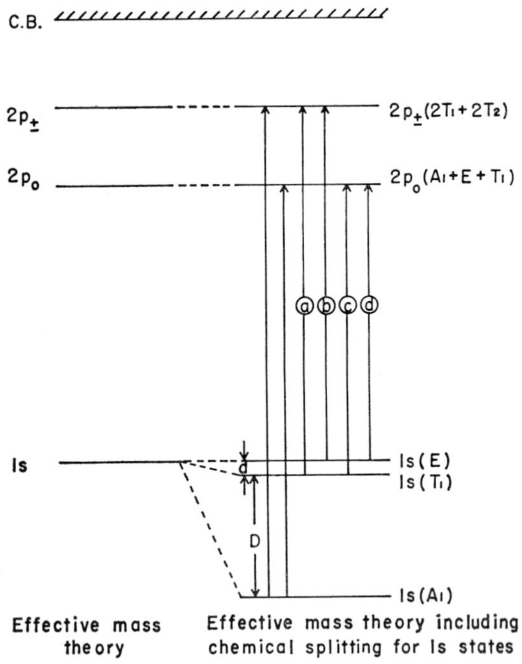

Fig. 1. The energy level scheme (not to scale) of Group V impurities in silicon for the transitions from the 1s states to the $2p_0$ and $2p_{\pm}$ states. The letters next to a level indicate the irreducible representations of the donor site symmetry, T_d, to which the level belongs. C.B. = conduction band.

II. DONOR STATES

1. *Zero stress*

The most comprehensive treatment of the energy states of Group V donors in silicon and germanium has been given by Kohn and Luttinger.[2,3] In this approach the wave functions of the impurity states are

$$\psi^{(i)} = \sum_{j=1}^{N} \alpha_j^{(i)} F_j(\mathbf{r}) \Phi_j(\mathbf{r}), \qquad i = 1, \dots, N \tag{1}$$

where $\Phi_j(\mathbf{r})$ is the Bloch wave function at the jth minimum, $\alpha_j^{(i)}$ are numerical coefficients which describe the relative contribution from each of the N

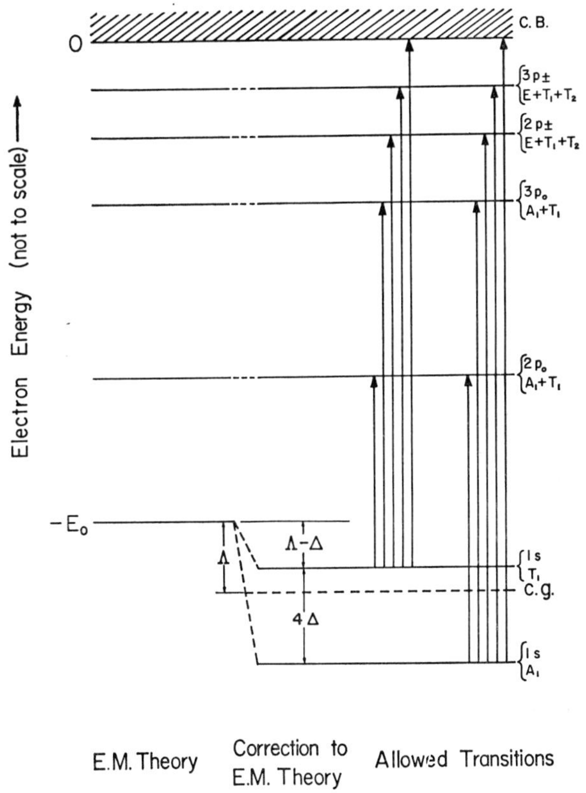

Fig. 2. The energy level scheme (not to scale) of a Group V impurity in germanium showing how a general perturbation affects the effective mass ground state. Shown also are the possible transitions between the two ground states and the excited p-states.

valleys, and the " hydrogen-like " envelope functions $F_j(\mathbf{r})$ satisfy the so-called effective mass wave equation

$$\left[-\frac{\hbar^2}{2m_t}\left(\frac{\partial^2}{\partial x_j^2}+\frac{\partial^2}{\partial y_j^2}\right)-\frac{\hbar^2}{2m_l}\frac{\partial^2}{\partial z_j^2}-\frac{e^2}{\kappa r}\right]F_j(\mathbf{r}) = EF_j(\mathbf{r}) \qquad (2)$$

In Eq. (2), m_t and m_l are the transverse and longitudinal effective masses respectively at the bottom of the conduction band, while κ is the static dielectric constant of the material. It is clear that the multi-valley nature of the conduction band results in an N-fold degeneracy. As is well known, Eq. (2) is not valid for small r and hence states derived from this equation which have large amplitudes near the impurity ion are not adequately described in the framework of the effective mass theory. It is customary to correct the effective mass equation by using a more appropriate expression for the impurity potential. Such a potential must, of course, conform to the site symmetry T_d of the donor ion. Thus the N-fold degeneracy of the effective mass ground state is lifted resulting in the splittings shown[4] in Figs. 1 and 2. In view of the form of the wave functions, it is expected[2] that the totally symmetric ground state, $1s(A_1)$, will be the most seriously

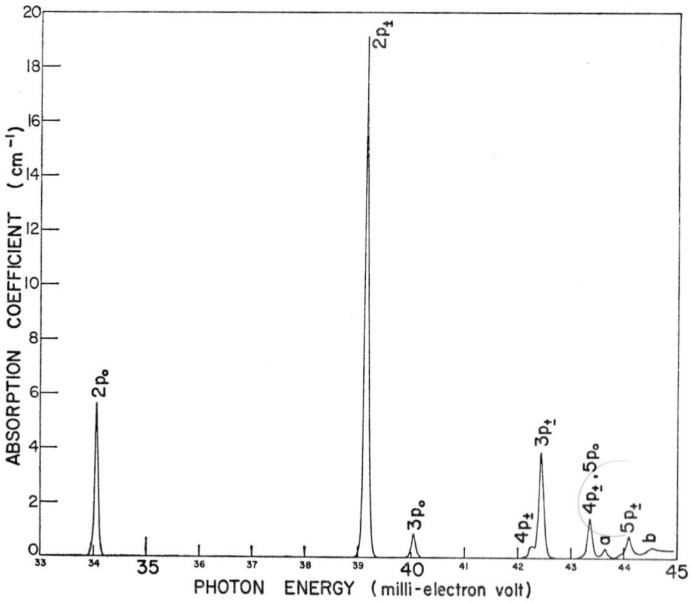

Fig. 3. The excitation spectrum of phosphorus impurity in silicon; $N_D \simeq 2\times10^{14}$ cm^{-3}. Liquid helium used as coolant. (See ref. 15).

affected by departure from the Coulomb potential of Eq. (2). In Figs. 1 and 2, the $1s(A_1)$ is shown depressed below the other ground states in agreement with electron spin resonance experiments,[3, 5–8] piezo-resistance measurements,[9] spectroscopic[10–12] and piezo-spectroscopic[12–14] observations. However, as will be seen later, in the case of interstitial lithium[15] in silicon the $1s(A_1)$ state lies above the others.

Lyman spectra arising from the multiple 1s-states to the p-states are shown for phosphorous in silicon in Figs. 3 and 4 and for arsenic in germanium in Fig. 5. Excitations from the upper ground states are observed by thermally populating these states.[10–12]

A comparison of the energy spacings of the corresponding lines of different impurities with those predicted by the effective mass theory is shown in Table I for silicon and Table II for germanium. As a consequence of the species-dependence of the ground state splitting, the spectra for the different impurities in a given material are shifted with respect to one another. The chemical splittings deduced from such measurements are also given in Tables I and II.

In addition to the Group V impurities, the excitation spectra of the

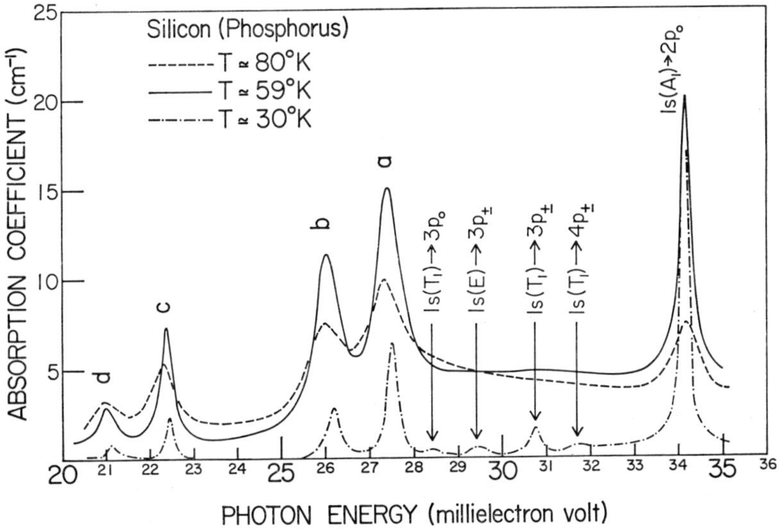

Fig. 4. Excitation spectrum of phosphorus donors in silicon. $N_D = 5.2 \times 10^{15}$ cm^{-3}. At ~30°K the peak of the line $1s(A_1) \rightarrow 2p_0$ could not be located since, for the sample used, the transmission was very low. (See ref. 12).

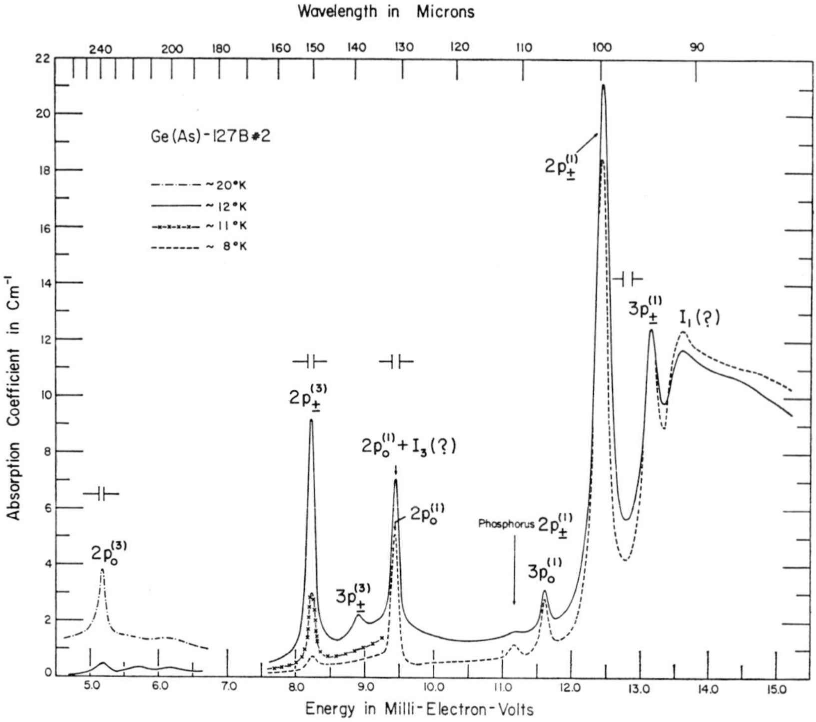

Fig. 5. The excitation spectrum of arsenic impurity in germanium for different sample temperatures. $N_D \simeq 1 \times 10^{15}$ cm^{-3}. (See ref. 10).

lithium donors in silicon and germanium have been investigated.[15] Figure 6 shows the spectrum of lithium in oxygen-free, floating-zone silicon. The upper curve is for phosphorus in silicon. In the figure, the $2p_0$ lines for both impurities have been brought into coincidence by displacing the energy scales. It is evident both from Table I and Fig. 6 that the energy spacings of the lines for lithium and for phosphorus are identical. Also, the relative intensities of the lines are very similar. It is particularly interesting to note that for lithium in floating zone silicon the excitation lines occur close to the positions predicted by the effective mass calculations, suggesting that the chemical splitting for the ground state is very small. The full implication of this feature will be discussed in detail in the following section.

The spectrum of lithium donors in crucible grown silicon,[15–17] having

TABLE 1. Spacings of energy states for donors in silicon (in meV).

States	Theory [a]	P [b]	As [b]	Sb [b]	Bi [c]	Li [d]	Li-O [d]
$2p_\pm - 2p_0$	5.0 ± 0.3	5.06	5.12	5.06	4.94	5.13	5.23
$3p_0 - 2p_\pm$	0.2 ± 0.7	0.93	0.86	0.95	0.93	0.88	0.81
$3p_\pm - 2p_\pm$	3.0 ± 0.15	3.27	3.25	3.34	3.31	3.28	3.18
$4p_0 - 2p_\pm$	—	3.11	2.6	?	2.61	3.09	3.02
$4p_\pm - 2p_\pm$	—	4.21	4.3	4.33	4.35	4.19	4.14
$5p_0 - 2p_\pm$	—	4.21	4.3	4.33	4.34	4.19	4.14
$5p_\pm - 2p_\pm$	—	4.95	4.9	?	5.26	4.93	—
$1s\,(T_1) - 1s\,(A_1)$ [e]	—	11.73	21.09	9.86	—	—	—
$1s\,(E) - 1s\,(T_1)$ [e]	—	1.37	1.41	2.42	—	—	—
$1s\,(E+T_1) - 1s\,(A_1)$	—	—	—	—	—	−1.8	7.7

a. See ref. 3.
b. See refs. 12 and 13.
c. S. Zwerdling, K. J. Button and B. Lax, Group Report M84-4 of the Lincoln Laboratory, Massachusetts Institute of Technology (unpublished). The spacings are calculated from the average zero field positions of the transitions.
d. See ref. 15.
e. A. Morita and H. Nara have recently calculated the chemical shifts of shallow donor levels in silicon. To be published in the *Proceedings of the International Conference on the Physics of Semiconductors, Kyoto, 1966* [Supplement to the Journal of the Physical Society of Japan, Vol. 21 (1966)].

TABLE II. Spacings of energy states for donors in germanium.[a] Units are millielectron volts.

States	Theory [b]	P	As	Sb	Bi
$3p_\pm - 2p_\pm$	0.75 ± 0.08	0.68	0.71	0.70	0.64
$2p_\pm - 3p_0$	0.75 ± 0.23	0.83	0.83	0.84	0.88
$2p_\pm - 2p_0$	2.90 ± 0.23	3.01	3.00	3.01	3.08
$1s\,(T_1) - 1s\,(A_1)$	—	2.83	4.23	0.32	2.87

a. See ref. 10 for experimental values.
b. See ref. 3.

a dispersed oxygen concentration of $\sim 5 \times 10^{17}$ per cm^{-3}, is shown in Fig. 7. This spectrum is also compared with that of phosphorus. As is the case for lithium in floating zone silicon, the two spectra of Fig. 7 are very similar. However, the spectra of lithium impurity in the two types of silicon are shifted relative to each other. The lithium spectrum in Fig. 6 is attributed to isolated lithium, while that in Fig. 7 is attributed to a lithium-oxygen (Li-O) complex. The latter has a chemical splitting comparable to that of antimony in silicon as can be seen from Table I.

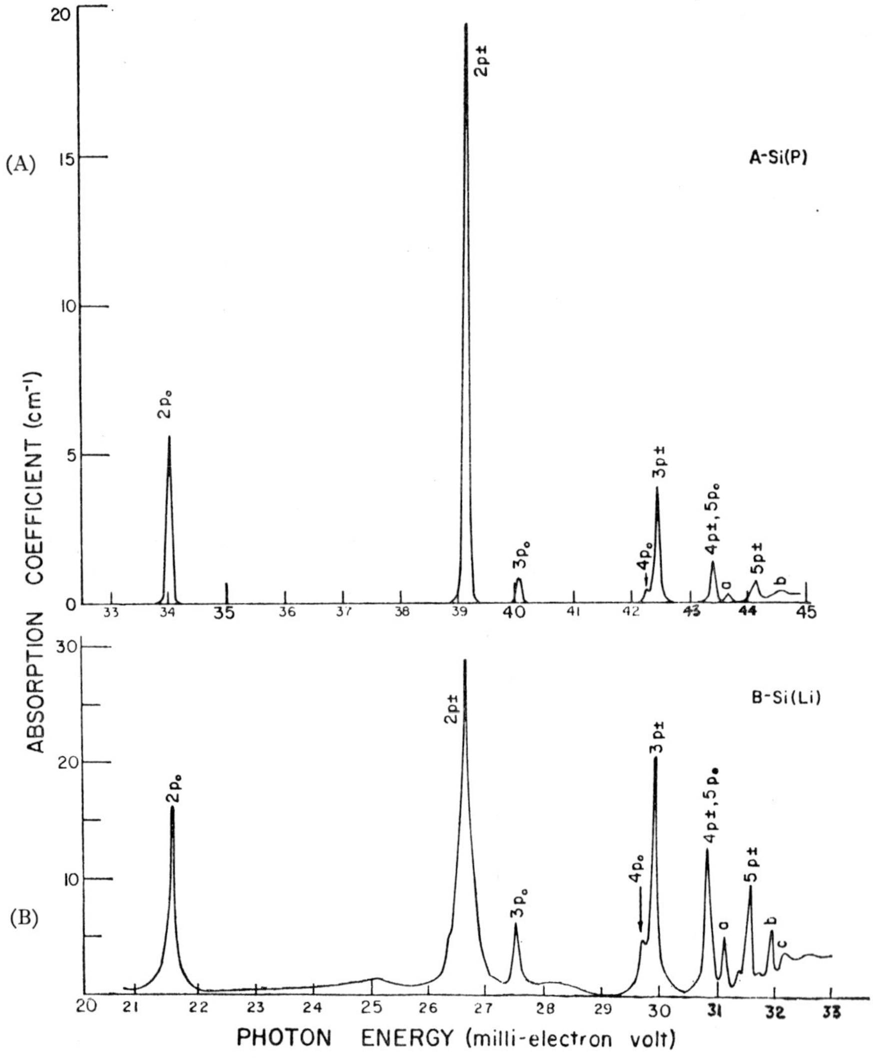

Fig. 6. (A) See Fig. 3.

(B) The excitation spectrum of lithium introduced into high resistivity floating-zone silicon; $N_D \simeq 1 \times 10^{15}$ cm^{-3}. Liquid helium used as coolant. The nature of the excited states giving rise to the lines designated by a, b, and c, have not yet been determined. (See ref. 15).

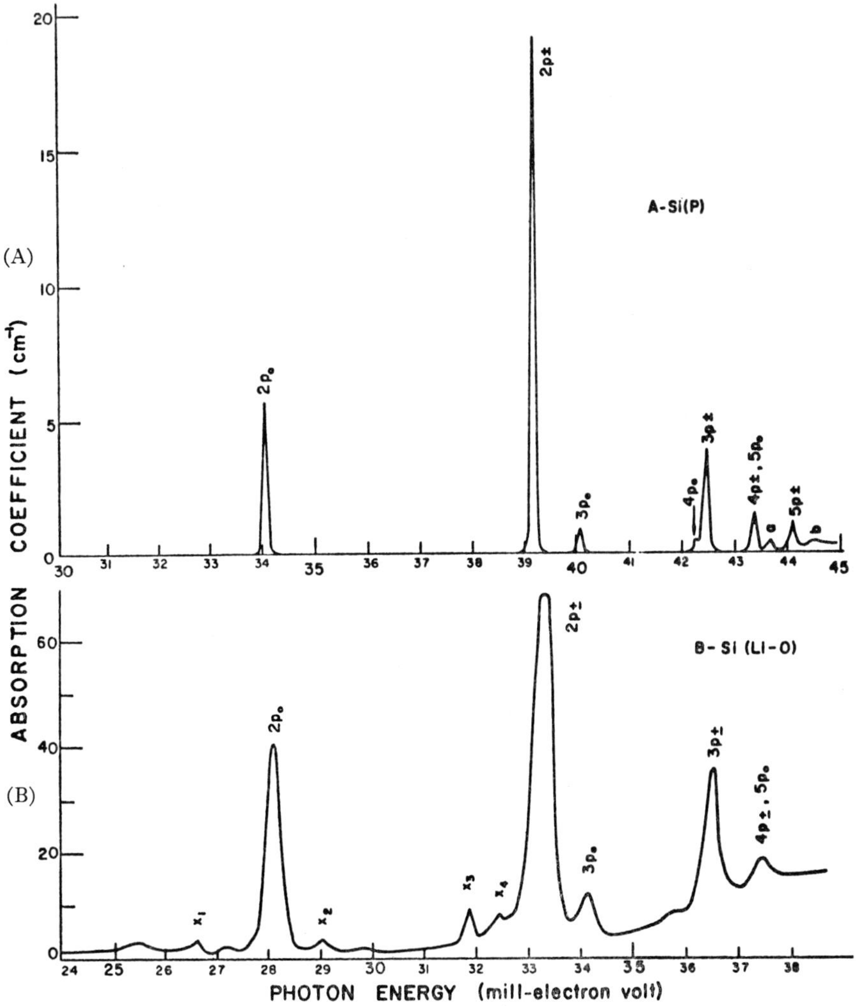

Fig. 7. (A) See Fig. 3.
(B) The excitation spectrum of lithium introduced into ~150 ohm-cm p-type crucible-grown silicon, with oxygen concentration of ~5 $\times 10^{17}$ cm^{-3}; $N_D \simeq 6 \times 10^{15}$ cm^{-3}. Liquid helium used as coolant. (See ref. 15).

2. *Effects of uniaxial stress*

A uniaxial stress, when applied to a multivalley semiconductor, shifts the conduction band minima with respect to one another.[18] Following Herring's deformation potential analysis one can write the shift in energy of the *j*th minimum as

$$\Delta E^{(j)} = [\Xi_d \delta_{\alpha\beta} + \Xi_u K_\alpha^{(j)} K_\beta^{(j)}] u_{\alpha\beta} \tag{3}$$

where $K_\alpha^{(j)}$ and $K_\beta^{(j)}$ are components of a unit vector pointing from the

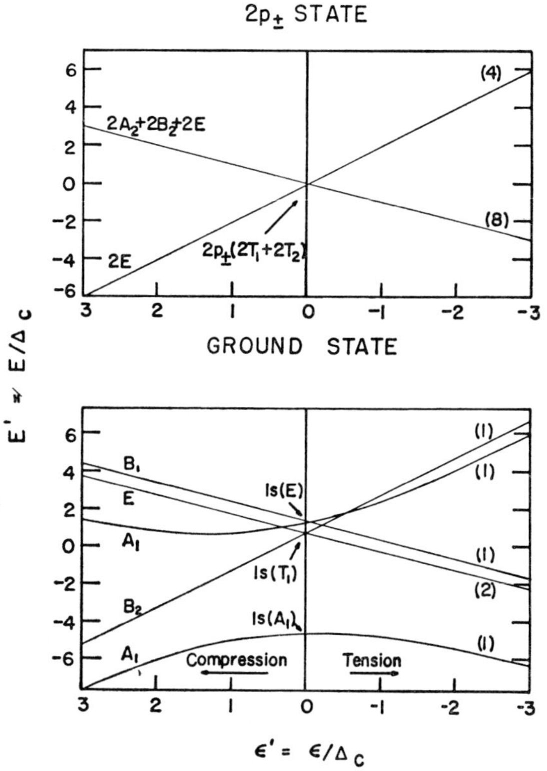

Fig. 8. Splitting of the 1s ground states and the 2p± state of Group V donors in silicon under strain for **F**||⟨100⟩, $E' = E/\Delta_c$, and $\varepsilon' = \varepsilon/\Delta_c$ where E is the energy of a given state, $\Delta_c = 1/6$ [spacing between 1s (E) and 1s(A_1)] and 3ε is the energy difference between the ⟨100⟩ conduction band minima. The numbers in parentheses indicate the degeneracies of the states while the letters denote the relevant irreducible representations of D_{2d}, the site symmetry of the donor for **F**||⟨100⟩. (See ref. 12).

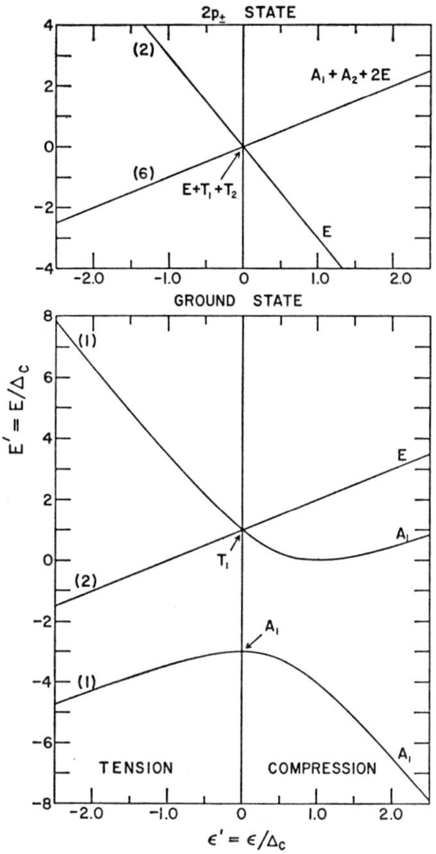

Fig. 9. Splitting of the 1s ground states and the 2p± state of Group V donors in germanium for **F**||⟨111⟩, $E'=E/\Delta_c$, and $\varepsilon'=\varepsilon/\Delta_c$ where E is the energy of a given state, $4\Delta_c$=the chemical splitting, and 4ε is the energy difference between the ⟨111⟩ conduction band minima. The numbers in parentheses indicate the degeneracies of the states while the letters denote the relevant irreducible representations of C_{3v}, the site symmetry of the donor for **F**||⟨111⟩.

center of the Brillouin zone towards the position in k-space of the jth minimum. The sub-index α or β designates a component along one of the cubic axes of the crystal. The symbols Ξ_d and Ξ_u are Herring's[18] deformation potential constants. In Eq. (3), a summation over α and β is implied. If it is assumed that both the dielectric constant and the effective masses characterizing the conduction band minima are unaltered by a small strain,

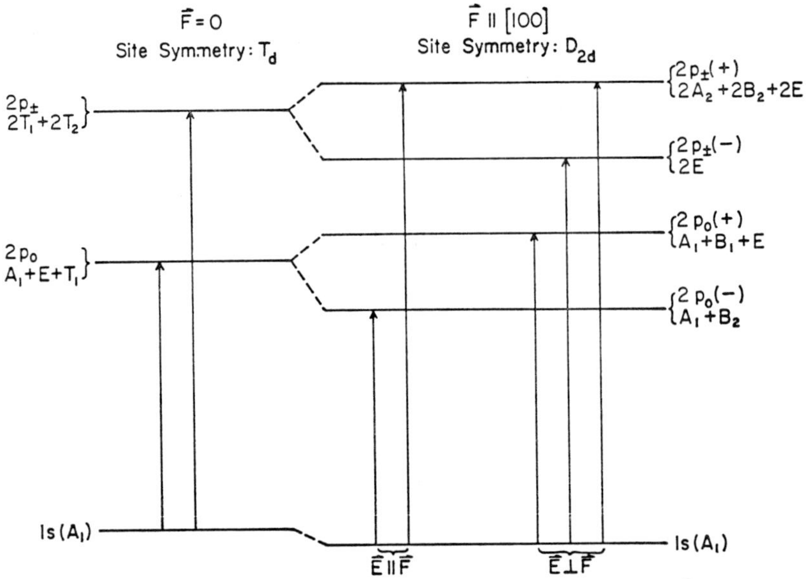

Fig. 10. Energy levels (not to scale) for $\mathbf{F}=0$ and $\mathbf{F}\|\langle 100\rangle$. The arrows indicate the allowed transitions with 1s (A_1) as the ground state. The letters next to a level denote the representations of the appropriate site symmetry. (See ref. 13).

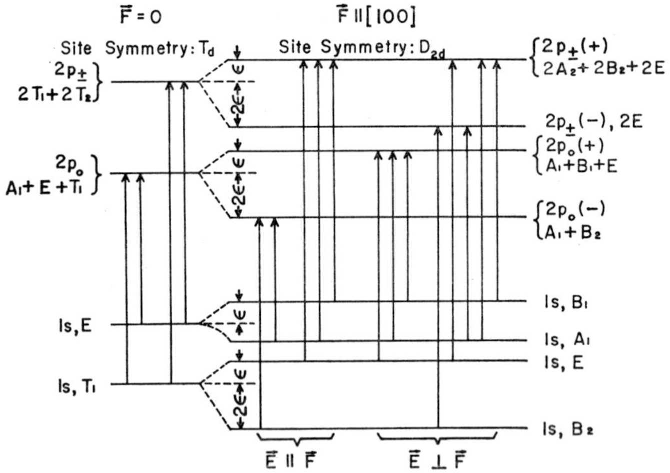

Fig. 11. Energy levels (not to scale) of a Group V donor for $\mathbf{F}=0$ and $\mathbf{F}\|\langle 100\rangle$. The arrows indicate the allowed transitions with 1s(E) and 1s(T_1) as the ground states. The letters next to a level denote the representations of the appropriate site symmetry. (See ref. 12).

then, for a given valley, the energy level scheme of a Group V donor will be unaffected by the stress since Eq. (2) will be unaltered. However, the energy level schemes bearing different valley labels will be shifted relative to one another by the amounts given in Eq. (3). This will be true for all states which are well described by the effective mass Hamiltonian, e.g., the p-states. For the ground states, Price[19] has considered the effect of strain for donors in germanium explicitly taking into account the chemical splitting. Following Price, similar calculations have been made by Wilson and Feher[20] for silicon.

The effect of uniaxial stress on a typical p-state and the ground state of a Group V donor is shown in Figs. 8 and 9 for silicon and germanium, respectively. The applied force, F, is along ⟨100⟩ for silicon and ⟨111⟩ for germanium. The labels on the stress-induced sub-levels denote the irreducible representations of the new site symmetry to which they belong.

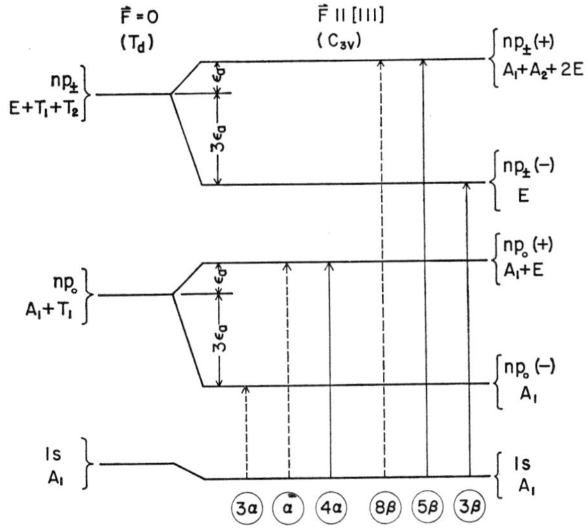

Fig. 12. Splitting of the shallow donor states in germanium under a ⟨111⟩ compression. The vertical arrows indicate the allowed transitions where the dashed arrows are for $E_{||}$ and the full arrows are for $E_{\perp}$. The capital letters labelling the levels denote the irreducible respresentations of T_d and C_{3v} to which the states belong. Note that T_1 here is the irreducible representation under which x, y, and z transform. The quantity ε_a is the same as ε in Fig. 9. The encircled quantity below each arrow refers to the intensity of the transition where $\alpha = A^2/12$ and $\beta = B^2/6$. The quantities A and B are defined in the Appendix, Eqs. (A6), of ref. 14.

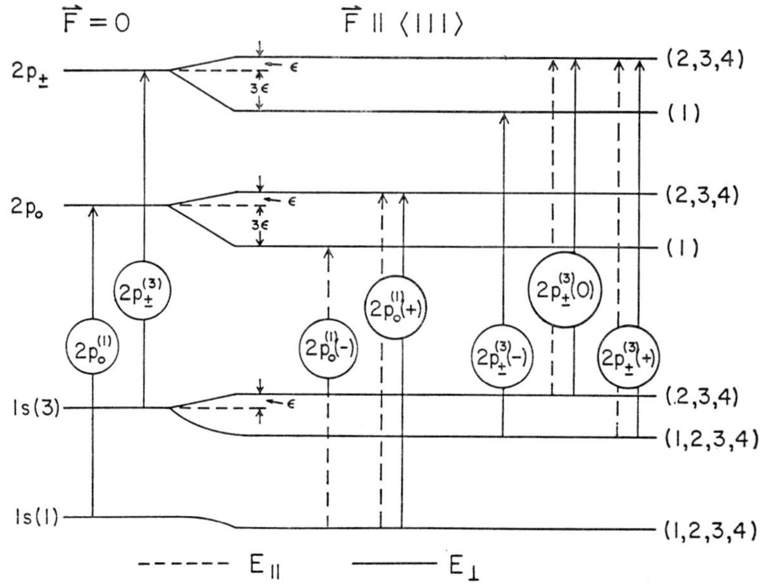

Fig. 13. The effect of a ⟨111⟩ compression on the energy states of a Group V impurity in germanium. The arrows show the allowed transition between the upper ground state and the 2p± state and between the lower ground state and the $2p_0$ state. The numbers in parentheses for the stress case indicate which valleys contribute to the linear combination of a given state for **F** along the axis of valley 1.

These assignments have been arrived at by the use of a group theoretical technique.[12–14, 21] The selection rules for electric dipole transitions are shown in Figs. 10–13. These can be deduced either by group theory or by an actual intensity calculation using linear combinations of wave functions appropriate for the new site symmetry of the impurity. From these figures, it is clear that the transitions which have $1s(A_1)$ as their ground state should, under stress, exhibit two components one on either side of the zero stress position. For silicon, transitions from $1s(T_1)$ occur at an energy which is coincident with the zero stress energy. On the other hand, transitions with $1s(E)$ as their ground state give rise to three components, one of which occurs at the zero stress value, the other two lying on either side of it. For germanium, transitions from the $1s(T_1)$ ground state give rise to three components with a similar pattern. Figs. 14–17 show the experimental results.[12–14, 22] As can be seen, close agreement between predictions and experimental observations is obtained. For most

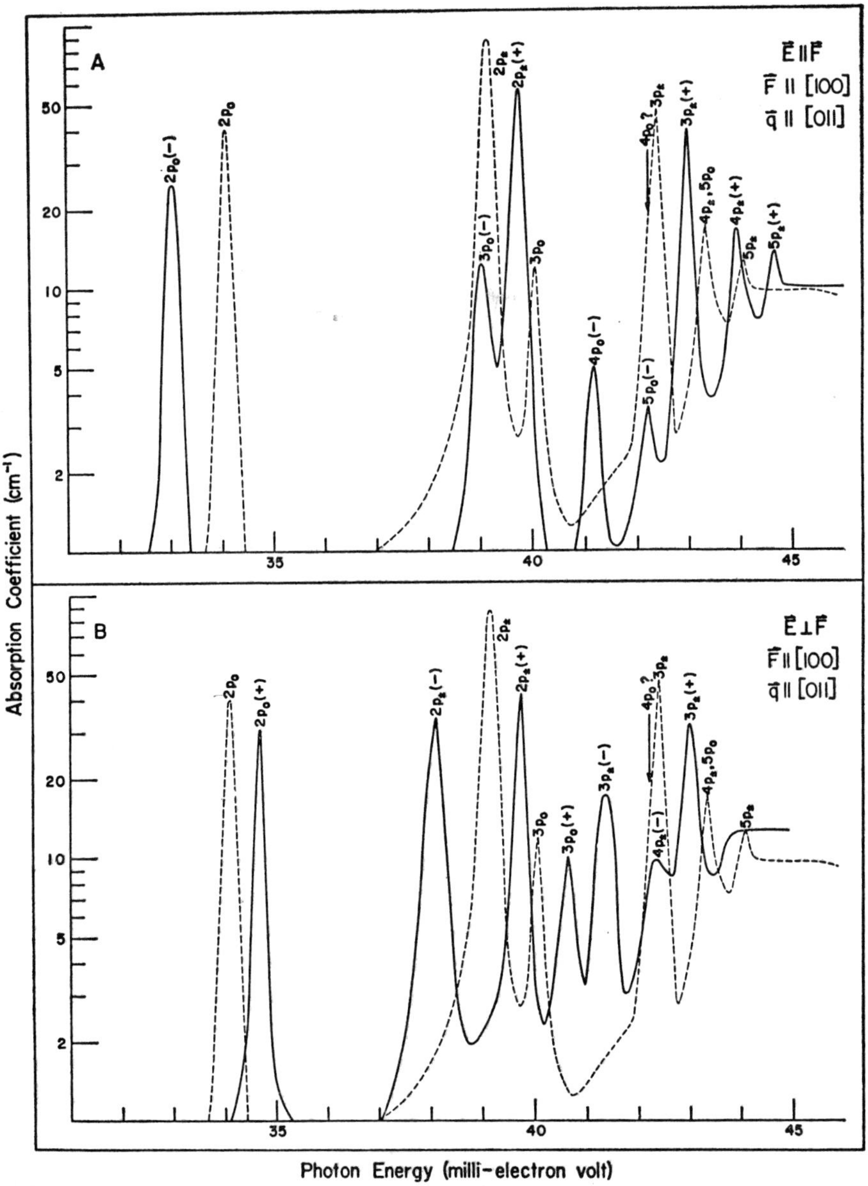

Fig. 14. Excitation spectrum of phosphorus in silicon [ρ(300°K) $=0.9\ \Omega\cdot$cm] under uniaxial compression, **F**, along ⟨100⟩. A: The electric vector **E** ∥ **F**. B: **E**⊥**F**. The dotted curve is for **F**=0. (See ref. 13).

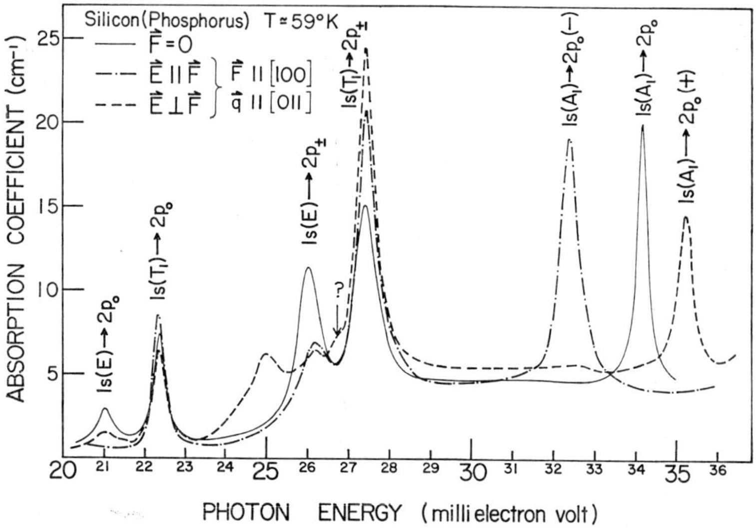

Fig. 15. The effect of a ⟨100⟩ compression on the excitation spectrum of phosphorus in silicon. $N_D = 5.2 \times 10^{15}$ cm^{-3}. (See ref. 12).

cases, observations and predictions have also been compared for both silicon and germanium for **F**||⟨110⟩. The noteworthy feature for this direction of applied force is that the theory predicts that the polarization properties of the spectra should be dependent on the direction of light propagation. This is confirmed in detail by the experimental observations.[12–14]

When the force is applied in such a direction that the energy equivalence of the conduction band minima is not destroyed, the excitation lines should not split. That this is the case has been demonstrated experimentally[13,14] for silicon with **F**||⟨111⟩ and for germanium with **F**||⟨100⟩.

The piezo-spectroscopic effects observed for Group V donors in silicon and germanium can thus be understood in the framework of the effective mass theory (together with the corrections to the ground state) in conjunction with the deformation potential theory. The technique is clearly useful for arriving at symmetries not only of the excited states but also of the ground states. The splitting of the excited states obtained from the observed line splittings can be used to determine the deformation potential constant Ξ_u of the conduction band minima.

The piezo-spectroscopic technique has been used to study the energy states of Li-O complex in silicon.[15] The results for Li-O with **F**||⟨100⟩

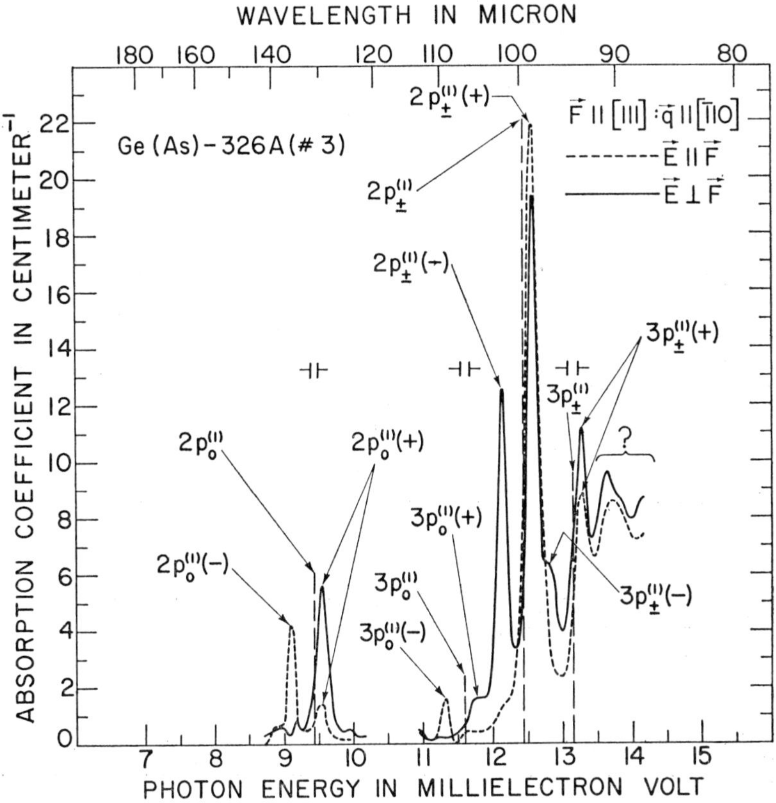

Fig. 16. The effect of a ⟨111⟩ compression on the excitation spectrum of arsenic in germanium at ~8°K. The dashed curve is for $E_{\parallel}$ while the full curve is for $E_{\perp}$. The vertical dashed lines indicate the energies of the zero stress transitions. N(As)~0.8×10^{15} cm^{-3}. (See ref. 14).

are shown in Fig. 18. A comparison of Figs. 14 and 18 shows that the behavior of the lines of the Li-O complex under stress is identical to the behavior of their counterparts in the phosphorus spectrum. From this we conclude that the lowest ground state for the Li-O complex must be the singlet 1s(A_1). Zero stress experiments at higher temperature have shown lines which have been attributed to transitions arising from the upper ground state; stress experiments have yet to be performed.

Similar observations have been made on the spectrum of isolated lithium in silicon.[15] The results for $\mathbf{F} \| \langle 100 \rangle$ are shown in Fig. 19. These

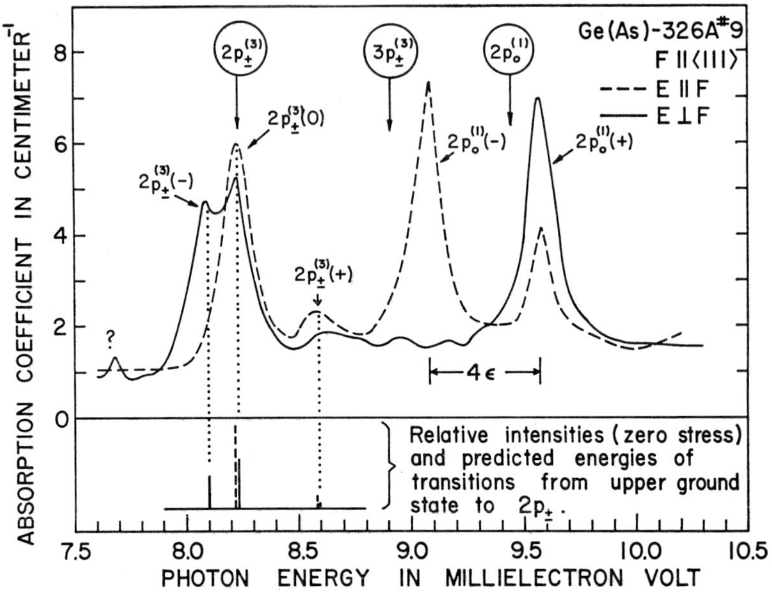

Fig. 17. The effect of a ⟨111⟩ compression on the transitions $1s(A_1) \to 2p_0$ and $1s(T_1) \to 2p_\pm$ for arsenic impurity in germanium. $N_D \simeq 2 \times 10^{15}$ cm^{-3}.

results are strikingly different from those for phosphorus in that zero stress components have been observed for all the lines in addition to the high and low energy components. This is true also for **F**∥⟨110⟩. For **F**∥⟨111⟩, however, the lines do not split just as is the case for Group V donors. The results for isolated lithium have been explained in detail assuming that the energy level scheme for lithium is identical to that of a Group V impurity, *except* that the ground states are inverted i.e. *the* $1s(A_1)$ *ground state of lithium lies above the* $1s(E+T_1)$ *ground states instead of below.*[15] The behavior of such a ground state for **F**∥⟨100⟩ is shown in Fig. 20 together with the selection rules.

The case of sulfur donors in silicon has been studied extensively by Krag and co-workers.[23] Excitation spectra associated with both neutral and singly ionized sulfur and sulfur complexes have been observed. The spacings of the excitation lines indicate that the excited p-states are effective-mass-like. Piezo-spectroscopic effects of these centers have demonstrated that some have symmetries lower than T_d. For example, some of the centers give spectra which exhibit splittings even for **F**∥⟨111⟩.

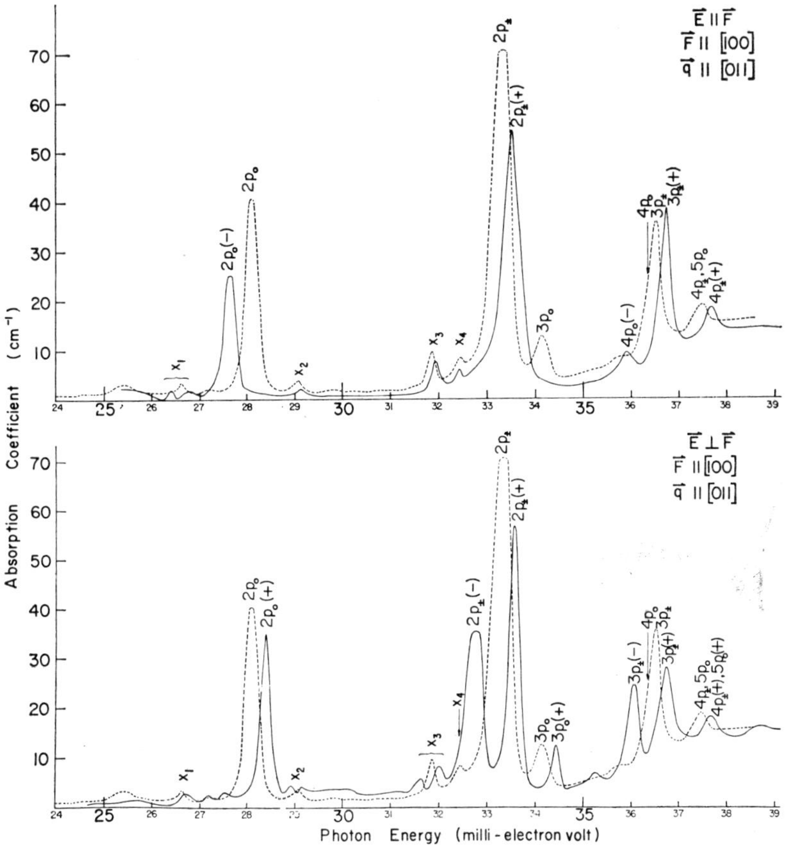

Fig. 18. The effect of a [100] compression on the excitation spectrum of Fig. 7(B) for **q** || [011]. The upper curve is for **E** || **F** and the lower is **E**⊥**F**. The dashed curve in each diagram is for **F**=0. (See ref. 15).

III. ACCEPTOR STATES

1. *Zero stress*

The valence band maximum of silicon or germanium occurs at the center of the Brillouin zone and is split by spin-orbit interaction into a four-fold branch ($p_{3/2}$) and a two-fold branch ($p_{1/2}$). When substitutional acceptors are present, each branch has associated with it a set of localized states. Figure 21 shows schematically a few typical states and the associated

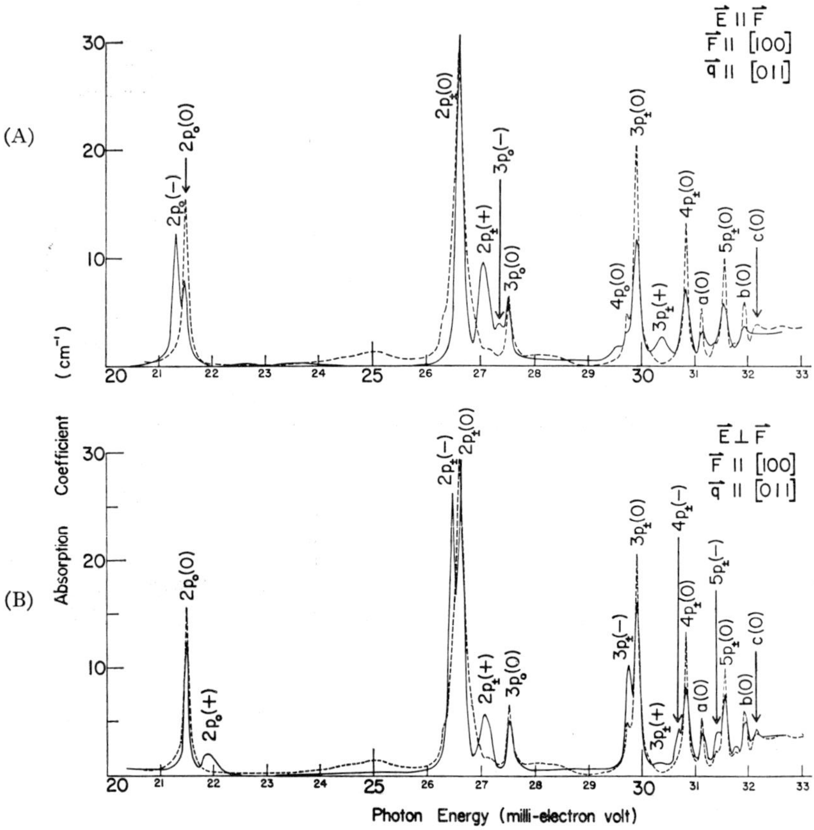

Fig. 19. The effect of a [100] compression on the excitation spectrum of Fig. 6(B) for **q** || [011]. The upper curve is fo **E** || **F** and the lower for **E**⊥**F**. The dashed curves are for **F**=0. (See ref. 15).

optical transitions. The effective mass description of the shallow acceptor states has been given by several authors.[3, 24, 25]

In the case of germanium, only the $p_{3/2}$ spectrum has been observed. For the Group III acceptors both the energy spacings and the relative intensities of corresponding lines for all the impurities are very similar. The energy spacings of the lines relative to the D line for the different Group III impurities are shown in Table III. The meaning of the nomenclature can be seen from Fig. 22 which shows the $p_{3/2}$ spectrum of a typical Group III impurity in germanium i.e. gallium.[26] The spectra are, however, displaced in energy with respect to one another due to a chemical shift of

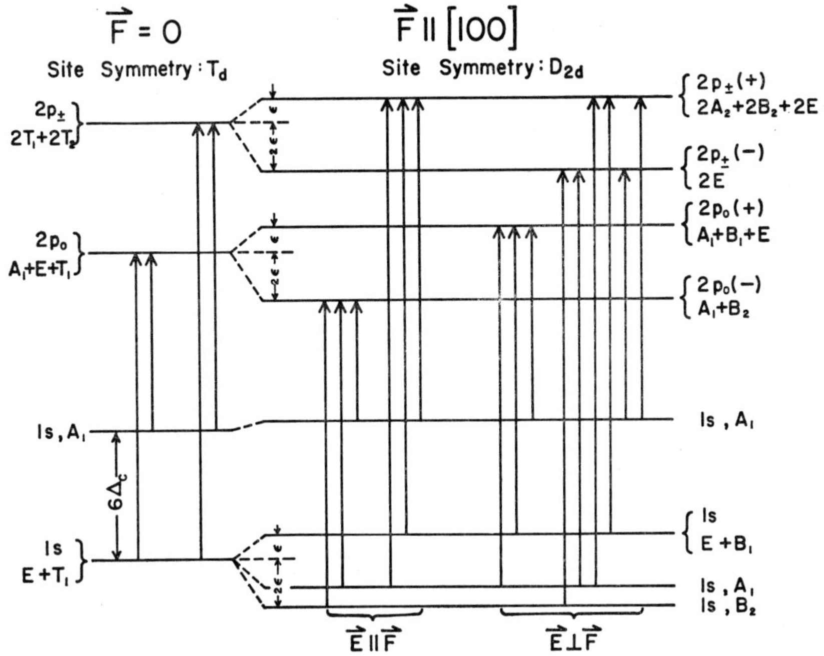

Fig. 20. Splitting of the donor levels in silicon (not to scale) under a ⟨100⟩ compression for a finite $6\Delta_c$ and with an "inverted" Group V-like ground state. The vertical arrows indicate the allowed transitions. The capital letters labelling the levels denote the irreducible representations of T_d and D_{2d} to which the states belong. (See ref. 15).

TABLE III. Energy spacings* of the excitation lines of neutral acceptor impurities in germanium. Units are millielectron volts.

	Boron[a]	Aluminum[a]	Gallium[a]	Indium[a]	Thallium[a]	Zinc[b]	Copper[b]	Mercury[c]
A′-D	1.87	1.88	1.87	1.88	1.86	1.91	—	1.94
A″-D	1.71	1.75	1.73	1.73	1.69	1.73	—	1.74
B-D	1.38	1.40	1.40	1.40	1.35	1.39	1.40	1.42
a-D	1.12	—	—	1.12	1.08	—	—	—
D-C	0.75	0.75	0.75	0.78	0.75	0.76	0.74	0.80
D-E	0.37	—	0.42	0.66	0.74	—	—	0.67
D-G	1.70	1.68	1.70	1.69	1.70	—	1.72	1.74

* The spacings are given relative to the D line.

a See ref. 26.

b See ref. 28, and R. L. Jones and P. Fisher (unpublished data).

c See refs. 29 and 30.

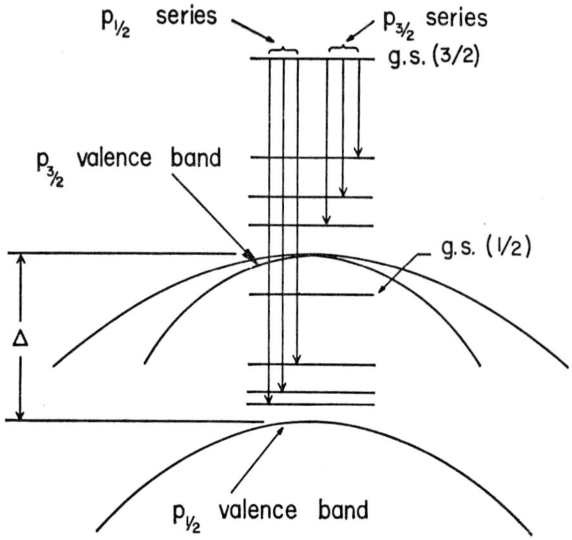

Fig. 21. The spin-orbit split valence band of silicon or germanium with associated acceptor states.

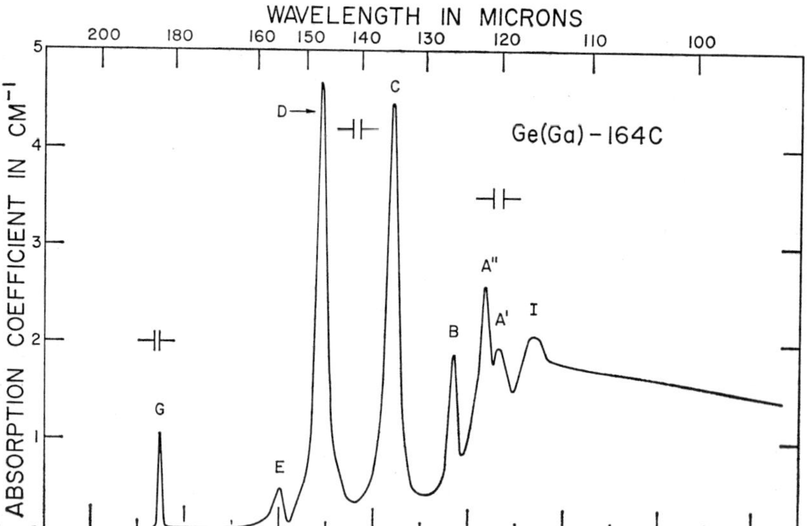

Fig. 22. The $p_{3/2}$ excitation spectrum of gallium impurity in germanium. Temp. $\simeq 9°K$; $N_A \simeq 2 \times 10^{14}$ cm^{-3}. (See ref. 26).

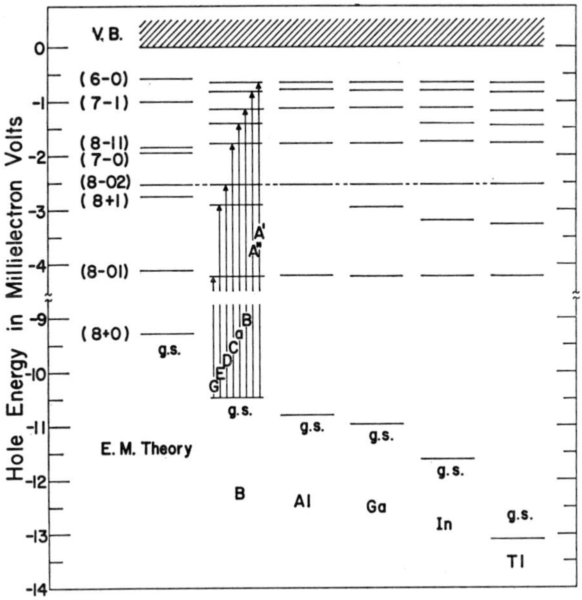

Fig. 23. Correlation of experimental and theoretical energy level schemes of Group III impurities in germanium. The final state of the D transition is assumed to correspond to the (8-02) state of Mendelson and James, ref. 25. (See ref. 26).

the ground state. Fig. 23 shows a correlation between the experimental results[26] and the predictions of the effective mass calculation of Mendelson and James.[25] Clearly, the agreement between experiment and theory is not as good as for Group V donors in germanium, although, as has been pointed out earlier, the spectra of the five Group III impurities are very similar. This similarity, especially with regard to the line spacings (see Table III), extends to the "deep" acceptors, beryllium,[27] zinc,[28, 29] copper[28, 29] and mercury.[29, 30] The implication of this is that the excited p-like states of all the acceptors in germanium so far studied are identical.

The spectra of substitutional Group III impurities in silicon have been observed and show both the $p_{3/2}$ series[31–35] and $p_{1/2}$ series[36, 37] of lines. However, for this case, unlike that of germanium, the $p_{3/2}$ spectra show some marked species dependence[35] other than the chemical shift. In view of this, the agreement with the effective mass theory is not expected to be even as good as for germanium. Further, the effective mass calculations for silicon[24] are not as complete as those for germanium.[25]

2. *Effects of uniaxial stress*

It is predicted that the energy states of the substitutional Group III impurities in silicon and germanium, including spin, belong to the double

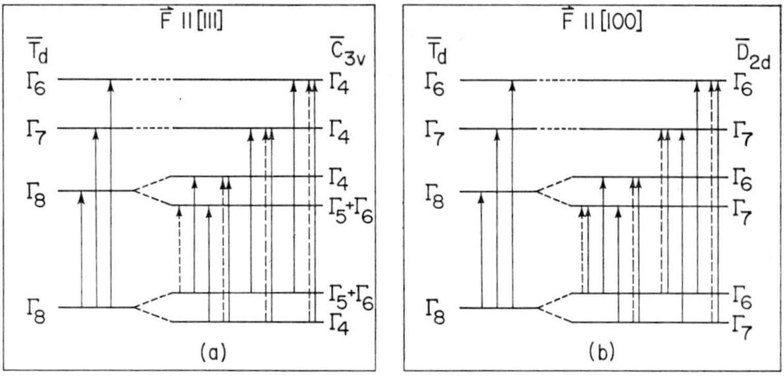

Fig. 24. The behavior of transitions from a Γ_8 ground state to Γ_6, Γ_7, and Γ_8 excited states of the double group $\overline{T}_d$ under uniaxial compression. The dashed arrows are for **E** || **F** while the full arrows are for **E** ⊥ **F**.

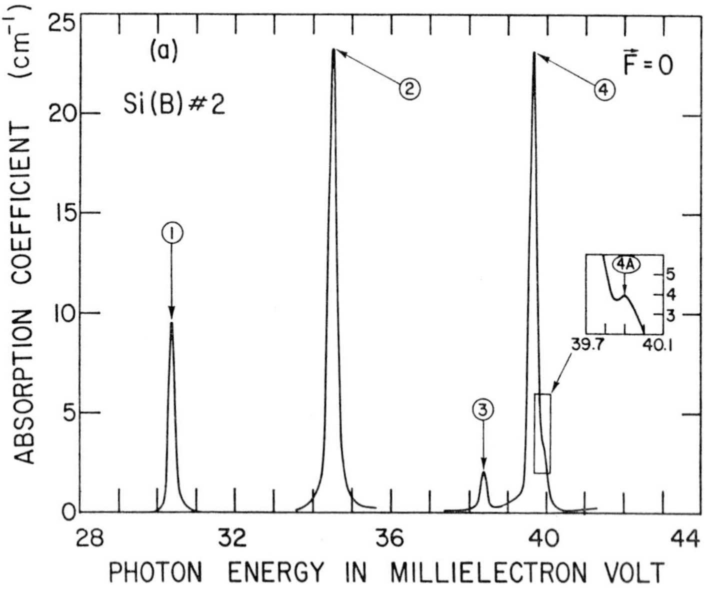

Fig. 25. The $p_{3/2}$ excitation spectrum of boron doped silicon with liquid helium as coolant. Room temperature resistivity ~12 ohm·cm. (See ref. 34).

tetrahedral group, $\overline{T}_d$. Thus, the wave functions of these impurities should conform to the irreducible representations[38] Γ_6, Γ_7, and Γ_8 of $\overline{T}_d$. Of these Γ_6 and Γ_7 are two-dimensional representations while Γ_8 is a four-dimensional representation. The effective mass theory predicts[24] that the ground state of these impurities is four-fold degenerate and hence is represented by Γ_8. A transition between the four-fold degenerate ground state and a two-fold degenerate state should, under uniaxial stress, give rise to at most two components, as neither Γ_6 nor Γ_7 can be split by strain whereas a Γ_8 state can split into two two-fold states.[3] On the other hand, a transition to a Γ_8 state can give rise to as many as four components. These results are summarized in Fig. 24 for $\mathbf{F}\|\langle 111\rangle$ and $\mathbf{F}\|\langle 100\rangle$.

A part of the $p_{3/2}$ spectrum of boron in silicon[34] is shown in Fig. 25. The effect of a $\langle 111\rangle$ compression on this spectrum as well as on the 2p′ line of the $p_{1/2}$ spectrum[37,39] is shown in Fig. 26. The insets show the effect of a higher stress on lines 1 and 2p′. Only two components have been observed for the 2p′ line for all directions and magnitudes of com-

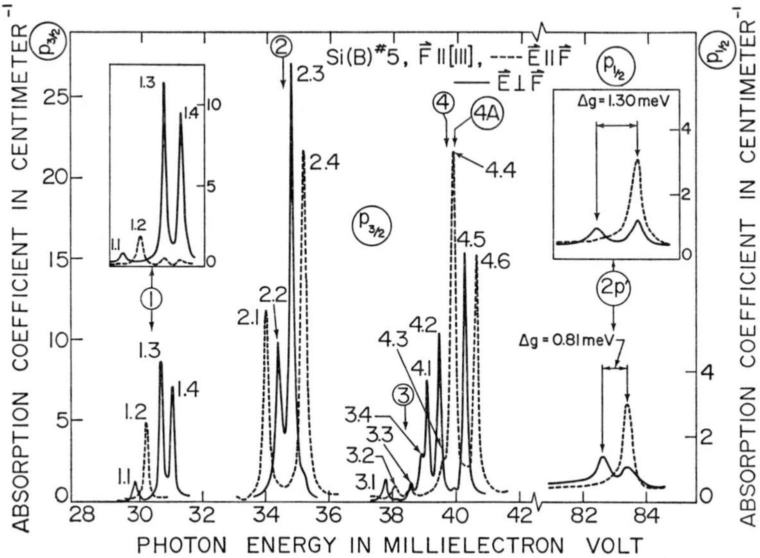

Fig. 26. The excitation lines 1-4A and 2p′ of neutral boron impurity in silicon under a $\langle 111\rangle$ compression. The encircled numbers together with the associated vertical arrows indicate the positions of the zero stress lines. The two insets show the effect of a larger stress on lines 1 and 2p′. Room temperature resistivity ~12 ohm·cm. (See ref. 39).

pression. This is consistent with the final state of this transition being either Γ_6 or Γ_7. Thus from the polarization features of the 2p′ line the sub-levels of the ground state for $\mathbf{F}||\langle 111\rangle$ can be ordered. Using this ordering good agreement has been obtained between the prediction of Fig. 24 and the experimental results of Fig. 26. It is found that the order of the stress induced sub-levels of the upper Γ_8 states of lines 1 and 3 is opposite to that of line 2. Similar studies have also been made of transitions to higher lying states. The spectra of aluminum, gallium and indium have also been studied under uniaxial compression.[39]

Since the excited states of neutral acceptors in germanium are identical, any difference in the stress induced spectra for the different types of acceptors must be attributed to the different natures of the ground states. Fig. 27 shows the effect of a ⟨100⟩ compression on some of the excitation

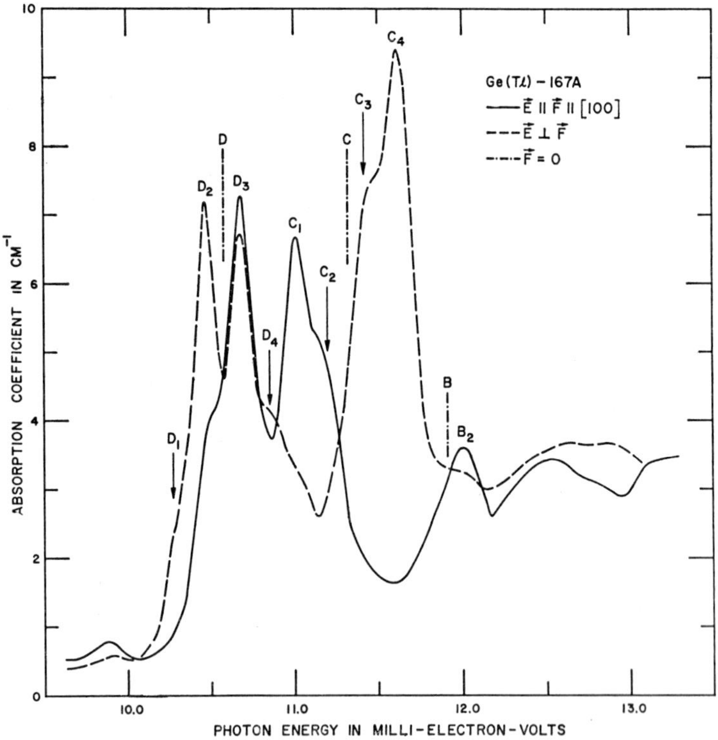

Fig. 27. The excitation spectrum of thallium impurity in germanium for $\mathbf{F} || \langle 100\rangle$. Sample temperature ~8°K. $N_A \simeq 4\times10^{14}$ cm^{-3}. The vertical dot-dashed lines denote the zero stress positions.

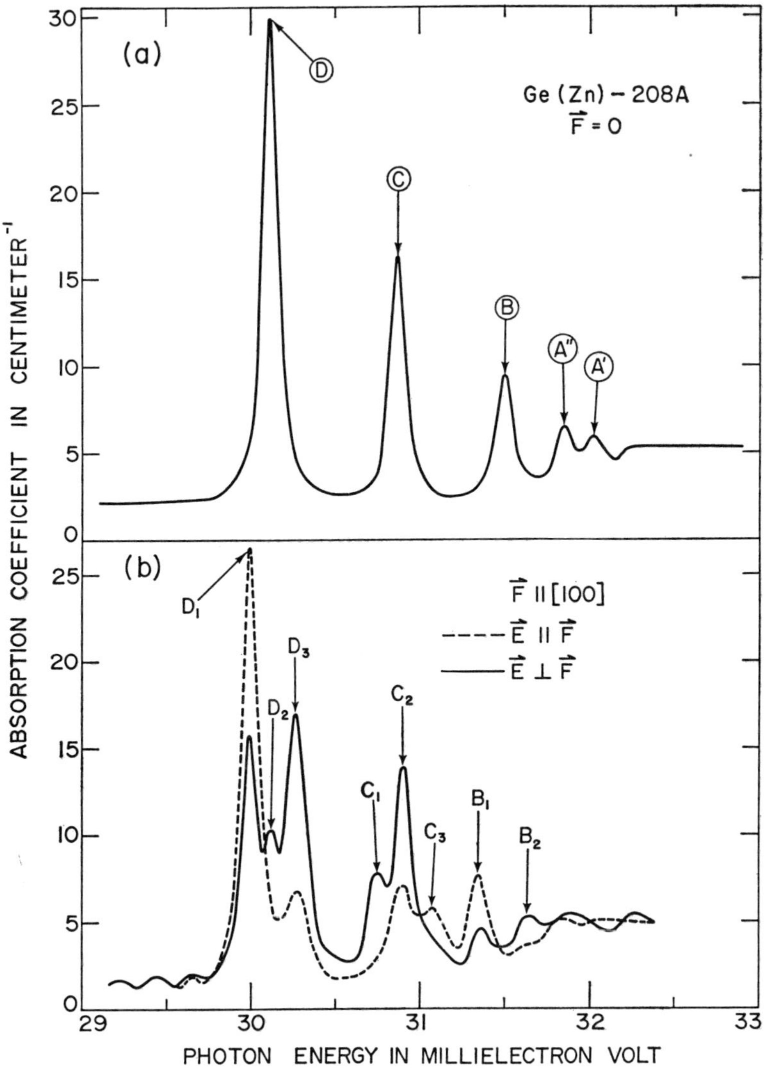

Fig. 28. The excitation spectrum of neutral zinc in germanium measured with and without uniaxial stress. Liquid helium used as coolant. Room temperature resistivity ~2.5 ohm·cm.

lines of thallium impurity in germanium.[40] The vertical dashed lines indicate the zero stress positions. Note that neither the C nor D line gives rise to a zero stress component while for light polarized parallel to

F, a high energy B component is observed. As can be seen from Fig. 28, this is not the case for neutral zinc in germanium.[41] For this case, lines C and D do give rise to zero stress components, while the B line gives a low energy component for parallel polarization. Similar but preliminary observations have been made for the case of neutral copper in germanium. The results indicate that the effect of uniaxial stress on the spectrum of copper in germanium is different from those for Group III and neutral zinc impurities. These differences may have their origin in the fact that the neutral impurities thallium, zinc and copper bind one, two and three holes, respectively. Evidence for a complex ground state has indeed been found for the neutral double acceptor mercury by Chapman *et al.*[42]

The spectra of thallium and neutral zinc in germanium have been examined for $\mathbf{F}||\langle 111\rangle$ and $\mathbf{F}||\langle 110\rangle$, also.[39] A complete understanding of all the experimentally observed features has not yet been obtained.

IV. CONCLUSIONS

It is clear from the above discussion that the piezo-spectroscopic technique when applied to impurity states in semiconductors gives considerable information not only about the excited states but also about the ground states. In this respect, the technique is more powerful than the Zeeman effect. For example, different acceptors in germanium show the same Zeeman effect[28, 43] but exhibit different piezo-spectroscopic effects. In addition, the splitting of the excitation lines for both acceptors[40] and donors[23] provides the most direct method of measuring some of the deformation potential constants of the host crystal.

Acknowledgements

Two of the authors (P.F. and A.K.R.) take pleasure in thanking R.L. Aggarwal, R. L. Jones, P. M. Lee and A. Onton for their enthusiastic participation in various aspects of the above work.

REFERENCES

1. The term " piezo-spectroscopic " was coined by A. A. Kaplyanskii who has made extensive studies of the effect of stress on the spectra of solids. See, for example, *Opt. i. Spectroskopiya*, **16**, 1301 (1964) [English transl.; *Opt. Spectry*, **16**, 557 (1964)].
2. W. Kohn and J. M. Luttinger, *Phys. Rev.*, **98**, 915 (1955).
3. W. Kohn, *Solid State Physics* (F. Seitz and D. Turnbull, eds.), Vol. 5, p. 257, Academic Press Inc., New York, 1957. This article contains a comprehensive bibliography of publications on shallow impurities in silicon and germanium prior to 1957. This

includes the pioneering work of Burstein and co-workers on the excitation spectra of these impurities in silicon.
4. The various levels are labeled in two ways (1) according to the magnetic quantum number, m, associated with the envelope wave functions and (2) according to the irreducible representations of the point group T_d to which the full wave functions belong. The notation for the representations follows E. B. Wilson, Jr., J. C. Decius and P. C. Cross, *Molecular Vibrations* (McGraw-Hill Book Company, Inc., New York, 1955) with $T_1 = F_2$ and $T_2 = F_1$.
5. W. Kohn and J. M. Luttinger, *Phys. Rev.*, **97**, 883 (1955).
6. G. Feher, *Phys. Rev.*, **114**, 1219 (1959).
7. D. K. Wilson and G. Feher, *Phys. Rev.*, **124**, 1068 (1961).
8. D. K. Wilson, *Phys. Rev.*, **134**, A265 (1964).
9. H. Fritzsche, *Phys. Rev.*, **115**, 336 (1959); *ibid.*, **120**, 1120 (1960).
10. P. Fisher, *J. Phys. Chem. Solids*, **23**, 1346 (1962); J. H. Reuszer and P. Fisher, *Phys. Rev.*, **135**, A1125 (1964).
11. R. L. Aggarwal, *Solid State Comm.*, **2**, 163 (1964).
12. R. L. Aggarwal and A. K. Ramdas, *Phys. Rev.*, **140**, A1246 (1965).
13. R. L. Aggarwal and A. K. Ramdas, in *Proceedings of the International Conference on the Physics of Semiconductors, Paris, 1964*, p. 797, Dunod Cie, Paris, 1964; R. L. Aggarwal and A. K. Ramdas, *Phys. Rev.*, **137**, A602 (1965).
14. J. H. Reuszer and P. Fisher, *Phys. Rev.*, **140**, A245 (1965).
15. R. L. Aggarwal, P. Fisher, V. Mourzine and A. K. Ramdas, *Phys. Rev.*, **138**, A882 (1965).
16. E. M. Pell, in *Solid State Physics in Electronics and Telecommunication* (M. Désirant and J. L. Michiels, eds.), Vol. 1, p. 261, Academic Press Inc., New York, 1960.
17. W. Kaiser, P. H. Keck and C. F. Lange, *Phys. Rev.*, **101**, 1264 (1956).
18. C. Herring, *Bell System Tech. J.*, **34**, 237 (1955); C. Herring and E. Vogt, *Phys. Rev.*, **101**, 944 (1956).
19. P. J. Price, *Phys. Rev.*, **104**, 1223 (1956).
20. D. K. Wilson and G. Feher, *Phys. Rev.*, **124**, 1068 (1961).
21. A. K. Ramdas, P. M. Lee and P. Fisher, *Phys. Letters*, **7**, 99 (1963).
22. J. H. Reuszer and P. Fisher, unpublished results on the effect of stress on $1s(T_1) \rightarrow np$ transitions.
23. W. E. Krag and H. J. Zeiger, *Phys. Rev. Letters*, **8**, 485 (1962); W. E. Krag, W. H. Kleiner, S. Fischler and H. J. Zeiger, *Bull. Am. Phys. Soc.*, **9**, 705 (1964); W. E. Krag, W. H. Kleiner, H. J. Zeiger and S. Fischler, to be published in the *Proceedings of the International Conference on the Physics of Semiconductors, Kyoto, 1966* [*Suppl. J. Phys. Soc. Jap.*, Vol. 21 (1966)].
24. W. Kohn and D. Schechter, *Phys. Rev.*, **99**, 1903 (1955); D. Schechter, *J. Phys. Chem. Solids*, **23**, 237 (1962).
25. K. S. Mendelson and H. M. James, *J. Phys. Chem. Solids*, **25**, 729 (1964).
26. R. L. Jones and P. Fisher, *J. Phys. Chem. Solids*, **26**, 1125 (1965).
27. W. J. Moore and R. Kaplan, *Bull. Am. Phys. Soc.*, **11**, 206 (1966).
28. P. Fisher and H. Y. Fan, *Phys. Rev. Letters*, **5**, 195 (1960).
29. W. J. Moore, *Solid State Comm.*, **3**, 385 (1965).
30. R. A. Chapman and W. G. Hutchinson, *Solid State Comm.*, **3**, 293 (1965).
31. E. Burstein, E. E. Bell, J. W. Davisson and M. Lax, *J. Phys. Chem.*, **57**, 849 (1953);

E. Burstein, G. Picus, B. Henvis and R. Wallis, *J. Phys. Chem. Solids*, **1**, 65 (1956); R. Newman, *Phys. Rev.*, **99**, 465 (1955); *ibid.*, **103**, 103 (1956).
32. H. J. Hrostowski and R. H. Kaiser, *J. Phys. Chem. Solids*, **4**, 148 (1958).
33. K. Colbow, *Can. J. Phys.*, **41**, 1801 (1963).
34. P. Fisher and A. K. Ramdas, *Phys. Letters*, **16**, 26 (1965); M. W. Skoczylas and J. J. White, *Can. J. Phys.*, **43**, 1388 (1965).
35. A. Onton, P. Fisher and A. K. Ramdas, *Bull. Am. Phys. Soc.*, **11**, 740 (1966).
36. S. Zwerdling, K. J. Button, B. Lax and L. M. Roth, *Phys. Rev. Letters*, **4**, 173 (1960).
37. A. Onton and A. K. Ramdas, *Bull. Am. Phys. Soc.*, **11**, 51 (1966).
38. G. F. Koster, J. O. Dimmock, R. G. Wheeler and H. Statz, *Properties of the Thirty-two Points Groups* (M.I.T. Press, Cambridge, Mass., 1963). In the case of the acceptors the nomenclature of this reference has been followed.
39. P. Fisher, R. L. Jones, A. Onton and A. K. Ramdas, to be published in the *Proceedings of the International Conference on the Physics of Semiconductors, Kyoto, 1966* [*Suppl. J. Phys. Soc. Jap.*, Vol. 21 (1966)].
40. R. L. Jones and P. Fisher, *Solid State Comm.*, **2**, 379 (1964); D. H. Dickey and J.O. Dimmock, *Bull. Am. Phys. Soc.*, **10**, 1086 (1965).
41. It should be noted that in Fig. 27 the dashed line represents the curve taken with light polarized perpendicular to **F** while in Fig. 28 it represents results for the parallel direction of polarization.
42. R. A. Chapman, W. G. Hutchinson and T. L. Estle, *Phys. Rev. Letters*, **17**, 132 (1966).
43. P. Fisher and H. Y. Fan, *Phys. Rev. Letters*, **2**, 456 (1959).

Spectra of U Centers in Alkali Halides*

SHASHANKA S. MITRA

Department of Electrical Engineering, University of Rhode Island, Kingston, Rhode Island

ABSTRACT

The infrared (300 to 800 cm^{-1}) absorption spectra of U centers (H^- ions) have been studied in several alkali halides. Extensive side band structure to the main local mode resonance has been noted. These are assigned to processes in which a photon is absorbed and a local mode phonon is created along with the creation or annihilation of one or more lattice phonons, belonging to appropriate critical points of the first Brillouin zone. U centers have also been formed in CsCl, CsBr and CsI which crystallize in the CsCl structure. The local mode resonance, associated with only moderate amounts of side band structure, occurs at 363 cm^{-1} (80°K) for CsBr.

The theory of the Mössbauer effect has been applied to the resonance absorption of light by impurity induced local modes. The resonance treated as modulation of the local mode by the host lattice phonons predicts the temperature dependence of peak position, half-width, and intensity of the absorption. Examples will be furnished from the local mode absorption by U centers in alkali and alkaline earth halides.

The first electronic transition of the U center has been studied by absorption spectroscopy in a number of alkali halides of the NaCl and CsCl types. The shape of the ultraviolet U band has been studied as a function of temperature (4 to 400°K) and explained in terms of a configurational coordinate model.

A variational calculation has been performed to determine the energy levels, wave functions and oscillator strengths of transitions of a U center in the various alkali halides. Both singlet and triplet states are predicted.

I. INTRODUCTION

The U center consists of an H^- ion occupying a vacant halogen ion site in an alkali halide, and shows the characteristic electronic absorption band

* Supported in part by the Air Force In-House Laboratory, Independent Research Fund under Codtract No. AF 19(628)-5144, 6042.

(U band) in the ultraviolet region. Martienssen[1] has reported the U band in several alkali halides of the NaCl structure. The H^- ions differ in mass, polarizability and interaction with their neighbors when compared to the halogen ions they replace. Such impurity centers thus affect the lattice vibration spectrum of the crystal. In addition to modifying the frequency distribution of the host lattice, certain new features known as the localized modes of vibration are introduced in certain cases. In a diatomic crystal, when the substituent is lighter than both the host lattice atoms, as is the case with the U centers, one expects resonance modes due to the localized vibrations of the impurity to occur with frequencies above the optic band of the host lattice.[2] The amplitude of a local mode is very large at the impurity site and decreases very rapidly with increasing distance from the impurity.

The U center local mode has an associated dipole moment and may thus interact with the electromagnetic field associated with incident radiation resulting in resonant absorption of infrared radiation. The near infrared absorption due to U center local modes for a number of alkali halides of the NaCl structure has been reported by Schaeffer.[3] Mitsuishi and Yoshinaga[4] have measured the local mode absorption due to D^- centers in the same crystals.

In the present communication, certain features of absorption of light by the U centers involving both vibrational and electronic transitions will be discussed.

II. RESULTS AND DISCUSSION

1. *Infrared-active multiphonon processes*

Schaeffer[3] found the local mode vibration bands in eight alkali halide crystals of NaCl structure in the spectral region between 17.75μ (NaCl: NaH) and 27.65μ (RbI: RbH). He had also observed one side band each on both the high and low frequency side of the local mode vibration band. Identification of these was not, however, attempted by him. The side bands were also noted by Mitsuishi and Yoshinaga[5] in certain deuterated alkali halides. The identical separation of the side bands from the main bands in KBr: KH and KBr: KD suggests that these may be due to interaction of the local mode with centro-symmetric lattice modes in which the impurity ion does not move. Fritz[6] has subsequently observed such side bands approximately symmetrically disposed about a central local mode peak in the infrared absorption spectra of NaCl, KCl, RbCl, KBr, RbBr, and KI containing U centers.

We have independently investigated the infrared side bands of the U center local mode absorption in KCl and KBr in great detail using thick samples and high concentration of U centers. Spectra were recorded as functions of temperature down to 4°K. In KBr: KH the side bands occurred at separations of 44, 82 and 114 cm^{-1} from the main band towards the high frequency side, with a prominent minimum at 103 cm^{-1}. For KCl: KH such side bands occurred at separations of 62, 85, 130, 167 and 199 cm^{-1}. The fine structure is assigned as combination modes in which a photon is absorbed and both a local mode phonon and a perturbed lattice phonon belonging to a critical point (c.p.) in the Brillouin zone are created. It may be remarked here that in the presence of impurities, the translational symmetry of the lattice is detroyed, and thus the requirement of wave vector conservation is relaxed. This, in principle, makes possible an abundance of such combinations. A comparison with the dispersion curves obtained by Woods *et al.*[8] for pure KBr enables one to assign the peaks to specific c.p. phonons. For KBr: KH the side bands thus correspond to TA(100), LA(0.7, 0.7, 0) and TO(000). Moreover, the minimum at 103 cm^{-1} may be identified with the gap between the optic and acoustic bands for this crystal. For KCl, comparison was made with the dispersion curves and the lattice vibration spectrum calculated by Hardy.[9] The assignment of the side bands is indicated in Fig. 1. The same figure also shows the

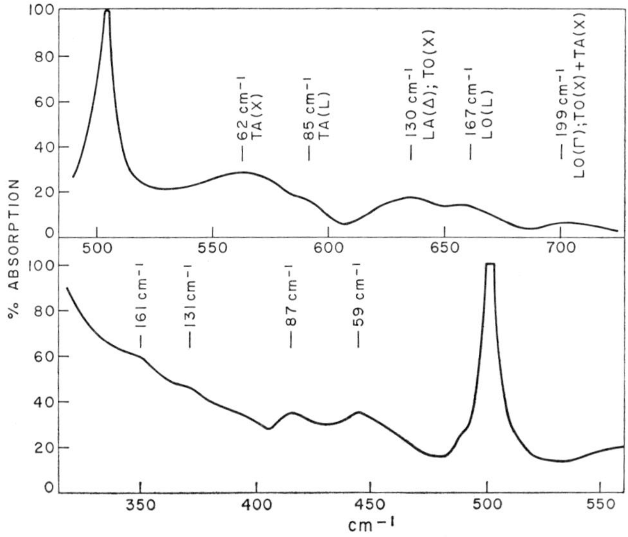

Fig. 1. High and low energy side bands of the U center local mode band in KCl at 80°K.

low energy side of the U center local mode band in KCl. At 300°K the tail of the KCl reststrahlen absorption mars any structure. At very low temperatures (~4°K) the spectrum, although practically free from the tail of the reststrahlen spectrum, does not show any fine structure, as is characteristic of difference bands in which the absorption of an infrared photon is accompanied by creation of a local mode phonon and the annihilation of a perturbed lattice phonon. For a spectrum recorded at 80°K, however, sufficient fine structure could be resolved. The spacings of the low frequency side bands from the main local mode band are essentially the same as was noted in the case of the sum bands. The slight measurable difference in the separations of the sum and difference bands may, however, be interpreted as indication of anharmonicity in the crystal potential.

2. *Temperature dependence of the local mode absorption*

Following a suggestion of Silsbee and Fitchen,[10] the theory of the Mössbauer effect is applied to the resonance absorption of light by impurity-induced local modes. The resonance treated as modulation of the local mode by the host-lattice phonons predicts the temperature dependence of peak position, half-width, and intensity of the absorption.

To establish the analogy we consider[11] the resonating system as consisting of two parts, the impurity center and the surrounding lattice. The width of the impurity local mode is determined mainly by the probability of decay of a local mode into crystal modes, and depends on the extent of coupling between the two. The infrared absorption by local mode, especially in the case of light substitutional impurities, is known to consist of a sharp line at low temperatures, and its intensity and width depend strongly on temperature.[12] Only a limited amount of multiphonon structure associated with the main resonance is usually observed. The local-mode resonance can therefore be regarded as belonging to a weak-to-intermediate-coupling situation with a relatively large f (recoilless fraction of Mössbauer effect). The main resonance may therefore, for the purpose of analogy, be termed a zero-lattice-phonon line.

The local mode or the zero-lattice-phonon line is known to be approximated by an Einstein oscillator,[13] whereas the host lattice will have a complete frequency spectrum of its own, albeit slightly perturbed by the presence of the impurity. In the case of non-Bravais lattices, the lattice phonons will belong to optic and acoustic branches. The Raman-like scattering of the local-mode phonon by the lattice phonons will be controlled by the requirements of momentum and energy conservation. For such a model of the local-mode resonance absorption, one may directly use the expressions developed by McCumber[14] (from the consideration of local lattice distortion in the neighborhood of the impurity) for the shift and width of zero-phonon lines in vibronic spectra. In the Debye approximation for the host-lattice phonons, with a Debye characteristic temperature θ chosen above the acoustic band but not exceeding the temperature corresponding to the maximum optic frequency, the shift and width of the zero-lattice-phonon lines are given by[15]

$$\Delta\omega(T) = \omega(T) - \omega(0) = \alpha\left(\frac{T}{\theta}\right)^4 \int_0^{\theta/T} \frac{x^3}{e^x - 1} dx \tag{1}$$

and

$$\Gamma(T) = \Gamma(0) + \beta\left(\frac{T}{\theta}\right)^7 \int_0^{\theta/T} \frac{x^6 e^x}{(e^x - 1)^2} dx \tag{2}$$

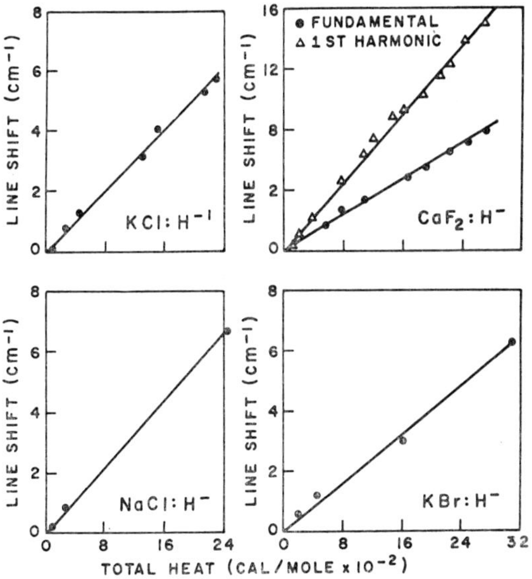

Fig. 2. Shift in local-mode peak position versus total heat. Line-shift data, KCl : H^- from our measurements; NaCl : H^- and KBr : H^- from G. Schaefer, *Phys. Chem. Solids*, **12**, 233 (1960); CaF_2 : H^- from R. J. Elliott, W. Hayes, G. D. Jones, H. F. Macdonald and C. T. Sennett, *Proc. Roy. Soc.*, **289**, 1 (1965). Heat capacity data, KCl and KBr from W. T. Berg and J. A. Morrison, *Proc. Roy. Soc.* (London, A **242**, 467 (1957); NaCl from T. H. K. Barron, A. J. Leadbetter and J. A. Morrison, *Proc. Roy. Soc.* (London), **A 279**, 62 (1964); CaF_2 from S. S. Todd, *J. Am. Chem. Soc.*, **71**, 4115 (1949) and D. R. Huffman and M. H. Norwood, *Phys. Rev.*, **117**, 709 (1960).

The constants α and β are associated with the scattering of phonons by the local mode and contain the unknown coupling factor.

The expression (1) for the shift of the peak position is the same, apart from a multiplicative constant, as that for the total heat. Thus a correlation between $\Delta\omega(T)$ and the total heat of the host crystal is expected. Such a correlation is shown in Fig. 2 for the local-mode absorption in NaCl: H^-, KCl: H^-, KBr: H^-, and CaF_2: H^-. For CaF_2: H^- the first overtone of the local mode is also infrared active, and a correlation with the total heat is possible in this case also.

For the correlation of the half-width with temperature, the residual width is obtained from the extrapolation of low-temperature data to 0°K. The integral of Eq. (2) is done numerically with assumed Debye temperature to give the best fit. θ obtained in such a fashion was slightly smaller than the calorimetric value but close to the temperature corresponding to the optic phonons. The calculated and observed half-widths are compared in Fig. 3. It may be noted that at high- and low-temperature

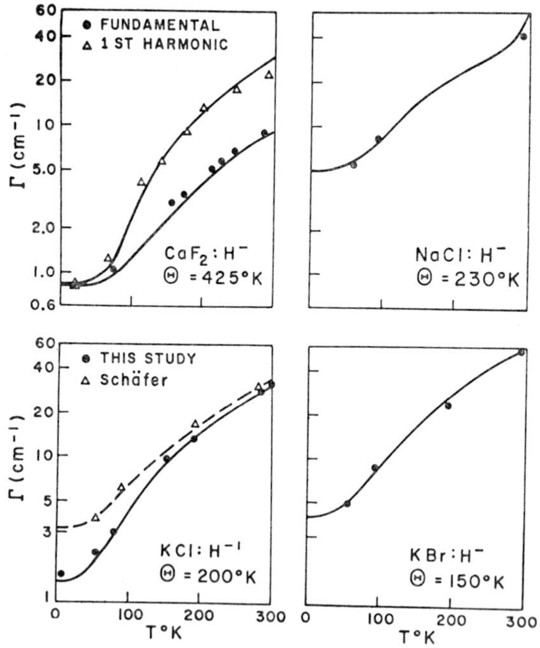

Fig. 3. Temperature dependence of U center local-mode half-width in several host-lattices. Lines are calculated curves with assumed Debye temperatures to give best fit with the experimental data (points). For sources of data see Fig. 2.

limits ($T \gg \theta$ and $T \ll \theta$), the temperature dependent portion of Eq. (2) becomes proportional to T^2 and T^7, respectively. The T^2 dependence of the half-width at high temperatures has already been noted by Mirlin and Reshina[16] in the local-mode resonance of U centers.

In the Debye approximation the theory of the Mössbauer effect[17] predicts that the fraction of transitions without change in the host-lattice states is given by

$$f = e^{-2w} \tag{3}$$

where the Debye-Waller factor w is given by

$$w = \frac{3R}{k\theta}\left[\frac{1}{4}+\left(\frac{T}{\theta}\right)^2\int_0^{\theta/T}\frac{x\,dx}{e^x-1}\right] \tag{4}$$

Fitchen, Silsbee, Fulton and Wolf[18] have used Eq. (4) successfully in the case of zero-phonon transitions of color centers in alkali halides. Equations (3) and (4) may be rewritten in a more convenient form:

$$\frac{I_L}{I_B} = \exp\left\{-S\left[1+4\left(\frac{T}{\theta}\right)^2\int_0^{\theta/T}\frac{x\,dx}{e^x-1}\right]\right\} \sim \exp\left\{-S\left[1+6.6\left(\frac{T}{\theta}\right)^2\right]\right\} \text{ for } T \ll \theta \tag{5}$$

where I_L represents the integrated intensity under the pure (zero-lattice-phonon) local-mode band whereas I_B represents the total band intensity including the multiphonon structure. The factor S in Eq. (5), known as the Huang-Rhys factor, may be identified with the most probable number of phonons involved in the transition, which in the absence of any coupling between the local mode and the lattice phonons should be equal to zero. In the other extreme, for very strong coupling, e.g. the F band, S can be as high as 30.

The total integrated intensity I_B is expected to be independent of temperature. It consists of the local-mode absorption, the difference bands, and the sum bands. The absorption coefficient of the high-frequency sum bands has a slight temperature dependence, but that of the low-frequency difference bands has a strong temperature dependence, vanishing completely at very low temperatures. The change in the side-band intensities thus produces a compensating change in integrated intensity of the local mode. The population of the vibrational ground (zero-lattice-phonon) state increases with decreasing temperature and increases the local-mode intensity. Thus, at very low temperatures

$$I_B = I_L + I_{\text{sum}} \tag{6}$$

and the factor S can be calculated from

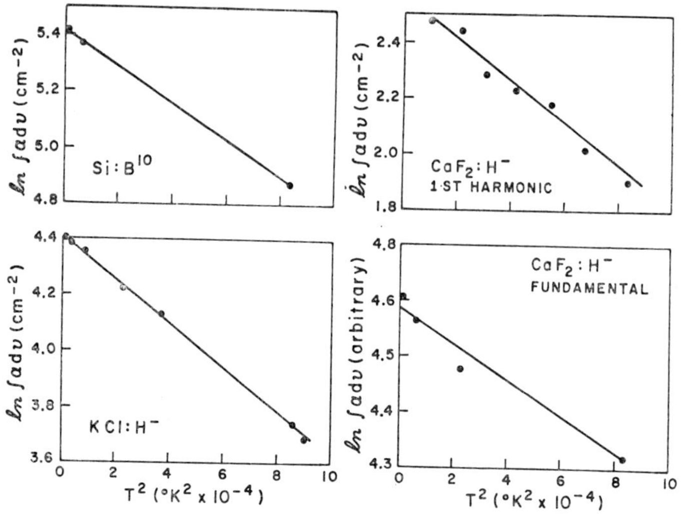

Fig. 4. Temperature dependence of integrated intensity under the local-mode band. Data, Si : ^{10}B (681 cm^{-1} band) from M. Balkanski and W. Nazarewicz, *to be published*; KCl : H^- from our measurements; CaF_2 : H^- from R. J. Elliott, W. Hayes, G. D. Jones, H. F. Macdonald and C. T. Sennett, *Proc. Roy. Soc.*, **289**, 1 (1965).

$$S = \ln (1+I_{sum}/I_L) \tag{7}$$

where I_{sum} indicates the integrated intensity under the sum bands.

In Fig. 4 are plotted $\ln I_L$ vs. T^2 for several impurity-host-lattice systems. As predicted by Eq. (5), linear dependence is indeed obtained. The U center (H^-) local mode in KCl was remeasured at 6°K. The value of S determined from Eq. (7) is 0.052, while that from the slope of the linear plot for this system (Fig. 4) using the value of θ obtained from the dependence of half-width on T is 0.048, in remarkably good agreement. The low value of S is indicative of relatively weak coupling between the local mode and the lattice modes.

Takeno and Sievers[19] have independently pointed out the T^2 dependence of the logarithm of the integrated intensity under the local mode. They did not, however, discuss the various parameters involved in Eq. (5).

3. *Electronic structure of the U center*[20]

Grourary[21] has given a variational method to calculate the position of the U-band in the NaCl type crystals. We have extended his method to include the CsCl structure also. In addition, the energies of the higher excited states of the U center have also been obtained. Finally, the possible triplet state energy levels for this center are predicted.

Following the procedure of Gourary[21] all ions are represented by point charges and polarization effects are neglected. This latter approximation may be justified since the center is electrically neutral. The Hamiltonian (in atomic units) for the U center may thus be written as

$$H = -\frac{1}{2}\nabla_1^2 - \frac{1}{2}\nabla_2^2 - \frac{1}{r_1} - \frac{1}{r_2} + \frac{1}{r_{12}} + V_L(r_1) + V_L(r_2) \tag{8}$$

Here the first two terms arise from the kinetic energy of the two electrons, the third and fourth terms from the Coulomb interaction between the electrons and the proton, the fifth is the electron-electron interaction and the last two terms arise from the interaction of the electrons with the rest of the ions in the lattice. The latter may be more explicitly written as

$$V_L(r) = \sum_{\xi\eta\phi} (-1)^{\xi+\eta+\phi}[(x-\xi a)^2 + (y-\eta a^2) + (z-\phi a)^2]^{-1/2} \tag{9}$$

where a is the nearest neighbor distance in the host lattice.

The simplest wave function that gives a stable ground state (singlet) for the H^- ion has two variational parameters[22]

$$\psi_{gs} = \frac{1}{2\pi}(\alpha\beta)^{3/2}(e^{-\alpha r_1}e^{-\beta r_2} + e^{-\alpha r_2}e^{-\beta r_1}) \tag{10}$$

This wave function gives an energy for the ground state which is 0.0143 au less than that obtained by using an eleven parameter wave function.[23] Therefore, in using the two parameter wave function for the singlet ground state of the U center, we shall make the above correction to the energies obtained. The energy functional obtained by using the wave function (10) and the Hamiltonian (8) is given by

$$E_{gs} = \frac{(\psi_{gs}|H|\psi_{gs})}{(\psi_{gs}|\psi_{gs})} = -\frac{2\alpha_M}{a} - \left[\frac{1}{2} + \frac{32(\alpha\beta)^3}{(\alpha+\beta)^6}\right]^{-1}\left\{\frac{1}{2}(\alpha+\beta) + \frac{22(\alpha\beta)^3}{(\alpha+\beta)^5}\right.$$
$$-\frac{1}{4}(\alpha^2+\beta^2) - \frac{32(\alpha\beta)^4}{(\alpha+\beta)^6} - \frac{\alpha\beta}{2}\frac{(\alpha^2+3\alpha\beta+\beta^2)}{(\alpha+\beta)^3}$$
$$-\sum_{\xi\eta\phi}(-1)^{\xi+\eta+\phi}\left[\frac{1}{2}\left(\alpha+\frac{1}{r_0}\right)e^{-2\alpha r_0} + \frac{1}{2}\left(\beta+\frac{1}{r_0}\right)e^{-2\beta r_0}\right.$$
$$\left.\left. +16\frac{(\alpha\beta)^3}{(\alpha+\beta)^6}\left(\alpha+\beta+\frac{2}{r_0}\right)e^{-(\alpha+\beta)r_0}\right]\right\} \tag{11}$$

Here $r_0=a(\xi^2+\eta^2+\phi^2)^{1/2}$ is the distance between the proton and the ion in question and α_M is the Madelung constant which is 1.7476 for the NaCl structure. The energy is obtained by minimization of Eq. (11) as a function of α and β. The difference between the calculation for NaCl type crystals and CsCl type crystals is, first of all, the difference in α_M and secondly, that for the CsCl structure the coordination number is eight while that for the NaCl structure is six.

The wave function used by Gourary[21] for the first excited singlet state is

$$\psi_{1es} = \frac{(\gamma\delta)^{3/2}}{8\pi}\gamma\,[r_1 e^{-\gamma r_1/2}\, e^{-\delta r_2}\cos\theta_1 + r_2\, e^{-\gamma r_2/2}\, e^{-\delta r_1}\cos\theta_2] \tag{12}$$

The energy function obtained using Eq. (12) is

$$E_{1es} = -2\frac{\alpha}{a}M + \frac{1}{2}\delta^2 - \delta + \frac{1}{8}\gamma^2 - \frac{\gamma^5}{64}\frac{\left(3\delta+\frac{1}{2}\gamma\right)}{\left(\delta+\frac{\gamma}{2}\right)^5} + \frac{7}{24}\frac{\gamma^5\delta^3}{\left(\delta+\frac{\gamma}{2}\right)^7}$$
$$+\sum_{\xi\eta\phi}(-1)^{\xi+\eta+\phi+1}\left[\left(\delta+\frac{1}{r_0}\right)e^{-2\delta r_0}+\frac{1}{24}\left(\gamma^3 r_0^2+6\gamma^2 r_0+18\gamma+\frac{24}{r_0}\right)e^{-\delta r_0}\right] \tag{13}$$

The energy (13) is next minimized as a function of δ and γ. As the H^- has no bound excited states,[24] no corrections need be made.

The transition energy calculated for the U band by both Gourary and ourselves as well as the experimental values available are shown in Table I. The calculated values are consistently about 16% lower than the observed values. This may be due to the fact that the two parameter wave functions used are not really a good fit to the true wave function. It may also be in part due to the neglect of the electron-phonon interaction which is very strong in the alkali halides. The calculated oscillator strengths for the transition are 1.65 for NaCl and 1.57 for KBr. For NaCl it is twice as large as the observed value.[25]

Similar variational calculations have been extended[20] to obtain the triplet ground state (1s2s) and the first triplet excited state energies. The transition energies calculated for the first two triplet states of the U center are shown in Table II. This transition occurs at a much lower energy than the singlet transition and should be observable in luminescence experiments. The calculated oscillator strengths for this transition is 1.00 in the case of NaCl.

The electrons in the second excited state have the configuration 1s3p.

TABLE I. Experimental and theoretical values of the U band energy in eV.

Substance	Theory	Experiment
LiF	8.11	
LiCl	6.12	
LiBr	5.66	
LiI	5.11	
NaF	6.88	
NaCl	5.52	6.46[a]
NaBr	5.14	5.90[a]
NaI	4.68	
KF	5.88	
KCl	4.84	5.79[a]
KBr	4.60	5.43[a]
KI	4.22	5.08[a]
RbF	5.52	
RbCl	4.60	5.41[a]
RbBr	4.38	5.12[a]
RbI	4.05	
CsF	5.11	
CsCl	4.38	
CsBr	4.16	4.99[b]
CsI	3.86	

a. Experimental values taken from J. W. Schulman and W. D. Compton, *Color Centers in Solids* (1962).
b. See next section.

The transition energies between the ground and second excited singlet and triplet states are shown in Table III. For NaCl the values of the oscillator strengths are 0.026 for the transition to the second excited singlet state and 0.720 for the transition to the second excited triplet state. Thus the contribution of the transition to the second excited singlet state to any band on the high energy side of the U band will be fairly small. In the triplet case, on the other hand, the contribution of this transition is of the same order as that of the first excited state. Therefore, a higher energy band should be more apparent in the case of transitions to higher excited triplet states than in the case of transitions to higher excited singlet states.

The theoretical values of the $1s^2 \rightarrow 1s2p$ singlet transition follow the Ivey formula very closely as can be seen in Fig. 5. The exponent of the lattice constant is $n=1.14$ for the NaCl type structure and $n=1.07$ for the CsCl type structure. This is in good agreement with the experimental values which give a value of $n=1.10$ for the exponent.[25]

Recently, calculations[26] and experimental observations[27, 28] have been

TABLE II. Theoretical values of the triplet transition for the U center eV.

Substance	Theoretical value
LiF	0.82
LiCl	1.39
LiBr	1.50
LiI	1.60
NaF	1.14
NaCl	1.52
NaBr	1.60
NaI	1.69
KF	1.44
KCl	1.66
KBr	1.71
KI	1.74
RbF	1.52
RbCl	1.69
RbBr	1.71
RbI	1.77
CsF	1.60
CsCl	1.71
CsBr	1.74
CsI	1.77

TABLE III. Theoretical values of the second singlet and triplet transitions for the U center in eV.

Host lattice	Singlet	Triplet
LiF	9.68	3.35
LiCl	7.76	3.64
LiBr	7.21	3.67
LiI	6.61	3.70
NaF	8.46	3.54
NaCl	7.07	3.70
NaBr	6.66	3.70
NaI	6.23	3.70
KF	7.42	3.67
KCl	6.39	3.70
KBr	6.12	3.70
KI	5.74	3.67
RbF	7.07	3.67
RbCl	6.15	3.70
RbBr	5.90	3.67
RbI	5.58	3.64
CsF	6.66	3.70
CsCl	5.85	3.67
CsBr	5.63	3.67
CsI	5.33	3.62

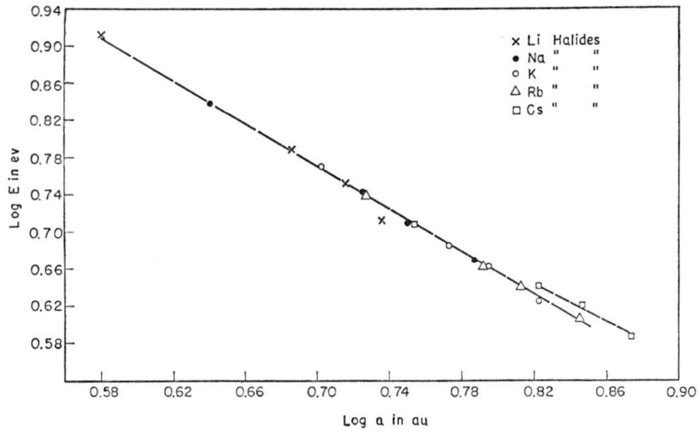

Fig. 5. The variation of calculated U band energy with lattice constant (Ivey plot).

made on the triplet state of the M center. The existence of the triplet state of the M center has been established by both optical[27] and electron spin resonance measurements.[28] It remains to be seen whether the U center triplet states can be investigated by similar techniques. Luminescence measurements from U centers, involving non-radiative transitions from singlet to triplet state with resulting emission from the triplet to triplet transition, could provide a means of observing the triplet state. This type of measurement although theoretically plausible, may not be feasible because of the bleaching properties of the U center. At high temperature (above $-100°C$), the irradiation of a crystal containing U centers with U band light bleaches the U band and generates the F band.[29] This is be cause, as pointed out by Gourary, the electron in the first excited singlet state of the U center has a slightly repulsive interaction with the remaining hydrogen atom which then will have a tendency to diffuse away leaving an F center behind. If this diffusion of the hydrogen atom occurs in a time shorter than that required for the singlet to triplet transition to occur, we would then expect the conversion of the U center to the F center to predominate over the formation of the triplet state. On the other hand, at temperatures below $-100°C$, the hydrogen atom does not tend to diffuse away from the extra electron[30] and the formation of the triplet state remains possible. Once the triplet state is formed, the diffusion of the hydrogen atom should not occur as the interaction between the F center and the hydrogen atom in the first excited triplet state is attractive.

To date no reliable measurements seem to have been done on the higher excited states of the U center to which comparison with the theoretical results can be made. However, some speculations have been made[31] regarding the effect of transitions involving these states on the high energy side of the U band in a number of alkali halides. Our calculations for the second excited singlet state indicate that because of its small oscillator strength, this state would only contribute to a small extent to any high energy tail of the U band.

It seems that transitions to the higher excited triplet states would be more interesting as the oscillator strength for the 1s2s→1s3p transition is of the same order as for the transition to the first excited state. This leads us to suspect that if the triplet transitions can be observed experimentally in the absorption, one would find two peaks, one lying on the red end of the visible spectrum between 1.5 and 1.7 eV and the other lying towards the ultraviolet end of the visible spectrum at around 3.7 eV. Other triplet-triplet transitions would not be expected to play an important role as these two transitions exhaust the sum rule.

4. *U centers in CsBr*[32]

The formation and optical properties of U centers in an alkali halide of the CsCl structure is reported. The U centers were produced by heating additively colored (with K) CsBr in an atmosphere of H_2 gas at about 500°C until all the F centers were converted into U centers. Initial attempts to produce additively colored CsBr using single crystals from Semi-Elements, Incorporated, were unsuccessful. Single crystals from Harshaw Chemical Company with much higher purity, were more amenable to additive coloration and to subsequent hydrogenation. By this method we obtained typical concentrations of 10^{17}–10^{18} U centers per cm^3. Compared to NaCl-type crystals the centers were relatively unstable at room temperature, but showed little decrease in concentration over a long period of time (six months) when stored in liquid N_2.

The infrared local-mode absorption was studied by a Perkin-Elmer Model 301 spectrophotometer. The spectra at 300 and 80°K are shown in Fig. 6. At 80°K the local-mode frequency occurs at 363 cm^{-1} (half-width 10 cm^{-1}). An Ivey plot of the peak position with lattice constant for all alkali halides with available data is shown in Fig. 7. Side bands due to interaction of local-mode phonons with perturbed lattice phonons in the case of NaCl type crystals have already been discussed. There are indications of a few side bands in the case of the CsBr U local mode (Fig. 6). The first side band observed on both high- and low-energy sides is probably due to summation and difference processes involving a $TA(X)$ mode with

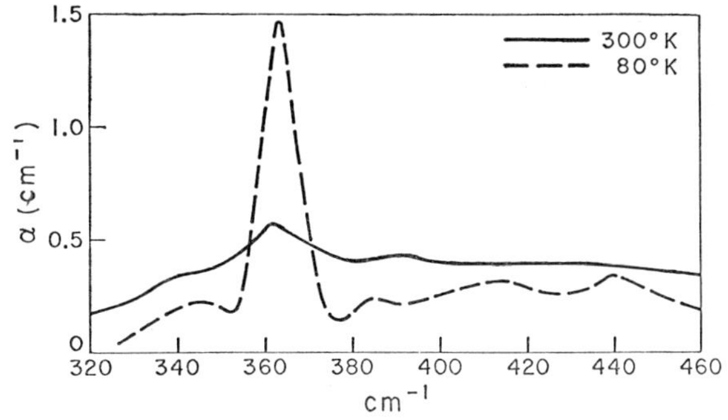

Fig. 6. Local-mode absorption spectrum due to U centers in CsBr.

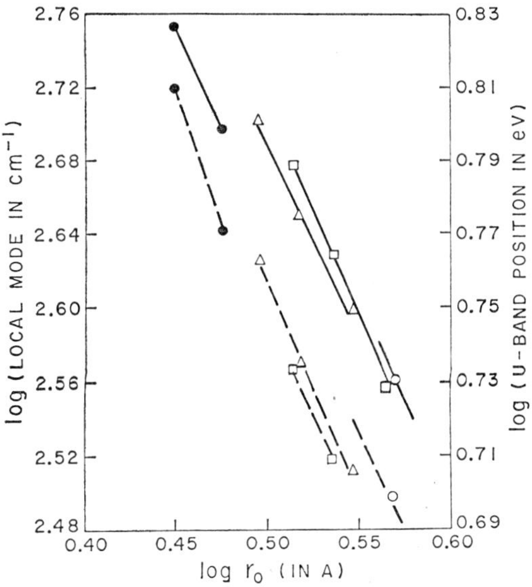

Fig. 7. Ivey plots of U center local mode (solid line) and U center electronic band (dashed line) in alkali halides as functions of nearest-neighbor distance. Na halides : ●, K halides : △, Rb halides : □, and Cs halides : ○.

an average value of 20 cm^{-1}. The slight difference in the separation of the sum and difference band from the main resonance may be due to anharmonicity in the crystal potential. This value compares well with the calculated value of 26 cm^{-1} for the same mode given by Krishnan.[33]

The first electronic transition of the U center in CsBr was studied at

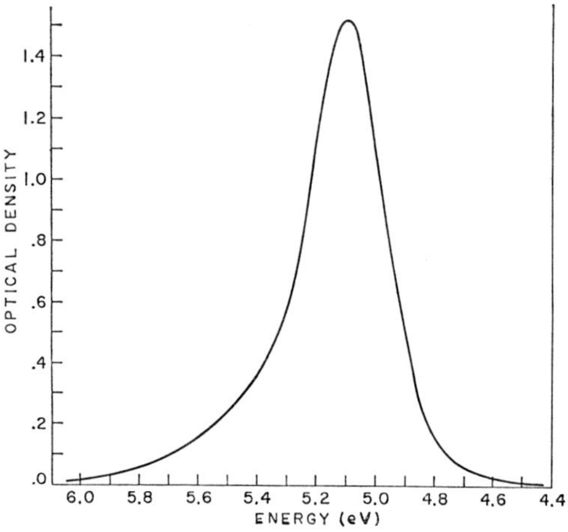

Fig. 8. U band in CsBr at 80°K.

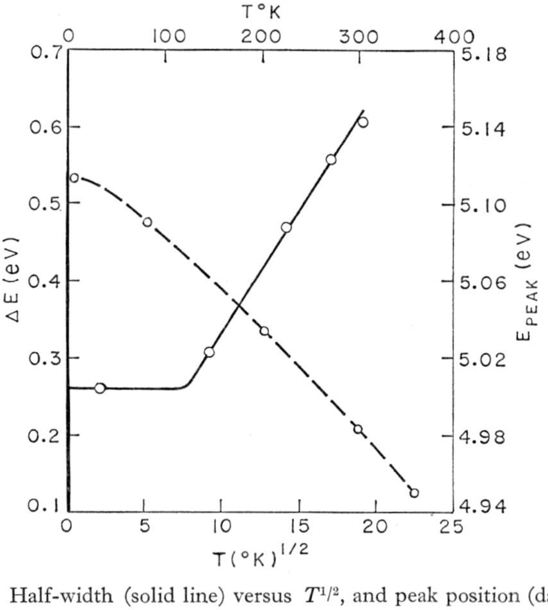

Fig. 9. Half-width (solid line) versus $T^{1/2}$, and peak position (dashed line) versus T for the U band in CsBr.

temperatures ranging from 6 to 360°K. Figure 8 shows the U band at 80°K after correction for the absorption edge of the host lattice. The Ivey plot of the U band position for the various alkali halides is shown in Fig. 7.

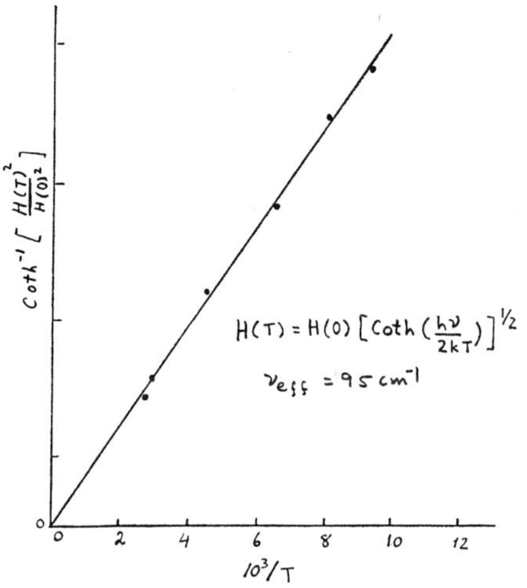

Fig. 10. Hebb-Williams plot for U band in CsBr.

The variation of the peak position and half-width with temperature is shown in Fig. 9. A Williams-Hebb[34] plot (Fig. 10) of the half-width with temperature yields an effective phonon frequency of 95 cm^{-1}. This value lies between the transverse and longitudinal optic branches of CsBr. A similar situation is known to exist in the case of F centers in alkali halides.[35]

5. *U centers in CsI*

Dötsch, Mitra Tanton and Shatas[36] have recently studied the formation and optical properties of U centers in CsI. By simultaneous additive coloration and hydrogenation, U center concentration to about 10^{19} cm^{-3} could be obtained both in CsI and CsBr. The infrared absorption spectrum induced by the localized vibrational mode of CsI:H^- has been investigated at temperatures ranging from 10° to 300°K. The local mode resonance occurs at 286 cm^{-1} at 80°K and shifts to higher frequencies with increasing temperature. Several multiphonon structures due to summation processes were noted at low temperatures. The first electronic transition of H^- and D^- centers in CsI was studied between 6° and 300°K. At 80°K the U-band position in CsI is measured at 269 mm (4.6 eV). Table IV gives the local mode phonon frequencies due to H^- and D^- centers in CsCl, CsBr and CsI.

Table IV. U center local mode phonon frequencies in cm^{-1} at 100°K

Impurity	Host Lattice		
	CsCl	CsBr	CsI
H^-	424	365	290
D^-	302	259	207

Acknowledgements

I am deeply indebted to Mr. T. N. Casselman for bringing to my attention the interesting properties of the U center, and for many helpful discussions. The work described in this article has resulted from the joint efforts of many. In particular, I would like to thank Dr. Y. Brada, Mr. R. S. Singh and Dr. H. N. Spector for their collaboration in various phases of the research.

REFERENCES

1. W. Martienssen, *Z. Physik*, **131**, 488 (1952).
2. R. F. Wallis and A. A. Maradudin, *Progr. Theoret. Phys.* (Kyoto), **24**, 1055 (1960).
3. G. Schaeffer, *J. Phys. Chem. Solids*, **12**, 233 (1960).
4. A. Mitsuishi and H. Yoshinaga, in *Proc. Intern. Symp. on Molecular Structure and Spectroscopy, Tokyo, Japan, 1962*, p. A 304.
5. A. Mitsuishi and H. Yoshinaga, *Progr. Theoret. Phys. Suppl.*, **23**, 241 (1962).
6. B. Fritz, *Intern. Conf. on Lattice Dynamics*, Copenhagen, Denmark, 1963, Paper D8.
7. S. S. Mitra and Y. Brada, *Phys. Letters*, **17**, 19 (1965).
8. A. D. B. Woods, B. N. Brockhouse, R. A. Cowley and W. Cochran, *Phys. Rev.*, **131**, 1025 (1963).
9. J. R. Hardy, *Phil. Mag.*, **7**, 315 (1962).
10. R. H. Silsbee and D. B. Fitchen, *Rev. Mod. Phys.*, **36**, 432 (1964).
11. S. S. Mitra and R. S. Singh, *Phys. Rev. Letters*, **16**, 694 (1966).
12. B. Fritz, U. Gross and D. Bäuerle, *Phys. Status Solidi*, **11**, 231 (1965).
13. J. A. D. Mathew, *J. Phys. Chem. Solids.*, **26**, 2067 (1965).
14. D. E. McCumber, *J. Math. Phys.*, **5**, 333 (1964).
15. D. E. McCumber and M. D. Sturge, *J. Appl. Phys.*, **34**, 1682 (1963).
16. D. N. Mirlin and I. I. Reshina, *Fiz. Tverd. Tela*, **6**, 945 (1964), [translation: Soviet Phys.-Solid State **6**, 728 (1964)].
17. H. Frauenfelder, *The Mossbauer Effect*, W. A. Benjamin, Inc., New York, 1962.
18. D. B. Fitchen, R. H. Silsbee, T. A. Fulton, and E. L. Wolf, *Phys. Rev. Letters*, **11**, 275 (1963).
19. S. Takeno and A. J. Sievers, *Phys. Rev. Letters*, **15**, 1020 (1965).
20. H. N. Spector, S. S. Mitra and H. N. Schmeising, *J. Chem. Phys.*, **46**, 2676 (1967).
21. B. S. Gourary, *Phys. Rev.*, **112**, 337 (1958); B. S. Gourary and F. J. Adrian, in *Solid State Physics* (F. Seitz and D. Turnbull, eds.), Vol. 10, p. 127, Academic Press, New York, 1960.

22. S. Chandrasekhar, *Astrophys. J.*, **100**, 176 (1944).
23. L. R. Heinrich, *Astrophysics. J.*, **99**, 59 (1943).
24. C. Pekeris, *Phys. Rev.*, **126** 1470 (1962).
25. J. W. Schulman and W. D. Compon, *Color Centers in Solids*, Pergamon Press, New York, 1962.
26. R. F. Wood and A. Meyer, *Solid State Comm.*, **2**, 225 (1964).
27. I. Schneider and M. E. Caspari, *J. Appl. Phys.* **33**, 3387 (1962); *Phys. Rev.*, **133**, A 1193 (1964); M. Ikezawa and M. Ueta, *J. Phys. Soc. Japan*, **18**, 145 (1963); R. T. McCall and L. I. Grossweiner, private communication.
28. H. Seidel, *Phys. Letters*, **7**, 27 (1963).
29. R. Hilsch and R. W. Pohl, *Trans. Faraday Soc.*, **34**, 883 (1938).
30. H. Thomas, *Ann. Physik*, **38**, 601 (1940).
31. G. Spinolo and D. Y. Smith, *Phys. Rev.*, **140**, A 2117 (1965); D. Y. Smith and G. Spinolo, *ibid.*, **140**, A 2121 (1965).
32. S. S. Mitra and Y. Brada, *Phys. Rev.*, **145**, 626 (1966).
33. N. Krishnamurthy and R. S. Krishnan, *Indian J. Pure and Appl. Phys.*, **1**, 239 (1963).
34. F. E. Williams and M. H. Hebb, *Phys. Rev.*, **84**, 1181 (1951); J. J. Markham, *Rev. Mod. Phys.*, **31**, 956 (1959).
35. G. E. Stungis, J. J. Markham and G. A. Noble, *J. Chem. Phys.*, **40**, 3634 (1964).
36. H. Dötsch, S. S. Mitra, G. A. Tanton and R. A. Shatas, *Bull. Am. Phys. Soc.*, **13**, 499 (1968).

Part II

Formation and Migration Energies and Annealing

Vacancies and Diffusion Mechanisms in Diamond-Structure Semiconductors

A. Seeger and M. L. Swanson*

Institut für Physik am Max-Planck-Institut für Metallforschung, Stuttgart, Germany

ABSTRACT

The formation energies of monovacancies $E_{1V}{}^{F}$ in Ge and Si are recalculated using a Morse type potential with the following improvements over earlier papers: (a) a generalized Morse potential, employing two adjustable exponents, is used; (b) the dissociation energy D appearing in the potential is taken as one-half of the sum of the sublimation energy and the promotion energy from the s^2p^2 to the sp^3 electronic configuration of the free atoms; (c) the relaxation around a vacancy for the three nearest neighbor shells is calculated.

It is found that the improvements (b) and (c) on the calculated values of $E_{1V}{}^{F}$ approximately cancel, giving $E_{1V}{}^{F}=1.97$ eV and $E_{1V}{}^{F}=2.35$ eV for Ge and Si respectively.

The present experimental knowledge concerning vacancies and diffusion mechanisms in Ge and Si is reviewed. Special attention is given to the inconsistencies between various results in relation to the role of divacancies and impurities. The concentration of point defects in thermal equilibrium is estimated and found to be too small to be studied by direct methods. It is shown that the self-diffusion and solute diffusion data suggest strongly different mechanisms of self-diffusion in Ge and Si. Whereas the Ge data are compatible with self-diffusion via strongly relaxed vacancies ("relaxions"), a considerable body of evidence indicates that in Si an interstitial mechanism might predominate, involving the migration of a complex interstitial configuration resembling locally melted regions. The vacancy formation energies $E_{1V}{}^{F}$ are thought to exhibit a substantial temperature dependence. The comparison with theory involves, therefore, considerable uncertainty; nevertheless it appears that the calculated values are too

* Permanent address: Metal Physics Branch, Chalk River Nuclear Laboratories, Chalk River, Ontario, Canada.

small. This is attributed to the neglect of the energy required for "bond-bending."

I. INTRODUCTION

There are three main experimental areas which have contributed to the understanding of vacancies in Ge and Si: diffusion, quenching, and irradiation damage. Many observations in each of these areas have been interpreted in terms of the formation and migration of single vacancies, but various inconsistencies among the data have shown that in general more complex explanations are required. For example, although diffusion of impurities in Ge and Si can often be interpreted by a vacancy mechanism,[1] the diffusion of Li, Cu, and Ni is very rapid and is due to an interstitial mechanism.[2,3] Quenched-in defects are usually thought to be single vacancies, but it is much more likely that most of them are divacancies, because of their high binding energy.[4] Similarly, defects introduced by irradiation at 300°K have been shown to consist largely of defect-impurity complexes,[5] rather than single vacancies and interstitials. It is thus clear that more experiments under carefully controlled conditions, and in particular low temperature irradiations of high purity semiconductors,[6] are required.

Only few theoretical calculations of formation and migration energies of vacancies in Ge and Si have been attempted, and these have utilized rather oversimplified models.[7–9] An extension of a simple valence bond model, using a generalized Morse potential, is presented in section II. Sections III and IV are devoted to the discussion of experimental diffusion, quenching, and irradiation data. The self-diffusion data on Si and Ge are considered in some detail. The differences from f.c.c. metals are discussed and analyzed in terms of models for high-temperature diffusion. This analysis suggests that the mechanisms of self-diffusion are different in Ge and Si. Whereas the Ge data are compatible with a vacancy type mechanism of self-diffusion, involving strongly relaxed vacancies, the Si data are best accounted for by a diffusion through complex interstitial configurations which may resemble locally melted regions.

II. CALCULATION OF VACANCY FORMATION ENERGIES

Attempts to calculate the formation energy of vacancies in diamond-structure semiconductors have been made by Swalin,[7] Scholz,[8] and Scholz and Seeger[9] by using a Morse potential function[10] for the energy of covalent bonds. Swalin considered only interactions between nearest neighbor atoms, so that the formation energy $E_{1V}{}^{F}$ of a vacancy was equated to the energy $4D$ required to break four bonds on removing an atom from a lattice site, minus the energy $2D$ gained by forming two new bonds on the surface of the crystal, minus the energy E_3 gained by forming two new bonds by the atoms nearest to the vacancy, plus the energy increase E_2 caused by stretching the bonds of other atoms surrounding the vacancy

$$E_{1V}{}^{F} = 2D+E_2-E_3 \tag{1}$$

It was assumed that the bond energy E could be well represented by the Morse potential[11,12]

$$E=D\{\exp[-2\alpha(r-r_0)]-2\exp[-\alpha(r-r_0)]\} \tag{2}$$

consisting of the sum of repulsive and attractive energy terms, which vary exponentially with the bond strain $(r-r_0)$. E has a minimum at $r=r_0$, the equilibrium separation of atoms at 0°K, where its value is $E=-D$, the bond dissociation energy (since the zero-point vibrational energy can be neglected).

The assumptions involved in this approach should be carefully considered:

(a) In view of the strong covalent bonds in diamond-structure semiconductors, the assumption that only nearest neighbor interactions are important is probably a good first approximation for the purpose of calculating defect energies.

(b) Since a potential of the Morse type has been justified to some extent by its use in the interpretation of vibrational spectra of diatomic molecules,[10,12] it would seem to offer a reasonable representation of the bond energy near the equilibrium separation of atoms. Its use for the large atomic displacements near a vacancy remains doubtful.

(c) The determination of the constants D and α is important and is open to some question. α can be found from the compressibility of the crystal at zero pressure, but the resulting potential is then generally incompatible with the experimental rate of change of compressibility with pressure. Thus a more general form of the Morse potential should be used.

The bond dissociation energy D was taken as half the sublimation energy by Swalin,[7] Scholz,[8] and Scholz and Seeger,[9] which would be true only if the electronic energy state of an atom in the crystal equalled that in the gas. In the present case, as pointed out by Lidiard,[13] the promotion energy E_p from the s^2p^2 to the sp^3 configuration, which is required for the tetrahedral bonds in diamond structures, must be considered, so that the sublimation energy is given by $E_s = 2D - E_p$. Values of E_p have been tabulated by Slater,[14] who stated that average values of the various energy levels corresponding to the configurations s^2p^2 and sp^3 were a more appropriate measure of E_p than the minimum values.

If the values of D in Eqs. (1) and (2) are equal, as is assumed, then the calculated value of $E_{IV}{}^{F}$ is roughly independent of D (as will be shown), so that an accurate value of D is not required. However, it must be mentioned that for the bonds near a vacancy the value of D in Eq. (2) might differ considerably from the value found from the sublimation energy, because the asymmetric distortions around the vacancy will alter the nature of the covalent bond.* In this case, the D in Eq. (2) would be less than that in Eq. (1), thus causing an appreciable increase in $E_{1V}{}^{F}$.

(d) The relaxation of atoms surrounding the vacancy was calculated in Swalin's paper[7] by prescribing an exponential decrease in strain with increasing distance from the vacancy, which is not in conformity with theory. In Scholz's and Seeger's papers[8, 9] the energy of the distortion around a vacancy was given by the Born lattice energy in the harmonic approximation, which may be inaccurate for large strains.

In view of these considerations, we shall calculate the formation energy of vacancies in Ge and Si by employing the following improvements over earlier papers:

(a) A generalized Morse potential will be used.

(b) The bond dissociation energy D will be put equal to $(E_s + E_p)/2$.

(c) A less restricted relaxation of atoms surrounding a vacancy will be allowed.

A generalized Morse potential of the form

$$E = D_1 \exp(-\alpha_1 r) - D_2 \exp(-\alpha_2 r) \tag{3}$$

will be used for the bond energies. By stipulating that $E(r_0) = -D$ and $(dE/dr)_{r_0} = 0$, this equation becomes

$$E = D\left\{\frac{\alpha_2}{\alpha_1 - \alpha_2}\exp[-\alpha_1(r - r_0)] - \frac{\alpha_1}{\alpha_1 - \alpha_2}\exp[-\alpha_2(r - r_0)]\right\} \tag{4}$$

where D is the bond dissociation energy and α_1 and α_2 are constants which can be determined from compressibility data. If $\alpha_1 = 2\alpha_2$ this equation is identical with the standard Morse potential, Eq. (2). The use of two

* Note that according to Pauling[12] even in the perfect diamond lattice the tetrahedral bonds contain some d and f character.

α's instead of one as in Eq. (2) makes possible a more exact fit to the two-constant compressibility equation found experimentally by Bridgman[15] using a static method, for pressures up to 3×10^4 kg/cm^2 (which corresponds to a linear contraction of 1%)

$$(V_0-V)/V_0 = aP-bP^2 \tag{5}$$

Here a is the isothermal compressibility at zero pressure and b is the pressure dependence of the compressibility. V_0 is the crystal volume at zero pressure and V is the volume at pressure P. Thus the isothermal compressibility is

$$K_T = -(1/V_0)(dV/dP)_T = a-2bP \tag{6}$$

This linear variation of compressibility with pressure has been verified in a more direct and presumably more accurate manner by measurements of ultrasonic wave propagation for somewhat smaller pressures. The values of a found by the two techniques are very similar, but different values of b are reported.[16] By the dynamic method the adiabatic compressibility K_S is measured. It is related to K_T by[17]

$$K_T = K_S+TV_m\gamma^2/C_p \tag{7}$$

where γ is the coefficient of thermal expansion, T is the absolute temperature, V_m is the molar volume, C_p is the specific heat at constant pressure, $K_S=3/(c_{11}+2c_{12})$ is the adiabatic compressibility for isotropic compression of a cubic crystal, and c_{11} and c_{12} are the appropriate elastic constants (Voigt's notation). The difference between K_T and K_S for Ge and Si is only 0.3% at 300°K.[18]

The values of α_1 and α_2 in Eq. (4) can be found from Eq. (5).[19] If a unit cell of initial volume $V_0=C^3r_0^3$ (where the geometric factor $C=4/\sqrt{3}$ for the diamond structure) is compressed at 0°K according to Eq. (5), the internal energy change is

$$\begin{aligned}(E+D)n = \Delta E &= -\int_{V_0}^{V} PdV \\ &= \frac{-C^3a}{2b}(r^3-r_0^3)-\frac{C^3r_0^3}{12b^2}\left\{a^3-\left[a^2+4b\frac{(r^3-r_0^3)}{r_0^3}\right]^{3/2}\right\}\end{aligned} \tag{8}$$

where E is given by Eq. (4) and n is the number of bonds in the unit cell ($n=16$ for the diamond structure). The second and third derivatives of this equation with respect to r, evaluated at $r=r_0$, give

$$\left.\begin{aligned} \alpha_1\alpha_2 &= 9C^3r_0/nDa = \beta_1 \\ \text{and} \qquad \alpha_1\alpha_2(\alpha_1+\alpha_2) &= 54C^3(b/a^2-1)/nDa = \beta_2 \end{aligned}\right\} \quad (9)$$

Thus

$$\left.\begin{aligned} 2\alpha_1 &= \beta_2/\beta_1+[(\beta_2/\beta_1)^2-4\beta_1]^{1/2} \\ \text{and} \qquad 2\alpha_2 &= \beta_2/\beta_1-[(\beta_2/\beta_1)^2-4\beta_1]^{1/2} \end{aligned}\right\} \quad (10)$$

The values of a, b, r_0, and $D=(E_s+E_p)/2$ at 0°K are required to calculate α_1 and α_2 (see Table I). It was found by McSkimin and Andreatch[20] (using the ultrasonic technique) that the values of a and b for Ge and Si were approximately 2–4% less at 77°K than at 298°K. Since lower temperature data are not available and since thermal expansion is usually small below 77°K, their 77°K values will be used. The interatomic spacings r_0 at 0°K for Ge and Si are shown from thermal expansion data[21] to differ little from the room temperature values.[22] It is of interest to note that the thermal expansion coefficient of Si is negative below 120°K, and that of Ge is negative below 40°K, in contrast to the usual behavior.[21] Values of E_s at 0°K and of E_p have been tabulated by Gschneidner[18] and Slater[14] respectively. The calculated values of α_1 and α_2 (Table I) lead to the following generalized Morse potentials:

TABLE I. Morse potential constants and calculated formation energies of vacancies in Ge and Si.

Physical quantity	Ge	Si
a (10^{-7} cm^2/kg)	12.80	9.88
b (10^{-12} cm^4/kg^2)	3.736	2.013
r_0 (Å)	2.448	2.350
E_s (eV)	3.85	4.66
E_p (eV)	8.1	7.2
$D = (E_s+E_p)/2$ (eV)	6.0	5.9
α_1 (Å^{-1})	2.625	1.728
α_2 (Å^{-1})	0.515	0.984
E_2 (eV)	1.52	1.80
E_3 (eV)	11.55	11.25
$E_{1V}{}^F = 2D+E_2-E_3$ (eV)	1.97	2.35
Results of previous calculations		
$E_{1V}{}^F$ (eV) [Swalin[7]]	2.07	2.32
$E_{1V}{}^F$ (eV) [Scholz[8,9]]	1.62–2.50	2.78

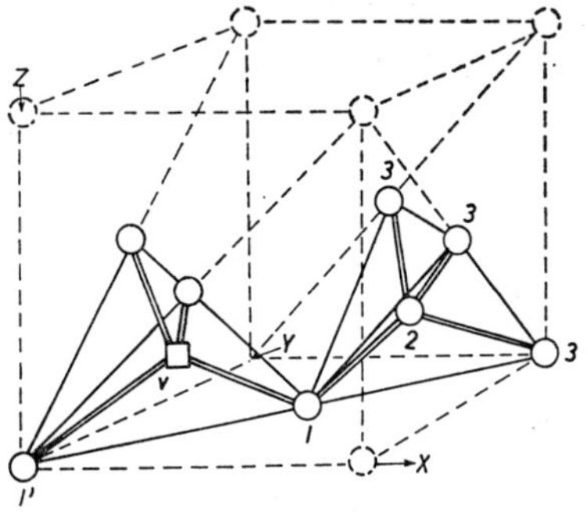

Fig. 1. Environment of a vacancy V in the diamond structure.

For Ge:

$$E = D\{0.244 \exp[-2.625(r-r_0)] - 1.244 \exp[-0.515(r-r_0)]\} \tag{11}$$

For Si:

$$E = D\{1.323 \exp[-1.728(r-r_0)] - 2.323 \exp[-0.984(r-r_0)]\} \tag{12}$$

The ratio $\alpha_1/\alpha_2=1.756$ for Si is much closer to that of the standard Morse potential $\alpha_1/\alpha_2=2$ than the ratio $\alpha_1/\alpha_2=5.097$ for Ge.

The relaxation model for atoms in the three nearest shells to the vacancy will now be described. The four nearest neighbours to the vacancy are assumed to form two new bonds perpendicular to one another, the relaxed configuration having tetragonal symmetry about one of the three ⟨100⟩ directions. The vacancy is put at position (1/4, 1/4, 1/4) in the unit cell of parameter $a_0=4r_0/\sqrt{3}$ (Fig. 1). The surrounding atoms at (0, 0, 0), (1/2, 1/2, 0), (1/2, 0, 1/2) and (0, 1/2, 1/2) are equivalent, so that only eon of them, atom 1 at (1/2, 1/2, 0), need be considered. The nearest neighbor atoms to the vacancy at (0, 0, 0) and (1/2, 1/2, 0) will be assumed to form one of the two new bonds, causing relaxation of atom 1 along the $[\bar{1}\bar{1}0]$-direction a distance $\varepsilon_1/2=\sqrt{2}\,a_0x_1$, where $(1/2-x_1, 1/2-x_1, 0)$ is its final position. Thus the bond distance between these atoms is $(1.633\,r_0-\varepsilon_1)$. Of the three nearest neigh- bors to atom 1, the two at (3/4, 1/4, −1/4) and (1/4, 3/4, −1/4) change their bond lengths little and are not shown in the figure. (However, the energy of these bonds is significant, and cannot be neglected as was done by Swalin.) The third, atom 2 at (3/4, 3/4, 1/4), is assumed to relax towards atom 1 in the $[\bar{1}\bar{1}\bar{1}]$-direction, having final position $(3/4-x_2, 3/4-x_2, 1/4-x_2)$. Thus the stretching of the bond between atoms 1 and 2 is

$$\varepsilon_2 = a_0[2(x_2-x_1-1/4)^2+(x_2-1/4)^2]^{1/2}-r_0 \tag{13}$$

The other three nearest neighbors of atom 2 (at positions 3) are assumed to remain fixed. Thus the bonds between these atoms and atom 2 are each stretched by

$$\varepsilon_3 = a_0[2(1/4+x_2)^2+(1/4-x_2)^2]^{1/2}-r_0 \tag{14}$$

The stretching of the bonds between atom 1 and the atoms at (3/4, 1/4, −1/4) and (1/4, 3/4, −1/4), which are also held fixed, is

$$\varepsilon_2' = a_0[(1/4+x_1)^2+(1/4-x_1)^2+1/16]^{1/2}-r_0 \tag{15}$$

The best values of x_1 and x_2 are found by minimizing

$$\begin{aligned} E_{1V}{}^{F} &= 2D-E_3+E_2 \\ &= 2D+\frac{2D}{\alpha_1-\alpha_2}\{\alpha_2 \exp[-\alpha_1(0.633r_0-\varepsilon_1)]-\alpha_1 \exp[-\alpha_2(0.633r_0-\varepsilon_1)]\} \\ &\quad +4D\left[1+\frac{\alpha_2 \exp(-\alpha_1\varepsilon_2)-\alpha_1 \exp(-\alpha_2\varepsilon_2)}{\alpha_1-\alpha_2}\right] \\ &\quad +8D\left[1+\frac{\alpha_2 \exp(-\alpha_1\varepsilon_2')-\alpha_1 \exp(-\alpha_2\varepsilon_2')}{\alpha_1-\alpha_2}\right] \\ &\quad +12D\left[1+\frac{\alpha_2 \exp(-\alpha_1\varepsilon_3)-\alpha_1 \exp(-\alpha_2\varepsilon_3)}{\alpha_1-\alpha_2}\right] \end{aligned} \tag{16}$$

The calculated values of $E_{1V}{}^{F}$ from this model for Ge and Si are 1.97 and 2.35 eV respectively (Table I). These values are quite close to those previously calculated,[7] so that the effects of the improvements in calculation have cancelled. The values of ε_1, ε_2, and ε_3 are 1.28, 0.20 and 0.14 Å respectively for Ge and 1.23, 0.20, and 0.13 Å for Si. The corresponding bond lengths for atoms 1–1′ and 1–2 are 1.11 r_0 and 1.08 r_0 for both Ge and Si.

If the compressibility data of Bridgman[15,18] is used, the results for Ge are essentially unchanged, but the ratio α_1/α_2 for Si is increased to 6.28 and $E_{1V}{}^{F}$ is decreased by about 0.20 eV. This difference is largely due to the change in b, and thus would not be found by using the standard Morse potential.

If the Swalin arbitrary relaxation relation is used, where it is assumed that $\varepsilon_3=\varepsilon_2/2$, and an approximate geometric relationship is assumed for the effects on ε_1 of strains out to the seventh nearest neighbor to the vacancy, then the calculated values of $E_{1V}{}^{F}$ for Ge and Si are 2.60 and 3.06 eV (using either the standard or generalized Morse potential). Thus it is clearly shown that a free relaxation model, even one as restricted as that used here, leads to lower (and thus presumably more accurate) activation energies than the model of Swalin. It is to be noted that these values calculated with the Swalin model are approximately 0.6 eV greater than found by him previously using lower values of D.[7] Thus the use of the standard Morse potential leads to values of $E_{1V}{}^{F}$ which are dependent on the choice of the dissociation energy D.

However, when the generalized Morse potential is used, this dependence of $E_{1V}{}^{F}$ on D is not realized, as is easily shown by expanding the exponentials of Eq. (16) in Taylor series, giving

$$E_{1V}{}^{F} \approx D\alpha_1\alpha_2\{\varepsilon_0{}^2[1-(\alpha_1+\alpha_2)\varepsilon_0/3]+2\varepsilon_2{}^2[1-(\alpha_1+\alpha_2)\varepsilon_2/3] \\ +4(\varepsilon_2')^2[1-(\alpha_1+\alpha_2)\varepsilon_2'/3]+6\varepsilon_3{}^2[1-(\alpha_1+\alpha_2)\varepsilon_3/3]\} \tag{17}$$

where $\varepsilon_0=0.633r_0-2\sqrt{2}\,a_0x_1$, and ε_2, ε_2' and ε_3 are given by Eq. (13)-(15).

The error in this approximation is only about 2–3% for the present values of the constants used. Since $\alpha_1\alpha_2$ varies as $1/D$ and $\alpha_1+\alpha_2$ is independent of D (see Eq. (9)), it is apparent that $E_{1V}{}^{F}$ is approximately independent of D.* With the standard Morse potential, however, where $a^2=9C^3r_0/2nDa$, terms involving $1/\sqrt{D}$ appear in the equation for $E_{1V}{}^{F}$, so that the calculated values of $E_{1V}{}^{F}$ depend on D.

Using the generalized Morse potential introduced above, the binding energies of divacancies $E_{2V}{}^{B}$ have been calculated by the approximation of Scholz and Seeger,[9] i.e. neglecting the relaxation of neighboring atoms. The numerical values are $E_{2V}{}^{B}=1.96$ eV for Ge and 2.17 eV for Si. These values are too high compared with the values of $E_{1V}{}^{F}$ given in Table I. A better comparison is obtained with the monovacancy formation energies calculated by the same approximation. These are given by

$$E_{1V}{}^{F} = 2D\left[1+\frac{\alpha_2}{\alpha_1-\alpha_2}\exp(-0.633\,r_0\alpha_2)-\frac{\alpha_1}{\alpha_1-\alpha_2}\exp(-0.633\,r_0\alpha_2)\right]$$

The numerical results are $E_{1V}{}^{F}=5.3$ eV for Ge and $E_{1V}{}^{F}=6.7$ eV for Si. Similar to the previous treatment,[9] the divacancy binding energy comes out between one third and one half of the monovacancy formation energy. These are rather high values compared with those encountered in metals.

---

* Since the sublimation energy E_s actually consists of not only the energy required to break nearest neighbour bonds, but also some longer range energy terms, the value $D=(E_S+E_P)/2$ is probably too large, but this error is shown by Eq. (17) to be unimportant.

III. POINT DEFECTS IN THERMAL EQUILIBRIUM

1. *Thermal equilibrium measurement of defect concentration*

Any experimental technique which is sensitive to the presence of point defects and which can be applied with sufficient accuracy can be used to study point defects in thermal equilibrium. From the point of view of simplicity and reliability of the interpretation, the most advantageous of these techniques is the simultaneous measurement of thermal expansion and X-ray lattice parameter as a function of temperature. The relative changes $\Delta l/l$ in length and $\Delta a/a$ in lattice parameter (with respect to a sufficiently low reference temperature, usually room temperature) are related to the concentrations C_{V} of vacant sites and C_{Int} of atoms on interstitial sites by

$$\frac{\Delta l}{l}-\frac{\Delta a}{a}=(C_{\mathrm{V}}-C_{\mathrm{Int}})/3 \tag{19}$$

Since the diamond structure is an open structure, it is not *a priori* clear whether the Gibbs free energy for the formation of single vacancies ($G_{1\mathrm{V}}{}^{\mathrm{F}}$) or interstitials ($G_{1\,\mathrm{Int}}{}^{\mathrm{F}}$) is lower, i.e. which of the two kinds of basic point defects dominates in thermal equilibrium. It would therefore be particularly desirable to obtain from an experimental determination of the left-hand side of Eq. (19) information about both the nature and the concentration of the defects that dominate in thermal equilibrium for Si and Ge, as has been done for metals,[23, 24] and also for AgCl.[25] Although Si and Ge are available as highly perfect crystals suitable for precision diffraction experiments, it appears quite unlikely that such an experimental determination could be carried out with sufficient accuracy. The reason is that all the available evidence (see sections IV-1 and IV-2) points to equilibrium concentrations of point defects in Si and Ge at melting points that are several powers of ten smaller than those in metals. In the discussion of the self-diffusion data, we shall show—provided that self-diffusion takes place by a vacancy mechanism—that the vacancy concentration $C_{1\mathrm{V}}$ at the melting point T_m of Si can be bracketed as follows:

$$3\times10^{-7} \gtrsim C_{1\mathrm{V}}(T_m) \gtrsim 5\times10^{-9} \tag{20}$$

Similar estimates are possible for Ge, and also for the case that an interstitialcy mechanism of self-diffusion prevails. Such concentrations are indeed too small to be studied by the techniques allowing the application of Eq. (19). Since the other standard techniques to study the equilibrium concentrations of defects, such as resistivity or specific heat measurements, do not appear to be more promising, the only equilibrium measurements capable of giving information on the nature and the concentrations of defects in thermal equilibrium are diffusion studies. Extracting this information is not possible without rather detailed and to some extent still speculative discussions since diffusion coefficients are, roughly speaking, the product of defect *concentrations* and defect *mobilities*.

The order of magnitude of the estimate Eq. (20) is plausible, since quenching of Ge from 1,100°K has yielded quenched-in acceptor concentrations of the order of 10^{-8}, and since vacancy type defects (during the quench the vacancies may have agglomerated into multiple vacancies) are expected to act as acceptors (see sections IV-1 and IV-2).

2. *Diffusion*

2.1. *Experimental results*

In recent years the availability of high purity Ge and Si and the use of radioactive tracers have facilitated accurate measurements of self-diffusion and dilute solute diffusion in these elements. The diffusion coefficient is usually expressed as

$$D = D_0 \exp(-Q/kT) \tag{21}$$

where Q is the activation energy for diffusion.

The *self-diffusion* coefficients of Ge and Si have been measured (over temperature ranges of only about 200°C), giving for Ge

$$D_0^{\mathrm{Sd}} = 10.8 \pm 2.4 \text{ cm}^2/\text{sec} \text{ and } Q^{\mathrm{Sd}} = 3.01 \pm 0.02 \text{ eV}$$ [26,27]

and for Si

$$D_0^{\mathrm{Sd}} = 1.8 \times 10^3 \text{ cm}^2/\text{sec} \text{ and } Q^{\mathrm{Sd}} = 4.8 \text{ eV}$$ [28]

The activation energies for self-diffusion Q^{Sd} are probably quite accurate, since considerable care was taken in using pure samples for the experiments. They are nevertheless higher than expected from quenching and irradiation data (sections IV-2 and IV-3). The values of D_0^{Sd} are also higher than expected from theory (section III-2.2).*

* The recent determination of the Si self-diffusion coefficient by Fairfield and Masters[36] employing the radioactive isotope ^{31}Si (the isotope also used by Peart[28]) gives between 1,100°C and 1,300°C

$$D_0^{\mathrm{Sd}} = 9 \times 10^3 \text{ cm}^2/\text{sec}; \qquad Q^{\mathrm{Sd}} = 5.14 \text{ eV}$$

i.e., even larger values for the pre-exponential factor and the activation energy than found by Peart.[28] In the temperature range in which the two determinations overlap (1,220–1,300°C) they give approximately the same values for the coefficients of self-diffusion, the Fairfield and Masters values being slightly lower. The difference between the two determinations is larger at the lower temperatures, where because of the short half-life of ^{31}Si the accuracy is least.

TABLE II. Diffusion in Ge and Si.

Group	Solute	Ge solvent D_0 (cm^2/sec)	Ge solvent Q (eV)	Proposed mechanism†	Si solvent D_0 (cm^2/sec)	Si solvent Q (eV)	Proposed mechanism†	Ref.‡
I	Li	9×10^{-3}	0.57	I	2.7×10^{-3}	0.63	I	29
I	Cu		0.33*	D		0.43	I	30
I	Au	$13\text{–}2.3\times10^{2}$	2.5	E	1.1×10^{-3}	1.1	D	
					1.7×10^{2}	2.4		
						0.40*		31
II	Zn	3.6	2.7			~1		
III	B	1.6×10^{9}	4.6		10.5	3.7		
					16	3.7		32
III	Al				8.0	3.5		
					2.8×10^{3}	3.8		
III	Ga	20	3.0	V	3.6	3.5		
		40	2.5–3.1			4.1		33
III	In	0.02	2.16–2.5	V	16	3.9	E	
		0.45	2.7					
		20	3.0					
III	Tl	0.06	2.7		16	3.9		
V	P	2.5–12	2.48–2.6	V	10.5	3.7		
V	As	0.7–12	2.25–2.5	V	0.32	3.6	V	
					68.6	4.2		34
					2.56	3.9		35
V	Sb	0.7–1.3	2.2–2.3	V	5.6	3.9	V	
		7	2.5					
V	Bi	3	2.5		1.0×10^{3}	4.6		
IV	Si				1.8×10^{3}	4.8		28
					9×10^{3}	5.1		36
IV	Ge	8	3.0		5×10^{5}	5.3		26, 37
		10.8	3.01					27

* For diffusion by the dissociation mechanism, no well-defined value of Q can be found. The starred values refer to the interstitial migration energy of solute atoms.

† I = Interstitial, D = Dissociation, E = Exchange, V = Vacancy.

‡ Only the latest references are given here. General references can found in ref. 1, 38–40.

Solute diffusion in Ge and Si has been studied extensively,[1, 38–40] and is summarized in Table II. Various diffusion mechanisms have been proposed to account for the large differences in diffusion coefficients for the various solute elements. The diffusion of Group III and V elements, which enter Ge and Si substitutionally, has been attributed to a vacancy mechanism.[40] In Ge, Group III elements diffuse slower than Group V, but both have approximately the same activation energies, $\sim$2.4 eV. The opposite is true for Si, in which Group III elements diffuse faster than Group V elements, but again both have approximately the same activation energies, $\sim$3.7 eV. The values of D_0 in both Ge and Si vary between 0.02 and 20. The diffusion of Bi, Si, and Ge in Si is associated with high values of $D_0(\sim 10^3\ \text{cm}^2/\text{sec})$ and $Q(\sim 4.8\ \text{eV})$.

Evidence for the vacancy mechanism of self-diffusion is given by the effect of Group III and V impurities on self-diffusion in Ge.[41] The solubility of an acceptor (e.g. vacancies) is greater in a donor environment than in an acceptor environment. Thus if the vacancy mechanism is valid one would expect that Ge which is heavily doped with a donor impurity ($\sim 10^{19}$ impurity atoms/cm^3) would have a higher vacancy concentration and thus a greater self-diffusion coefficient than undoped Ge or Ge doped with an acceptor impurity. This was in fact found,[41] thus providing circumstantial evidence for the vacancy mechanism. An analogous argument can be applied to explain the difference mentioned above for solute diffusion of Groups III and V in Ge, but it fails to account for the observations on solute diffusion in Si. Further evidence that vacancies are responsible for self-diffusion in Ge has been provided by the analysis of the precipitation kinetics of supersaturated Cu in Ge. Since Cu diffuses by the dissociative mechanism,[2] involving equilibrium between interstitial Cu atoms, substitutional Cu atoms, and vacancies, the rate of Cu precipitation in dislocation-free samples is determined by the diffusion of vacancies. By assuming a vacancy mechanism* for self-diffusion, Penning[3] calculated self-diffusion coefficients for Ge at temperatures down to 700°K which agreed well with the higher temperature values measured by standard tracer techniques.[26, 27]

For less soluble elements in Ge and Si (which also have deeper energy levels in the forbidden gap), the vacancy mechanism is no longer valid. The diffusion of In in Si and of Au in Ge has been attributed to a replacement mechanism, in which a substitutional atom jumps into an interstitial position, a neighboring solvent atom jumps into the resulting vacancy, and the interstitial atom moves into the new vacancy.[42] The diffusion of Zn and Au is much more rapid in Si than in Ge, apparently because these solutes have an appreciable interstitial solubility in Si. The diffusion of insoluble elements in Ge and Si, such as Li, Cu, and Ni, has been shown

* The role of divacancies, which were not considered in these calculations, may be important in these quasi-equalibrium conditions because of the high divacancy binding energy (*cf.* section IV).

to be due to interstitial diffusion, sometimes via the dissociative mechanism.[2,3] Li diffuses as a singly charged interstitial ion in Ge and Si with activation energies 0.57 and 0.63 eV respectively.[29] The ratios of substitutional to interstitial Cu solubility in Ge and Si are 6 and 10^{-4} respectively at 700°C.[30] In general, interstitial solubility is greater in Si than in Ge; consequently the interstitial diffusion mechanism is favored in Si.[42]

Calculations of the migration energy of interstitial Ge atoms in Ge or Si atoms in Si give approximately 0.2 eV.[43] As this is also the same order of magnitude as the migration energy of vacancies in Si, the interstitial mechanism of self-diffusion in Si cannot be disregarded.

The contribution of divacancies to diffusion in Ge and Si is small because the divacancy must partially dissociate before it can migrate. Thus the migration energy of divacancies, $E_{2V}{}^{M}$, is approximately equal to the binding energy, $E_{2V}{}^{B}$. But if the divacancy mechanism predominated, $E_{2V}{}^{M}+E_{2V}{}^{F}<E_{1V}{}^{M}+E_{1V}{}^{F}$ (where $E_{1V}{}^{F}$ and $E_{2V}{}^{F}$ are the formation energies, and entropy terms are neglected). Since

$$E_{2V}{}^{F} = 2E_{1V}{}^{F}-E_{2V}{}^{B}$$

then

$$E_{1V}{}^{F} < E_{1V}{}^{M}$$

and also

$$E_{2V}{}^{F} < E_{1V}{}^{F}$$

Both of these inequalities are contrary to theory and experiment.

2.2. *Discussion of self-diffusion*

Let us compare self-diffusion in Si and Ge with that in f.c.c. metals such as Ag, Au, Cu, Ni, etc., which have melting points similar to those of Si and Ge. This group of metals happens to contain those elements in which the mechanism of self-diffusion is best understood. In them self-diffusion was found to be dominated by the migration of monovacancies with possibilities of small contributions by divacancies and larger vacancy clusters.[44,45] When comparing quantitatively the self-diffusion data of these metals with those of Si and Ge, the following significant differences come to mind:

(i) Extrapolated to the melting points the self-diffusion coefficients of Si and Ge are a factor of about 10^3 smaller than those of the above mentioned group of metals.

(ii) The activation energy for self-diffusion, Q^{Sd}, and the pre-exonential factor $D_0{}^{Sd}$ in Eq. (21) are considerably larger for Si and Ge than for f.c.c. metals with similar melting points. (Very large values of D_0 are also found for diffusion of some solute elements in Ge and Si, e.g. B in Ge, Bi and Ge in Si.)

As we shall see presently, the explanation of these differences, particularly of the large values of D_0, between the diamond structure valence crystals and the f.c.c. metals, presents considerable difficulties.

It is true that the mechanism of self-diffusion in Si and Ge may be quite different from the vacancy mechanism that has been established for the f.c.c. metals with similar melting points. Since the diamond structure is a very open structure, the energy of formation of an interstitial atom should not be very different from that of a vacancy, and since the energy of migration of an interstitial atom is expected to be fairly low,[43] an interstitial mechanism is a serious competitor for the mechanism of self-diffusion. In the following discussion we shall nevertheless start out considering vacancy mechanisms of self-diffusion, for the following reasons:

(a) Some of the difficulties to be discussed are the same for vacancy and interstitial mechanisms, e.g. those connected with an anomalously large D_0^{Sd}.

(b) If self-diffusion were mainly due to a non-vacancy mechanism, say an interstitial or an exchange mechanism, vacancies would contribute only little to the measured self-diffusion coefficients, which means that difficulties encountered in explaining (i) will be aggravated.

The contribution of single vacancies to the self-diffusion coefficient is given by

$$D_{1V}^{Sd} = D_{1V} \cdot f_{1V} \cdot C_{1V} \tag{22}$$

Here D_{1V} is the diffusion coefficient for monovacancies, f_{1V} is the correlation coefficient for single vacancies ($f_{1V}=0.5$)and

$$C_{1V} = \exp(S_{1V}^F/k)\exp(-E_{1V}^F/kT) \tag{23}$$

is the concentration of monovacancies in thermal equilibrium (S_{1V}^F, E_{1V}^F $=$entropy and energy of formation of monovacancies).

If the migration of a monovacancy occurs by jumps between nearest neighbour sites only, the diffusion coefficient of monovacancies is given by

$$D_{1V} = \frac{1}{8}a^2\nu_{1V}^0 \exp(S_{1V}^M/k)\exp(-E_{1V}^M/kT) \tag{24}$$

where a is the edge length of the elementary cube of the diamond structure and $\nu_{1V}^0 \exp(S_{1V}^M/k)\exp(-E_{1V}^M/kT)$ is the frequency factor for jumps from one site to a neighboring site. Inserting Eq. (23) and (24) into Eq. (22) gives us for the pre-exponential factor of the self-diffusion coefficient

$$D_0^{Sd} = \frac{1}{8} a^2 \nu_{1V}^0 f_{1V} \exp\left[(S_{1V}^M + S_{1V}^F)/k\right] \quad (25)$$

In Eq. (25) the only unknowns are ν_{1V}^0, S_{1V}^M, and S_{1V}^F. Assuming a sinusoidal potential energy curve for an atom jumping from its crystal site into a neighboring vacant site (height of the potential barrier E_{1V}^M), one may estimate (m=atomic mass)

$$\nu_{1V}^0 = (16 E_{1V}^M / 6 m a^2)^{1/2} \quad (26)$$

From the studies of monovalent metals it is known that the entropies S_{1V}^M and S_{1V}^F are both positive, the latter having magnitudes between 0.5 k and 1.5 k. Similar magnitudes are expected for the contributions of the lattice vibrations in the diamond structure. An additional contribution may arise if the defects are capable of taking up different orientations in the crystal (for example divacancies). According to Scholz and Seeger[9] the single vacancy in the diamond structure is capable of three different orientations on account of the degeneracy of the electronic configuration; this contributes an additional factor of 3 to $\exp(S_{1V}^F/k)$.

Similar considerations apply to other crystal structures and other defect mechanisms of self-diffusion, e.g., for the vacancy mechanism in the f.c.c. structure the factor 1/8 in Eq. (24) has to be replaced by unity.

The preceding considerations allow us to make a numerical estimate of D_0^{Sd} and to compare this with experiments. For a monovacancy mechanism in the above-mentioned group of f.c.c. metals one gets values for D_0^{Sd} between about 0.08 cm²/sec and 0.40 cm²/sec. These estimates are in excellent agreement with the experiments, if the contribution of divacancies, etc., is corrected. Such corrections are necessary for Ni[45] and Pt,[46] for example. The estimates for Si and Ge give similar magnitudes. It may thus be concluded that the experimental values of D_0^{Sd} are too large for a vacancy mechanism (and also for a mechanism involving single interstitials) by at least one order of magnitude for Ge, and by several orders of magnitude for Si. A consideration of various so-called direct mechanisms of self-diffusion, such as the direct exchange of nearest neighbor atoms or the various ring mechanisms, leads to the same conclusion.

None of the " simple " mechanisms of self-diffusion can therefore explain the self-diffusion data on Si (and, to a lesser extent, also on Ge). A superposition of various " simple " mechanisms does not help either, since at least one of them would than have to have a pre-exponential factor about as large as that determined experimentally, apart from the unlikely exception that several mechanisms happen to have almost exactly the same activation energy of self-diffusion.

An anomalously high pre-exponential factor may arise from a temperature dependence of the activation energy of self-diffusion. In order to demonstrate this, we rewrite Eq. (21) as

$$D = D_0' e^{-G/kT} \tag{27}$$

$$G = H - TS \tag{28}$$

where D_0' is a constant of dimensions cm²/sec. G is the Gibbs free energy of self-diffusion, and H and S are the activation enthalpy (≈energy) and the entropy of self-diffusion. If the activation enthalpy is temperature independent, we have $H = Q^{Sd}$. A temperature dependence of H entails a temperature dependence of S on account of the thermodynamic relation

$$T\frac{\partial S}{\partial T} = \frac{\partial H}{\partial T} \tag{29}$$

which follows from Eq. (28) and $S = -(\partial G/\partial T)_p$. Eq. (29) may be integrated to give

$$S(T) = \int_{T=0}^{T} \frac{1}{T}\frac{\partial H}{\partial T}\, dT \tag{30}$$

We see that an increase of the activation enthalpy H with temperature leads to an increase of the entropy and thus of the pre-exponential factor. As is well-known, the slope of an Arrhenius plot of Eq. (27) gives the activation energy $H(T)$ according to

$$\frac{d \ln D(T)}{d(1/kT)} = -H(T) \tag{31}$$

irrespective of the details of the temperature variation of H and S. In the case of Ge, no significant temperature variation in the measured self-diffusion coefficient can be detected.[3,26,27] Due to the lack of long-lived radio-isotopes the experimental accuracy of the present self-diffusion data on Si is less than that on Ge; within the experimental accuracy no variation of the activation energy of self-diffusion is found within the temperature range investigated.[28] Any significant temperature dependence would therefore have to occur at lower temperatures.

As a working hypothesis we assume that the activation energy changes from a "low-temperature" value Q_0 to a "high-temperature" value

$(Q_0+\Delta Q)$ (which would be equal to the experimentally determined Q^{Sd}) in a relatively narrow temperature interval centered at a temperature T_0. We may divide the integrand of Eq. (30) into two contributions, viz., one due to the "transformation," which is non-vanishing only around T_0, and a second one which contains the weakly temperature-dependent terms associated with the differences of the vibrational spectrum of the defect crystal and the perfect crystal. We write the total entropy accordingly as

$$S(T) = \frac{\Delta Q}{T_0} + S_{\text{Vibr.}} \tag{32}$$

$S_{\text{Vibr.}}$ increases from zero to a finite (positive) value, which is reached near the Debye temperature. At higher temperatures a slight further increase of $S_{\text{Vibr.}}$ due to anharmonicity effects is feasible. This effect is negligible in the present discussion, since it would give rise to a (non-observed) curvature of the ln D vs. $(kT)^{-1}$-plots.

Let us tentatively assume for Si that $T_0=100°$K and that a factor of 10^4 in D_0^{Sd} has to be accounted for by the "excess entropy" $\Delta S=\Delta Q/T_0$. This gives

$$\Delta Q = kT_0 \ln 10^4 = 0.08 \text{ eV}$$

We see that in principle a surprisingly small anomaly in the energy of self-diffusion can explain the anomalously large pre-exponential factor in the self-diffusion coefficient, provided the anomaly in the activation energy occurs at sufficiently low temperatures.

The preceding considerations are completely general and not confined to any particular mechanism of self-diffusion. We shall now apply them to the vacancy mechanism of self-diffusion. If self-diffusion is dominated by monovacancies, the activation energy of self-diffusion Q^{Sd} is related to the energies of formation and migration of single vacancies by

$$Q^{Sd} = E_{1V}^{F} + E_{1V}^{M} \tag{33}$$

By annealing studies on p-type silicon at about 160°K to 180°K, Watkins[47] has determined the energy of migration of a vacancy as

$$E_{1V}^{M} = (0.33 \pm 0.03) \text{ eV} \tag{34}$$

He suggests that this value pertains to the electrically neutral vacancy, and that a negatively charged vacancy has an even lower energy of migration. These very small energies raise a problem with respect to the vacancy concentration in thermal equilibrium. Inserting into Eq. (33) the measured activation energy of self-diffusion for Si of 4.8 eV[28] and Eq. (34) gives us a vacancy formation energy of at least 4.4 eV. At the melting point of silicon we have $\exp(-4.4 \text{ eV}/kT) \approx 10^{-13}$. If the entropy of formation of monovacancies were in the range usual for metals, as mentioned above, the vacancy concentration at the melting point would be of the order of 10^{-12}. This would be an implausibly low value, which would make it virtually impossible to carry out well-defined experiments involving vacancies in thermal equilibrium on account of vacancy-impurity interactions.

The estimate of the preceding paragraph disregards the "entropy anomaly" discussed above. This anomaly may occur in the entropy of formation or in the entropy of migration, or in both. Detailed inspection shows that the high-temperature concentration of vacancies is higher when the anomaly in the entropy and energy of *formation* is larger. Let us assume that the energy of migration is given by Eq. (34) independent of the temperature (this means that the anomaly is attributed entirely to the formation quantities), and that the entropy of migration is of the order of magnitude $S_{1V}{}^{M} = k$. Inserting the experimental parameters for Si into Eqs. (25) and (26) gives us as high temperature value $\exp(S_{1V}{}^{F}/k) = 3 \times 10^{6}$, yielding a monovacancy concentration at the melting point $C_{1V}(T_m) = 3 \times 10^{-7}$. In view of the assumptions made this is an approximate *upper limit* to the vacancy concentration at the melting point. This upper limit also applies if self-diffusion is not dominated by a vacancy mechanism, since the contribution of the monovacancies to self-diffusion have then to be less than the measured self-diffusion coefficient.

If self-diffusion of Si and Ge is indeed due to the migration of vacancies, a *lower limit* for the vacancy concentration in thermal equilibrium may be obtained as follows (a similar estimate for interstitials is possible if the self-diffusion is due to an interstitialcy mechanism): by combining Eq. (22) with (24) the concentration of single vacancies may be written as

$$C_{1V} = \frac{16D^{Sd}}{a^2 \nu_{1V}{}^{0}} \exp\left(\frac{E_{1V}{}^{M} - TS_{1V}{}^{M}}{kT}\right) \tag{35}$$

where a is the lattice constant (Si: $a = 5.43$Å; Ge: $a = 5.66$Å[22]). In order to obtain a lower limit for C_{1V}, we put the free energy of migration of a vacancy, $E_{1V}{}^{M} - TS_{1V}{}^{M}$, equal to zero, i.e., we assume that the vacancy migration takes place without an appreciable energy barrier. The attempt frequency for vacancy migration, $\nu_{1V}{}^{0}$, is presumably not larger than 10^{13} sec^{-1}. For the coefficient of self-diffusion D^{Sd} we insert the value at the melting point T_m (Si: $T_m = 1{,}690$°K, $D^{Sd}(T_m) = 1 \times 10^{-11}$ cm^2/sec; Ge:

$T_m=1{,}210°K$, $D^{Sd}(T_m)=3\times10^{-12}$ cm²/sec). This gives us as estimates for the concentration of vacant sites at the melting point

$$\text{for Si:}\quad C_{1V}(T_m) \gtrsim 5\times10^{-9}$$

$$\text{for Ge:}\quad C_{1V}(T_m) \gtrsim 1.5\times10^{-9}$$

These estimates have been made use of in section III-1.

The interpretation of the self-diffusion data in Si given above in terms of an anomalously high entropy of formation requires that the nature of the monovacancy at high temperatures be different from that at low temperatures. If the low-temperature modification has an entropy of formation $S_{1V}{}^F=2k$, the " transformation " from the low temperature form to the high-temperature form must be accompanied by an entropy increase of $\Delta S_{1V}{}^F=12.9k$ and by an energy increase of $\Delta E_{1V}{}^F=T_0\Delta S_{1V}{}^F$, where T_0 is the transformation temperature. Putting more or less arbitrarily $T_0=500°K$, we get $\Delta E_{1V}{}^F=0.56$ eV and as the low temperature value of the formation energy of monovacancies $E_{1V}{}^F=3.9$ eV. The transition must be of such a nature that the high-temperature modification is considerably more disordered than the low-temperature modification. This suggests that the high-temperature form of the vacancy is spread out as a " relaxion," [48, 49] e.g. over the sixteen nearest and next-nearest sites of the original vacancy. Such a configuration would have a considerably higher entropy of formation than an " ordinary " vacancy. Since the energy of such a configuration is presumably substantially higher than that of the simple, only slightly relaxed vacancy, the high-temperature configuration could not be obtained by the conventional theoretical studies of defect configurations and energies such as those of section II, which are based on essentially zero-temperature theories.

The idea of the relaxion mechanism of self-diffusion was originally introduced in an attempt to relate self-diffusion in metals to the melting process.[48] Both Si and Ge have a much higher melting entropy per atom $S_m=L_m/T_m$ than metals with similar melting points[18] (L_m=latent heat of melting; Si: $S_m=3.6\,k$; Ge: $3.1\,k$; Ag: $1.1\,k$; Cu: $1.2\,k$; Au: $1.1\,k$; Ni: $1.2\,k$). If the relaxion in the diamond structure is pictured as consisting of a total of about 16 atoms in a melt-like configuration, there would be no difficulty in accounting for the high excess entropy $S_{1V}{}^F=13k$ in terms of the melting data. However, it must be remembered that Si and Ge are known to contract on melting (relative volume changes on fusion −9.6% [Si] and −5.0% [Ge]),[50] so that a locally melted region is certainly not a good model for a strongly relaxed *vacancy*. Rather, one might expect that in the diamond structure an *interstitial* atom with a strongly distorted environment might be related to such locally melted regions.

It is tempting to carry this line of thought further. We would than, reversing our previous viewpoint, attribute the self-diffusion in Si to a relaxed interstitial mechanism. In this way we could account for the anomalously high pre-exponential factor in the diffusion coefficient. In Ge the anomaly in the entropy of self-diffusion is less than half that of Si. It might be possible to account for this smaller entropy value without resorting to the analogy to melting and to explain it on the basis of formation and migration of strongly relaxed vacancies. The order-of-magnitude differences of D_0^{Sd} for Si and Ge could thus be explained by different high-temperature self-diffusion mechanisms, namely a relaxed interstitial mechanism for Si and a relaxed vacancy mechanism for Ge.

The preceding assignment postulates that the mechanisms of self-diffusion in Si and Ge are different. In addition to the different orders of magnitude of the pre-exponential factors of the self-diffusion coefficients a number of further experimental facts support such an interpretation:

(i) Although Si has a smaller lattice parameter than Ge, it forms a more open crystal structure because of the smallness of the Si atoms. Table III compares the atomic radii as derived by Slater[51] from interatomic bonds with the nearest neighbor distances in the diamond structure crystals. It is seen that the ratio of the interatomic distance to the atomic radius is about 10% larger for Si than for Ge. Whereas in Si the atomic diameter is smaller than the nearest neighbor separation in the crystal, the opposite is true for Ge. It appears therefore entirely feasible that an interstitial type mechanism dominates in Si and a vacancy type mechanism in Ge.

(ii) As discussed in section III-2.1, the interstitial solubility of solute atoms is larger in Si than in Ge, which favors the interstitial mechanism of solute diffusion in Si. This supports also the idea that self-diffusion in Si might occur by an interstitial mechanism.

TABLE III. Comparison between nearest-neighbor interatomic distance r_0 and Slater atomic radii r_a (ref. 51) for diamond-structure elements.

Element	Atomic radius r_a (Å)	Interatomic distance r_0 (Å)	$r_0/2\ r_a$
C	0.70	1.54	1.1
Si	1.10	2.35	1.07
Ge	1.25	2.45	0.98
Sn	1.45	2.80	0.97

(iii) In Ge, quenching and precipitation data (to be discussed in section IV-2) show that vacancies dominate in Ge at high temperatures, with a concentration of approximately 10^{-7} at the melting point. Moreover, as shown by Penning,[3] the assumption of a vacancy mechanism as applied to the precipitation kinetics of Cu in Ge leads to diffusion coefficients which agree with self-diffusion data. No such evidence for the vacancy mechanism has appeared for Si.

(iv) As indicated in section III-2.1, the differences between diffusivities of Group III and Group V solutes can be accounted for by a vacancy type diffusion mechanism in Ge, but not in Si. The evidence that in Si the solute diffusion of Group III elements is faster than that of Group V elements could be explained in terms of an interstitial mechanism, since interstitial solute atoms have been shown to be donors.

We see that quite a number of hitherto unrelated experimental facts are explained by our proposal. Figure 2 summarizes this proposal schematically. It gives the temperature dependences of self-diffusion in Si by the relaxed interstitial mechanism and by the vacancy mechanism, and in Ge by the vacancy mechanism. The curvature of the latter curve is meant to indicate the transition from the low-temperature vacancy configuration to the more relaxed high-temperature configuration. Since, however, no such curvature has been noted over the temperature interval over which self-diffusion

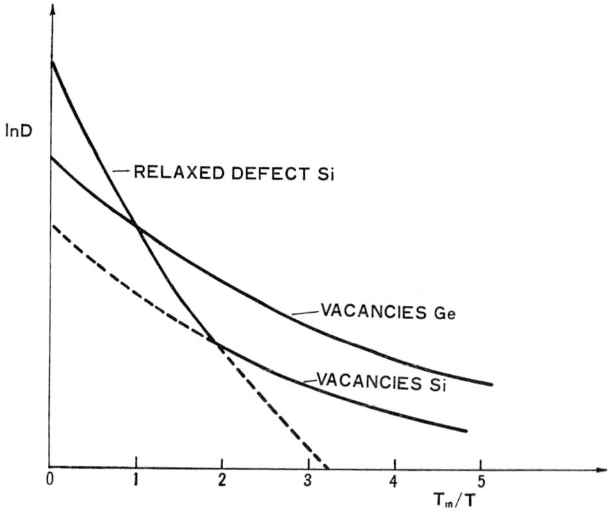

Fig. 2. The self-diffusion coefficient D of Si and Ge as a function of reciprocal absolute temperature T^{-1}(T_m=melting temperature).

and precipitation experiments have been carried out, we must conclude that it is fairly small over that range.

The full lines give the temperature dependence that would be found if self-diffusion could be observed down to very low temperatures. An estimate of the lower curve in Si (vacancies Si) will be given in section IV-3. It should be noted that from the present view the approximate agreement of the self-diffusion coefficients of Si and Ge as extrapolated to the melting point appears fortuitous, since they pertain to different mechanisms of self-diffusion.

IV. QUENCHING AND IRRADIATION RESULTS

1. *General considerations*

The lattice defects introduced by quenching or irradiation of solids (as well as the defects responsible for diffusion) may be interstitial atoms, vacancies, or combinations of these with one another or with impurities and dislocations. Information about the mobility of isolated defects from annealing experiments is thus not always unambiguous. The thermally activated diffusion of defects can best be discussed in terms of the general equation for the diffusion coefficient in a cubic crystal[44]

$$D = \frac{1}{6}\Gamma d^2 \tag{36}$$

where Γ is the jump frequency of the defect and d is the unit jump distance.

An effective diffusion coefficient for migration of vacant lattice sites can be written[52]

$$(D_\mathrm{V})_\mathrm{eff} = \frac{1}{6}\,\frac{d_{1\mathrm{V}}^2\Gamma_{1\mathrm{V}}C_{1\mathrm{V}}+2d_{2\mathrm{V}}^2\Gamma_{2\mathrm{V}}C_{2\mathrm{V}}+d_{i\mathrm{V}}^2\Gamma_{i\mathrm{V}}C_{i\mathrm{V}}}{C_{1\mathrm{V}}+2C_{2\mathrm{V}}+C_{i\mathrm{V}}} \tag{37}$$

where $C_{1\mathrm{V}}$, $C_{2\mathrm{V}}$ and $C_{i\mathrm{V}}$ are the atomic fractions of single vacancies, divacancies and vacancy-impurity pairs, and $d_{1\mathrm{V}}$, $d_{2\mathrm{V}}$, $d_{i\mathrm{V}}$, $\Gamma_{1\mathrm{V}}$, $\Gamma_{2\mathrm{V}}$, and $\Gamma_{i\mathrm{V}}$ are the corresponding jump distances and frequencies. In thermal equilibrium or quasi-equilibrium the concentrations $C_{2\mathrm{V}}$ and $C_{i\mathrm{V}}$ are given by[53]

$$C_{2\mathrm{V}} = \frac{z}{2}C_{1\mathrm{V}}^2 \exp(E_{2\mathrm{V}}{}^\mathrm{B}/kT)\exp(S_{2\mathrm{V}}{}^\mathrm{B}/k \tag{38}$$

and

$$C_{i\mathrm{V}} = C_i\left[1+\frac{1}{zC_{1\mathrm{V}}\exp(E_{i\mathrm{V}}{}^\mathrm{B}/kT)\exp(S_{i\mathrm{V}}{}^\mathrm{B}/k)}\right]^{-1}$$

$$\approx C_i\cdot z\cdot C_{1\mathrm{V}}\exp(E_{i\mathrm{V}}{}^\mathrm{B}/kT)\exp(S_{i\mathrm{V}}{}^\mathrm{B}/k) \tag{39}$$

since $E_{1V}^{F} > E_{iV}^{B}$.

Here $E_{2V}^{B} = 2E_{1V}^{F} - E_{2V}^{F}$ denotes the binding energy of a divacancy, $S_{2V}^{B} = S_{2V}^{F} - 2S_{1V}^{F}$ is the association entropy of a divacancy, E_{1V}^{F}, E_{2V}^{F}, S_{1V}^{F}, S_{2V}^{F} are the energies and entropies of formation of single vacancies and divacancies, C_i is the impurity concentration, z is the coordination number, and the binding energy and association entropy of an impurity-vacancy pair are $E_{iV}^{B} = E_{1V}^{F} - E_{iV}^{F}$ and $S_{iV}^{B} = S_{iV}^{F} - S_{1V}^{F}$.

The recovery process for vacant lattice sites can usually be closely approximated by

$$\frac{dC_V}{dt} = A \cdot C_V \cdot (D_V)_{eff} \tag{40}$$

for diffusion to fixed sinks[52] (e.g. dislocations). Here

$$C_V = C_{1V} + 2C_{2V} + C_{iV} \tag{41}$$

is the total concentration of vacant sites; A depends on the nature, distribution and concentration of sinks. Then the effective migration energy of vacancies $(E_V^M)_{eff}$ can be found from Eq. (37) by the usual procedures.

Let us first consider only the migration of vacancies and divacancies. In contrast to metals, the binding energy of divacancies in diamond structure valence crystals is large,[4, 54] and the migration energy of single vacancies is much less than that of divacancies, because a divacancy must first partially dissociate in order to migrate. Several interesting cases can be discussed, which may be applicable to annealing studies after irradiation or quenching of Ge and Si:

(a) If single vacancies and divacancies are in thermal equilibrium, then the effective migration energy of vacant lattice sites is dependent on the concentration of free vacancies and thus on the binding energy of divacancies, giving $(E_V^M)_{eff} = E_{1V}^M + E_{2V}^B/2$ for a suitable large binding energy.

This can be shown as follows: the equilibrium relation between single vacancies and divacancies is given by Eq. (38). Thus, in terms of the total vacancy concentration, $C_V = C_{1V} + 2C_{2V}$,

$$C_{1V} = \frac{[1 + 4z \exp(E_{2V}^B/kT) C_V]^{1/2} - 1}{2z \exp(E_{2V}^B/kT)} \tag{42}$$

(neglecting the entropy factor).

The effective diffusion coefficient is approximately given by

$$(D_V)_{eff} \approx \frac{C_{1V} D_{1V}}{C_{1V} + 2C_{2V}}, \qquad D_{1V} = \frac{1}{6} \Gamma_{1V} d_{1V}^2 \tag{43}$$

since $C_{1V} \approx C_{2V}$ at the annealing temperature, and $D_{1V} \gg D_{2V}$. Therefore,

$$\frac{dC_V}{dt} = A \cdot (C_{1V} + 2C_{2V}) \cdot (D_V)_{\text{eff}} = AC_{1V}D_{1V} \tag{44}$$

where C_{1V} is given by Eq. (42). For large divacancy binding energy or large C_V,

$$\frac{dC_V}{dt} \approx \{C_V/[z \exp(E_{2V}{}^B/kT)]\}^{1/2} \cdot A \cdot D_{1V} \tag{45}$$

Thus the effective migration energy

$$(E_V{}^M)_{\text{eff}} = E_{1V}{}^M + E_{2V}{}^B/2 \tag{46}$$

and the apparent order of the reaction is one-half. (See Damask and Dienes[53] for a discussion of effective migration energies in metals.)

(b) If the rate at which single vacancies anneal out is larger than the rate at which divacancies dissociate, then thermal equilibrium between C_{1V} and C_{2V} is not achieved, and the rate-determining process for annealing out of vacant sites is the dissociation of divacancies. That is, if

$$3C_{1V}C_{\text{sink}}\Gamma_{1V} > 6C_{2V}\nu_{2V} \exp(-E_{2V}{}^{\text{Dis}}/kT) \tag{47}$$

then the effective activation energy $(E_V{}^M)_{\text{eff}}$ equals $E_{2V}{}^{\text{Dis}}$, the energy of dissociation. If one considers that in order to dissociate the divacancy must first separate into vacancies at next-nearest neighbor positions, it is clear that the dissociation energy equals $E_{1V}{}^M + E_{2V}{}^{B+}$, where $E_{2V}{}^{B+}$ is almost as large as $E_{2V}{}^B$.

(c) If the binding energy of divacancies is very high (≈ 2 eV) but the energy to dissociate partially is appreciably smaller, the annealing out of vacant sites will take place by migration of divacancies, with an effective migration energy

$$(E_V{}^M)_{\text{eff}} = E_{1V}{}^M + E_{2V}{}^{PB}$$

where $E_{2V}{}^{PB}$ is a partial binding energy roughly equal to $(3/4)E_{2V}{}^B$, depending on the saddle point energy for partial dissociation (or approximately on the binding energy of the separated divacancy).

These different possibilities will be considered later when discussing the recovery of quenched-in vacant sites and of irradiation-induced divacancies.

Similarly, if only the interaction of vacancies with impurities is important, the concentration of vacancy-impurity pairs is given by Eq. (39) and the total concentration of vacant sites is $C_V = C_{1V} + C_{iV}$.

Then

$$(D_V)_{eff} = \frac{C_{1V}D_{1V}}{C_{1V}+C_{iV}} \tag{48}$$

since

$$C_{1V} \approx C_{iV} \quad \text{and} \quad D_{1V} \gg D_{iV}$$

Thus

$$\frac{dC_V}{dt} = A\,C_{1V}D_{1V} = \frac{C_{iV}}{zC_i}\exp(-E_{iV}{}^B/kT)A\,D_{1V} \tag{49}$$

For a large binding energy, but with the thermal equilibrium condition (39) satisfied,

$$C_{iV} \approx C_V \quad \text{and} \quad (E_V{}^M)_{eff} = E_{1V}{}^M + E_{iV}{}^B \tag{50}$$

However, if as with divacancies the binding energy is so high that the migration is by movement of the vacancy-impurity complex, then

$$(E_V{}^M)_{eff} = E_{1V}{}^M + E_{iV}{}^{PB} \tag{51}$$

where $E_{iV}{}^{PB}$ represents a partial binding energy corresponding to migration of the complex by partial dissociation. Such a recovery process has been observed for vacancy-phosphorous pairs in Si.[56]

A further complication in semiconductors is that defects having several different charge states may be involved in diffusion. James and Lark-Horovitz[57] considered that vacancies are acceptors and introduced two energy levels in the forbidden gap at 0.05 eV and ≈0.2 eV above the valence band, being charged negatively when the Fermi level is between the two levels. Blout[58] calculated, however, that the vacancy is amphoteric (either a donor or acceptor), with similar energy levels as found by James and Lark-Horovitz, but being neutral for the Fermi level between the two energy levels. The experiments of Watkins[5] have supported the second model, since the vacancies produced in electron-irradiated Si are neutral in p-type material with the Fermi level 0.05 eV above the valence band but become negative when the Fermi level rises. They found that a further energy level may be present near the conduction band, above which the vacancies are assumed to be doubly charged. Callcott and MacKay[59] have presented evidence for the existence of a doubly charged vacancy in electron-irradiated Ge. Energy levels at ~0.04 eV and ~0.18 eV above the valence band in quenched Ge have been attributed to the single vacancy.[60,61] After irradiation of Ge, many energy levels have been found, none of which could be associated definitely with the vacancy.[62, 63]

Corbett and Watkins[63] have found three energy levels for divacancies in Si at ~0.25 eV above the valence band, approximately in the center of the gap, and 0.4 eV below the valence band.

2. *Quenching results*

In metals, since formation and migration energies of single vacancies are approximately equal to 1 eV,[52] large numbers of vacancies are in thermal equilibrium at high temperatures, and they can easily be quenched in because of their relative immobility. In semiconductors the formation energy is considerably higher (2–4.5 eV in Ge and Si) and the migration energy is lower (~0.3 eV in Si).[47] Thus not only are few vacancies in thermal equilibrium near the melting point, but they anneal out rapidly during the quench. Consequently the few vacancies which are quenched in are largely in association with impurities or other vacancies.

Nevertheless, useful quenching experiments have been performed on Si and Ge. Mayburg[64] could not detect any reversible changes in Si radiation quenched from near the melting point that could be attributed to frozen-in defects which had been in thermal equilibrium at high temperatures. He concludes that less than 5×10^{12} donors or acceptors per cm^3 (corresponding to a defect concentration of less than 10^{-10}) have been frozen in. Although with this relatively slow quench many of the vacancies would anneal out during the quench, the experiment indicates that $E_{1V}{}^{F}$ is high (probably >3 eV) and is consistent with the low value of $E_{1V}{}^{M}$ found by Watkins.[47] (An analogous argument holds if the dominant defect in high-temperature equilibrium is not the vacancy, but an interstitial type defect.) Mayburg found an irreversible reduction in resistivity of p-type Si samples near the surface after annealing at 1,100°C, but no change for n-type. He interpreted this by the annealing out of frozen-in interstitial Si atoms, which would imply a very high interstitial migration energy. An impurity effect is much more likely, and indeed it has since been found that a high temperature anneal of Si containing O-vacancy complexes (which act as donors) causes their disappearance by clustering,[65] thus clearly explaining this effect as due to diffusion in of oxygen from the surface.

Up to 5×10^{14} acceptors/cm^3 (atomic fraction 10^{-8}) have been quenched into high purity Ge from near the melting point,[60, 66–68] giving an apparent activation energy of formation of vacancies equal to 1.9–2.0 eV. Since the scatter of points over the small quenching temperature range of ~130°K was rather large, this energy is open to some doubt. Furthermore, even in the fastest quench[60] the presence of only 10^6 dislocations/cm^2 provided a sufficiently high sink concentration to prevent vacancies from being quenched in. As this dislocation concentration corresponds to an atomic fraction of only about 10^{-9}, it is clear that most of the observed quenched-in vacancies in dislocation-free, high purity samples must have encountered another vacancy during the quench. Thus the observed acceptor concentration after quenching is probably due largely to divacancies (or possibly larger vacancy clusters). The effect of impurities can also not be neglected, since the presence of 6×10^{19} Sn atoms/cm^3 lowered the apparent formation energy to 1.3 eV.[60] If vacancy-impurity pairs predominate at the quenching temperature, then from Eq. (39) the concentration of quenched-in vacancies is

$$C_{\mathrm{V}} \approx C_{i\mathrm{V}} \approx zC_i \exp{(E_{i\mathrm{V}}{}^{\mathrm{B}} - E_{1\mathrm{V}}{}^{\mathrm{F}})/kT} \tag{52}$$

so that the effective formation energy of vacancies would equal $E_{1\mathrm{V}}{}^{\mathrm{F}} - E_{i\mathrm{V}}{}^{\mathrm{B}}$.

In the case of the high purity quenching samples it seems that impurities alone do not determine the quenched-in acceptor concentration, since the experimental pre-exponential factor is $\sim 2\times10^{13}$/cm^3, which is not in agreement with Eq. (52). However, it is easily shown that a relatively small impurity concentration could affect the quenching results: the total concentration of vacancies in equilibrium at the quenching temperature can be written

$$C_{\mathrm{V}} = C_{1\mathrm{V}}[1 + zC_i \exp{(E_{i\mathrm{V}}{}^{\mathrm{B}}/kT)}] \equiv C_{1\mathrm{V}}[1+y] \tag{53}$$

The effective formation energy of vacancies is

$$(E_{\mathrm{V}}{}^{\mathrm{F}})_{\mathrm{eff}} = \frac{-\partial(\ln C_{\mathrm{V}})}{\partial(1/kT)} \tag{54}$$

if isolated vacancies and vacancy-impurity pairs act as equally effective acceptors.

For small y,

$$\ln C_{\mathrm{V}} = -E_{1\mathrm{V}}{}^{\mathrm{F}}/kT + y - \frac{y^2}{2} + \frac{y^3}{3} - \dots$$

and

$$(E_{\mathrm{V}}{}^{\mathrm{F}})_{\mathrm{eff}} \approx E_{1\mathrm{V}}{}^{\mathrm{F}} - zC_i \exp{(E_{i\mathrm{V}}{}^{\mathrm{B}}/kT)} E_{i\mathrm{V}}{}^{\mathrm{B}} \tag{55}$$

For an impurity concentration of only 10^{-7}, and $E_{i\mathrm{V}}{}^{\mathrm{B}} = 1.2$ eV, a quench from 1,100°K will give an effective formation energy which is 0.2 eV smaller than the formation energy of single vacancies. Such values are not unreasonable for the Ge quenches, especially since the samples are often annealed in an oxygen atmosphere to avoid Cu contamination.[67, 68]

The concentration of divacancies which is at equilibrium at the quenching temperature is not expected to be significant, since

$$\frac{2C_{2V}}{C_{1V}} = z \exp(S_{2V}{}^{F}/k) \exp(-S_{1V}{}^{F}/k) \exp(E_{1V}{}^{F} - E_{2V}{}^{F})/kT \tag{56}$$

and $E_{1V}{}^{F}$ is presumably less than $E_{2V}{}^{F}$ by at least 0.5 eV. However, for a quenched sample, the quasi-equilibrium condition (from Eqs. [38] and [42]) is

$$\frac{2C_{2V}}{C_{1V}} = \frac{[1+4z \exp(S_{2V}{}^{B}/k) \exp(E_{2V}{}^{B}/kT)C_V]^{1/2}-1}{2} \tag{57}$$

where the binding entropy term $S_{2V}{}^{B}$ has now been included.

From this equation it is seen that at the annealing temperature of 800°K, divacancies will predominate in samples containing 10^{-8} atomic fraction of quenched-in lattice sites, if $E_{2V}{}^{B} \gtrsim 1.1$ eV (which is much less than the experimental value for Si).

Recovery of quenched-in defects in Ge occurs from 300–550°C (varying greatly among the different experiments) with an effective migration energy of 1.2–1.3 eV.[60, 66–68] Since divacancies are likely to predominate at the annealing temperature in high purity samples and since the single vacancy migration energy is expected to be less than these values (from other evidence), this migration energy should be considered in terms of the quasi-equilibrium between single vacancies and divacancies. If the vacant sites themselves form the main sinks for annealing, then from Eq. (57)

$$\frac{C_{2V}}{C_{1V}} \approx [C_V \exp(E_{2V}{}^{B}/kT)]^{1/2} \tag{58}$$

and the inequality (47) becomes

$$\frac{(C_V)^{1/2}}{2} > \exp(-E_{2V}{}^{B}/2kT) \tag{59}$$

It follows that equilibrium between single vacancies and divacancies occurs above a critical temperature

$$T_c \approx 600 E_{2V}{}^{B}\ (°\mathrm{K}), \text{ where } E_{2V}{}^{B} \text{ is in eV.}$$

It seems therefore that $(E_V{}^{M})_{\mathrm{eff}}$ cannot be given by Eq. (46), because equilibrium would not be achieved at the annealing temperature. Moreover, annealing by the dissociative mechanism (b) of section IV-1 would lead to an $(E_{\mathrm{V}}^{\mathrm{M}})_{\mathrm{eff}}$ which is larger than observed. It is concluded that the divacancies probably migrate as a unit (similar to the case of irradiated Si—see section IV-3).*

If the impurity concentration of quenched samples $C_i \gtrsim 10^{-7}$ and $E_{i\mathrm{V}}^{\mathrm{B}} \gtrsim 1.2$ eV, the effective migration energy may be that of the vacancy-impurity complexes. From precipitation studies of Li in Ge, $E_{i\mathrm{V}}^{\mathrm{B}}$ was found to be ~ 0.6 eV for a possible vacancy-oxygen pair[69] but a value of ~ 2 eV is deduced for the vacancy-oxygen pair in Si.[82]

The migration energy of single vacancies can also be estimated from the fact that almost all vacancies annealed out during a rapid quench from 1,100°K if only 10^6 dislocations/cm² were present.[60] The estimated value is $E_{1\mathrm{V}}^{\mathrm{M}} \sim 0.6$ eV, and is almost certainly not as great as 1 eV. Thus in Ge, as in Si, $E_{1\mathrm{V}}^{\mathrm{M}} \ll E_{2\mathrm{V}}^{\mathrm{M}}$.

Further evidence that the migration energy of vacancies in Ge is less than 1 eV was given by the experiments of Tweet[70] on diffusion of Cu by the dissociative mechanism into dislocation-free Ge crystals. He found that $D_{\mathrm{V}} > 10^{-6}$ cm²/sec at 750°C. If this diffusion is largely due to single vacancies, $\frac{1}{6}\,\Gamma_{1\mathrm{V}} d^2 > 10^{-6}$ cm²/sec, which leads to

$$E_{1\mathrm{V}}^{\mathrm{M}} < 0.8 \text{ eV} \quad \text{if} \quad \Gamma_{1\mathrm{V}} \approx 10^{14} \exp(-E_{1\mathrm{V}}^{\mathrm{M}}/kT)^{\dagger}$$

For radiation-quenched Ge samples containing an initial concentration of 6×10^{12} acceptors/cm³, Mayburg[64] found approximately five times more acceptors/cm³ than did Logan,[60] who used a faster quench. Mayburg then observed from electrical resistivity measurements at 195°K that annealing below 516°C produced an increase in acceptor concentration, while annealing above 516°C gave an initial rapid decrease in acceptor concentration. Measurements of resistivity at 28°C did not show an increase in acceptor concentration caused by annealing. These results were interpreted by Letaw[71] in terms of the formation of divacancies during the anneal, and their subsequent removal by clustering or by migration to surfaces. He found it necessary to assume that divacancies introduced more than twice the number of acceptors than did single vacancies, and

* However, if dislocations formed the main sinks for migrating defects, as is indicated by annealing studies,[66–68] equilibrium between vacancies and divacancies might be achieved at the annealing temperature, and $(E_{\mathrm{V}}^{\mathrm{M}})_{\mathrm{eff}}$ might be given by Eq. (46).

† Recent experiments by Stojic and Hasiguti, and by Swanson (*Radiation Effects in Semiconductors*, Plenum Press, New York, 1968) have indicated that quenched-in defects are mobile below 300°K in Ge, with migration energy < 0.5 eV.

that a long range repulsion between single vacancies effectively added 0.7 eV to the migration energy of single vacancies. These two assumptions are not in accordance with theory. A more likely explanation is that many of the acceptors introduced by the quench were vacancy-impurity complexes, as has also been recently suggested by Fuller and Wolfstirn.[73] If some of these complexes were donors (as in the case of vacancy-oxygen complexes) which annealed out below 516°C, an apparent increase in acceptor concentration would be observed.

3. *Irradiation results*

Fast particle irradiation at low temperatures offers the best method of introducing isolated defects into semiconductors, but only recently have techniques advanced to the stage where quantitative results on defect configurations and migration energies have been obtained. This is because the annealing kinetics of defects in doped samples is very complex, as has been shown by electron paramagnetic resonance (EPR) and infra-red absorption studies of irradiated Ge and Si.[63] Numerous vacancy-impurity complexes are formed and disappear during the annealing process from 60–600°K after electron or neutron irradiation.[62, 63] Single vacancies, divacancies, complexes of a vacancy associated with O, P and Al and other unidentified complexes have been observed by EPR after electron irradiation of Si.[5] Often infra-red absorption bands have been associated with these defects.[72, 74] In Ge the EPR studies are lacking but similar infra-red bands to those observed in Si have been observed.[75] Measurements of spectral response of photo-conductivity after electron and neutron irradiation have also shown that many defect levels appear in Ge and Si during annealing above 80°K.[76, 77]

Watkins and coworkers have reported that the migration energies of negatively charged and neutral vacancies in Si electron-irradiated at 4°K are ~0.13 and 0.33 eV, respectively.[5, 47] Spectra associated with divacancies and various vacancy-impurity complexes appear as the vacancy spectra disappear. The neutral vacancy annealing stage at ~160°K has also been reported by electrical studies after γ-irradiation at 80°K.[78] (In addition, some evidence has been given for the migration of interstitials below 4°K in Si.[5])

In Ge an annealing stage associated with single vacancies has not been positively identified. Early work on recovery of electrical properties from 100–200°C after electron irradiation was interpreted as migration of single vacancies,[79] but subsequent studies have shown that many recovery stages associated with vacancies occur below 300°K in irradiated Ge. The main ones are at 35°, 65°, 90° and 180°K.[6,79] The first two have been rather arbitrarily attributed to recombination of Frenkel pairs,[80] but recent infra-red absorption studies of electron-irradiated Ge doped with 10^{17} oxygen atoms per cm^3 have shown that an oxygen-vacancy complex appears when annealing between 45° and 80°K,[75] very much as in Si.

Thus one might conclude that the migration of vacancies (perhaps negatively charged) occurs in this range for Ge. The 90°K stage has been shown by photoconductivity measurements to be associated with the production of defect-impurity complexes or defect clusters.[76] The 180°K recovery stage might well be due to the migration of single vacancies.[6] Measurements of minority carrier lifetime in γ-irradiated As-doped Ge have shown a recovery stage near 300°K with migration energy 0.64 eV,[81] which is attributed to motion of a defect-impurity complex.

Many defect-levels in the energy gap have been observed by electrical measurements of irradiated Ge and Si above 300°K, and several of these have been shown to be caused by impurity-defect complexes.[62,63] The oxygen-vacancy[72,82] and phosphorous-vacancy[56] pairs have been identified in Si by EPR. They form when the vacancy migrates in suitably doped Si. (The oxygen-vacancy pair has also been observed in floating-zone refined Si, thus indicating that the oxygen content of such material is by no means negligible.) The oxygen-vacancy pair consists of an interstitial oxygen atom bound to a normal vacancy[72] with an energy level 0.16 eV below the conduction band. The oxygen atom bonds with two of the four nearest neighbors to the vacancy and the bonds can switch to the other two atoms with an activation energy of 0.38 eV. The oxygen-vacancy pair anneals out with an activation energy 2.1 eV[82] which may correspond to the dissociation energy.

The phosphorous-vacancy pair consists of a substitutional phosphorous atom bound to a normal vacancy.[56] Its energy level is 0.4 eV below the conduction band. The activation energy of " bond switching " of the unpaired electron between the three equivalent Si atoms is only 0.06 eV. The activation energy for alignment of the P-vacancy direction (which equals the migration energy of the P-vacancy pair) is 0.93 eV. The P-vacancy pair did not completely dissociate during the alignment studies, indicating that the binding energy was at least 1 eV.

The activation energy for diffusion of phosphorous in Si is 3.66 ± 0.18 eV.[83] Watkins and Corbett[56] argue that on account of the high-binding energy of the solute phosphorous atoms to vacancies the migration of phosphorous is enhanced over the self-diffusion of Si. They point out that it is the binding between a P atom and a vacancy at a third neighbor site that is operative in enhancing the P-diffusion, since the vacancy of a vacancy-oxygen pair has to go through such a site in order for the phosphorous to diffuse. By estimating the corresponding binding energy, Watkins and Corbett[56] arrive at an activation energy for self-diffusion of

$$Q^{Sd} = (3.94 \pm 0.33) \text{ eV} \tag{60}$$

The value Eq. (60) is substantially lower than the value $Q^{Sd}=4.8$ eV found by tracer measurements.[28] This may be the following reason: the derivation of Eq. (60) is based on the assumption that the diffusion of substitutional phosphorus occurs through the migration of single, well defined vacancies. In view of the discussion of self-diffusion of silicon (section III-2.2) this assumption may be wrong. However, since P acts as a donor, it is more likely that it stabilizes the vacancy mechanism of self-diffusion.* Such an interpretation would account for the fact that in silicon group V elements diffuse less rapidly than group III elements (see section III-2.1). Eq. (60) would then give the activation energy of self-diffusion by the vacancy mechanism. Let us assume that the corresponding pre-exponential factor (allowing for $f_{1V}=0.5$) is 0.2 cm^2/sec. At the lowest temperature at which Peart[28] determined the tracer self-diffusion coefficient of Si, 1,496°K, this gives a vacancy self-diffusion coefficient of $10^{-14\pm1}$ cm^2/sec compared to the measured value 10^{-13} cm^2/sec. It appears that the contribution of single vacancies to tracer self-diffusion is indeed negligible over the temperature range investigated so far by tracer measurements. Our proposed interpretation is thus self-consistent. However, if the tracer-measurements were extended to substantially lower temperatures, the vacancy contribution to self-diffusion should become noticeable and result in a curvature of Arrhenius plot and a lowering of the apparent activation energy of self-diffusion.

EPR spectra which have been attributed to positively and negatively charged divacancies appear during low temperature electron bombardment of Si[54] and also during the annealing stages where the single vacancy spectra disappear. Recovery of the stress-induced alignment of the divacancy occurs at ~130°C with an activation energy of 1.3 eV (assuming first order kinetics[54]). This energy must also equal the migration energy of a divacancy. No loss of divacancies occurred during these annealing

* The detailed interpretation of the EPR spectra shows that the phosphorous is indeed associated with a well defined vacancy and not with a " relaxion " type defect. However, this evidence pertains to low temperatures, and need not necessarily reflect the situation during the high-temperature migration of the phosphorus atoms.

studies. Divacancies annealed out at ~290°C in Si containing ~10^{-5} atomic fraction of oxygen, and at ~350°C in floating zone refined Si. This indicates that the binding energy of divacancies was sufficiently high to prevent their dissociation up to at least 350°C. That is, the probability of dissociation at 350°C is less than

$$10^{-3}/\text{sec} = \frac{1}{C_{2V}} \cdot \frac{dC_{2V}}{dt} = 6\nu_{2V} \exp\left(-E_{2V}{}^{\text{Dis}}/kT\right)$$

This gives a lower limit for the dissociation energy $E_{2V}{}^{\text{Dis}} \gtrsim 2.07$ eV or a binding energy $E_{2V}{}^{\text{B}} \gtrsim 1.7$ eV (see the discussion of the inequality [47]).

V. CONCLUSIONS

The survey of the experimental data on *quenching* and *irradiation* of Ge and Si has shown that:

(i) Monovacancies migrate well below room temperature in Si and probably also in Ge.

(ii) Since the binding energies of divacancies and of vacancy-impurity pairs are very large, quenching experiments cannot give direct evidence about monovacancies; the same holds for irradiation experiments unless carried out at very low temperatures.

(iii) Definitive irradiation experiments on high purity Ge at low temperatures are needed for a full understanding of the nature and the properties of monovacancies in Ge.

(iv) Quenching experiments on Si (preferably into liquid He) are desirable in order to establish whether the defects dominant under high-temperature equilibrium conditions are vacancy or interstitial type defects.

The discussion of the *high-temperature data* on Ge and Si has shown that:

(v) The concentration of the thermal equilibrium defects is presumably too small (estimated to lie between about 5×10^{-9} and 5×10^{-7} at the melting points) to permit a direct determination of the nature and the concentration of the high-temperature equilibrium defects.

(vi) The data on self-diffusion and solute diffusion in Ge are best accounted for by a vacancy mechanism of self-diffusion.

(vii) The magnitude of the pre-exponential factor of the self-diffusion coefficient suggests a strongly relaxed configuration (" relaxion ") for the vacancy in Ge at high temperatures.

(viii) Significant differences between Si and Ge exist both for self-diffusion and solute diffusion, suggesting different mechanisms for self-diffusion in these elements.

(ix) Quite a number of hitherto unrelated experimental facts can be explained by postulating an interstitial type self-diffusion mechanism at high temperatures in Si; it is suggested that self-diffusion occurs through an extended interstitial complex bearing some similarity to a locally melted region.

(x) The vacancy mechanism of self-diffusion in Si (which would have both a smaller activation energy and a smaller pre-exponential factor and which would therefore predominate at intermediate temperatures) could be investigated through studies of solute diffusion and strongly bound solute-vacancy pairs (e.g., phosphorus-vacancy pairs).

(xi) Since with regard to the "openness" of the crystal structure diamond is similar to Si and grey tin is similar to Ge, an interstitial mechanism for self-diffusion may be expected for diamond and a vacancy mechanism for grey tin (hypothetical at high temperatures).

The *calculations of vacancy formation energies* pertain to the absolute zero of temperature. The analysis of the experimental data indicates that the vacancy parameters are temperature dependent; the energies of formation (not the *free* energy!) increase presumably with increasing temperature. A direct comparison between theoretical values and self-diffusion data can therefore not be carried out. The calculations of the present paper, using an improved Morse potential model, give for the monovacancy formation energy $E_{1V}{}^{F}=1.97$ eV (Ge) and $E_{1V}{}^{F}=2.35$ eV (Si). The corresponding experimental values can be estimated only very roughly; reasonably values might be 2.5 eV (Ge) and 3.0 eV (Si). We see that the theoretical values are too low. This is presumably due to the fact that in the defect calculations the strength of the valence bonds was assumed to be independent of their directions with respect to the crystal structure. Taking into account the energy required for "bond-bending" might bring the experimental values to the right order of magnitude.

Acknowledgement

The authors are grateful to Dr. Gerd Lorenz, Pinneberg, for a discussion of diffusion in Si and Ge and for supplying a list of recent references on that subject.

REFERENCES

1. B. I. Boltaks, *Soviet Phys.-Tech. Phys.*, **1**, 443 (1956).
2. F. C. Frank and D. Turnbull, *Phys. Rev.*, **104**, 617 (1956).

3. P. Penning, *Phys. Rev.*, **110**, 586 (1958); *Philips Res. Rep.*, **13**, 17 (1958).
4. A. Seeger, in *Solid State Physics in Electronics and Telecommunications* (M. Désirant and J. L. Michiels, eds.), Vol. 1, p. 61, Academic Press, London-New York, 1960.
5. G. D. Watkins, in *Radiation Damage in Semiconductors* (Paris-Royaumont 1964), p. 97, Dunod, Paris, 1965.
6. M. L. Swanson, *Can. J. Phys.*, **44**, 2181 (1966).
7. R. A. Swalin, *J. Phys. Chem. Solids*, **18**, 290 (1961).
8. A. Scholz, *Phys. Stat. Sol.*, **3**, 42 (1963).
9. A. Scholz and A. Seeger, *Phys. Stat. Sol.*, **3**, 1480 (1963); in *Radiation Damage in Semi conductors* (Paris-Royaumont 1964), p. 315, Dunod, Paris, 1965.
10. P. M. Morse, *Phys. Rev.*, **34**, 57 (1929).
11. J. C. Slater, in *Quantum Theory of Matter*, p. 206, McGraw-Hill, New York, 1951; *Quantum Theory of Molecules and Solids*, Vol. 1, p. 17, McGraw-Hill, New York, 1963.
12. L. Pauling, in *The Nature of the Chemical Bond*, p. 596 and p. 126, Cornell University Press, Ithaca, New York, 1960.
13. A. B. Lidiard, in *Radiation Damage in Semiconductors* (Paris-Royaumont 1964), p. 336, Dunod, Paris, 1965.
14. J. C. Slater, *Phys. Rev.*, **98**, 1039 (1955).
15. P. W. Bridgman, *Physics of High Pressure*, Bell, London, 1949.
16. C. A. Swenson, *Solid State Physics*, **11**, 41 (1960).
17. H. B. Huntington, *Solid State Physics*, **7**, 213 (1958).
18. K. A. Gschneider, Jr., *Solid State Physics*, **16**, 275 (1964).
19. J. Waser and L. Pauling, *J. Chem. Phys.*, **18**, 747 (1950).
20. H. J. McSkimin and P. Andreatch, Jr., *J. Appl. Phys.*, **34**, 651 (1963); **35**, 2161 (1964).
21. R. H. Carr, R. D. McCammon and G. K. White, *Phil. Mag.*, **12**, 157 (1965).
22. *American Institute of Physics Handbook*, McGraw-Hill, New York, 1963.
23. R. Feder and A. S. Nowick, *Phys. Rev.*, **109**, 1959 (1958).
24. R. O. Simmons, in *Radiation Damage in Solids* (Proc. Int. Summer School Phys. E. Fermi), XVIII, p. 568, Academic Press, London-New York, 1962.
25. R. D. Fouchaux and R. O. Simmons, *Phys. Rev.*, **136**, A 1664 (1964).
26. H. Letaw, W. M. Portnoy, and L. Slifkin, *Phys. Rev.*, **102**, 636 (1956).
27. H. Widmer and G. R. Gunther-Mohr, *Helv. Phys. Acta*, **34**, 635 (1961).
28. R. F. Peart, *Phys. Stat. Sol.*, **15**, K 119 (1966).
29. B. Pratt and F. Friedmann, *J. Appl. Phys.*, **37**, 1893 (1966).
30. R. N. Hall and J. H. Racette, *J. Appl. Phys.*, **35**, 379 (1964).
31. W. R. Wilcox and T. J. La Chapelle, *J. Appl. Phys.*, **35**, 240 (1964).
32. E. L. Williams, *J. Electrochem. Soc.*, **108**, 795 (1961).
33. A. D. Kurtz and C. L. Gravel, *J. Appl. Phys.*, **29**, 1456 (1958).
34. W. J. Armstrong, *J. Electrochem. Soc.*, **109**, 1065 (1962).
35. P. S. Raju, N. R. K. Rao, and E. V. K. Rao, *Indian J. Pure Appl. Phys.*, **2**, 353 (1964).
36. J. M. Fairfield and B. J. Masters, *Appl. Phys. Letters*, **8**, 280 (1966).
37. D. A. Petrow, J. M. Schaschkow and J. R. Akimchenko, *Chem. Abstr.*, **54**, 17190 b (1960).
38. B. I. Boltaks, *Diffusion in Semiconductors*, Academic Press, New York, 1963.

39. D. A. Petrow, J. M. Schaschkow and A. S. Belanovski, in *Halbleiter und Phosphore*, p. 652, Vieweg und Sohn, Braunschweig, 1958.
40. H. Reiss and C. S. Fuller, in *Semiconductors*, p. 222, Reinhold Publishing Corp., New York, 1960.
41. M. W. Valenta and C. Ramasastry, *Phys. Rev.*, **106**, 73 (1957).
42. M. F. Millea, *J. Phys. Chem. Sol.*, **27**, 309, 315 (1966).
43. R. R. Hasiguti, *J. Phys. Soc. Japan*, **21**, 1927 (1966).
44. P. G. Shewmon, *Diffusion in Solids*, McGraw-Hill, New York, 1963.
45. A. Seeger, G. Schottky and D. Schumacher, *Phys. Stat. Sol.*, **11**, 363 (1965); A. Seeger and D. Schumacher, *Mater. Sci. Eng.*, in press.
46. D. Schumacher, A. Seeger and O. Härlin, *Phys. Stat. Sol.*, **25**, 359 (1968).
47. G. D. Watkins, *J. Phys. Soc. Jap.*, **18**, Suppl. II, 22 (1963).
48. N. H. Nachtrieb and G. S. Handler, *Acta Met.*, **2**, 797 (1954).
49. N. H. Nachtrieb, *Proc. Conf. React. of Solids*, Madrid, 1956.
50. J. R. Wilson, *Metallurg. Rev.*, **10**, 381 (1965).
51. J. C. Slater, in *Quantum Theory of Molecules and Solids*, Vol. 2, p. 55, McGraw-Hill, New York, 1965.
52. A. Seeger and D. Schumacher, in *Lattice Defects in Quenched Metals*, p. 15, Academic Press, New York, 1965.
53. A. C. Damask and G. J. Dienes, *Point Defects in Metals*, Gordon and Breach, New York, 1963.
54. G. D. Watkins and J. W. Corbett, *Disc. Farad. Soc.*, **31**, 86 (1961); *Phys. Rev.*, **138**, A 543, A 555 (1965).
55. M. W. Lomer and A. H. Cottrell, *Phil. Mag.*, **46**, 711 (1955).
56. G. D. Watkins and J. W. Corbett, *Phys. Rev.*, **134**, A 1359 (1964).
57. H. M. James and K. Lark-Horovitz, *Z. physik. Chem.*, **198**, 107 (1951).
58. E. I. Blount, *J. Appl. Phys.*, **30**, 1218 (1959).
59. T. A. Callcott and J. W. MacKay, in *Radiation Damage in Semiconductors* (Paris-Royaumont 1964), p. 27, Dunod, Paris, 1965.
60. R. A. Logan, *Phys. Rev.*, **101**, 1455 (1956).
61. A. Hiraki and Y. Inuishi, *Tech. Rept. Osaka Univ.*, **15**, No. 647, 65 (1965).
62. J. H. Crawford, Jr., in *The Interaction of Radiation with Solids*, p. 421, North-Holland, Amsterdam, 1964.
63. J. W. Corbett, *Electron Radiation Damage in Semiconductors and Metals* (Solid State Physics, Suppl. 7), Academic Press, New York-London, 1966.
64. S. Mayburg, *Acta Met.*, **4**, 52 (1956).
65. W. Kaiser, *Phys. Rev.*, **105**, 1751 (1957).
66. S. Mayburg, *Phys. Rev.*, **95**, 38 (1954).
67. A. Hiraki and T. Suita, *J. Phys. Soc. Jap.*, **18**, Suppl. III, 254 (1963).
68. A. Hiraki, *J. Phys. Soc. Jap.*, **21**, 34 (1966).
69. R. Weltzin and R. A. Swalin, *J. Phys. Soc. Jap.*, **18**, Suppl. III, 135 (1963).
70. A. G. Tweet, *Phys. Rev.*, **106**, 221 (1957); *J. Appl. Phys.*, **30**, 2002 (1959).
71. H. Letaw, *J. Phys. Chem. Sol.*, **1**, 100 (1956).
72. G. D. Watkins and J. W. Corbett, *Phys. Rev.*, **121**, 1001 (1961).
73. C. S. Fuller and K. B. Wolfstirn, *J. Phys. Chem. Sol.*, **26**, 1463 (1965); **27**, 1431 (1966).
74. R. E. Whan, *Appl. Phys. Lett.*, **8**, 131 (1966).

75. R. E. Whan, *Appl. Phys. Lett.*, **6**, 221 (1965); *Phys. Rev.*, **140**, A 690 (1965).
76. R. F. Konopleva, S. R. Novikov, E. E. Rubinova, and S. M. Ryvkin, in *Radiation Damage in Semiconductors* (Paris-Royaumont 1964), p. 287, Dunod, Paris, 1965.
77. K. Matsui, *Thesis, University of Paris*, Dec. 1965.
78. K. Matsui and R. R. Hasiguti, *J. Phys. Soc. Jap.*, **20**, 487 (1965).
79. R. R. Hasiguti and S. Ishino, in *Radiation Damage in Semiconductors* (Paris-Royaumont), p. 259, Dunod, Paris, 1965.
80. J. W. McKay and E. E. Klontz, in *Radiation Damage in Semiconductors* (Paris-Royaumont 1964), p. 11, Dunod, Paris, 1965.
81. O. L. Curtis, Jr. and J. H. Crawford, Jr., in *Radiation Damage in Semiconductros* (Paris-Royaumont, 1964), p. 143, Dunod, Paris, 1965.
82. J. W. Corbett, G. D. Watkins, and R. S. McDonald, *Phys. Rev.*, **135**, A 1381 (1964).
83. C. S. Fuller and J. A. Ditzenberger, *J. Appl. Phys.*, **27**, 544 (1956).

The Formation and Binding Energies of a Divacancy in Diamond Lattices

RYUKITI R. HASIGUTI

Department of Metallurgy, Faculty of Engineering, University of Tokyo, Bunkyo-ku, Tokyo

ABSTRACT

The formation and binding energies of a divacancy in diamond lattices were calculated assuming the Morse type potential energy between two covalently bonded atoms. The calculated values of divacancy formation energies are 4.07 eV(Ge), 4.56 eV(Si) and 8.83 eV(diamond), and the calculated values of divacancy binding energies are 0.84 eV (Ge), 0.93 eV(Si) and 1.67 eV(diamond). The relaxations of six nearest neighbor atoms of a divacancy were also calculated.

I. INTRODUCTION

Previously it was considered that a single vacancy in germanium has a migration energy of about 1 eV. (See, for example, refs. 1 and 2 and references therein.) But recently many authors have come to consider that a single vacancy in germanium migrates with a migration energy from about 0.2 eV to about 0.3 eV depending on electric charge states.[2,3]

On the other hand, the formation energy of a vacancy in germanium is calculated to be about 2 eV.[4] The sum of the migration and formation energies is then 2.2 to 2.3 eV, which is far smaller than the activation energy of 3.0 eV of self-diffusion in germanium.[5] If it is assumed that self-diffusion in germanium takes place by means of single vacancies, the discrepancy between the energies of 3.0 eV and of 2.2–2.3 eV is too large to be within the experimental error.

Some considerations must be made in order to solve this paradox. One of the possible solutions is to consider that divacancies may be contributing to self-diffusion together with single vacancies, because a divacancy is considered to have a migration energy far larger than that of a single vacancy. The contribution of divacancies can occur, if the binding energy of a divacancy is very large, so that the concentration of divacancies at self-diffusion temperatures is sufficiently large.

Of course, the problem of the binding energy of divacancy is of interest apart from the above mentioned self-diffusion problem. The aim of the present paper is to show the results of calculation of the formation and binding energies of a divacancy in germanium, silicon and diamond based on a simple model with Morse interaction potentials.

II. THE MODEL

Figure 1 shows the atomic configuration of a divacancy. Two dotted circles show the divacancy. By elimination of two atoms, two short bonds and one long bond are formed as shown in Fig. 1. The two newly formed short bonds are quite similar to the bonds resulting from formation of a single vacancy.[6–8] The long bond was described by Watkins and Corbett when they made an electron spin resonance study of a divacancy.[9,10]

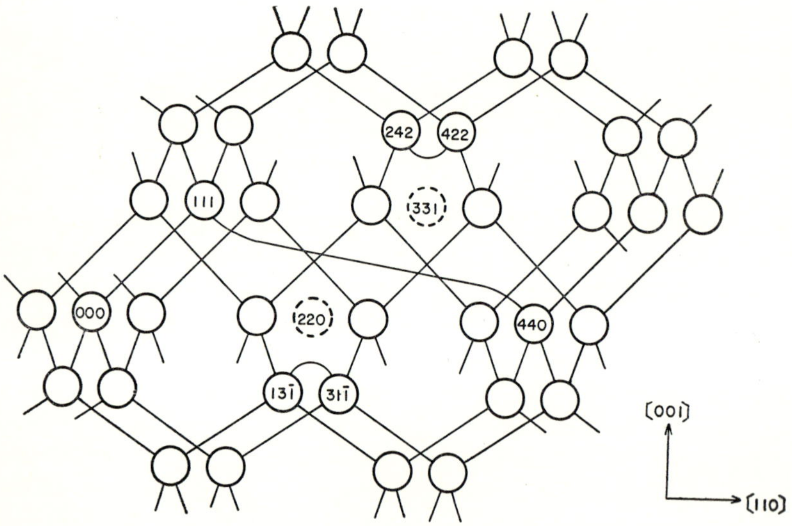

Fig. 1. A divacancy in the diamond lattice.

The relaxation of atoms around a divacancy was performed in the present calculation as follows. The six nearest neighbor atoms of a divacancy were completely relaxed. The eighteen next nearest neighbor atoms were not relaxed, but were fixed at their regular sites. This gives an over-estimation of the divacancy formation energy. This point will be discussed more fully later.

TABLE I. The constants for the Morse potentials.[6]

Material	D(eV)	α(Å^{-1})	d_0(Å)
Ge	1.63	1.50	2.44
Si	1.83	1.56	2.34
Diamond	3.61	2.30	1.54

As the potential energy between two covalently bonded atoms, we shall use the Morse type potential, for which the constants are shown in Table I.

III. CALCULATIONS

The Morse type potential ΔH is expressed as follows:

$$\Delta H = D\{1+\exp[-2\alpha(d-d_0)]-2\exp[-\alpha(d-d_0)]\} \tag{1}$$

where D and α are the constants, d is the distance between two covalently bonded atoms, and d_0 is the value of d at the equilibrium separation of nearest neighbor atoms in the perfect lattice. The constants D, α and d_0 are listed in Table I.[6]

The energy contribution from the twenty-one bonds inside the fixed eighteen next nearest neighbor atoms of a divacancy was calculated to minimize the energy of a divacancy.

The energy which includes two short bonds (or the " short bond energy ") and the energy which includes a long bond (or the " long bond energy ") are separately calculated, because there is no interaction between these two kinds of energies in the present approximation. Thus the short bond energy ΔH_S is

$$\begin{aligned}\Delta H_S = {} & 4D\left\{1+\exp\left[-2\alpha\left(\frac{1}{\sqrt{3}}\sqrt{2(1+x)^2+(1-y)^2}-1\right)d_0\right]\right.\\ & \left.-2\exp\left[-\alpha\left(\frac{1}{\sqrt{3}}\sqrt{2(1+x)^2+(1-y)^2}-1\right)d_0\right]\right\}\\ & +8D\left\{1+\exp\left[-2\alpha\left(\frac{1}{\sqrt{3}}\sqrt{2+2x^2+(1+y)^2}-1\right)d_0\right]\right.\\ & \left.-2\exp\left[-\alpha\left(\frac{1}{\sqrt{3}}\sqrt{2+2x^2+(1+y)^2}-1\right)d_0\right]\right\}\\ & +2D\left\{\exp\left[-2\alpha\left(\frac{2\sqrt{2}}{\sqrt{3}}(1-x)-1\right)d_0\right]\right.\\ & \left.-2\exp\left[-\alpha\left(\frac{2\sqrt{2}}{\sqrt{3}}(1-x)-1\right)d_0\right]\right\}\end{aligned} \tag{2}$$

and the long bond energy ΔH_L is

$$\begin{aligned}\Delta H_L = 2D\Big\{&1+\exp\Big[-2\alpha\Big(\frac{1}{\sqrt{3}}\sqrt{2(1+u)^2+(1+v)^2}-1\Big)d_0\Big]\\&-2\exp\Big[-\alpha\Big(\frac{1}{\sqrt{3}}\sqrt{2(1+u)^2+(1+v)^2}-1\Big)d_0\Big]\Big\}\\+4D\Big\{&1+\exp\Big[-2\alpha\Big(\frac{1}{\sqrt{3}}\sqrt{2+2u^2+(1-v)^2}-1\Big)d_0\Big]\\&-2\exp\Big[-\alpha\Big(\frac{1}{\sqrt{3}}\sqrt{2+2u^2+(1-v)^2}-1\Big)d_0\Big]\Big\}\\+D\Big\{&\exp\Big[-2\alpha\Big(\frac{1}{\sqrt{3}}\sqrt{2(3-2u)^2+(1+2v)^2}-1\Big)d_0\Big]\\&-2\exp\Big[-\alpha\Big(\frac{1}{\sqrt{3}}\sqrt{2(3-2u)^2+(1+2v)^2}-1\Big)d_0\Big]\Big\}\end{aligned} \tag{3}$$

Here x, y, u and v are the displacements of atoms expressed as follows: Let us take the coordinate as shown in Fig. 1. Two vacant sites are at $(2a/4, 2a/4, 0)$ and $(3a/4, 3a/4, a/4)$ where a is the lattice constant. (The coordinates will be abbreviated hereafter as (2, 2, 0) and (3, 3, 1), i.e. they will be expressed in units of $a/4$.) The displacements of atoms in units of $a/4$ will be denoted as x and y for the directions of [100] and [001], respectively, in the case of atoms associated with the short bonds, and denoted as u and v for the directions of [100] and [001], respectively, in the case of atoms associated with the long bond. The displacements in the [010] direction are the same as those in the [100] direction because of the crystal symmetry.

The formation energy of a divacancy E_{FV_2} is

$$E_{FV_2} = 3D+(\Delta H_S+\Delta H_L) \tag{4}$$

The energy ΔH_S can also be used to calculate the formation energy of a single vacancy E_{FV_1}, because the short bonds are the same for a divacancy and for a single vacancy in the present approximation, so that we obtain

$$E_{FV_1} = 2D+\Delta H_S \tag{5}$$

Then the binding energy of a divacancy E_{BV_2} is obtained from

$$E_{BV_2} = 2E_{FV_1}-E_{FV_2} = D+\Delta H_S-\Delta H_L \tag{6}$$

The results of calculations of E_{FV_1}, E_{FV_2} and E_{BV_2} are shown in Table II, and the displacements of the six nearest neighbor atoms are shown in Table III. The atoms at (1, 1, 1) and (4, 4, 0), i.e. the atoms at both

TABLE II. The single vacancy formation energy E_{FV_1}, the divacancy formation energy E_{FV_2} and the divacancy binding energy E_{BV_2} in eV.

Material	E_{FV_1}	E_{FV_2}	E_{BV_2}
Ge	2.45	4.07	0.84
Si	2.75	4.56	0.93
Diamond	5.25	8.83	1.67

TABLE III. Displacements of nearest neighbor atoms, of which the coordinates are shown in units of $a/4$.

Material	Before displacement	After displacement
Ge and Si	(1, 3, −1)	(1.12, 2.88, −0.95)
	(3, 1, −1)	(2.88, 1.12, −0.95)
	(2, 4, 2)	(2.12, 3.88, 1.95)
	(4, 2, 2)	(3.88, 2.12, 1.95)
	(1, 1, 1)	(1, 1, 1)
	(4, 4, 0)	(4, 4, 0)
Diamond	(1, 3, −1)	(1.16, 2.84, −0.95)
	(3, 1, −1)	(2.84, 1.16, −0.95)
	(2, 4, 2)	(2.16, 3.84, 1.95)
	(4, 2, 2)	(3.84, 2.16, 1.95)
	(1, 1, 1)	(1, 1, 1)
	(4, 4, 0)	(4, 4, 0)

ends of the long bond, show no displacement.

IV. DISCUSSION

Scholz and Seeger[11] calculated the formation and binding energies of a divacancy in germanium, silicon and diamond without allowing any relaxation of atoms.[12] The completely free relaxation of the nearest neighbor atoms in the present calculation gives some improvement in the approximation.

As was pointed out in Section II, the present approximation neglecting the displacements of next nearest neighbor atoms gives some over-estimation of the formation energy of a divacancy. This is also true for the formation energy of a single vacancy. As can be seen from Eqs. (4), (5) and (6), all three energies, namely E_{FV_1}, E_{FV_2} and E_{BV_2}, show the same degree of over-estimation as that of ΔH_S, because ΔH_L is very small compared with ΔH_S.

Scholz and Seeger[11] give larger values of all three energies for all three crystals compared with those of the present calculation. This is due to the greater degree of over-estimation in their calculations.

The degree of over-estimation by Scholz and Seeger[11] is 0.20 eV larger than that of the present calculation in the case of germanium. This is obtained from the difference between 2.65 eV (E_{FV_1}, Scholz-Seeger) and 2.45 eV (E_{FV_1}, Hasiguti). This value (0.20 eV) comes from neglecting the relaxation of nearest neighbor atoms of a divacancy.

Neglecting the relaxation of next nearest neighbor atoms of a divacancy results in an over-estimation of energies, the degree of which is considered to be far smaller than 0.20 eV, because the relaxation of next nearest neighbors is far smaller than the relaxation of nearest neighbors. This shows that the over-estimation in the present calculation is not serious at all. The numerical evaluation of the degree of our over-estimation will not be made here, because further calculations which include the next nearest neighbor relaxations are presently being performed in our laboratory. For silicon and diamond the differences between E_{FV_1} (Scholz-Seeger) and $E_{FV_1}{}^*$ (Hasiguti) are 0.20 eV and 0.50 eV, respectively.[13]

The contribution from the long bond is rather small in the present approximation. But it is possible that the interaction in the long bond is larger than that expected from the Morse potential. If this is the case, there will be a smaller divacancy formation energy ($+\Delta H_L$ in Eq. [4]) and a larger divacancy binding energy ($-\Delta H_L$ in Eq. [6]), because ΔH_L has a negative value. It is difficult to evaluate the degree of mis-estimation of ΔH_L at this moment.

These two kinds of mis-estimations of binding energies are in opposite directions and are both supposed to be very small compared with the binding energies themselves, so that the present calculation is considered to give nearly correct values of binding energies.

As seen from the results shown in Table II, the binding energies of a divacancy are fairly large. This suggests that the contribution of divacancies to self-diffusion in germanium, silicon and diamond must be considered.

Acknowledgement

The author wishes to thank Miss S. Motomiya for her assistance in making calculations.

REFERENCES

1. R. R. Hasiguti and S. Ishino, in *Proc. Symp. on Radiation Damage in Semiconductors*, p. 259, Dunod, Paris; Academic Press, New York, 1965.

2. R. R. Hasiguti, *J. Phys. Soc. Japan*, **21**, 1927 (1966).
3. R. E. Whan, *Appl. Phys. Letters*, **6**, 221 (1965).
4. S. Ishino, F. Nakazawa and R. R. Hasiguti, *J. Phys. Soc. Japan*, **20**, 817 (1965).
5. H. Letaw, W. M. Portnoy and L. M. Slifkin, *Phys. Rev.*, **102**, 636 (1956).
6. R. A. Swalin, *J. Phys. Chem. Solids*, **18**, 290 (1961).
7. A. Scholz and A. Seeger, *Phys. Status Solidi*, **3**, 1480 (1963).
8. M. L. Swanson and A. Seeger, in this book.
9. J. W. Corbett and G. D. Watkins, *Phys. Rev. Letters*, **7**, 314 (1961).
10. G. D. Watkins, in *Proc. Symp. on Radiation Damage in Semiconductors*, p. 97, Dunod, Paris; Academic Press, New York, 1965.
11. A. Scholz and A. Seeger, in *Proc. Symp. on Radiation Damage in Semiconductors*, p. 315, Dunod, Paris; Academic Press, New York, 1965.
12. M. L. Swanson and A. Seeger also made calculations of the energies of a divacancy in germanium and silicon without allowing any relaxation (Scholz-Seeger method). See ref. 8.
13. Scholz and Seeger (ref. 11) use somewhat different values of the constants for the Morse potentials in the case of silicon and diamond. In obtaining the defferences between E_{FV_1}(Scholz-Seeger) and $E_{FV_1}{}^*$(Hasiguti), the values of $E_{FV_1}{}^*$(Hasiguti), which are obtained using the Scholz-Seeger values of the Morse constants, rather than E_{FV_1}(Hasiguti) were used in the case of silicon and diamond, although the differences between $E_{FV_1}{}^*$(Hasiguti) and E_{FV_1}(Hasiguti) are small, i.e. 0.02 eV(Si) and 0.04 eV(diamond).

Further Investigation on Quenched-in Vacancies in Germanium

Akio Hiraki and Yoshio Inuishi

Department of Electrical Engineering, Osaka University, Osaka, Japan

ABSTRACT

The detailed investigation on quenched-in vacancies in Ge has been continued. The effects of oxygen doping and electronic level associated with quenched-in defects were investigated by the Hall effect and lifetime measurement. These results give $E_V+0.04\sim0.06$ eV and $E_V+0.2$ eV double acceptor level.

I. INTRODUCTION

The investigation on quenched-in vacancies in Ge crystals has been continued by the authors with the Hall effect and minority carrier life-time measurement. The results obtained in the previous papers[1] can be summarized as follows.

(i) The density of a quenched-in vacancy N_V obtained from the Hall effect measurement can be given by

$$N_V = 1.8\times10^{23} \exp(-1.9\,\text{eV}/kT_Q) \quad (1/\text{cc}) \tag{1}$$

where T_Q is the quenching temperature. The formation energy of a quenched-in vacancy is then estimated to be 1.9 eV, in good agreement with Seeger's recent calculation[2] for a single vacancy in Ge.

(ii) From the isochronal annealing curve for nearly intrinsic Ge samples with different amounts of dislocation density, the diffusion constant D of quenched-in vacancies was obtained as

$$D = 2 \exp(-1.2\,\text{eV}/kT) \quad (\text{cm}^2/\text{sec}) \tag{2}$$

In the present paper, the effect of impurities such as oxygen and antimony on the introduction of quenched-in vacancies and the multiple level associated with quenched-in vacancies will be discussed.

II. PRODUCTION OF A VACANCY BY QUENCHING

To avoid contaminations due to Cu and other impurities, the procedure shown in Fig. 1 was used to produce quenched-in vacancies. After soaking in HNO_3 solution, Ge specimens were put into the furnace with running wet oxygen atomosphere. The oxidized surface thus obtained seems to have gettering effect to impurities (especially Cu) as was found by Weiser.[3] By this technique, we were able to keep Cu contamination under 10^{13}/cc judging from lifetime measurement. After increasing the temperature of the furnace slowly up to quenching temperature T_Q (700–800°C), the samples were dropped into an air or ethylene glycol bath, the cooling rate of which was ca. 800°C/sec.

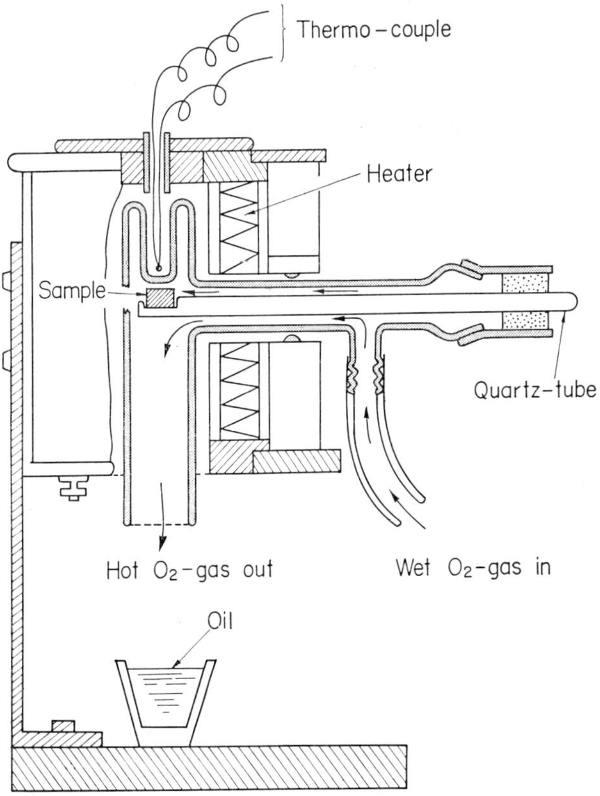

Fig. 1. Sketch of the furnace used for thermal treatments.

III. EXPERIMENTALS

(a) *Effect of impurities on a quenched-in vacancy*

As will be mentioned later, quenched-in vacancies in nearly intrinsic Ge act as double acceptors which become completely ionized at room temperature.[4] If the quenched-in vacancy interacts with an oxygen impurity to form a complex defect as A-center in Si, the hole concentration may be reduced due to the deeper acceptor level of Ge: A-center (E_c—0.09 eV).[5] Intrinsic specimens of about 50 Ω·cm at room temperature doped with 3×10^{16}/cc oxygen were quenched from various temperatures as mentioned previously. After quenching all specimens become p-type, and a number of acceptors due to quenched-in defects were obtained from resistivity and the Hall effect measurement at room temperature. The results obtained are shown in Fig. 2 along with the data for nearly oxygen free Ge for comparison. As shown in the figure, no appreciable effect of oxygen impurity on the number and formation energy of quenched-in defects was found. This is noteworthy since the oxygen concentration 3×10^{16}/cc was more than 50–70 times larger than that of a quenched-in vacancy. The effect of oxygen doping on the annealing of a quenched-in vacancy was also studied. In the case of nearly oxygen-free specimens,

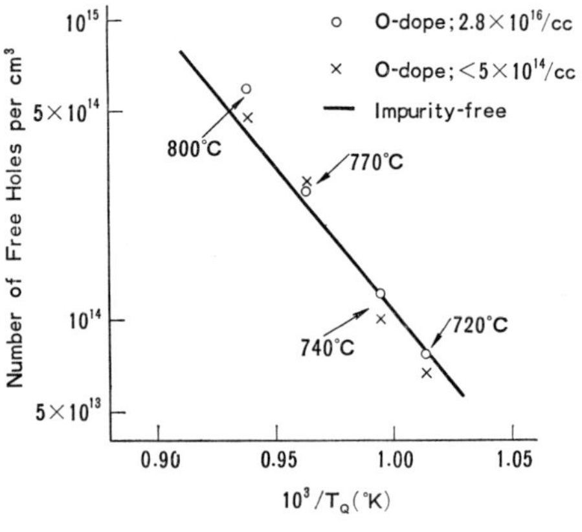

Fig. 2. Number of free holes per cm³ due to quenched-in vacancy in oxygen-doped Ge as a function of quenching temperature (T_Q). The straight line is a solubility curve of quenched-in vacancy in the impurity-free Ge. × indicates the experimental points of oxygen doped Ge.

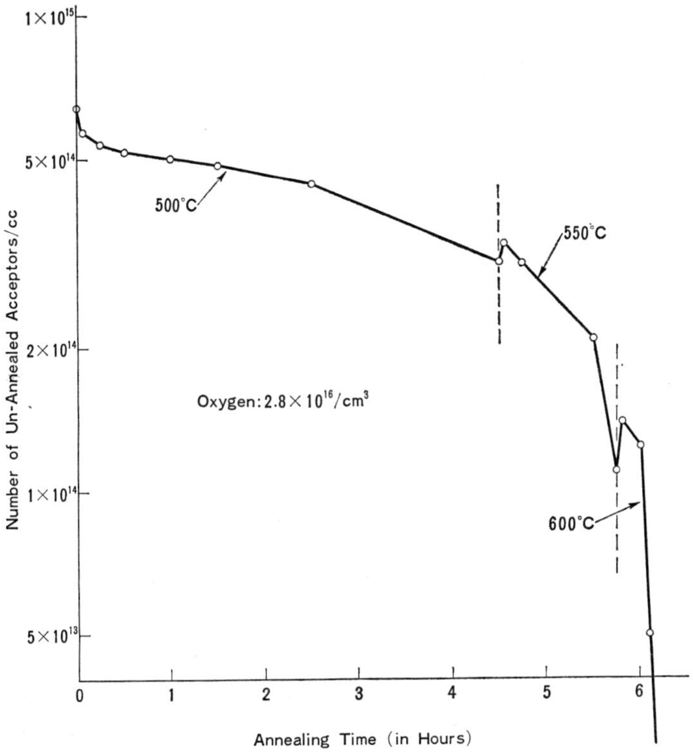

Fig. 3. Annealing behavior of quenched-in acceptors in oxygen doped Ge. Annealing temperatures are indicated in °C.

the annealing curve can be expressed by simple exponential decay[1] except for dislocation-free Ge. The annealing curves for oxygen-doped Ge shown in Fig. 3 were rather complex and similar to that for dislocation free specimens,[6] suggesting a cluster formation.

(b) Acceptor level due to a quenched-in vacancy

The position of acceptor levels due to quenched-in vacancies was obtained from the temperature dependence of Hall coefficient R_h(25∼330°K). Taking the existence of both light and heavy holes into account,[7] a high magnetic field limit (5KG) for lower temperature (25∼100°K) and a low magnetic field limit (200 G) for higher temperature (77∼330°K) were used. In the high field limit (5KG) Hall coefficient R_h is given by

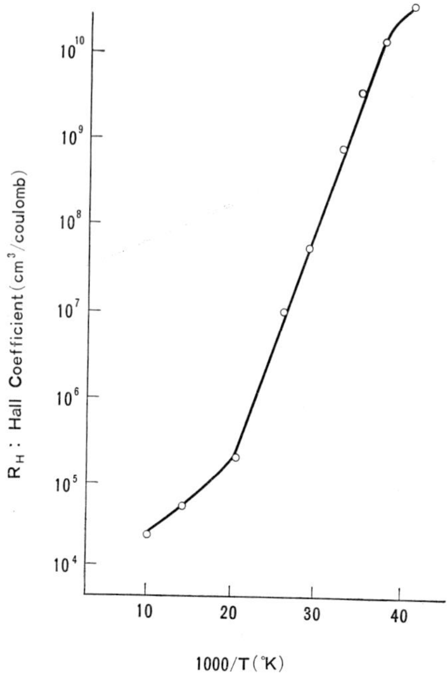

Fig. 4. Hall coefficient vs. temperature for the sample containing quenched-in vacancy. Magnetic field used is 5KG (strong field-limit).

$$R_h = \frac{1}{|e|} \cdot \frac{1}{P_h + P_l} \tag{1}$$

where P_h and P_l represent the density of heavy and light holes respectively. At the lower magnetic field (100–500 G) in a high temperature range

$$R_h = \frac{3\pi}{8|e|P_h} \cdot \frac{1+P_l/P_h(\mu_l/\mu_h)^2}{[1+P_l/P_h(\mu_l/\mu_h)]^2} \tag{2}$$

where μ_l and μ_h represent the mobility of light and heavy holes respectively.

(i) *Acceptor level in originally intrinsic specimens*

All specimens originally intrinsic (ca. 60 Ω·cm at room temperature) become p-type after quenching from various temperature. As shown in Fig. 4, the acceptor level at $E_V + 0.03 \sim 0.05$ eV was obtained from the slope of R_h vs. $1/T$ curve at the lower temperature range of these specimens. The quenching temperature was 830°C.

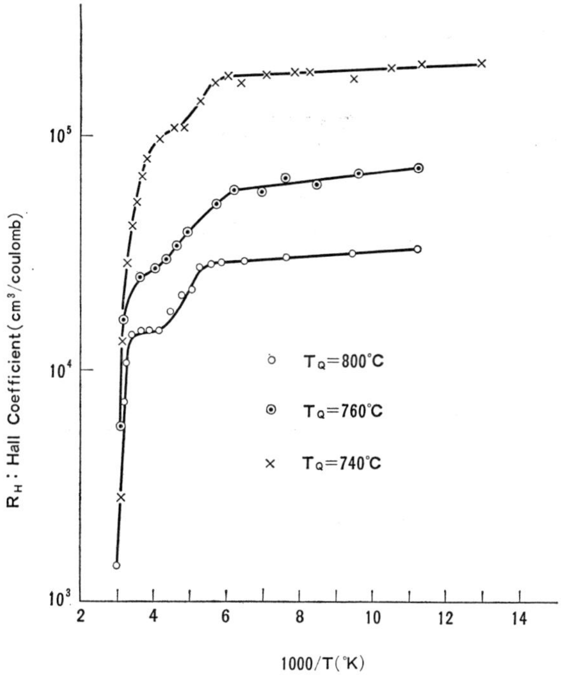

Fig. 5. Hall coefficient vs. temperature for samples containing different numbers of quenched-in vacancies. Magnetic field used are 100–500 G (low field-limit).

The R_h vs. $1/T$ curves for higher temperature and low magnetic field show two plateau as shown in Fig. 5. The lower plateau corresponds to the exhaustion region of $E_V+0.03\sim0.05$ eV, and the second plateau corresponds to the $E_V+0.2$ eV level. The latter facts can be understood from the curve of Fermi level as the function of temperature as shown in Fig. 6, where pinning of the level at $E_V+0.2$ eV is seen. Accordingly the quenched-in vacancy provides a double acceptor level.

(ii) *Acceptor level in originally n-type Sb doped samples*

First, similar experiments were done for n-type Ge doped with Sb ($1.5\times10^{15}\sim5\times10^{14}$/cc). The quenching temperature (760°C) was chosen so as to keep the specimens n-type after quenching. The R_h vs. $1/T$ curve after quenching is shown in Fig. 7, which suggests the existence of a level at $E_c-0.2$ eV. An $E_c-0.2$ eV level was also found by γ irradiation of Sb doped Ge and ascribed to the formation of Sb-divacancy or Sb-vacancy complex.[5] When the quenching temperature is higher (780°C) the

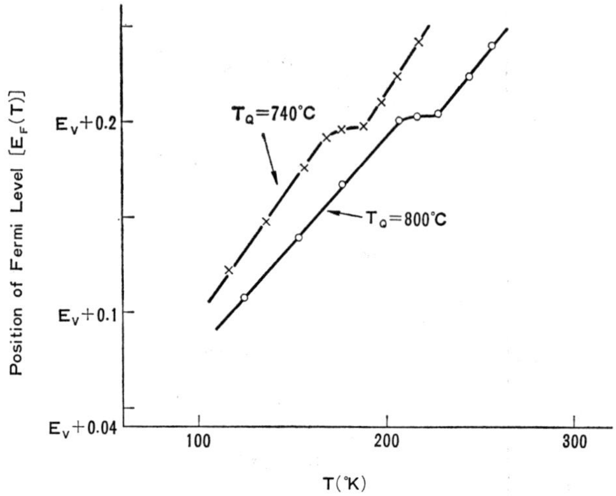

Fig. 6. Fermi level plot vs. temperature for samples quenched at 740° and 800°C.

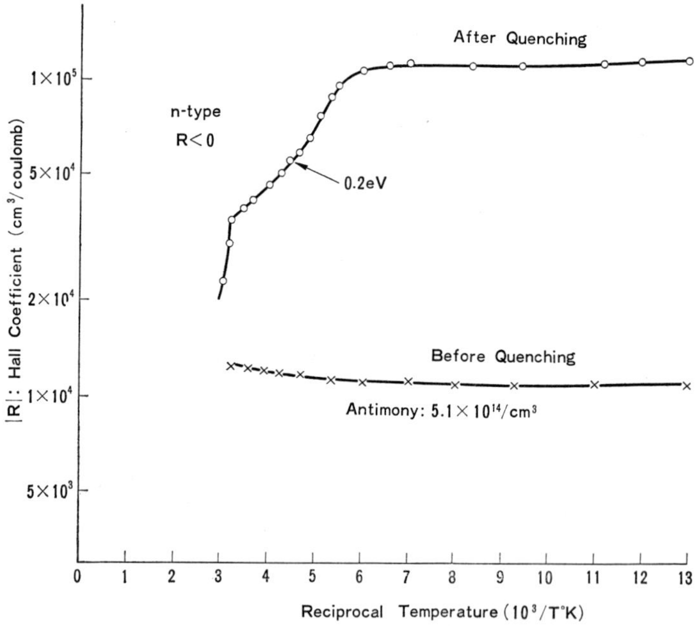

Fig. 7. Hall coefficient vs. absolute temperature of Sb doped n-type specimens before and after quenching.

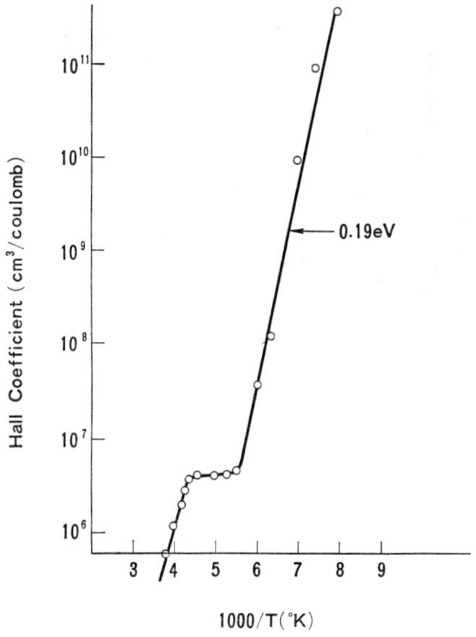

Fig. 8. Hall coefficient vs. temperature for the Sb doped sample quenched at higher temperature (780°C).

originally n-type samples converted to p-type after quenching due to the acceptor level of the vacancy. The R_h vs. $1/T$ curve of these samples for a higher temperature range is shown in Fig. 8. The $E_V+0.19$ eV acceptor level is also seen in this case to be in good agreement with that of an intrinsic sample.

(iii) *Acceptor level of a quenched-in vacancy in oxygen doped specimens*

The same experiment was repeated for oxygen doped Ge. From the Hall effect measurement both for lower temperature high field and for higher temperature low field, the $E_V+0.04\sim0.06$ eV acceptor level was confirmed as seen in Fig. 9. This level is slightly higher than the level for intrinsic Ge. The higher level was difficult to estimate, but seemed to lie above $E_V+0.2$ eV. Although Ge: A-center (vacancy-oxygen complex) is thought to give an $E_c-0.09$ eV level, no confirmation was obtained in this experiment.

IV. DISCUSSIONS

By keeping the sample at a high quenching temperature (700–800°K), the equilibrium number of single vacancies should be created, since

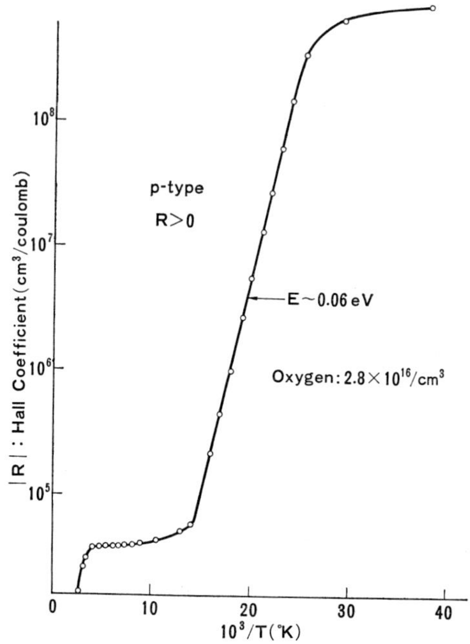

Fig. 9. Hall coefficient vs. absolute temperature of quenched-in acceptors in oxygen doped Ge.

divacancies and the whole vacancy-impurity complex seems to be unstable at these temperatures. Hence the number of total single vacancies (Schottky defect) N_V can be given by

$$N_V = N_0 \exp(-W/kT_Q) \tag{3}$$

where W is the formation energy of a single vacancy. N_V should not be sensitive to impurities such as oxygen. These single vacancies might freeze by quenching the sample down to room temperature. However, since the vacancy can move even at room temperature, these single vacancies should move to other defects such as oxygen, substitutional impurity, and other vacancies in order to stabilize themselves by forming an A-center dopant-vacancy complex and divacancy respectively. The total number of vacancies should be preserved except for those which annihilate at dislocations. Each of these complex defects may provide multiple acceptor levels as shown in the previous experiment. Hence the total number of holes produced by quenching should be a good measure of the total number

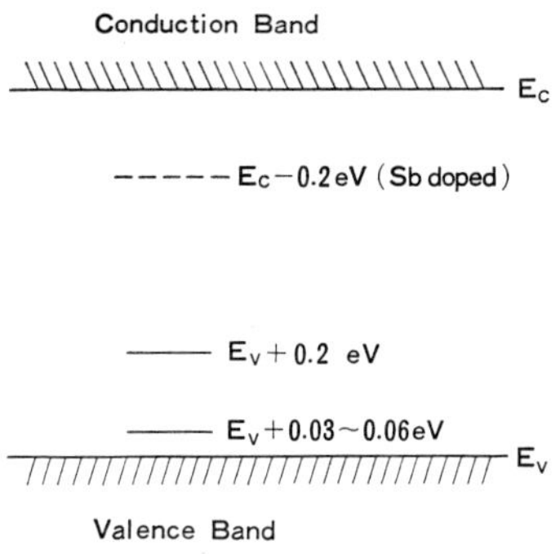

Fig. 10. Energy diagram of the double acceptor level due to quenched-in vacancies.

of the defect or total number of vacancies created at quenching temperatures.

From these considerations the activation energy obtained from Fig. 2 should be the formation energy of a single vacancy in Ge. In fact Seeger's theoretical calculations are in good agreement with this value. The electronic levels obtained are summarized in Fig. 10. From the insensitivity of the $E_V+0.2$ eV and $E_V+0.03\sim0.05$ eV levels to impurities, these levels seem to be associated with divacancy. The $E_c-0.2$ eV level, however, may be an Sb-vacancy or Sb-divacancy complex. These interpretations seem to be consistent with the carrier lifetime measurement published elsewhere.[4]

REFERENCES

1. A. Hiraki, *J. Phys. Soc. Japan*, **21**, 34 (1966).
2. A. Seeger, preprint of *Int. Symp. Latt. Def. Semicond.* (*Tokyo*), *Sep. 1966.*
3. K. Weiser, *J. Appl. Phys.*, **28**, 271 (1957).
4. A. Hiraki and Y. Inuishi, *Tech. Rep. Osaka Univ.*, **15**, 65 (1965).
5 J C. Pigg and J. H. Crawford, Jr., *Phys. Rev.*, **135**, A 1141 (1964).
6. A. Hiraki and T. Suita, *J. Phys. Soc. Japan*, **18**, Supple. III, 254 (1963).
7. R. K. Willardson, T. C. Harman and A. C. Beer, *Phys. Rev.*, **96**, 55 (1954).

Dissociative Diffusion of Nickel in Silicon and Self-diffusion of Silicon*

MASAYUKI YOSHIDA AND KEN SAITO

Electrical Communication Laboratory, Midoricho, Musashinoshi, Tokyo, Japan

ABSTRACT

The behavior of nickel acceptors in silicon was studied. The solubility of $N_S = 10^{26} \exp[-(3.1 \pm 0.2)/kT] \mathrm{cm}^{-3}$ and the reciprocal product of the time constant of a supersaturated state by a dislocation density of $1/\tau n_d = 1.3 \times 10^{-2} \exp[-(1.4 \pm 0.2)/kT] \mathrm{cm}^2 \mathrm{sec}^{-1}$ were obtained by the Hall coefficient and resistivity measurements. From the results it was concluded that the nickel acceptors locate at substitutional sites, that the substitutional and the interstitial nickel diffuse dissociatively and that the self-diffusion constant of silicon is given by $30 \exp[-(4.5 \pm 0.4)/kT] \mathrm{cm}^2 \mathrm{sec}^{-1}$.

I. INTRODUCTION

One of the authors has studied the behavior of nickel as an impurity in silicon and found interstitial nickel.[1] It has also been found that nickel atoms act as acceptors in silicon and that the concentration of the acceptors is extremely small compared with that of the interstitial nickel atoms. As their diffusion constant is large, the interstitial nickel atoms, which are saturated at high temperatures, easily precipitate at the dislocation and surface of silicon specimens by rapid cooling and only at the surface by slow cooling. The nickel acceptors, on the other hand, remain unprecipitated by the rapid and slow coolings. This indicates that they do not associate with the interstitial nickel atoms or nickel-precipitates but associate with the nickel atoms which occupy more stable sites than the interstitial sites in a silicon crystal lattice.

It has been reported that copper[2] and nickel atoms[3] in germanium and gold atoms[4] in silicon occupy both the interstitial and substitutional sites

* Details have been published in the *Japanese Journal of Applied Physics*, **6**, 573 (1967).

of the host crystal lattice. The substitutional atoms dissociate to the interstitial atoms and the vacancies. The resultant diffusion and precipitation of impurity atoms are called "dissociative."

As the nickel acceptors occupy more stable sites than the interstitial sites, it is natural to assume, referring to the cases of the impurity atoms mentioned above, that they occupy the substitutional sites and that the diffusion and precipitation of nickel atoms in silicon are dissociative. In the present work the annealing characteristics of supersaturated nickel acceptors are examined on the basis of dissociative mechanism.

II. THEORETICAL[3]

In order to obtain the differential equation of annealing by the dissociative mechanism the following was assumed:

1) During annealing the interstitial nickel atoms are always in thermal equilibrium.

2) Dislocations act as the sink and source of the vacancies.

3) As the diffusion constant of the substitutional nickel atoms is small, their migration is neglected.

From this we obtain the following equations:

$$\left.\begin{aligned} \frac{dN_S}{dt} &= -k_1 N_S + k_2 N_I N_V \\ \frac{dN_V}{dt} &= k_1 N_S - k_2 N_I N_V - n_d D_V{}^m (N_V - N_V{}^\circ) \\ N_I &= N_I{}^\circ \end{aligned}\right\} \qquad (1)$$

and at thermal equilibrium

$$k_1 N_S{}^\circ = k_2 N_I{}^\circ N_V{}^\circ \qquad (2)$$

where k_1 and k_2 are reaction constants, N_S, N_I and N_V are the concentration of the substitutional and interstitial nickel and the vacancy, n_d is the dislocation density, $D_V{}^m$ is the diffusion constant of the vacancy and suffiix ○ indicates thermal equilibrium value.

If the dislocation density and the diffusion constant of the vacancy are small or $k_1 + k_2 N_I > n_d d_V{}^m$ (assumption 4), the annealing process is controlled by the diffusion of vacancies. After a long period of annealing the concentration of the substitutional nickel atoms decays exponentially with a time constant of

$$\frac{1}{\tau} = n_d D_V{}^m \frac{N_V{}^\circ}{N_S{}^\circ} \qquad (3)$$

if N_S° is larger than N_V° (assumption 5).

With the measurements of τ, n_d and N_S°, $D_V^m N_V^\circ$ can be determined. $D_V^m N_V^\circ$ divided by the concentration of silicon atoms gives the self-diffusion constant of silicon. If the diffusion constant is reasonable, this can be taken as strong evidence supporting the assumptions described in the introduction.

III. EXPERIMENTAL

The silicon crystals we used were grown by the floating zone method. They contain phosphorus as a donor impurity. The silicon specimens were cut and polished to slices of about $15\times8\times1.5$ mm³. Surface layers

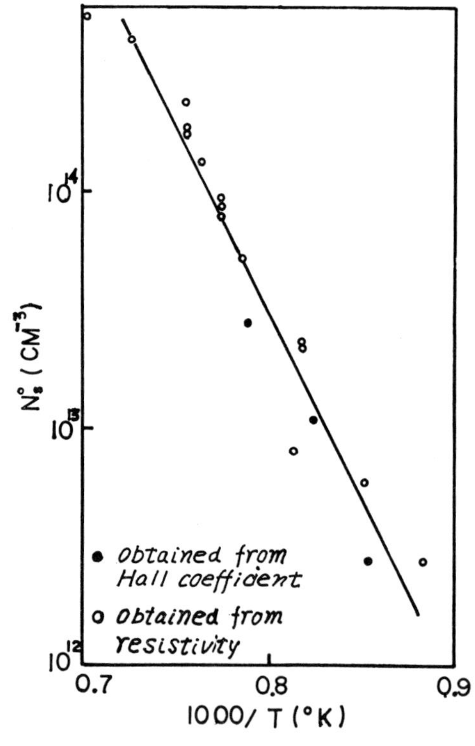

$$N_S^\circ = 10^{26} \exp\left\{\frac{-(3.1\pm0.2)}{KT}\right\}$$

Fig. 1. Solubility of the substitutional nickel atoms (nickel acceptors) in silicon.

about 200 microns thick were removed by etching to make the specimens free from surface damage. Nickel chloride was used as the source of nickel. A few drops of aqueous solutions of $NiCl_2$ were spread over the surface of the specimens and dried. The specimens were then heated in a 5×10^{-5} mmHg vacuum using a quartz tube long enough to be saturated with nickel. They were then cooled rapidly (within a few seconds) by dropping to a cold part of the quartz tube.

The specimens were always prepared in such a way that the concentration of the nickel acceptors did not exceed the net concentration of donors, in order to avoid the large variation of Fermi level caused by the diffusion of nickel acceptors.

By measuring the change of resistivity and Hall coefficient of the speci-

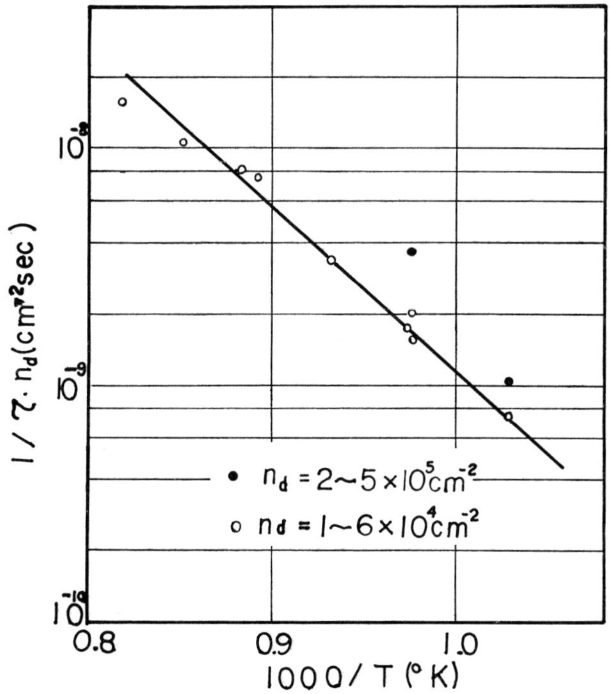

$$\frac{1}{\tau\cdot n_d}=1.3\times10^{-2}\exp\left\{\frac{-(1.4\pm0.2)}{KT}\right\}$$

Fig. 2. The reciprocal product of the time constant of the supersaturated substitutional nickel atoms (nickel acceptors) by the dislocation density.

mens before and after the nickel diffusion, the solubilities of the nickel acceptors N_S were given by

$$N_S = 10^{26} \exp[-(3.1 \pm 0.2)/kT] \quad \text{cm}^{-3} \tag{4}$$

The solubilities are shown in Fig. 1; open and solid circles indicate the values obtained by the resistivity and Hall coefficient measurements respectively.

The original specimens for annealing of the nickel acceptors were saturated with nickel at 1,000°C in general. Annealing of the supersaturated nickel acceptors was investigated by following the rate of change of the resistivity for a known time of heating of the specimens in the temperature range of 700° to 900°C. The dislocation desntiy was obtained by measuring the etch pit density. The reciprocal products of the time constant τ of the annealing and the dislocation density n_d are shown in Fig. 2; open and solid circles indicate the values for the dislocation density of 10^4 and 10^5 cm^{-2} respectively. They are given by

$$1/\tau n_d = 1.3 \times 10^{-2} \exp[-(1.4 \pm 0.2)/kT] \quad \text{cm}^2\,\text{sec}^{-1} \tag{5}$$

Substitution of Eqs. (4) and (5) into Eq. (3) yields

$$D_V{}^m N_V{}^\circ / N^\circ = 30 \exp[-(4.5 \pm 0.4)/kT] \quad \text{cm}^2\,\text{sec}^{-1} \tag{6}$$

where N° is the concentration of silicon atoms, 5×10^{22} cm^{-3}. Eq. (6) will be discussed in the next section.

IV. DISCUSSION AND CONCLUSIONS

Although the self-diffusion constant of silicon has not yet been measured directly, the values obtained by theory or by indirect measurement have been published. The values for Eq. (6) were also obtained by indirect measurement. They are tabulated in Table I.

In the theoretical calculation of Swalin,[5] the potential between nearest neighbor atoms has been assumed to have the form of the Morse function. Benneman[6] has theoretically calculated the change of the electrostatic energy of the system of ions arising from the point defect plus the change in energy of the system of valence electrons due to their redistribution associated with the point defect. It is interesting that their values are in fairly good agreement, although there is a discrepancy between the experimental and theoretical values as seen in Table I. Watkins and Corbett[7] have obtained the values from the activation energy of phosphorus

TABLE I.

	$E_V{}^m$	$E_V{}^f$	E_d	notice
	Self diffusion of silicon			
Swalin	1.06	2.32	3.38	theoretical
Benneman	1.09	2.13	3.22	,,
Watkins and Corbett	0.33±0.2	3.6±0.5	3.94±0.33	experimental
Wilcox and LaChapelle			4.9	,,
Yoshida and Saito			4.5±0.4	,,
	Chemical diffusion in silicon			
germanium			5.48	experimental
tin			4.2*	,,

$E_V{}^m$: Activation energy for migration of vacancy of silicon.
$E_V{}^f$: Activation energy for formation of vacancy of silicon.
E_d : Activation energy for self and chemical diffusion. (energy in eV)

* Obtained from ref. 4, Fig. 9.

diffusion and the Coulomb interaction potential of vacancy and phosphorus in silicon, assuming that the vacancy is negatively ionized. The question of whether the Coulomb interaction with an impurity plays a dominant role in the behavior of the vacancy or not is important. Although Swalin[8] has tried to explain the difference in the activation energy of diffusion of III and V group elements in germanium and silicon by the coulomb interaction and lattice strain, there is as yet no general explanation for it that includes the self-diffusion of germanium. Considering only the coulomb interaction, the activation energies for diffusion of germanium[9] and tin[10] in silicon cannot be different from the self-diffusion of silicon. They are shown in Table I; they are not in good agreement with each other.

In determining the self-diffusion constant by the dissociative diffusion experiment, it is unnecessary to correct it by the coulomb interaction and others. Wilcox and LaChapelle[4] have obtained the constant by using the dissociative diffusion of gold in silicon.

In the present work, the five assumptions in Section II were fairly well satisfied or were consistent with the experimental results. As seen in Table I, the present result is in fairly good agreement with the values of Wilcox and LaChapelle. Based on Table I and the above discussion the authors conclude that Eq. (6) is reasonable for the self-diffusion constant of silicon, that the nickel acceptors in silicon locate at substitutional sites and that nickel atoms diffuse dissociatively.

REFERENCES

1. M. Yoshida and K. Furusho, *Japan. J. Appl. Phys.*, **3**, 521 (1964).
2. F. v. d. Maesen and J. A. Brenkman, *J. Electrochem. Soc.*, **102**, 229 (1955).
3. P. Penning, *Philips Res. Rep.*, **13**, 17 (1958).
4. W. R. Wilcox and T. J. LaChapelle, *J. Appl. Phys.*, **35**, 240 (1964).
5. R. A. Swalin, *J. Phys. Chem. Solids*, **18**, 290 (1961).
6. K. H. Benneman, *Phys. Rev.*, **137**, A 1497 (1965).
7. G. D. Watkins, *J. Phys. Soc. Japan*, **18**, Sup. II 22 (1963). G. D. Watkins and J. W. Corbett, *Phys. Rev.*, **134**, A 1359 (1964).
8. R. A. Swalin, *J. Appl. Phys.*, **29**, 670 (1958).
9. D. A. Petrov, J. M. Schaschkow and A. S. Belanovski, in *Semiconductors and Phosphors* (M. Schon and H. Welker eds.), p. 652, Interscience Publishers, Inc., New York, 1958; D. A. Petrov, J. M. Schaschkow and I. P. Akimchenko, *Chem. Abst.*, **54**, 17190 b (1960).
10. M. F. Millea, *Ph. D. Dissertation*, University of Illinois, 1959 AFDSR-TN-59-660, ASTIA No. AD-215, 265.

Interactions of Lithium with Impurities and Defects in Silicon

P. H. FANG

National Aeronautics and Space Administration, Goddard Space Flight Center, Gerrnbelt, Maryland, U.S.A.

ABSTRACT

Two new types of defects were observed in electron irradiated silicon that contains lithium. These defects were observed in Si in the presence of either boron or phosphorus as long as lithium has the higher concentration. The first type of defect is observed in silicon with low oxygen content and anneals at room temperature. The second type of defect is observed in Si with high oxygen content and anneals at 150°C. In both cases, no other higher temperature annealing stages, up to 400°C, were observed.

I. INTRODUCTION

In a regular Si lattice, when lithium is introduced, one can expect a local mode. Such an effect has not been observed in the pure state (Li-Si) because of the presence of the excessive free carriers from shallow lying Li^+ donors. However, this has been observed in heavily B (boron) doped Li-compensated Si.[1] The pairing of (Li^+B^-) has been studied by Reiss *et al.*,[2] and of $(LiO)^-$ by Pell,[3] mainly in the thermodynamical diffusion properties. From the work described in refs. 2 and 3, the high mobility of Li in Si is established. A more complicated system is that of Vavilov,[4] who reported the reactivation of B by the interaction of Li with V (vacancies) in Si where (Li^+B^-) pairs were formed. The vacancies in his experiment were introduced by a high energy electron irradiation.

Recently, Wysocki *et al.*[5] studied a system consisting of Li, P (phosphorus) and V in Si and found and interesting property: the defect formed by electron irradiation anneals spontaneously at room temperature. They also found that in a system consisting of Li, P, O and V in Si, a major annealing stage occurs at 150°C.

In the present work, we propose to interpret the defect which anneals at room temperature as a Li-V complex and that which anneals at 150°C as a Li-O-V complex.

II. Li-V COMPLEX

The experimental method for investigating the nature of this defect is to measure the photovoltaic conversion efficiency of the Si p-n junction device in the form of solar cells.[6] The experimental results have been reported elsewhere.[7] The essence is that in solar cells made from float zone grown Si, with Li diffused in the bulk n-region, an electron-irradiation introduced minority carrier traps. But these traps are not stable at the room temperature and are spontaneously annealed in a period of hours. This annealing property was observed when the bulk region either contains P of various concentrations (from 1 to 2,500 ohm·cm before Li diffusion), or without P, but an n-type which was converted from an initial pl type by Li overcompensation. Therefore, we conclude that this defect is not associated with the type of the initial doping impurities. Furthermore, a drastic reduction of the room temperature annealing is observed in the presence of high oxygen concentration. From electron paramagnetic resonance, Goldstein[8] reported an observation of a new resonance line which was associated with the Li and V. But this finding disagrees with our observation in one respect: the reported resonance line was observed in float zone, as well as in crucible grown Si. On the other hand, in the presence of oxygen, we observed a very different annealing behavior, as will be discussed in Section III below.

Another speculative suggestion[5] was that the role of Li in the annealing is to neutralize a trapping state of an unspecified complex which was formed during the electron irradiation. An association of Li with A- and E-center can be excluded from the preceding experimental results. An association with C-center is excluded because of the extremely small defect production rate (0.01 e/cm^2) and the small total integrated electron flux (10^{14} e/cm^2). In such a small concentration of defects, no damage compatible to what we observed is expected. Another argument is that a heat treatment up to 400°C was applied to the specimen which was irradiated and subsequently annealed. Damages which would be expected if there is a dissociation between Li and the defect-complex were not observed. Such a high stability is not consistent with the high mobility of Li at 400°C.[2]

The above deduction leads us to a defect model whose elements are Li^+ and V^-. The charge state of these two individual point defects are known while Li^+ is in interstitial position and V^- in lattice position.

Therefore, there is a Coulomb force between Li^+ and V^-. However, an instant collapse of this configuration is prevented by the repulsive force of the surrounding Si coordination. This coordination is related to the preferential Li in the interstitial (111) orientation instead of the lattice position.[1] The difference between the Coulomb attraction and the lattice repulsion apparently is sufficiently small that the defect is only metastable at room temperature.

The actual structure has yet to be investigated by a more basic measurement. One interesting problem would be to study the local modes of Li-Si system where the free carriers were removed by an electron-irradiation. It was found that the carrier removal rate in Si diffused with Li is one order of magnitude higher than the case of equivalent concentration with P as donors.[9]

III. Li-V-O COMPLEX

By the same method of observation on systems with different impurities and different oxygen contents,[5,7] it was found that in Si with high oxygen content, when Li is present at a concentration in the range of 10^{16} atoms/cc, a major annealing stage is observed at 150°C and the usual major annealing stage observed near 400°C[7] is greatly inhibited. This effect, similar to the high rate of carrier removal discussed in the preceding section, indicates a favorable pairing of the mobile vacancies with either the interstitial Li or double interstitial Li-O pairs rather than the pairing to form A-, E-, C-centers, etc.

Since there is a favorable pairing between Li and V or Li-O and V on the one hand, and the unambiguous difference in the annealing properties of these two types of pairings on the other hand, a combination of Li and V could provide a tool to measure the oxygen content in Si. In this case, since the measurement will be the electrical properties instead of the optical properties,[3] greater sensitivity and consequently a coverage of a large range of oxygen content could be achieved.

IV. CONCLUSION

The above discussion presents some preliminary results and speculation about two new defects which promise important technical applications[5] as well as giving rise to some interesting problems in the kinetics of annealing. This latter aspect will be discussed elsewhere.[10]

REFERENCES

1. M. Balkanski and W. Nazarewicz, *J. Phys. Chem. Solids*, **25**, 437 (1964); W. G. Spitzer and M. Waldner, *J. Appl. Phys.*, **36**, 2450 (1965).
2. H. Reiss, C. S. Fuller, and F. J. Morin, *Bell Syst. Tech.* J., **35**, 535 (1956).
3. E. M. Pell, in *Solid State Physics in Electronics and Telecommunications, Brussels 1958*, Vol. 1, p. 261, Academic Press, Inc., New York, 1960.
4. V. S. Vavilov, in *Proc. 7th International Conference on the Physics of Semiconductors*, Vol. 3, p. 115, Dunod, Paris, 1965.
5. Wysocki, Rappaport, Davison, Hand and Loferski, *Appl. Phys. Letters*, **9**, 44 (1966).
6. P. H. Fang, *Phys. Letters*, **20**, 343 (1966).
7. P. H. Fang and Y. M. Liu, *Appl. Phys. Letters*, **9**, 364 (1966).
8. B. Goldstein, *Phys. Rev. Letters*, **17**, 428 (1966).
9. J. Carter, private communication.
10. P. H. Fang, *Phys. Letters*, **23**, 7 (1966).

Effect of Impurities on Annealing Behavior of Irradiated Silicon

MITSUJI HIRATA, MASAKO HIRATA, HARUO SAITO

Department of Physics, College of General Education, Osaka University
Toyonaka, Osaka, Japan

and

JAMES H. CRAWFORD, JR.

*Solid State Division, Oak Ridge National Laboratory**
Oak Ridge, Tennessee, U.S.A.

ABSTRACT

Both isochronal and isothermal annealing of n-type silicon specimens doped with $\sim 10^{14}$ atoms of phosphorous, arsenic, antimony, and bismuth have been studied after exposure to ^{60}Co gamma rays at room temperature. Two annealing stages in which the activation energies and frequency factors are regular functions of the covalent radius of the dopant have been observed in the range 100° to 240°C. The stage which occurs at the lower temperature (Stage I) has been previously reported and the anomalously large apparent activation energies and frequency factors for the larger atoms (Sb and Bi) have been attributed to a high temperature dependence of the process of E-center (vacancy-donor complex) reorientation for these dopants. Since reorientation of the center also produces displacement by one lattice distance, Stage I is thought to be motion of the E center to capturing sites without breakup. Stage II also exhibits activation energies and frequency factors whose magnitude depends in a regular way upon the size of the dopant atom, though their dependence upon covalent radii is not nearly so large as in Stage I. Evidently, this stage cannot be attributed simply to the breakup of E centers since it does not obey the simple first-order law, and the frequency factors for the different dopants are several orders of magnitude smaller than vibrational frequencies. Consequently, it is concluded that two distinct defect-dopant complexes are responsible for annealing in each of these two stages.

* Oak Ridge National Laboratory is operated by Union Carbide Corporation for the U.S. Atomic Energy Commission.

I. INTRODUCTION

The pioneering electron spin resonance studies of Watkins and Corbett[1] on electron-bombarded silicon first revealed the profound influence impurity exerts on radiation effects in this material. In n-type silicon, they found that the dominant radiation defects stable at room temperature contain oxygen (oxygen-vacancy complex called the A center) or donor atoms (phosphorus-vacancy complex called the E center). We have been investigating the influence of donor impurity on the thermal stability of the donor-vacancy complex in irradiated, oxygen-free, n-type silicon, as reflected by the annealing of minority carrier lifetime. The first paper[2] (hereafter referred to as I) treated the annealing in the temperature range 100° to 180°C. It was reported that the activation energy and frequency factors are dependent on both impurity type and concentration. Recently it was found that there is another annealing stage at about 200°C and the activation energies for the annealing stage are also dependent on impurity. It is this second stage that forms the topic of this paper. In this paper, the dependence of activation energies on impurity will be discussed, and the results of the paper I will be briefly summarized as needed for this discussion.

II. EXPERIMENTAL PROCEDURE

Samples used in this study were grown by the floating zone method and their dopants are phosphorus, arsenic, antimony, and bismuth. Impurity concentrations ranged from 7×10^{13} to 2×10^{15} atoms/cm^3. Measurements of minority carrier lifetime were carried out by detecting the decay of photoconductivity by means of microwave energy absorption when the specimen is inserted in a wave guide. No ohmic contact on the specimen was necessary and possible difficulties which might have resulted from preparing contacts were thereby avoided. The microwave frequency was about 10 Gc/s and the TE_{10} mode was used. The specimens were always mounted in the cavity at the position of maximum electric fields.

It is well known that surface states affect lifetime. To avoid this problem the surface of the samples were carefully etched with a mixture of 5 parts of HNO_3 and 1 part of 48% HF. A xenon flash tube was used as the light source and the duration of the light pulse was about 0.5 μsec. The samples were irradiated with ^{60}Co gamma rays. The dose rate was 9.5×10^{14} photons cm^{-2} hr^{-1}, and the total dose was 1.4×10^{15} photons cm^{-2}. Annealing was performed in a silicon-oil bath with a stirrer providing circulation of the oil. The temperature was controlled to ± 0.5°C. All lifetime measurements were made at 300°K.

Previous measurements have shown that in floating zone grown n-type crystals the lifetime after irradiation is dominated by minority carrier capture at a recombination center in the upper half of the gap.[3,7] The minority carrier lifetime τ_M is then related to the integrated flux ϕ, according to the following expression:

$$\frac{1}{\tau_M} = \frac{1}{\tau_0} + \Delta(E_B)\sigma_c v f_B(E_r - E_F)\phi_\gamma$$

where τ_0 = initial lifetime,

$\Delta(E_B)$ = the probability that an incident gamma ray will produce a recombination center,

σ_c = the cross section for minority carrier capture by the occupied center,

v = a thermal velocity of minority carrier,

ϕ_γ = integrated flux,

E_r = energy level of recombination center,

E_F = Fermi level, and

$f_B(E_r - E_F)$ = probability (Fermi function) that the recombination center is occupied by majority carrier.

Figure 1 summarizes the results obtained on crystals doped with arsenic,

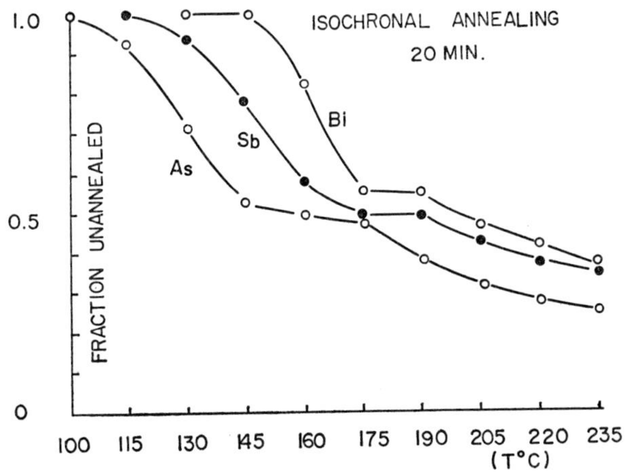

Fig. 1. Isochronal annealing curves for three specimens containing 10^{14} cm^{-3} arsenic, antimony, or bismuth atoms.

antimony, and bismuth. In this figure the fraction unannealed was taken to be

$$f = \left(\frac{1}{\tau_M} - \frac{1}{\tau_0}\right) \bigg/ \left(\frac{1}{\tau_{AB}} - \frac{1}{\tau_0}\right)$$

where τ_{AB} is the post-irradiation lifetime, and τ_M is the lifetime following anneal. Two annealing stages were observed, one at about 130°C and the other at about 200°C. It is interesting to note that the annealing temperature increased with increasing atom size of the impurity in both stages.

The level 0.4 eV below the conduction band is known to be effective as a recombination center in floating zone silicon containing phosphorous,[3,4] and presumably the other donor-vacancy centers as well. It was observed that in both stages the density of the level $E_c - 0.4$ eV decreased after annealing. Unfortunately, the E center is not the only level located at 0.4 eV below the conduction band but the divacancy[5] and the B center[6] have their levels at about the same position. This accounts for observations from Hall coefficient measurements[7] that the annealing of the 0.4 eV level was not complete but about 30% of these states remained after isothermal annealing at about 150°C. The activation energy for this annealing has been reported to be 0.94 eV. Sonder and Templeton[8] confirmed this observation and the thermally stable $E_c - 0.47$ eV center was found after annealing at about 150°C.

In studying the recovery from stress-induced alignment of the Si-E center, Watkins *et al.*[9] determined the activation energy for the reorientation of the phosphorous-vacancy axis to be 0.93 eV. Their more recent study[10] reveals that the reorientation energy for the antimony-vacancy axis is 1.29 eV.

III. RESULTS AND DISCUSSION

(a) *Lower temperature stage (Stage I)*

Isothermal annealing curves of the fraction unannealed are shown in Figs. 2 and 3. Superposition of curves in Figs. 2 and 3 are shown in Figs. 4 and 5, respectively, where the dashed curves indicate a simple exponential first-order decay.

The characteristic recovery time τ is given by

$$\tau^{-1} = \nu_0 \exp(-E_a/kT)$$

where E_a is the activation energy for the annealing, and ν_0 is the frequency factor. The activation energies for the annealing were calculated to be

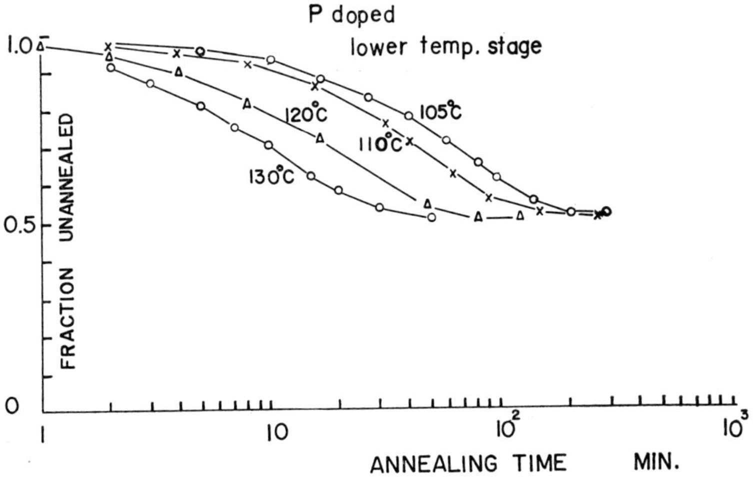

Fig. 2. Isothermal annealing curves for Stage I in phosphorus-doped specimens after irradiation with ^{60}Co gamma rays.

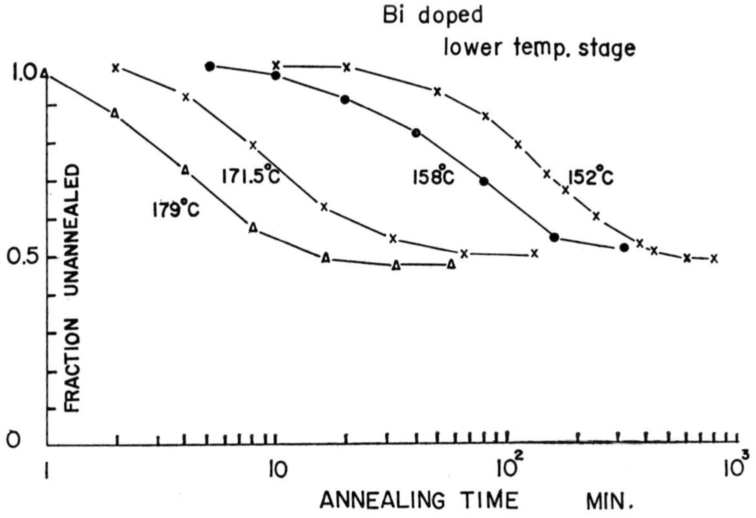

Fig. 3. Isothermal annealing curves for Stage I in bismuth-doped specimens after irradiation with ^{60}Co gamma rays.

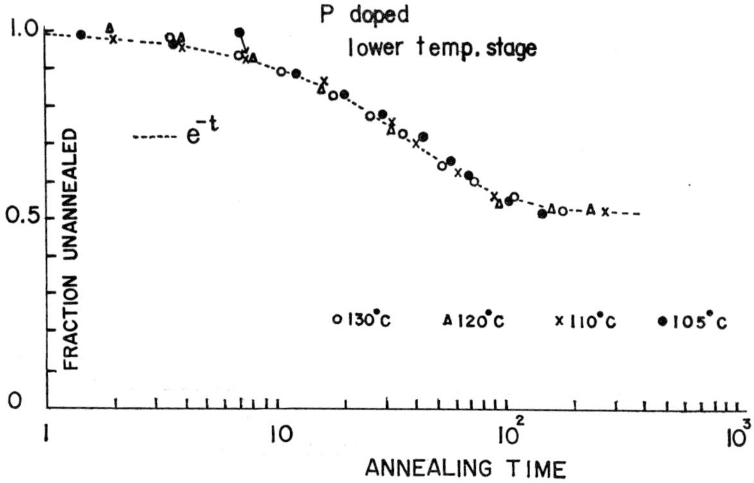

Fig. 4. A composite annealing curve obtained by adjusting the time scales for superposition of curves in Fig. 2.

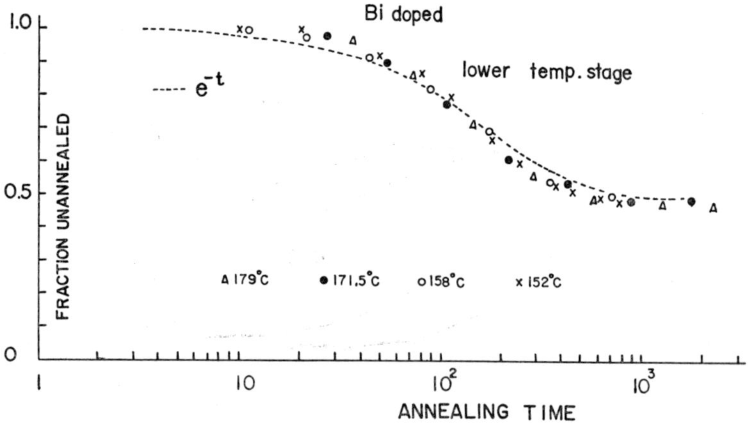

Fig. 5. A composite annealing curve obtained by adjusting the time scales for superposition of curves in Fig. 3.

0.93, 1.27, 1.84, and 2.22 eV, and the frequency factors were determined to be 6×10^{8}, 4×10^{12}, 8×10^{18}, and 2×10^{22} cps for the impurities phosphorous, arsenic, antimony, and bismuth, respectively.

The activation energy for the annealing increased with increasing tetrahedral covalent radii of the impurity atoms. The values are listed in Table I together with data accumulated so far.

The relation between activation energies and the covalent radii can be expressed as a linear function as follows:

$$E_a = 3.6r - 3.0 \text{ eV}$$

where r is the covalent radius of impurity atom in Å. Both E_a and ν_0 increased dramatically with increasing donor atom size. An increase in E_a which reflects an increase in the tightness of binding the vacancy to the impurity is expected on the basis of elastic consideration alone. However, the drastic increase in ν_0 was completely unexpected and has been discussed[2] on the basis of elastic interactions between the oversized impurity atoms and the vacancy.

The activation energy for the annealing of phosphorous E center corresponds to that obtained by Watkins and Corbett[9] for the reorientation of this center determined from electron spin resonance techniques. They point out that in the diamond lattice relaxation of a prefered orientation of the E center occurs by the motion of a vacancy from its position adjacent to an impurity atom around a six-membered ring to the opposite side of the impurity. According to their observation, the vacancy most probably remains coupled to the donor atom during the interchange and the reorientation energy of 0.93 eV is smaller than the total binding energy of the center. Nevertheless, during reorientation the E center moves one lattice spacing and this process corresponds to migration of the complex without breakup. The frequency factor of 6×10^8 seems appropriate for the migration of the E center as a unit to sinks and one possible sink is the oxygen impurity at which a new complex might be formed. It does not appear that a similar relaxation process can be occurring in the case of the larger radii donor atoms unless the inordinately large values of ν_0 and E_a can be accounted for in a reasonable way. One way of rationalizing the unexpectedly large values of these parameters is to assume that when the covalent radius of the substituent impurity is larger than that of silicon, the energy required for a vacancy to move around the impurity atom is strongly temperature dependent. The crowding of the lattice by the oversized atom could be appreciably relieved by the thermal expansion of the lattice. Such a temperature dependence would have the effect of increasing the effective magnitude of ν_0. According to this view, the observed phenomenon that relaxtion occurred at a much lower temperature than normally would be expected for such apparently high activation energy is not too surprising.

(b) *Higher temperature stage (Stage II)*

Annealing curves in the higher temperature stage are shown in Figs. 6, 7 and 8. These curves cannot be expressed as a simple exponential function. But, each of these curves possesses a characteristic decay time. Values of E_a and ν_0 obtained from the temperature dependence of the time for 70% complete annealing are listed in Table I. In this stage too the activation energy and the frequency factor were found to depend on impurity though the dependence is not nearly so strong as for the lower temperature stage. Once again the relation between the activation energies and the covalent radii can be expressed as a linear function

$$E_a = 1.5r - 0.7 \text{ eV}$$

It is interesting to note that, even though the measured activation energies for Stage II annealing are smaller than those for corresponding impurities in Stage I annealing, Stage II nevertheless occurs at a higher temperature. The reason for this behavior is obviously connected with the anomalously large values of ν_0 for the Stage I processes. As noted above we have attempted to explain these large values on the basis of a very large linear temperature dependence of the energy of reorientation for the E centers with large donor impurity.

Since both E_a and ν_0 for Stage II processes are regularly dependent upon

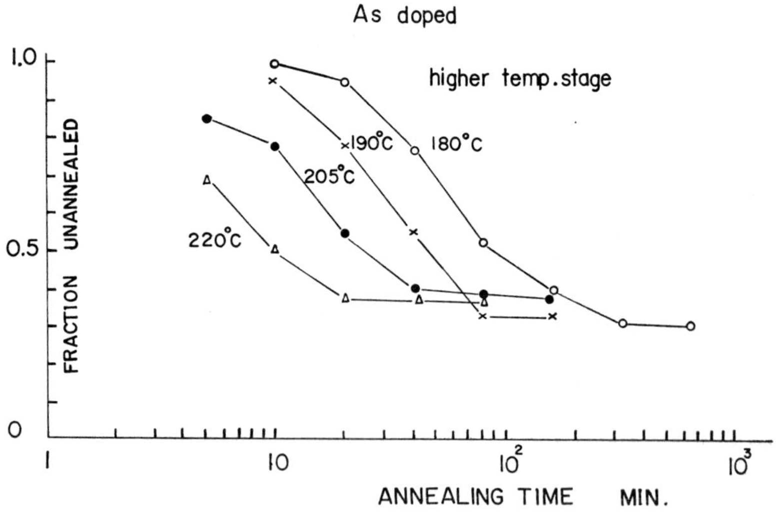

Fig. 6. Stage II annealing curves for a group of samples containing arsenic atoms.

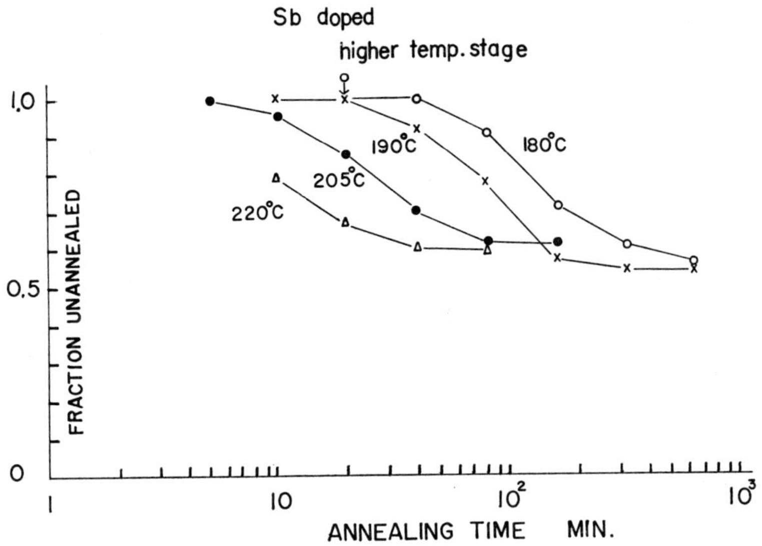

Fig. 7. Stage II annealing curves for a group of samples containing antimony atoms.

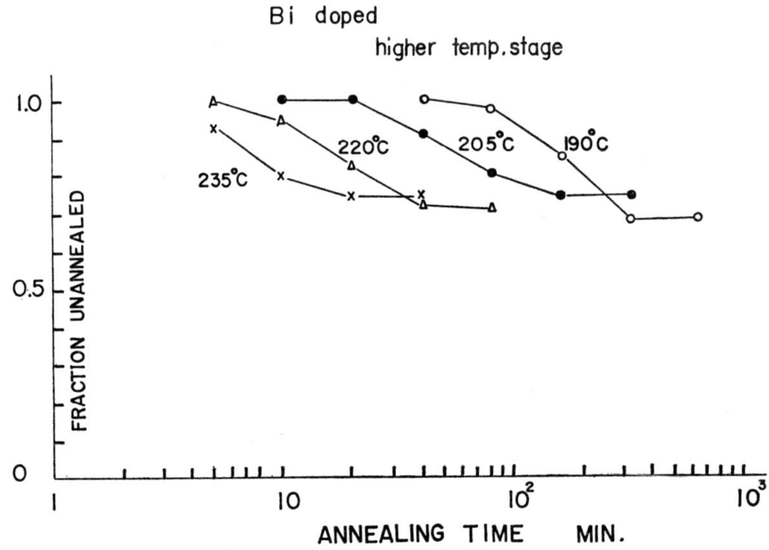

Fig. 8. Stage II annealing curves for a group of samples contining bismuth atoms.

TABLE I.

Dopant	Stage I (130°C)		Stage II (200°C)		E_a (eV)	ν_0 (cps)
	E_a (eV)	ν_0 (cps)	E_a (eV)	ν_0 (cps)		
P (1.1) [a]	0.93	6×10^{8}			0.93 [b]	1.6×10^{13} [b]
As (1.18) [a]	1.27	4×10^{12}	1.07	2×10^{8}		
Sb (1.36) [a]	1.84	8×10^{18}	1.28	1×10^{10}	1.29 [c]	9×10^{14} [c]
Bi (1.46) [a]	2.22	2×10^{22}	1.46	5×10^{11}		

a Covalent radius in Å.
b Watkins and Corbett.
c Watkins, private communication.

covalent radii of the impurity, it is concluded that these processes are also associated with a defect-donor complex such as the E center. This point of view is supported by a recent measurement by Watkins[10] who finds that the energy required to reorient the Sb-vacancy complex (see Table I),

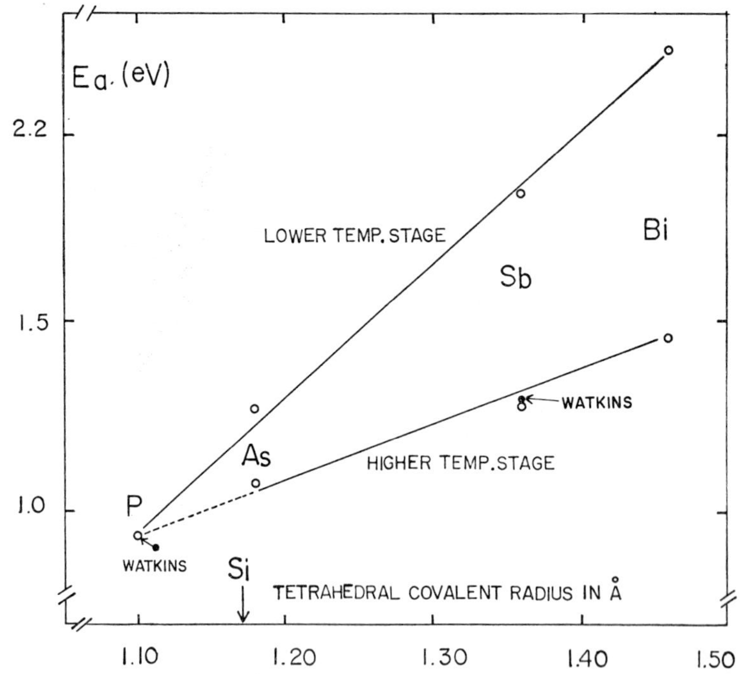

Fig. 9. Activation energies as a function of tetrahedral covalent radii for both Stage I and Stage II annealing. The arrows indicate the results from uniaxial stress studies by Watkins *et al.*

as detected by electron paramagnetic resonance methods, is the same as the activation energy for Stage II annealing of E centers in Sb-doped Si.

It is tempting to explain the higher temperature stage as being due to the actual breakup of E centers. However, one would expect this to be a straightforward first-order process with a frequency factor comparable to atomic vibrational frequencies. Since the observed values of ν_0 are smaller than 10^{13} sec^{-1} for all dopants studied and, since the first-order law is not obeyed, one must conclude that a more complicated relation process such as diffusion to sinks is taking place. Therefore, one is led to the view that two different entities may be involved in the two annealing stages under consideration, but it is difficult to conceive of two forms of the E center which would move with sufficiently different jump frequencies to produce annealing in these two different temperature ranges. Additional work is necessary to elucidate the details of the kinetic processes underlying the annealing stages reported here.

Acknowledgements

The authors are greatly indebted to Professor M. Kawanishi and other members of Osaka University Radiation Laboratory for their kind help in irradiating the samples with ^{60}Co gamma rays.

REFERENCES

1. G. D. Watkins and J. W. Corbett, *Phys. Rev.*, **121**, 1001 (1961); *ibid.*, **134**, A 1359 (1964).
2. M. Hirata, M. Hirata, H. Saito and J. H. Crawford, Jr., *J. Appl. Phys.*, **38**, 2433 (1967).
3. M. Hirata, M. Hirata and H. Saito, *J. Appl. Phys.*, **37**, 1867 (1966).
4. R. H. Glaenzer and C. J. Wolf, *J. Appl. Phys.*, **36**, 2197 (1965).
5. J. W. Corbett and G. D. Watkins, *Phys. Rev. Letters*, **7**, 314 (1961).
6. G. D. Watkins and J. W. Corbett, *Disc. Faraday Soc.*, **31**, 86 (1961).
7. H. Saito, M. Hirata and T. Horiuchi, *J. Phys. Soc. Japan*, **18**, Suppl. **3**, 246 (1963); *Japanese J. Appl. Phys.*, **2**, 768 (1963).
8. E. Sonder and L. C. Templeton, *J. Appl. Phys.*, **34**, 3295 (1963).
9. G. D. Watkins and J. W. Corbett, *Phys. Rev.*, **134**, A 1359 (1964).
10. G. D. Watkins, private communication.

Reverse Annealing of Conductivity in Electron-Irradiated n-Type Silicon

REGINA V. TAUKE and BRUCE J. FARADAY

U.S. Naval Research Laboratory, Washington, D.C., U.S.A.

ABSTRACT

Isochronal annealing of radiation-induced changes in the conductivity of n-type silicon has been studied by means of the four-point probe technique over the temperature range from 30° to 600°C. Pulled and float-zone single crystals (phosphorus $\sim 6\times10^{14}/\text{cm}^3$) were exposed to 1 MeV electrons in air at 30°C to integrated fluxes from 7.5×10^{15} to 1×10^{18} e/cm^2. A prominent reverse annealing stage was observed in the pulled silicon samples in the temperature range between 250° and 375°C. The magnitude of the reversal and the temperature range of the reverse annealing stage were seen to be functions of the integrated flux. The largest reversal was observed for a flux of 3×10^{16} e/cm^2 where the concentration of observed defects rose to 1.3 times the post-irradiation value. Similar annealing studies performed on float-zone specimens showed no evidence of a reverse annealing stage as exhibited by the pulled samples having a much higher oxygen content. These results suggest that the reverse annealing stage is due to the dissociation of A centers (oxygen atom+vacancy) to form some (as yet unidentified) defect center or complex which acts as an electron trap. From Fermi level considerations, the level associated with this defect should lie below $E_c - 0.28$ eV. Similar reverse annealing stages in n-type silicon have been reported in the minority carrier lifetime at lower temperatures following ^{60}Co γ-irradiation and in the impurity band optical absorption at higher temperatures after bombardment by 1.5 MeV electrons. Disagreement in the temperature range of the reverse annealing stage as observed in the three experiments is ascribed to differences in the defect concentrations produced at the flux levels utilized in each study.

I. INTRODUCTION

The irradiation of silicon by high energy electrons (>150 keV) introduces defect centers which give rise to localized levels in the forbidden energy gap. Despite a considerable number of electron radiation damage studies in silicon, the state of knowledge of the exact dependence of these levels upon the nature and concentration of dopant and residual impurities as well as upon the elctron energy and the irradiation temperature is far from complete. One firmly established feature of these studies is the fact that irradiation of phosphorus-doped n-type silicon containing oxygen as a residual impurity by electrons of a few MeV at room temperature may result in the formation of recombination centers with associated levels at 0.17 eV and at 0.40 eV below the conduction band. These have been designated as the A center and the E center, respectively. The defect structures determined by spin resonance studies show that the A center is a vacancy containing an oxygen atom[1–4] while the E center consists of a vacancy associated with a substitutional phosphorus atom.[1,2] The production of these defect centers is not restricted to electron bombardments. For example, the formation of A centers and of E centers in n-type silicon has been reported following neutron[5,6] and ^{60}Co γ-irradiations.[5–8]

It has also become well-established that both E centers and A centers in silicon anneal over a wide range of temperatures. Thus in a review of annealing studies in silicon, Hasiguti and Ishino[9] have suggested the range from 370°K to 470°K as the stage for the dissociation of E center and the range from 470° to 600°K as the annealing stage for A centers. In a temperature region overlapping the latter stage, i.e., between 210° and 450°C, reverse annealing stages have been reported in two studies of n-type silicon irradiated at room temperature. In the first experiment, Corbett, Watkins and McDonald[10] observed the annealing of defects produced in crystals by 1.5 MeV electrons as manifested by changes in the impurity band optical absorption α_{I}. In the second, Nakano and Inuishi[11] studied the annealing of damage in similar crystals exposed to ^{60}Co γ-rays by observing changes in the minority carrier lifetime τ.

The purpose of the present paper is to report a similar reversal observed in the annealing of conductivity changes induced in n-type silicon crystals by 1 MeV electrons and to compare the characteristics of this reverse annealing stage with the behavior of α_{I} and of τ. Another purpose is to report the extension of conductivity annealing data in n-type silicon up to 600°C.[12]

II. EXPERIMENTAL

The silicon samples utilized in this study were n-type (phosphorus-doped). They were cut as 0.5 mm thick slices of approximately 3 cm^2 area from single crystal ingots which had been grown by the Czochralski (pulled)[13] and the float-zone techniques.[14] The pre-irradiation resistivity of the pulled samples ranged between 7.8 and 8.5 ohm-cm whereas that of the float-zone specimens was measured as 11.4 ohm-cm. The oxygen concentration in the pulled samples was about $5\times10^{17}/cm^3$ and less than $10^{16}/cm^3$ in the float-zone material. The dislocation densities as determined by the respective suppliers were approximately $1\times10^3/cm^2$ for the pulled crystals and less than $2\times10^4/cm^2$ for the float-zone specimens. The samples were irradiated equally on opposite faces in air at 25°C by 1 MeV electrons from the NRL 2 MeV van de Graaff accelerator to eleven levels of integrated flux between 7.5×10^{15} and 1×10^{18} e/cm^2. The bombardment current was $\approx 0.3\ \mu A/cm^2$; the maximum temperature rise in each sample was about 20°C.

Resistivities were measured using a standard four-point probe technique.[15] The advantage of the contact-less method was that it avoided the possibility of contact deterioration or diffusion of contact material at the high annealing temperatures utilized in this investigation. After irradiation, isochronal annealing was performed by heating the specimens for 20-minute intervals in an argon atmosphere. The temperature was increased progressively in increments of 25°C and the resistivity was measured at room temperature following each annealing step.

III. RESULTS AND DISCUSSION

(a) *Radiation-induced conductivity decrease*

The conductivity of the pulled specimens was measured as a function of the integrated electron flux ϕ. The fractional decrease in the electron concentration, $(n_0-n)/n_0$, with increasing bombardment by 1 MeV electrons is shown in Fig. 1. The solid circles represent measurements obtained from six samples which had been irradiated to different levels of integrated flux. The data were based on the assumption that radiation-induced changes in the mobility were insignificant in comparison with corresponding changes in the conductivity. The curve is linear of unity slope up to a value of ϕ of $5\times10^{16} e/cm^2$ showing that over this range acceptor defect levels are being filled in proportion to the integrated flux. For heavier bombardments, the Fermi level shifts toward the center of the forbidden gap thereby causing a reduction in the fractional filling of acceptors. This is illustrated by the gradual departure from linearity and the eventual leveling of the 1 MeV curve in Fig. 1.

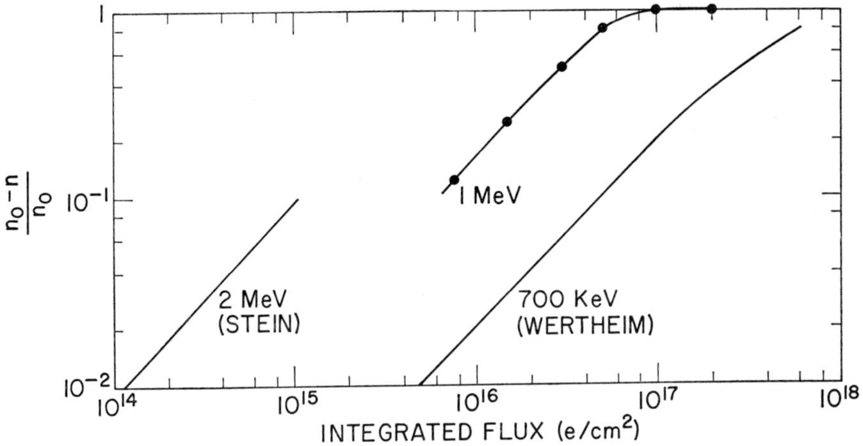

Fig. 1. Fractional decrease in electron concentration as a function of 1-MeV electron bombardment of n-type pulled silicon (8 ohm-cm). Comparative data for electron bombardment energies of 700 keV (ref. 16) and of 2 MeV (ref. 12) are shown.

The linearity of the conductivity data (up to $\phi=5\times10^{16}$ e/cm^2) as well as the internal consistency of measurements made on different specimens at various values of ϕ lend support to the validity of the four-point probe technique. In addition, the present data are seen to fall between results reported by Wertheim[16] and by Stein[12] for electron irradiations of n-type silicon at 700 keV and at 2 MeV, respectively, as shown in Fig. 1. The introduction rate of acceptors increases with increasing bombardment energy as expected.

(b) *Annealing effects*

The variation of the room-temperature resistivity measurements as a function of the annealing temperature in pulled and in float-zone specimens of n-type silicon is shown in Figs. 2 and 3, respectively. Data pertaining to six of the eleven levels of radiation exposure are presented in Figs. 2 and 3; they are representative of the others as well. These data are included to illustrate certain features of the annealing curves which do not appear so prominently in the manner of presentation adopted below. For example, it may be seen that the peak in the reverse annealing stage for the pulled samples (near 350°C) shown in Fig. 2 is present in all specimens except the one irradiated to the lowest level, i.e., $\phi=7.5\times10^{15}$e/cm^2. This observation also applies to the data not shown for the samples irradiated to the

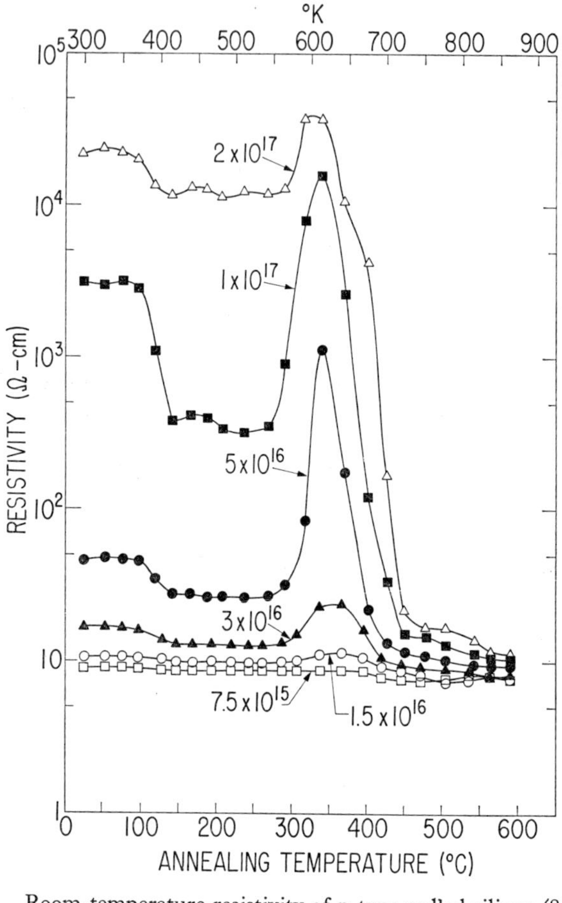

Fig. 2. Room-temperature resistivity of n-type pulled silicon (8 ohm-cm) as a function of annealing temperature for integrated flux levels of 1 MeV electrons between 7.5×10^{15} and 2×10^{17} e/cm^2.

other five levels of exposure (all in excess of 7.5×10^{15} e/cm^2). The annealing data for the float-zone specimens given in Fig. 3 show the presence of at least two annealing stages, the first between 100° and 150°C and the second between 300° and 425°C. These features will be referred to in the subsequent discussion.

In the treatment of annealing data, it is customary to introduce a parameter f which is the unannealed fraction of the irradiation-incuded conductivity decrease. Here f is defined as $(\sigma_0 - \sigma_a)/(\sigma_0 - \sigma_f)$, where σ_0, σ_f, and σ_a represent the conductivity before irradiation, after irradiation, and

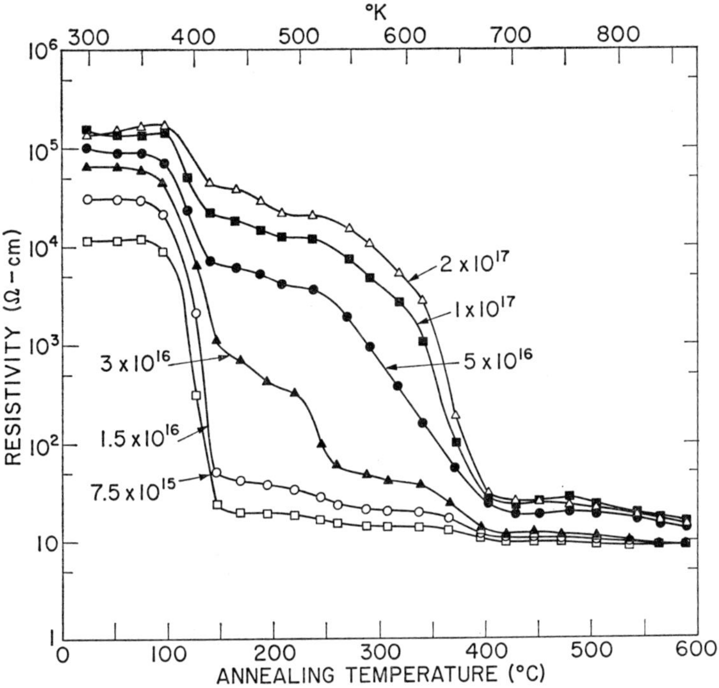

Fig. 3. Room-temperature resistivity of n-type float-zone silicon (11 ohm-cm) as a function of annealing temperature for integrated flux levels of 1 MeV electrons between 7.5×10^{15} and 2×10^{17} e/cm^2.

after annealing to a temperature T_a. The value of f can be shown to be directly proportional to the concentration of electrons trapped by radiation defects in the crystal at the temperature T_a, on the assumption that the electron mobility remains constant.

Annealing curves showing the manner in which f varies as a function of T_a for the pulled samples are shown in Figs. 4, 5 and 6. Note that each curve represents data for a separate sample, the sample having been cut from adjacent segments of the same ingot. Data for the pulled samples irradiated to the three lowest levels of ϕ in this study appear in Fig. 4. Three prominent recovery stages are apparent. The first, in the temperature range between 100° and 160°C, has been ascribed to the thermal dissociation of E centers.[17] This is followed by a large reverse annealing stage peaking at 350°C which is observed for samples irradiated to ϕ—values of 1.5 and 3×10^{16} e/cm^2. The measurements for the specimen irradiated to $\phi = 7.5 \times 10^{15}$ e/cm^2 are not definitive since the four-point

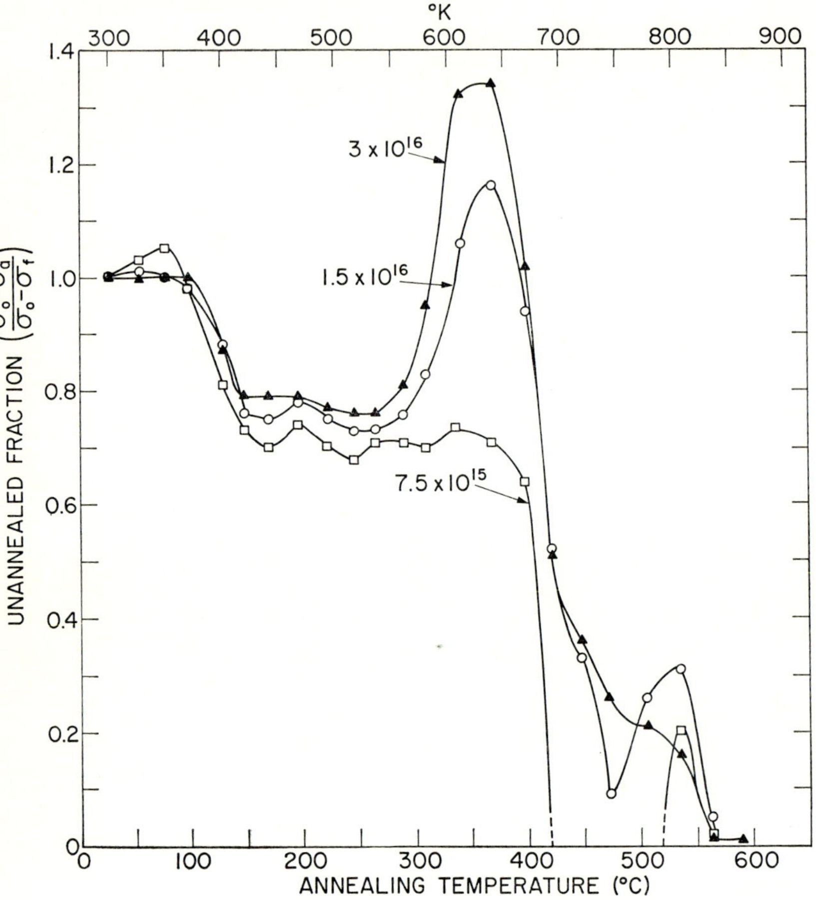

Fig. 4. Unannealed fraction of conductivity decrease in n-type pulled silicon (8 ohm-cm) as a function of annealing temperature for inte- grated flux levels of 1 MeV electrons between 7.5×10^{15} and 3×10^{16} e/cm^2.

probe is not sensitive to extremely small changes in the resistivity. Consequently, from the data exhibited in Figs. 2 and 4, it is not clear whether there is a reversal in the annealing curve for the lowest level of irradiation in the study. Stein[12] has reported a reverse annealing stage peaking at about 320°C in n-type silicon irradiated by 2 MeV electrons to a level of $\phi=9.4\times10^{14}$ e/cm^2.

A second prominent annealing stage can be seen in the range from 350° to approximately 450°C. Some additional remarks pertaining to this region will be presented below. The third recovery stage between 500°

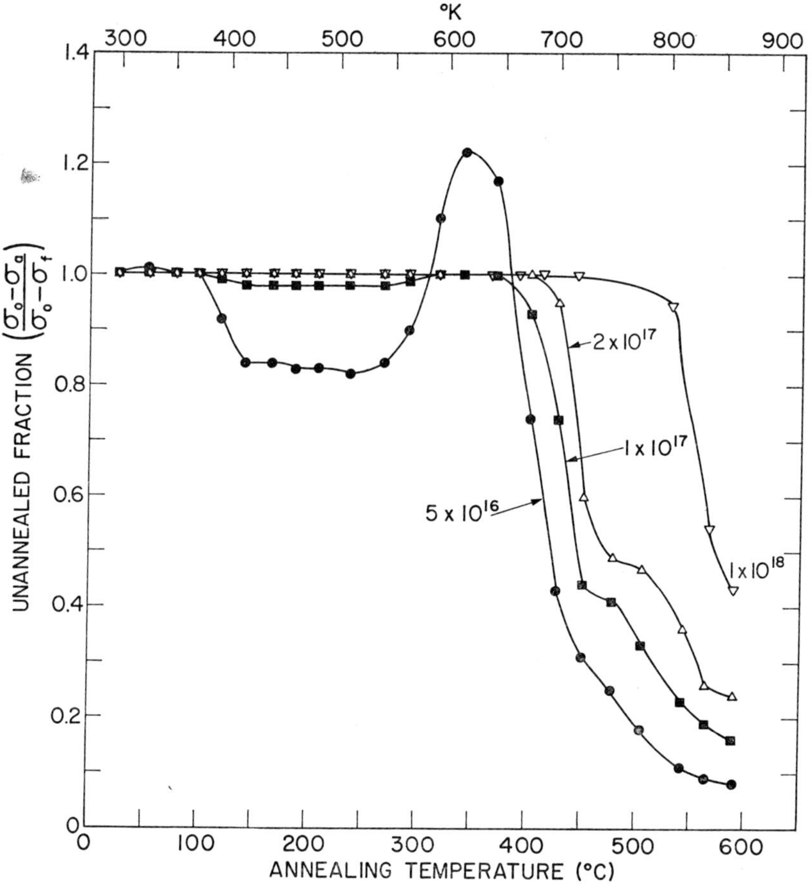

Fig. 5. Unannealed fraction of conductivity decrease in n-type pulled silicon (8 ohm-cm) as a function of annealing temperature for inte- grated flux levels of 1 MeV electrons between 5×10^{16} and 1×10^{12} e/cm^2.

and 550°C is preceded by a reversal for the two lightest exposures. This also appears to be a region of relative insensitivity for the four-point probe. In view of the fact that the third annealing stage is not readily observed for the heavier exposures, no further details regarding the third stage will be presented at this time.

Data for four other levels of irradiation (including the three highest levels) are shown in Fig. 5. Here the reverse annealing stage is not apparent for the two heaviest bombardments. Reference to Fig. 2 shows that the reversal is quite prominent for the resistivity measurements carried

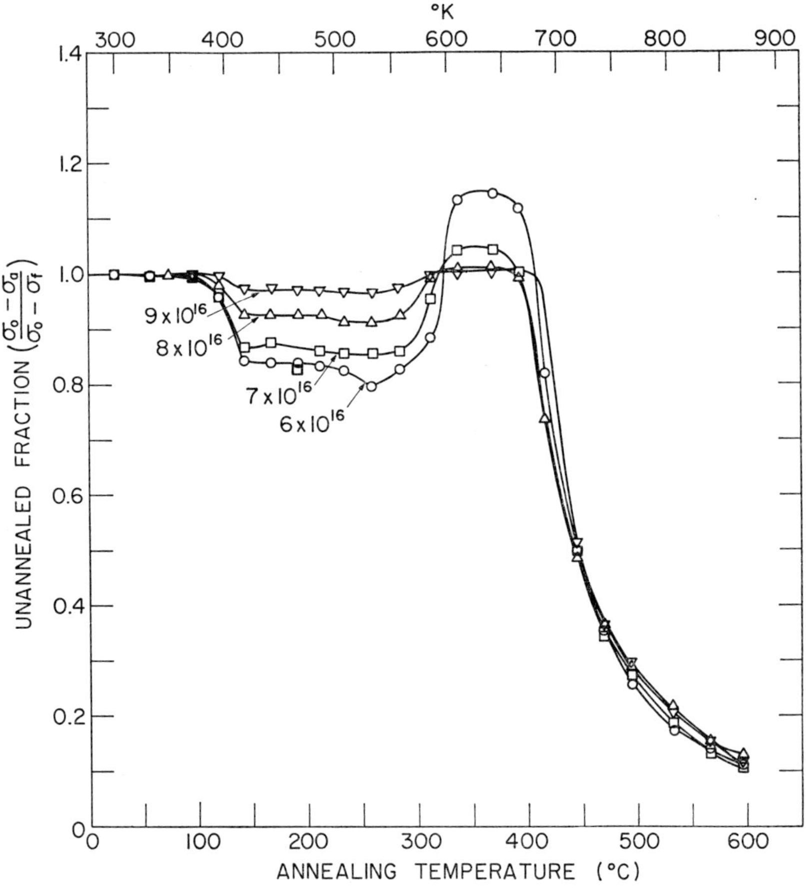

Fig. 6. Unannealed fraction of conductivity decrease in n-type pulled silicon (8 ohm-cm) as a function of annealing temperature for integrated flux levels of 1 MeV electrons between 6×10^{16} and 9×10^{16} e/cm^2.

out for the pulled sample irradiated to a flux of 2×10^{17} e/cm^2. This is also true (though not shown) for the specimen irradiated to a flux of 1×10^{18} e/cm^2. The reverse annealing stage is not evident for the heavy irradiations when the data are presented in terms of the parameter f because the conductivity values σ_f and σ_a are both negligible in comparison with σ_0. For this condition, f can be represented as $f \approx 1 - a + b$, where $a = \sigma_a/\sigma_0$, $b = \sigma_f/\sigma_0$, and $a < b \ll 1$. Thus the value of f is approximately equal to, though slightly larger than unity.

The data presented in Fig. 6 are included to show the annealing behavior of specimens irradiated to a series of ϕ-values between 6×10^{16} and 9×10^{16} e/cm². In the temperature range between about 100° and 220°C, the data for these four intermediate exposures are seen to be bracketed between the curves for the two irradiations given in Fig. 5. Since the annealing stage for E centers falls in this temperature range, these curves (together with all of the curves shown in Figs. 4 and 5) indicate that the unannealed fraction of E centers increases with increasing flux in the range of flux levels examined in this study. For the highest level, i.e., $\phi=1\times10^{18}$ e/cm², while a large number of E centers are annealed in the above temperature range, the fraction annealed is negligible.

Earlier, annealing curves for the resistivity in float-zone specimens of n-type silicon (phosphorus-doped) having approximately the same initial resistivity as the pulled samples were presented in Fig. 3. Data for the same six values of ϕ are replotted to exhibit the annealing behavior for the conductivity, as shown in Fig. 7. Neither in Fig. 3 nor in Fig. 7 is there

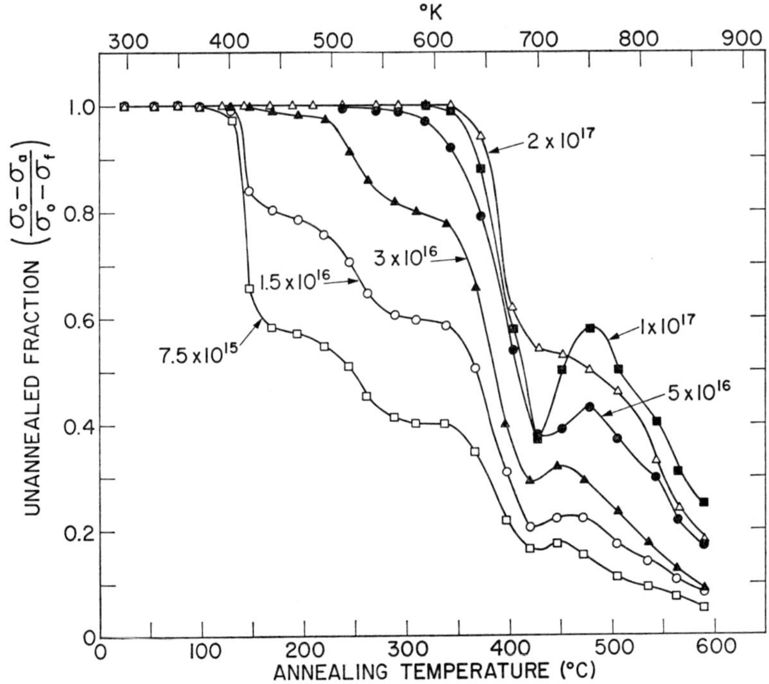

Fig. 7. Unannealed fraction of conductivity decrease in n-type float-zone silicon (11 ohm-cm) as a function of annealing temperature for integrated flux levels of 1 MeV electrons between 7.5×10^{15} and 2×10^{17} e/cm².

evident a reverse annealing stage similar to the reversals described for the pulled samples. These observations are in agreement with those of Nakano and Inuishi[11] and of Stein[12] who reported the absence of any reverse annealing in carrier lifetime or conductivity, respectively, for float-zone n-type phosphorus-doped silicon.

Some additional comments pertaining to the annealing behavior of the float-zone specimens are in order. Reference to Fig. 3 suggests the presence of at least two annealing stages. The first occurs in the temperature range between 100° and 150°C and is presumably due to the thermal dissociation of E centers.[17] The second annealing stage is much broader, falling into the range between about 220° and 390°C. Stein[12] has observed these same annealing stages in his study and has suggested that the second stage may be due to the dissociation of the divacancy. This defect is known to anneal near 350°C in the float-zone silicon.[18]

(c) *Reverse Annealing Stage*

The principal results relating to the reverse annealing stage in conductivity are now summarized. First, there is a broad overlap between T_{ra}, the temperature range of the reverse stage, and the A-center annealing stage. Second, the reverse annealing stage is observed only in the pulled silicon and is absent in the float-zone crystals. Third, there is an increase in the concentration of observed defects as indicated by a decrease in the conductivity. These results suggest that the reverse annealing stage is due to the dissociation (or modification) of A centers to form some (as yet unidentified) defect center or complex which acts as an electron trap. For the n-type material used here, the level associated with such a defect should lie below the pre-irradiation Fermi level, i.e., below $E_c-0.28$ eV. The data also suggest that the defect is a *secondary* defect since annealing is required for its formation.

As pointed out earlier, similar reverse annealing stages were reported for α_I and for τ. A comparison of these reversals with that in σ reported in the present study is presented in Table I. The reversal in α_I was manifested by the disappearance of the absorption band at 820 cm^{-1} (the absorption associated with the A center) and the corresponding growth of a band at 887 cm^{-1}.[19] Corbett, Watkins, and McDonald[10] suggest that the defect responsible for the growth of the latter band is an A center to which an oxygen atom has migrated, or alternatively, a vacancy and two associated oxygen atoms. The reversal in the annealing in τ, on the other hand, has been ascribed by Nakano and Inuishi to the thermal dissociation of the A center and to the simultaneous production of vacancies. These view-

TABLE I. Summary of comparative data for reverse annealing stages observed in n-type silicon following room-temperature irradiation.

	τ [a]	σ [b]	α_I [c]
Resistivity (ohm-cm)	80	8	0.6
Phosphorus (per cm^3)	6×10^{13}	6×10^{14}	10^{16}
Oxygen (per cm^3)	6×10^{17}	5×10^{17}	10^{18}
$E_{\text{irradiation}}$ (MeV)	1	1	1.5
Integrated Flux	$3\text{–}9\times10^7$R	10^{16}–10^{17} e/cm^2	$\sim10^{19}$ e/cm^2
Defect concentration calc. (per cm^3) [d]	$\sim10^{14}$	$\sim10^{17}$	$\sim10^{19}$
T_{ra} (°C)	210–330	250–380	300–450

T_{ra} refers to the temperature range for which the observed defect concentration increases with increasing temperature.

a See ref. 11.
b This study.
c See ref. 10.
d See ref. 20.

points are consistent insofar as the dissociation or modification of A centers is required for the production of the defect giving rise to the reverse annealing stage. Beyond this, no evidence is available in any of the three experiments to formulate an unambiguous description of the defect model.

Reference to Table I shows that the range of T_{ra} in τ differs from and is generally lower than that observed in σ. Furthermore, Nakano and Inuishi[11] reported that the temperature at which the reverse annealing stage peaks is seen to shift to higher temperature with increasing γ-ray fluxes. In fact, the peak shifts from 240° to 330°C as ϕ is varied from 3 to 8.6×10^7R. By contrast, the temperature at which the reverse annealing stage is observed in the present study decreases with increasing electron flux, as shown in Fig. 8. The curve is drawn to best fit the data. Such a dependence of T_{ra} (and presumably the peak) is consistent with a model in which the A center is modified by forming a complex with another species, e.g., an oxygen atom,[10] as suggested by the following simple argument. The lower the flux, the smaller is the concentration of A centers formed. Thus the separation between A centers is larger and the higher is the temperature required for the migration of an oxygen atom to combine with the A center. A similar argument may be used to indicate that the higher the flux, the smaller is the separation between A centers and hence the lower is the temperature required to form the defect or complex.

The data showing the second prominent annealing stage of the irradiated pulled samples (between roughly 350° and 450°C) in Figs. 4–6 is pertinent to the discussion. This stage appears to be due to the dis-

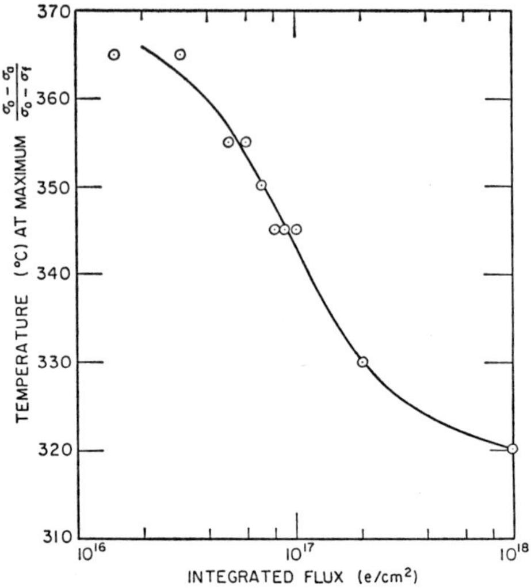

Fig. 8. The annealing temperature at which the peak of the reverse annealing stage is observed in n-type silicon (8 ohm·cm) as a function of the 1 MeV electron integrated flux.

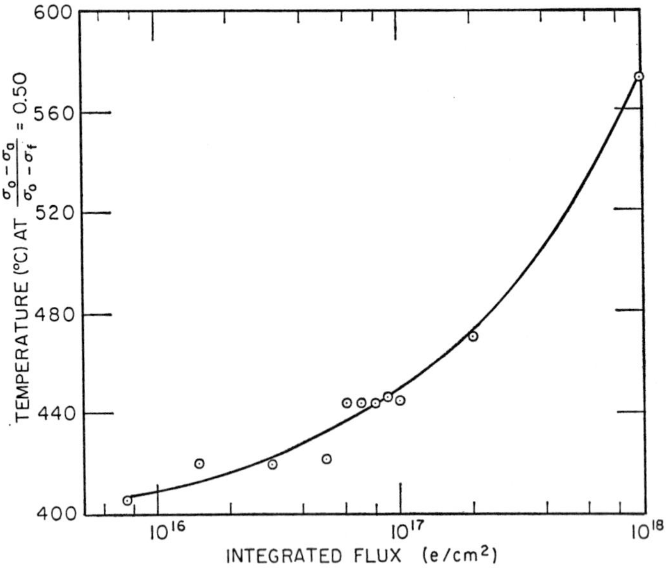

Fig. 9. The annealing temperature at which the unannealed fraction of conductivity decrease equals 0.5 (on the high temperature side of the peak of the reverse annealing stage) in n-type pulled silicon (8 ohm·cm) as a function of the 1 MeV electron integrated flux.

sociation of the unidentified defect or complex which gives rise to the reverse annealing stage. For convenience, the results presented in Fig. 9 show the temperature at which the unannealed fraction of conductivity decrease is equal to 0.5 plotted against the electron flux. The temperature in each case is very nearly that at which the value of f decreases to half its value at the peak of the reverse annealing stage. The curve is again drawn to best fit the available data. The results indicate that as the flux level of irradiation is increased, progressively higher temperatures are required to anneal the reverse annealing stage. Such behavior is unexpected in view of the fact that the peak of the reverse annealing stage was observed to shift to lower temperatures for increasing flux levels, as shown above in Fig. 8. It appears that for the higher flux levels, a second unknown defect or complex is being formed in the approximate temperature range between 450° and 600°C. It is suggested that the formation of this defect competes with the dissociation of the defect responsible for the reverse annealing stage. As the flux is increased, the former process becomes more important and so a higher temperature is required for the latter process. Reference to the annealing curves shown previously in Figs. 2, 4 and 5 suggests the formation of a defect or complex between about 450° and 600°C. The crossovers which are clearly evident in the annealing curves exhibited in Fig. 6 also support this viewpoint.

A few comments can be made regarding the position of the level associated with the unknown defect. First, the defect level should lie below the pre-irradiation Fermi level, i.e., $E_c-0.28$ eV. Second, if the absorption band associated with the reversal in α_I, i.e., the band at 887 cm^{-1}, arises from an electronic transmission rather than a vibrational mode as Corbett *et al.* suggest,[10] it should have associated with it a level at 0.11 eV from an unspecified band edge. Also, if the reverse annealing stages in α_I and in σ are the same, from Fermi level considerations, the associated defect level should be located at $E_V+0.11$ eV. If, however, the 887 cm^{-1} band is indeed due to a vibrational mode of two oxygen atoms in a vacancy, as proposed by Corbett *et al.*, the optical absorption band associated with the electronic level described here has not as yet been reported. Finally, the defect responsible for the reversal in τ appears to give rise to a hole recombination level in the lower half of the energy gap. This is not inconsistent with a level located at $E_V+0.11$ eV.

The lack of agreement in the observed values of T_{ra} in the three experiments can be ascribed to differences in radiation defect concentrations involved in each study.[20] This results from the fact that radiation-induced changes in τ saturate for values of ϕ below those required to observe changes in σ which in turn saturate for values of ϕ below those required in α_I studies. There are no values of ϕ suitable for any two of the measurements and, therefore, comparisons of annealing of two or more parameters cannot be readily effected.

Note added in proof: An analysis of the isochronal annealing behavior of the conductivity of n-type silicon irradiated by 1 MeV electrons has been carried out. The results indicate that the defect level associated with the reverse annealing stage is located at $E_c - 0.41$eV.[21]

Acknowledgement

The authors are deeply grateful to Professor J. J. Loferski of Brown University for many helpful discussions and suggestions.

REFERENCES

1. G. D. Watkins, J. W. Corbett, and R. M. Walker, *J. Appl. Phys.*, **30**, 1198 (1959).
2. G. D. Watkins, J. W. Corbett, *Discussions Faraday Soc.*, **31**, 86 (1961).
3. G. Bemski, *J. Appl. Phys.*, **30**, 1195 (1959).
4. G. D. Watkins and J. W. Corbett, *Phys. Rev.*, **121**, 1001 (1961).
5. Y. Inuishi and K. Matsuura, *J. Phys. Soc. Japan*, **18**, Suppl. III, 240 (1963).
6. T. Tanaka and Y. Inuishi, *J. Phys. Soc. Japan*, **19**, 167 (1964).
7. E. Sonder and L. C. Templeton, *J. Appl. Phys.*, **34**, 3295 (1963).
8. H. Saito and M. Hirata, *Japan. J. Appl. Phys.*, **2**, 678 (1963).
9. R. R. Hasiguti and S. Ishino, in *Proceedings of the Symposium on Radiation Damage in Semiconductors, Paris, 1964*, p. 268, Dunod, Paris, 1965.
10. J. W. Corbett, G. D. Watkins, and R. S. McDonald, *Phys. Rev.*, **135**, A 1381 (1964).
11. T. Nakano and Y. Inuishi, *J. Phys. Soc. Japan*, **19**, 851 (1964).
12. Annealing data up to 710°K for n-type silicon irradiated at 300°K by 2-MeV electrons have been presented by H. J. Stein, Sandia Report SC-R-65-938 (1965).
13. Supplied by Mallinckrodt Chemical Works.
14. Supplied by Dow Corning Corporation.
15. L. Valdes, *Proc. IRE*, **42**, 420 (1954).
16. G. K. Wertheim, *Phys. Rev.*, **105**, 1730 (1957).
17. H. Saito, M. Hirata, and T. Horiuchi, *J. Phys. Soc. Japan*, **18**, Suppl. III, 246 (1963).
18. See for example G. D. Watkins and J. W. Corbett, *Phys. Rev.*, **138**, A 543 (1965).
19. Similar annealing data for α_I in 40-MeV electron-irradiated silicon have been presented by J. C. Corelli, G. Oehler, J. F. Becker, and K. J. Eisentraut, *J. App. Phys.*, **36**, 1787 (1965).
20. J. H. Cahn, *J. Appl. Phys.*, **30**, 1310 (1959); G. J. Dienes and G. H. Vineyard, in *Radiation Effects in Solids*, p. 46, Interscience Publishers, Inc., New York, 1957; V. S. Vavilov, in *Effects of Radiation on Semiconductors*, p. 149, 155, Consultants Bureau, New York, 1965.
21. B. J. Faraday, R. V. Tauke and J. J. Loferski, *Bull. Am. Phys. Soc.*, **12**, 1070 (1967).

Low Temperature Electron Irradiation of n-Type Germanium

S. Ishino* and E. W. J. Mitchell

J. J. Thomson Physical Laboratory, University of Reading, Whiteknights Park, Reading, Berkshire, England

ABSTRACT

Two kinds of antimony doped n-type germanium (1–2 Ω·cm and 25 Ω·cm at room temperature) were irradiated with 2 MeV electrons mostly at 7–10°K.

Annealing around 60°K is the only annealing stage observed between 7° and 80°K. The temperature at which this annealing occurs depends on the doping level. In the higher resistivity samples, the central temperatures are between 55° and 60°K, whereas in the lower resistivity samples, they are around 65°K.

The changes in conductivity per unit amount of irradiation depend on the irradiation temperature. The rate decreases rapidly as the irradiation temperature increases and becomes nearly constant above 50°K. The overall carrier removal rate for irradiations between 7° and 17°K is found to be 2.6 ± 0.8 cm^{-1}.

The effect of infrared illumination on annealing has been investigated. No effect of illumination is observed below about 25°K. However, " optical bleaching " effects occur above 25°K. The process takes place most efficiently above 30°K. After the optical bleaching is completed no 65°K annealing stage is found.

These experimental evidences seem to give further support to MacKay-Klontz mechanism.

I. INTRODUCTION

In silicon the relatively complex behavior of " simple " lattice defects has been clarified in recent years by Watkins and co-workers[1] using the technique of electron spin resonance. On the other hand in germanium

* Now at Department of Nuclear Engineering, University of Tokyo, Tokyo, Japan.

the behavior of defect is apparently different but owing to the absence of spectroscopic data unambiguous microscopic models of defects in germanium have not been obtained.

As an alternative approach MacKay and Klontz[2–4] have studied the effects of low temperature irradiation and the low temperature annealing stages. In particular these authors have been able to account for many properties of the low temperature (10°K) irradiation of germanium by assuming that the annealing stage at 60°K arises from the recombination of vacancy-interstitial close pairs. They assume that such pairs are metastable at low temperatures if they have trapped electrons, thereby explaining the most important difference between the effects of low temperature fast electron irradiation on n- and p-type samples.[3]

As a further test of this model the experiments reported in this paper have been carried out to study the effect of infrared illumination on the occurrence of the 65°K annealing stage. Results are also presented of warm-up measurements and the variation of carrier removal rate with temperature.

II. EXPERIMENTAL PROCEDURE

1. *Samples*

Two kinds of n-type germanium were used. The Sb-M samples have resistivities of about 1–2 Ω·cm at room temperature and Hall coefficients in the exhaustion range of about $2\text{–}5 \times 10^3$ cm³/coulomb. The other samples (Sb-H) have room temperature resistivities of about 25 Ω·cm and Hall coefficients in the exhaustion range of 1.5×10^5 cm³/coulomb. In most of the experiments, the samples were cut with four or six arms, with the current orientation parallel to ⟨110⟩. The electron beam was incident on (111) plane of the samples. The samples were etched with CP-4, and gold lead wires were welded directly to the sample.

2. *Cryostat*

The cryostat used in the experiment was similar to that reported by MacKay and Klontz.[2] The lower part of the cryostat is schematically shown in Fig. 1. In order to introduce infrared light, the sample chamber was equipped with a thin polyethylene window sealed with indium. This was successfully used down to about 7°K. The outer case had a rotatable collar on which two windows were mounted: one, a thin aluminium foil for the electron beam; and the other, a thick germanium filter (~1 mm) for the infrared illumination.

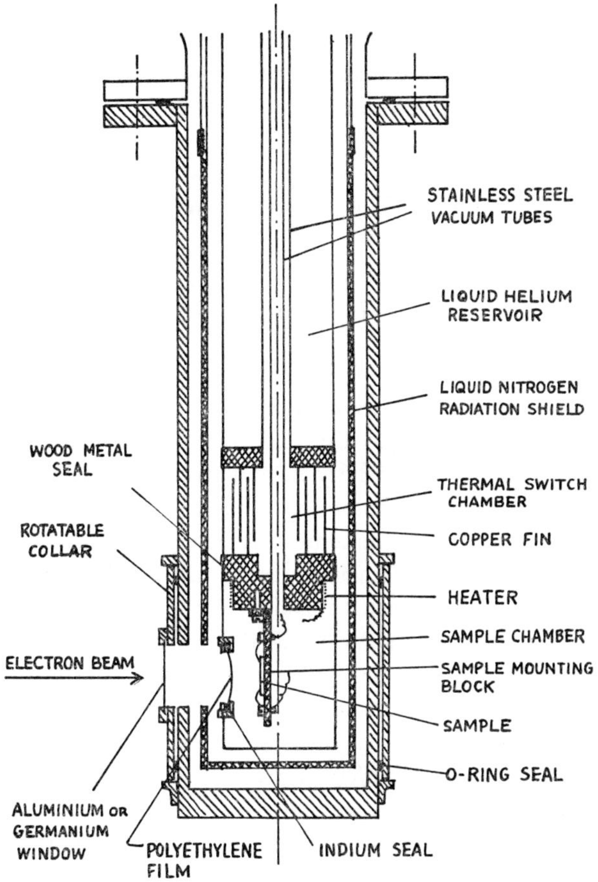

Fig. 1. Details of cryostat for low temperature irradiations and measurements.

The temperature of the sample was controlled by using a heater and by adjusting the helium pressure in a stainless steel can. The temperature was measured using a p-type germanium crystal thermometer, the resistivity of which was calibrated against an Allan-Bradley Co. carbon resistor, which in turn was calibrated at 4.2°, 77° and 319°K. It is assumed that the resistivity of the p-type germanium did not change with irradiation.[3] This assumption is well established and reproducible temperature measurements can be made by this method.

3. *Irradiation*

Irradiations were performed with 2 MeV electrons from the University of Reading accelerator. Since there was no space to put a Faraday cup in the sample chamber, the flux was calibrated using a Faraday cup which simulated the geometry of the cell immediately before the irradiation run. The time of irradiation was then used to obtain the dose. To check the stability of the beam, a current integrator was used to measure electrons collected by the sample. The sensitivity of the integrator was 6×10^{10} electrons per digit.

III. RESULTS

1. *Temperature dependence of Hall coefficient and resistivity before and after irradiation and after annealing*

Since the samples used in the experiments were not degenerate, the Hall coefficient and the resistivity have a strong temperature dependence, especially below 25°K. Therefore, a fixed reference temperature could not be used for the measurements. Instead, the variation of resistivity and Hall coefficient was measured as a function of temperature in the temperature range below each annealing temperature. If there are no annealing stages, the resistivity and the Hall coefficient curves should coincide with those obtained before the warm-up. In this way, we found that the annealing around 60°K is the only annealing stage between 7° and 80°K. In these experiments, typical electron doses were of the order of 10^{13} e/cm^2 for Sb-H, and of 10^{14} e/cm^2 for Sb-M samples.

In many of the samples, it was found that the resistivity after the 60°K annealing almost coincides with that before annealing (after irradiation) in the temperature range below 10°K. This might be caused by introduction of an acceptor level near the chemical donor level during the 60°K annealing or by a change in the degree of compensation, but the measurements below 10°K were not sufficiently accurate to comment in more detail on this point.

2. *Annealing in dark*

As mentioned in the previous section, no annealing was observed until the sample was warmed up to about 60°K.

Two Sb-M and two Sb-H samples were irradiated at temperatures near to 10°K and the resistivity and the Hall coefficient were measured as a function of temperature and time during warming up to liquid nitrogen temperature. An interesting difference in annealing temperature in the two kinds of samples was found.

Figure 2 shows the temperature dependence of resistivity in one of the Sb-M samples. The curve *a* was obtained before irradiation. After

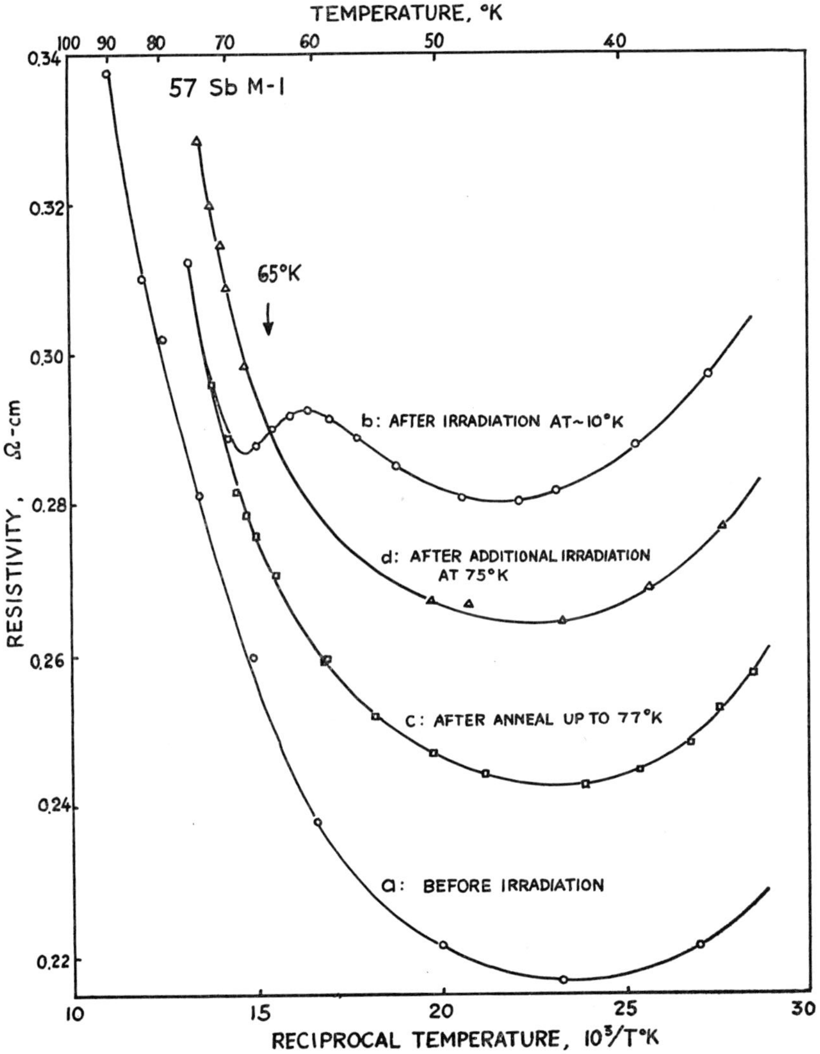

Fig. 2. Temperature dependence of resistivity of a Sb-M sample. Curve *a*, before irradiation; curve *b*, during warm-up after irradiation of 6.3×10^{13} e/cm² at 10°K; curve *c*, after warming up to 77°K; curve *d*, after second irradiation of 6.9×10^{13} e/cm² at 75°K.

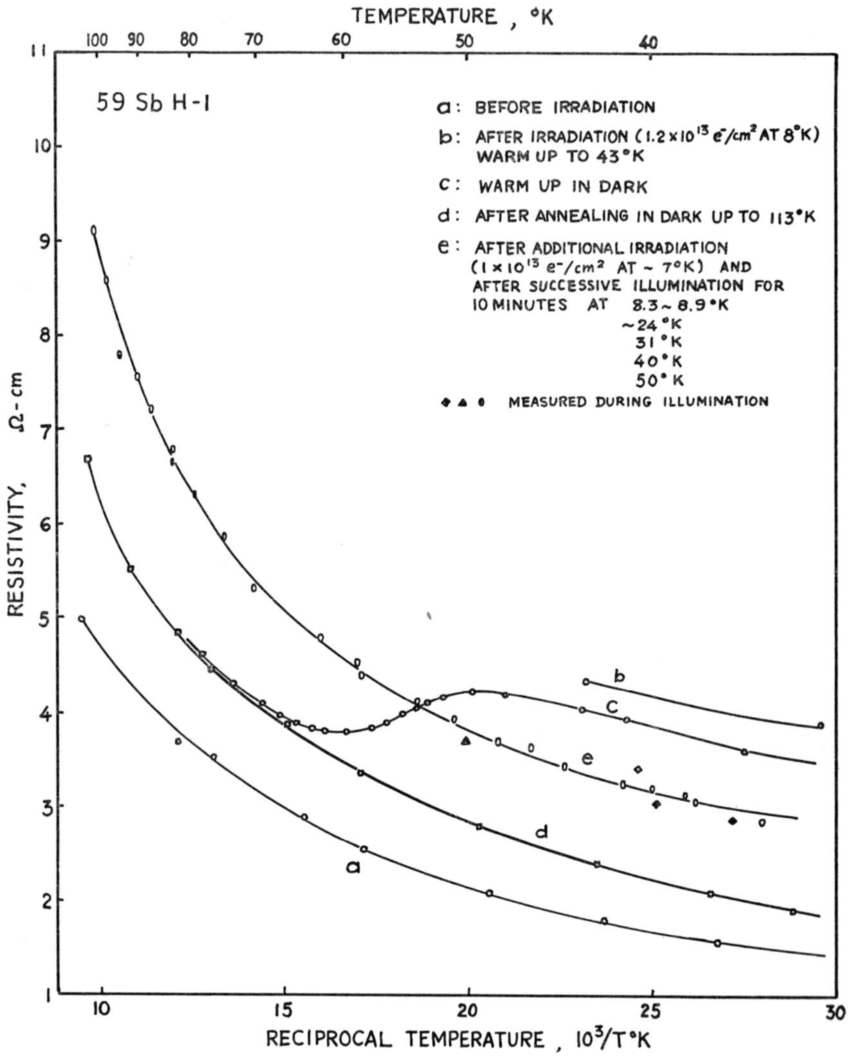

Fig. 3. Temperature dependence of resistivity of a Sb-H sample. Curve *a*, before irradiation; curve *b*, after irradiation of 1.2×10^{13} e/cm² at 8°K and warm-up to 43°K; curve *c*, during warm-up in dark; curve *d*, after annealing in dark up to 113°K; curve *e*, after additional irradiation of 1×10^{13}e/cm² at 7°K and during and after successive illuminations at various temperatures; curve *f*, after warm-up to 118°K.

irradiation with 6.3×10^{13} e/cm^2 at about 10°K, the sample was allowed to warm up slowly to liquid nitrogen temperature. Curve *b*, obtained during this warming up, shows that the center of the annealing stage is 65°K. After keeping the sample at liquid nitrogen temperature for about 8 hours, the temperature dependence of resistivity was as shown in curve *c*. From this curve it is obvious that the step on the curve *b* around 65°K is caused by removal of the defect rather than by a change in charge state of a defect. Curve *d* was obtained after an additional irradiation of a similar amount (6.9×10^{13} e/cm^2) at 70°K.

Figure 3 shows a similar set of curves obtained with one of Sb-H samples. Hall coefficient data obtained in the same sample are given in Fig. 4.

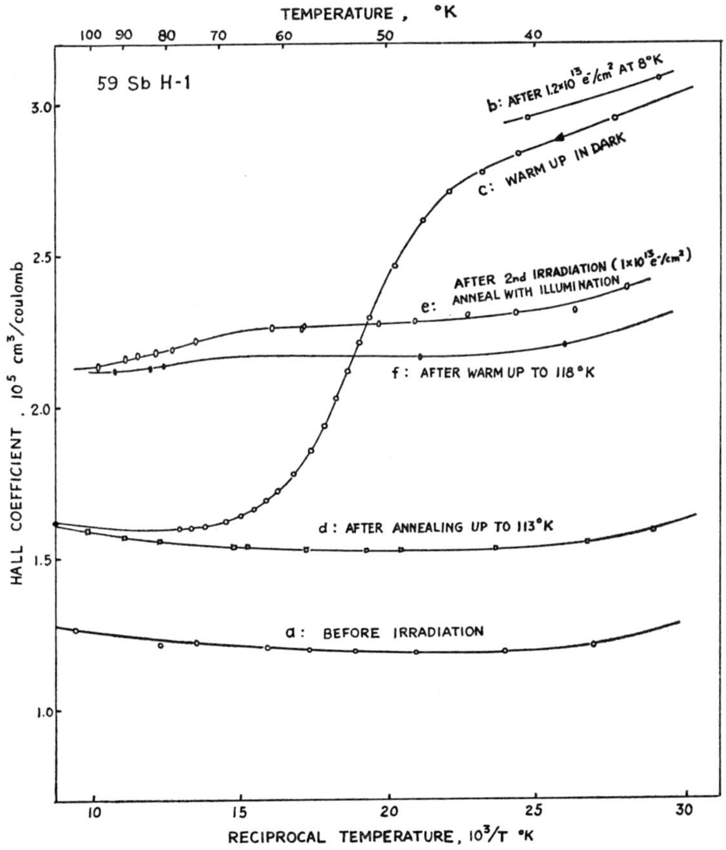

Fig. 4. Temperature dependence of Hall coefficient of the same sample as that of Fig. 3.

It is clearly seen that the annealing takes place at much lower temperature than in the Sb-M sample. The center of the annealing stage in the case of Figs. 3 and 4 is 55°K. In this sort of experiment, the rate of warming up is obviously important. In order to confirm that the difference in annealing temperature described above is due to the difference in the doping levels, two samples, 58Sb-M-6 and 58Sb-H-3, were warmed up at almost the same rates. The central temperatures of the annealing stage were 66.7° and 60.6°K in 58Sb-M-6 and 58Sb-H-3, respectively. In this case the effect of heating rate on the temperature of the annealing stage is eliminated and the observed change must arise from the difference of doping level.

3. *The effect of infrared illumination on annealing*

According to MacKay and Klontz,[3] the defects responsible for the 60°K annealing stage become metastable when electrons are trapped, being otherwise unstable. The annealing process is governed by the thermal release of these electrons. If this model is correct, it is of interest to study the annealing when the trapped electron is excited by illuminating the crystal with infrared light. The effect of infrared illumination was examined on several samples. The illumination was carried out by using an ordinary 30 watt tungsten lamp at about 4 cm from the specimen. The light was incident on the samele through a thick germanium filter and a thin polyethylene film. During illumination, the sample temperature was controlled by adjusting the vacuum in the thermal switch chamber—a stainless steel can connecting the helium reservoir with the sample chamber—and by applying a heater current so that the sample temperature did not change significantly.

Figure 5 shows a result obtained using one of Sb-H samples. Curve *a* represents the resistivity before irradiation.* The sample was irradiated with 1×10^{13}e/cm^2 at 7°K and was illuminated at 8.3°K for 10 minutes. No annealing was observed. Then the sample was allowed to warm up to about 24°K, Hall and resistivity measurements being made (curve *b*). The second illumination was carried out at 24.0–26.5°K for ten minutes. The point indicated by an inverse full triangle shows the resistivity measured during illumination. After illumination the sample was cooled down to 9.4°K and then warmed up to 32°K (curve *c*). The

* This curve refers to the resistivity before the second irradiation of the sample which had previously been irradiated to 1.2×10^{13}e/cm^2 followed by annealing to 113°K. The curve is the same as curve *d* in Fig. 3.

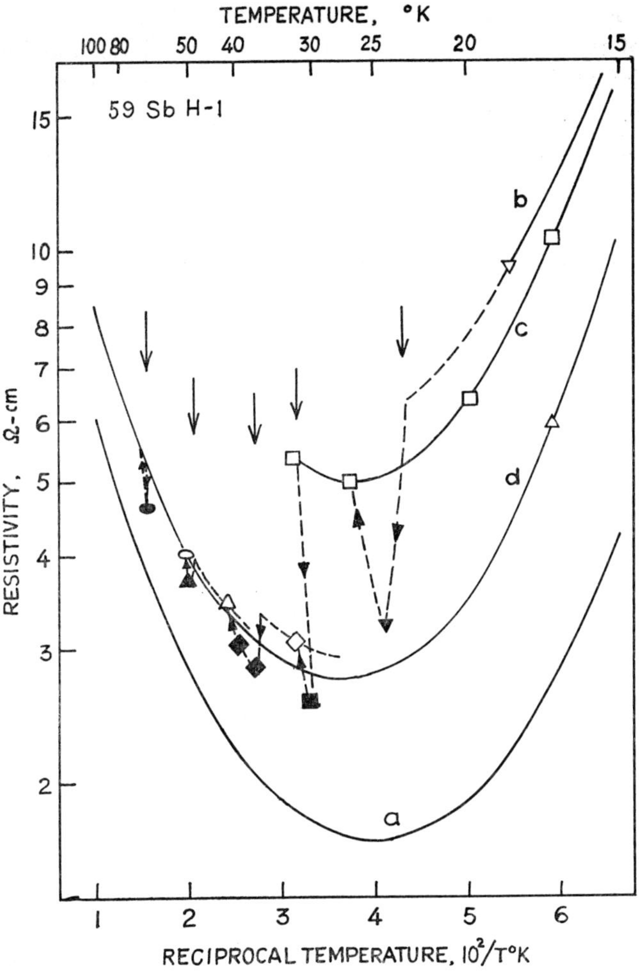

Fig. 5. Temperature dependence of resistivity of a Sb-H sample, showing the effect of illumination on annealing. Curve *a*, before second irradiation (same as curve *d* in Fig. 3); curve *b*, after irradiation of 1×10^{13} e/cm² at 7°K and illuminated for 10 minutes at 8.3°–8.9°K; curve *c*, after illumination for 10 minutes at 24.0°–26.5°K; curve *d*, after illuminations for 10 minutes at 31°–32°, 40°, 50° and 72°K. Points shown by black symbols are obtained during illumination.

third illumination was performed at 31–32°K for ten minutes. A substantial change was induced by the third illumination. Similarly, ten minute illuminations were further carried out at 40°, 50° and 72°K. After the annealings with illumination were completed, the temperature dependence of resistivity was as shown in curve *d*. It should be noticed that with infrared illumination the annealing takes place at much lower temperature than it does in the dark. One can completely eliminate the 60°K annealing stage by illumination at lower temperatures.

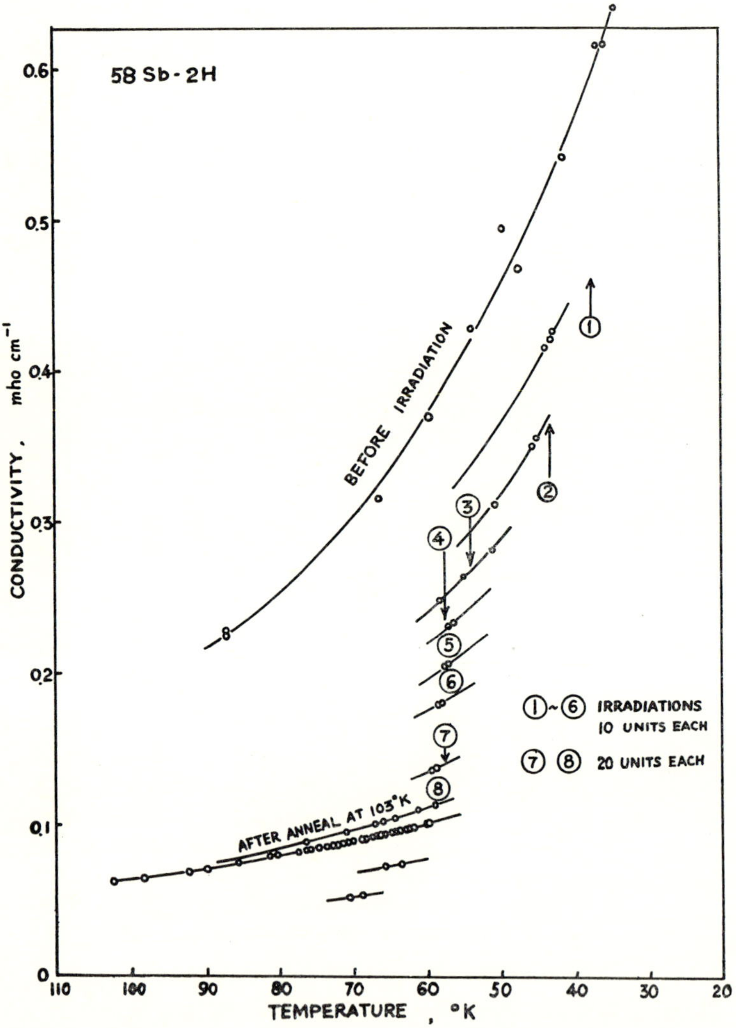

Fig. 6. Temperature dependence of conductivity before irradiation and after successive irradiations at various temperatures.

4. *Defect introduction*

In several samples, Hall coefficient data before irradiation, after irradiation at low temperature and after annealing up to liquid nitrogen temperature are all available. If the defect responsible for the 60°K annealing is called a " primary defect " and if one electron is removed per one primary defect, a cross section for the production of the primary defect as well as a total cross section for the carrier removal can be calculated.

In Table I, these cross sections are tabulated. They were obtained from the Hall data at 40°K. Although the dose measurements in these runs were not made to better than 10%, the calculated crosssection values are in reasonable agreement with each other. The macroscopic cross section is defined by $N\sigma_d$ where σ_d is the microscopic crosssection and N the number of atoms per unit volume. From simple displacement theory the micro- and macroscopic cross sections for producing primary displacements are 61×10^{-24} cm^2, 2.7 cm^{-1} and 45×10^{-24} cm^2, 2.0 cm^{-1} for the displacement energies of 25 eV and 31 eV, respectively at an electron energy of 2 MeV.

The temperature dependence of the carrier removal rate is also of interest. Figure 6 shows the reduction of conductivity after successive irradiations at various temperatures. The data were extrapolated to 50°K

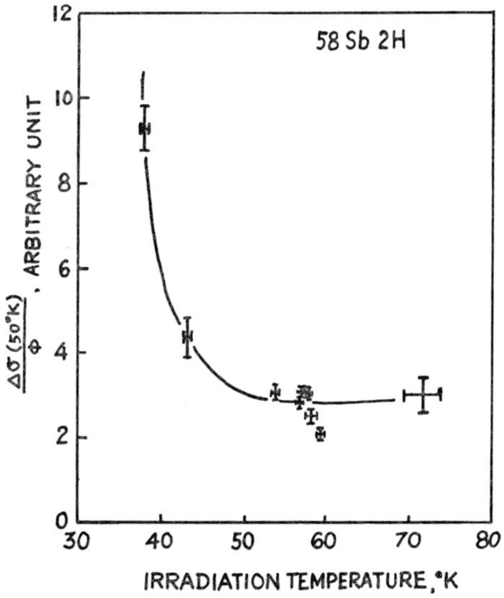

Fig. 7. Temperature dependence of rate of change in conductivity by irradiation with 2 MeV electrons.

TABLE I. Calculation of macroscopic

Sample	Irradiation temperature (°K)	Carrier concentration (cm^{-3})		
		Before irradiation	After irradiation	After annealing
58Sb-M-3	11	2.13×10^{15}	1.64×10^{15}	1.84×10^{15}
58Sb-M-6	11.5	1.09×10^{15}	7.69×10^{14}	9.0×10^{14}
58Sb-H-3	17	5.14×10^{13}	3.30×10^{13}	4.48×10^{13}
59Sb-H-1	8	5.25×10^{13}	2.11×10^{13}	4.07×10^{13}

and the rate of change in conductivity per unit amount of irradiation was calculated for the various irradiation temperatures. These values are plotted in Fig. 7. Unfortunately, the Hall coefficient was not measured in this sample, so that one could not obtain the carrier removal rate from this measurement. However, it is clearly shown that the rate of change in conductivity decreases rapidly around 40°K with increasing irradiation temperature and becomes nearly constant above about 50°K.

IV. DISCUSSION

As mentioned earlier, MacKay and Klontz[3,4] have proposed the model that the 65°K annealing stage is of a primary defect and that the defect is stabilized by trapping an electron. They also suggested that the annealing mechanism is limited by the thermal detrapping of the trapped electron.[3] All of the present experimental results seem to support this proposed mechanism.

A) First, we shall consider the results of doping level dependence of annealing temperature. According to MacKay and Klontz,[3] the depth of the trapping level is estimated to be 0.1 eV from the bottom of the conduction band based on the assumption that the 65°K annealing is partially limited by the rate of excitation of the trapped electron from the defect state. It is therefore expected that the higher resistivity samples will have lower annealing temperatures than the lower resistivity samples, because the Fermi level lies deeper in the higher resistivity samples, for the same temperature. This is exactly the case with our experiment. The Fermi level of Sb-M samples at 65°K, i.e. at the central annealing temperature, is at 0.04 eV from the conduction band edge, whereas in Sb-H samples it is about 0.05 eV from the conduction band edge at about 55°–

cross sections from Hall data at 40°K.

Δn Total carrier removal (cm^{-3})	Δn_2 Permanent defect (cm^{-3})	Δn_1 Primary defect (cm^{-3})	ϕ Dose $\left(\frac{10^{13}}{e/cm^2}\right)$	Macroscopic cross section (cm^{-1})	
				Total $\Delta n/\phi$	Primary defect $\Delta n_1/\phi$
4.9×10^{14}	2.9×10^{14}	2.0×10^{14}	15	3.4	1.3
3.2×10^{14}	1.9×10^{14}	1.3×10^{14}	14	2.3	0.93
1.84×10^{13}	7.6×10^{12}	1.08×10^{13}	1.0	1.84	1.1
3.14×10^{13}	1.18×10^{13}	1.96×10^{13}	1.2	2.64	1.63
			Average	2.6 ± 0.8	1.3 ± 0.4

60°K. If it is considered that the occupation of a given energy level is appreciably affected if the Fermi level comes within about $2\,kT$ of the level, the acceptor level in question would be situated about 0.06 eV from the conduction band. The possibility will be discussed later that the rate of annealing of the defect may be limited by the lifetime of the defect when no electrons are trapped.

B) Secondly, we shall discuss the effect of infrared illumination. The fact that the " optical bleaching " is observed seems to provide the most direct evidence of the validity of the MacKay-Klontz mechanism, because, if the annealing is due to the excitation of the trapped electron by thermal means, the same phenomenon should occur when the electron is excited by other means, i.e. by optical means.

Not only in such a qualitative manner, but also the experiment can lead to quantitative information about the barrier height when no electrons are trapped (E_R) which can be estimated by the equation

$$E_R = -kT \ln \frac{n_j}{\nu_0 t}$$

where n_j, the necessary number of jumps for the annealing to occur, which is assumed to be 1, ν_0 the vibrational frequency of the order of 10^{13}/sec, t, the annealing time, and k and T have their usual meanings. In the present experiment, annealing took place effectively at 30°K with illumination. This implies that E_R should be of the order of 0.08 eV, based on the assumption that about half of the total anneal proceeds during illumination for 10^3 sec at 25°K. In this calculation we also assumed that the infrared illumination is sufficiently intense so that the rate of annealing is determined by overcoming the barrier, E_R, and not by an excitation rate of the trapped electron.

Although the experimental results give strong evidence for the mechanism given above, alternative mechanisms are possible, for example, some modifications of the " primary defect " by infrared light leading to the production of a 35°K annealing stage. However, since only one fairly sharp annealing step (at 30°K) by illumination is found, the above possibility does not seem to be likely. To reach a final conclusion, it seems necessary to study the spectral response of annealing behaviour.

C) The temperature dependence of rate of change in conductivity is qualitatively consistent with the MacKay-Klontz mechanism, since during irradiation excitation of electrons is frequently taking place so that the annealing occurs together with the defect production above about 30°K. No such anomalous temperature dependence of production rate of defects as seen in low-energy electron irradiations was observed in the relatively high energy irradiation of 2 MeV in the present experiment.

The mean total displacement cross section observed in the present experiment at about 10°K is about 58 barn at 2 MeV which roughly corresponds to the displacement energy of 25 eV.

Although the present experiments give some insight into the mechanism of annealing at low temperature, the nature of the defect responsible for the 60°K annealing stage remains to be clarified.

Acknowledgements

The authors would like to thank Dr. C. D. Clark for his interest and many helpful discussions. We also wish to thank Mr. B. Hayter and Mr. R. Winters for their assistance with the actual experiments. One of the authors (S.I.) wishes to thank the Ramsay Memorial Fellowships Trust for a fellowship which enabled him to spend two years at the J. J. Thomson Physical Laboratory, University of Reading, England.

REFERENCES

1. G. D. Watkins, in *Proc. Symp. on Radiatin Damage in Semiconductors, Paris-Royaumont, 1964*, p. 97 Dunod, Paris, 1965.
2. J. W. MacKay and E. E. Klontz, *J. Appl. Phys.*, **30**, 1269 (1959).
3. J. W. MacKay and E. E. Klontz, in *Radiation Damage in Solids III* (Proc. Symp. on Radiation Damage in Solids and Reactor Materials, Venice, 1962), IAEA, p. 27, Vienna, 1963.
4. E. E. Klontz and J. W. MacKay, *J. Phys. Soc. Japan*, **18**, Supp. III, 216 (1963).

Annealing of p-Type Germanium Bombarded with 1.5 MeV Electrons

HARUO SAITO, NOBORU FUKUOKA, HIRONORI HATTORI

Department of Physics, College of General Education, Osaka University
Toyonaka, Osaka, Japan

and

JAMES H. CRAWFORD, JR.

*Solid State Division, Oak Ridge National Laboratory**
Oak Ridge, Tennessee, U.S.A.

ABSTRACT

Annealing of changes in Hall coefficient and resistivity induced by 1.5-MeV electron irradiation at 77°K in gallium-doped germanium was studied by isochronal and isothermal annealing techniques. Initial carrier concentrations of the samples ranged from 10^{13} to 5×10^{15} cm^{-3}, and the measurements were made at 77°K. The formation of traps was evident after irradiation. Five annealing stages were observed in the range 77°–400°K. In Stage III, which occurs in the range 210° to 270°K, the traps annealed out and the donor density increased. The traps which annealed in this stage were found to be associated with a level 0.16 eV above the valence band. The activation energy for this stage was found to be 0.31 eV with traps empty and 0.23 eV with traps full. The donor level 0.08 eV above the valence band increased in density as annealing proceeded. Characteristics of the other stages are also discussed, correlated with previous investigations.

I. INTRODUCTION

Although radiation effects in germanium have been extensively investigated, most of this effort has been devoted to n-type material. Not only has much attention been given to the position of energy levels in n-type germanium but the influence of the type of chemical donor upon both

* Oak Ridge National Laboratory is operated by Union Carbide Corporation for the U.S Atomic Energy Commission.

the nature of the effect and the course of thermal annealing has been examined. These studies have been reviewed in the recent monograph by Corbett.[1] In contrast, p-type germanium has been accorded much less attention. One reason for this is the fact that the yield of defects for low temperature ($\leq 10°$K) is much smaller than for n-type.[2] The most extensive examination of the annealing behavior was carried out by Brown *et al.*[3] on material irradiated with 1 MeV electrons at 77°K. They found that the annealing processes were markedly different from that in n-type germanium. More recently, Klontz and coworkers[4] have exposed p-type germanium to 1.1 MeV electrons at 0°C and by rapid quenching after the cessation of radiation have been able to study complex defect structures which transform reversibly with temperature cycling in a way that strongly modifies the energy level structure. Whitehouse[5] has chosen to investigate the annealing of defect structures at a much lower temperature, namely, those produced by 4.5 MeV electron bombardment at 4.2°K, and he finds that the thermal relaxation of the defect structure can be strongly altered with infrared illumination after irradiation. In each of these studies there is not sufficient evidence to permit identification of the type of complexes which are responsible for the change in electronic behavior and which anneal during thermal treatment.

This investigation of radiation damage in p-type germanium was undertaken with the aim of covering a wider range of annealing conditions than was covered by Brown and coworkers.[3] In particular, we wished to determine the influence of acceptor density on the types of defects created during irradiation and subsequent annealing. Consequently, the initial gallium density ranged from 10^{13} to 5×10^{15} cm^{-3}, all irradiations were performed at 77°K, and the temperature range for the annealing extended from 77° to 400°K. Hall coefficient and resistivity were the properties monitored after irradiation and during the annealing process, and 77°K was taken as the reference temperature for both isochronal and isothermal anneals. Although a number of annealing stages have been identified and correlated in part with previously reported investigations, much work remains to be done in this general study, and this report should be considered preliminary in nature.

II. EXPERIMENTAL PROCEDURE

The crystals used in this study were gallium doped and with impurity concentrations in the range 10^{13}–10^{15} atoms/cm^3. The bridge-shaped specimens were made by slicing discs perpendicular to the ⟨111⟩ axis of the crystal ingot and cutting specimens from these by means of ultrasonic die. After the surface of the specimen was etched with CP-4, ohmic contacts were made by using indium solder containing 5% lead. The specimen was placed in a slot at one end of a sample holder. The holder is a bakelite stick of dimensions $2 \times 12 \times 370$ mm. Both sides of the slot are covered with phosphor-bronze strips with a small hole allowing the electron beam to enter. Electrical insulation between the sample and the phosphor-bronze was provided by a layer of cellulose acetate tape placed on the phosphor-bronze strip. This arrangement protects the sample from breakage without undue thermal insulation.

Irradiation with 1.5 MeV electrons from a Van de Graaff accelerator was carried out at 77°K. During irradiation the sample stick is floated in liquid nitrogen with the surface of the sample at the same level as that of the liquid. Total dose of irradiation ranged from 7.5×10^{14} to 2.6×10^{16} e/cm². An integrating digital voltmeter was used to measure the Hall coefficient and resistance of the sample. The magnetic field used in the Hall measurement was 500 gauss. Isochronal annealing was done in a small furnace. The temperature was controlled with a copper-constantan thermocouple and an electronic controller of supply voltage to the heater. All measurements were made at 77°K.

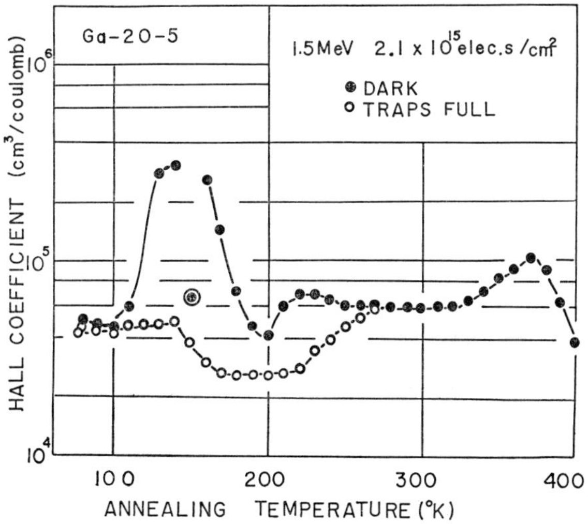

Fig. 1. 20-min pulse annealing of Hall coefficient after irradiation at 77°K.

III. EXPERIMENTAL RESULTS

1. *Isochronal Annealing*

Figures 1 and 2 show the results of the 20-minute isochronal annealing of Hall coefficient and resistivity after irradiation of 2.1×10^{15} e/cm^2. The initial carrier concentration of the sample was 3.1×10^{14} cm^{-3} and after the exposure it was 1.8×10^{14} cm^{-3}, indicating that the dominant effect of the 77°K irradiation is introduction of net donors. The annealing was performed in the dark. Hall coefficient and resistivity were measured at liquid nitrogen temperature. The change in mobility during the annealing is shown in Fig. 3. From these figures it is evident that there are five annealing stages in this temperature range. Stage I is in the range 110°–150°K; Stage II is in the 150°–170°K range; Stage III is in the 210°–270°K range; Stage IV is in the 320°–370°K range; and Stage V is above the 380°K range.

Evidently the first stage is associated with emptying of electron traps[6] since illumination after recooling to 77°K reestablishes the original value until the temperature exceeds 140°K. Also, the sharp mobility decrease in this stage indicates that the decrease in hole concentration is accompanied by an increase in scattering center density, which is expected for the release of minority carriers. This sample shows conversion from p- to n-type at the end of this stage, suggesting that the stage is not to be

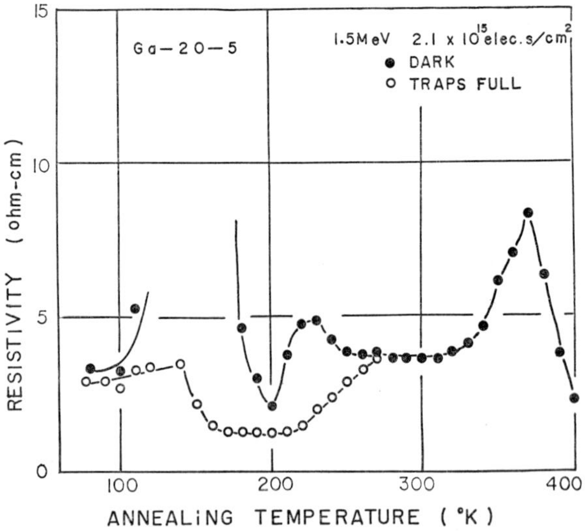

Fig. 2. 20-min pulse annealing of resistivity after irradiation at 77°K.

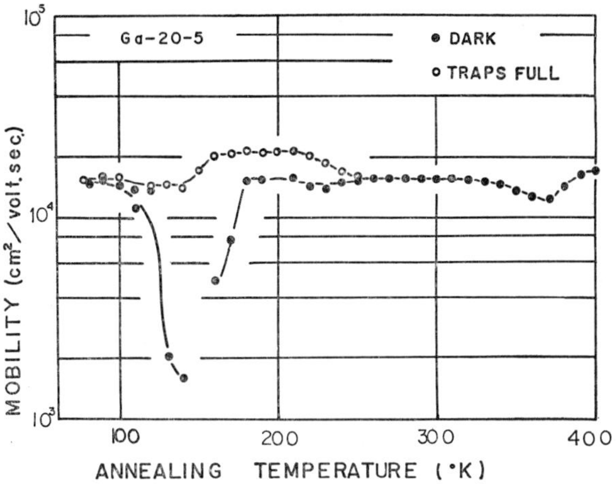

Fig. 3. 20-min pulse annealing of mobility after irradiation at 77°K.

interpreted merely as release of electrons from the traps, but formation of shallow donor levels by some rearrangement of defects apparently occurs as well. The type conversion was observed to be more pronounced and to cover a wider temperature range in heavily irradiated specimens.

Stage II was observed in the range 150°–170°K. In this stage the total donor density decreased and the traps began to disappear. Mobility also recovered.

In Stage III which appeared in the range 210°–270°K, traps completely annealed and the donor density increased once again. Mobility with traps empty showed no change in this stage, whereas mobility with traps full decreased until it reached the same value as that with traps empty. This suggests that the traps are transformed into stable donors which in their ionized state scatter carriers as effectively as the former empty traps.

Stage III decreases in importance with decreasing gallium concentration, which indicates that this stage may have some connection with a gallium-defect complex.

Similar isochronal annealings were made on samples containing 3.1×10^{13} and 3.8×10^{15} atoms/cm³ of gallium. Qualitatively the annealing behavior was the same as that shown in Figs. 1–3. Stages I and II were observed in every sample.

Figure 4 shows the result of isochronal annealing made under illumination by visible light. The specimen was cut from the same crystal as was the specimen described in Figs. 1–3. Stage I cannot be seen here, but a process considered to correspond to the former Stage II seems to

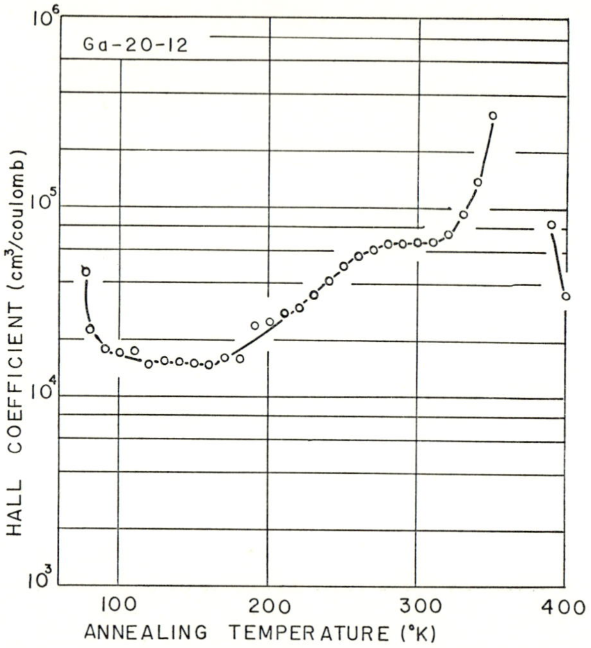

Fig. 4. Isochronal annealing of Hall coefficient. The annealing was performed under illumination by visible light, and the measurements were made in the dark.

be occurring in the range 77°–90°K. Stage III, where the traps apparently transform into stable donors, also seems to occur at a lower temperature than when annealing was performed in the dark.

Temperature dependence of carrier concentration was determined after isochronal annealing to 280°K (here the Stage III is completed), 330°K, and 400°K as is shown in Fig. 5. From these curves it can be concluded that a donor level at $E_V + 0.08$ eV exists after annealing at 280°K, which cannot be seen after completion of annealing up to 400°K. On comparing curves after 280° and 330°K, it seems that the donor at 0.08 eV is first transformed into a deeper lying donor state which also is removed by annealing at 400°K (Fig. 5).

In Stage IV (320°–370°K) an increase in donor concentration or a decrease in acceptor density was observed. In this stage too, type conversion was observed in heavily irradiated specimens.

In State V (above 380°K) annealing of donors seems to be occurring. Stages IV and V decreased in importance with decreasing irradiation dose.

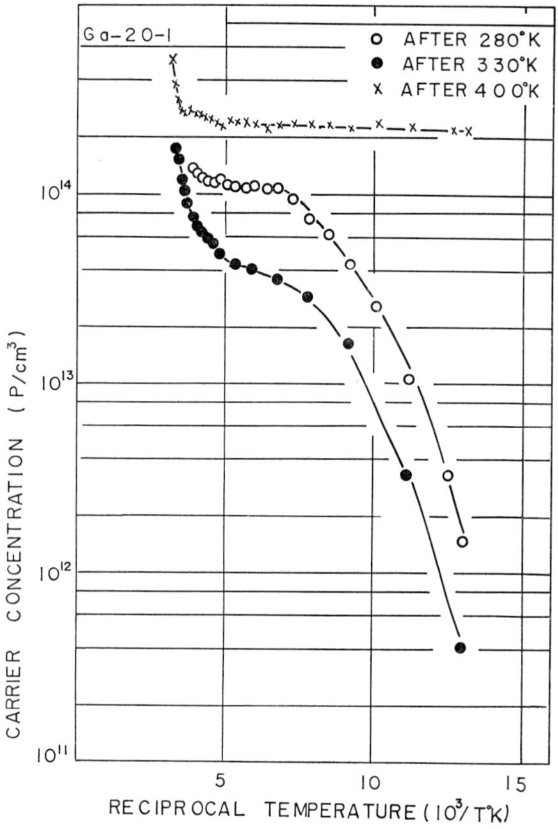

Fig. 5. Temperature dependence of carrier concentration after successive annealings.

There is no illumination effect on annealing in these stages since traps do not exist anymore.

2. *Isothermal Annealing*

In order to investigate Stage III in greater detail, three isothermal annealings were made at 230°, 240°, and 260°K. These annealing curves obtained on three samples from the same ingot, which have received approximately the same irradiation, are shown in Fig. 6. The reason for the different amounts of change in this stage for the three specimens is not known. The amount of recovery was followed through Hall coefficient measurements at 77°K after illumination by visible light to fill the traps.

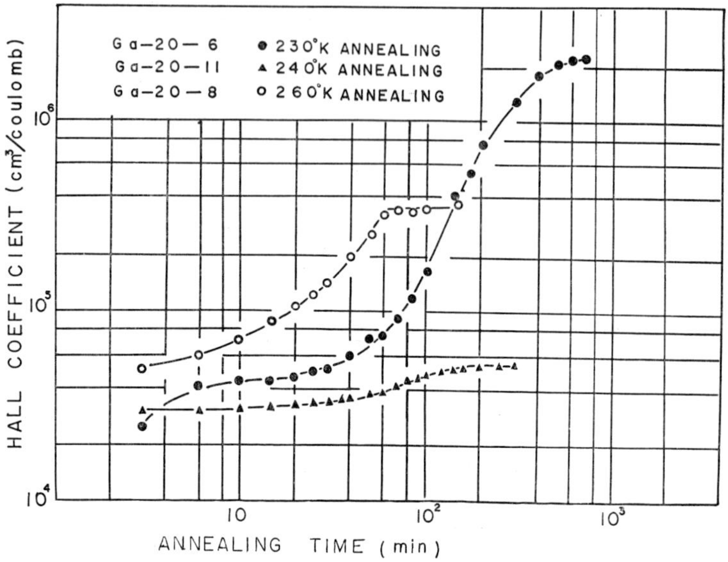

Fig. 6. Isothermal annealings of Hall coefficient at different temperatures.

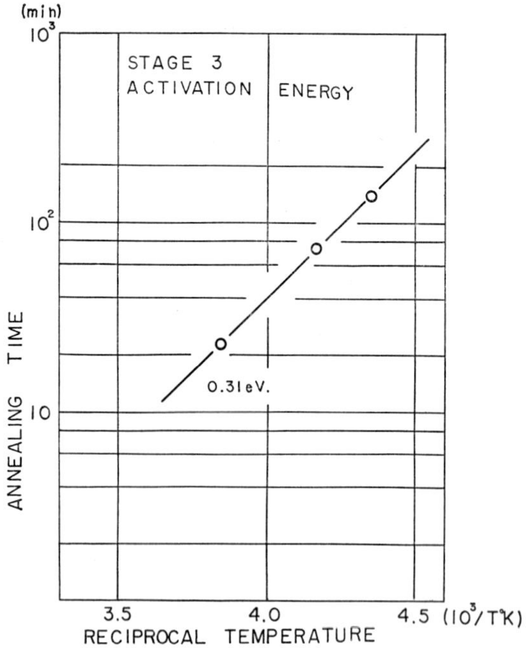

Fig. 7. Time for 50% anneal vs. reciprocal temperature.

Figure 7 shows the temperature dependence of the time for 50% anneal, and from these data the activation energy is estimated to be ~0.3 eV with traps empty. From two isothermal annealings made at 230° and 260°K under illumination, the activation energy for the traps full condition appears to be much smaller (~0.2 eV).

The temperature dependence of Hall coefficient after successive annealings at 230°K is shown in Fig. 8. In the curve plotted after 6-minutes anneal and with traps empty, a step in Hall coefficient at about 6 in abscissa is evident. This step is found to be caused by a level at about 0.16 eV above the valence band, and the level is considered to be present prior to the anneal since the step in the curve after 100 minutes shows little difference from that after 6 minutes. This level is apparently associated with the trapping level because it is not evident under illumination by visible light and, furthermore, it is removed simply by continuing the annealing at the same temperature.

The increase in Hall coefficient below 100°K is considered to be caused by a donor level lying ~0.08 eV above the valence band. It may be concluded that the density of this level increases as annealing proceeds.

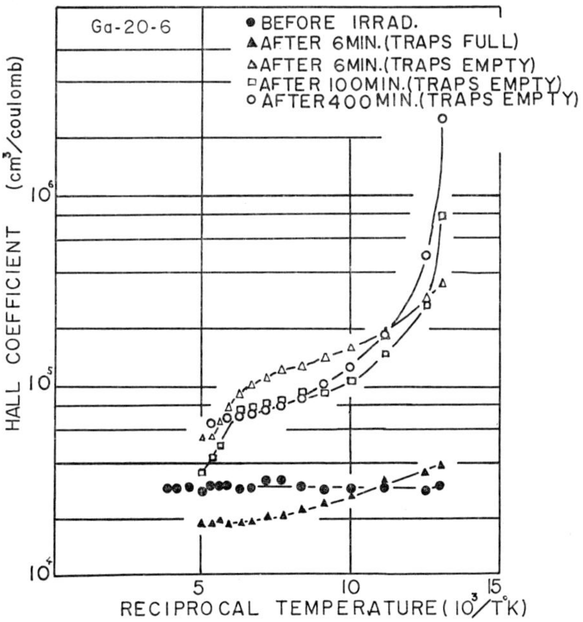

Fig. 8. Temperature dependence of Hall coefficient after successive annealings at 230°K.

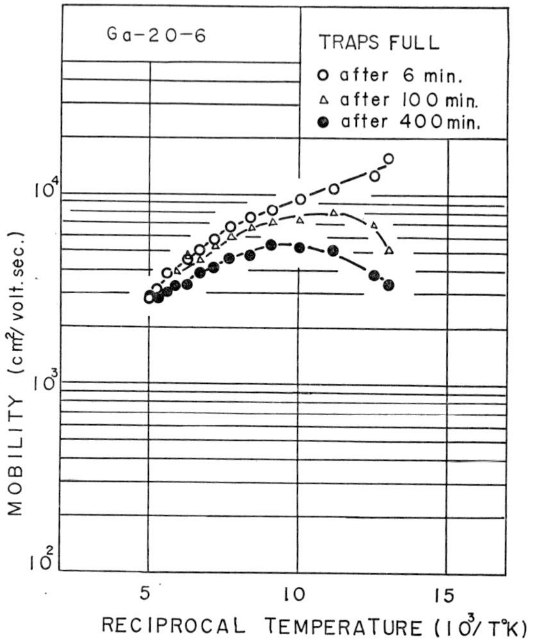

Fig. 9. Temperature dependence of mobility after successive annealings at 230°K.

Temperature dependence of mobility with traps full after successive annealings at 230°K is shown in Fig. 9. The decrease in mobility at low temperature may be caused by the ionized $E_V+0.08$ eV center.

IV. DISCUSSION

Although it is much too early to attempt to identify the defect structures responsible for the various energy states and annealing stages in irradiated p-type germanium, it would nevertheless be helpful if the stages observed in this study could be correlated at least in part with previous observations. In view of the distinct possibility that the location of these stages in temperature are dependent upon hole concentration (Fermi level or quasi-Fermi level) and on gallium concentration, and, in view of the wide range of acceptor concentration and irradiation dose and temperature used in the different studies, it is not easy to correlate the various stages with confidence. Even so, we feel that some conclusions can be drawn.

Stage I which is complete by 140°K apparently corresponds with the stage Whitehouse[5] labels 2P. He finds that this stage is largely suppressed if the metastable states (traps) are emptied by infrared light with $h\nu < E_g$. Upon recooling from 120°K, he also finds that the post-irradiation hole concentration can be restored by a short burst of ionizing radiation. This behavior of course is similar to the response of our specimens (Figs. 1–3) recooled from temperatures up to 140°K and exposed to band gap light. This behavior also corresponds to that observed by Brown and coworkers[3] who determined the rate of donor introduction for both traps empty and traps filled and found the rate of hole removal to be an order of magnitude larger for the traps empty condition. Evidently, as was also pointed out by Whitehouse,[5] the process is a more subtle one than simply emptying and refilling a metastable trap, since type conversion occurs in this stage which would not be expected if minority trapping were alone involved. Whitehouse finds that the activation energy for the process is ~0.16 eV, which is in accord with the isochronal behavior observed here.

In Stage II (150° to 170°K) the total donor density appears to decrease and a significant fraction of the traps are lost. This is clearly evident in the mobility annealing curves. This stage seems to have no clearcut counterpart in the observations of previous workers which may be due to the different experimental conditions employed.

Stage III (210° to 270°K) corresponds to the complete destruction of minority carrier traps and their conversion into stable donors. This donor apparently is responsible for a level lying ~0.08 eV above the valence band. This stage is evidently similar to that reported by Brown *et al.*[3] which occurs in the range 200° to 220°K, and which exhibits markedly different activation energies, depending upon whether the traps are full (0.19 eV) or empty (0.56 eV). In our study the corresponding energies are ~0.2 and 0.3 eV, respectively. In addition, we find that illumination with band gap light not only enhances Stage III, but Stage II as well (Fig. 4.).

Stage IV (320° to 370°K), which leads to an increase in donor concentration, was not observed by Brown and coworkers.[3] However, Stage V (>370°K) undoubtedly corresponds to the high temperature stage (389° to 404°K) which they explored and in which nearly complete recovery of the original hole concentration occurs.

It seems fruitless at the present to speculate about the defect structures responsible for each stage. By analogy with defect structures observed in silicon[1] through electron spin resonance studies, the number of possibilities is known to be high.

Vacancies, divacancies, interstitial impurity (gallium) and, in the range where vacancies are mobile, vacancy impurity complexes are all expected. Moreover, Whan[6] has recently found evidence that vacancies in germanium may move as low as 65°K. Finally, Klontz *et al.*[4] have presented results which they interpret as evidence for interstitial germanium atoms loosely bound to acceptor atoms. Any model of defect structures would have to account for the observed annealing behavior in terms of these various structures but the number of possibilities is quite large, and additional experimental information is necessary to decide between various possibilities.

Acknowledgements

The authors are thankful to Messrs. M. Matsuda and T. Takagaki of the Japanese Association for Radiation Research on Polymers, Osaka Laboratory, for their kind help in making the electron irradiations.

REFERENCES

1. J. W. Corbett, *Electron Radiation Damage in Semiconductors and Metals*, in *Solid State Physics* (F. Seitz and D. Turnbull, eds.), Suppl. 7, Academic Press, New York, 1966.
2. J. W. MacKay and E. E. Klontz, *J. Appl. Phys.*, **30**, 1269 (1959).
3. W. L. Brown, W. M. Augustyniak, and T. R. Waite, *J. Appl. Phys.*, **30**, 1258 (1959).
4. E. E. Klontz, J. W. MacKay, and T. M. Flanagan, *Radiation Damage in Semiconductors*, p. 41, Academic Press, New York, 1965.
5. J. W. Whitehouse, *Phys. Rev.*, **143**, 520 (1966).
6. R. E. Whan, *Phys. Rev.*, **140**, A690 (1965).

Annealing of γ-Irradiated p-Type Germanium

I. HIGASHINAKAGAWA,* S. ISHINO,† F. NAKAZAWA‡
AND R. R. HASIGUTI

Department of Metallurgy, University of Tokyo, Tokyo, Japan

ABSTRACT

Isochronal and isothermal annealings of p-type germanium doped with gallium and with indium were studied above room temperature after ^{60}Co γ-irradiation at 20°C. The annealing behaviors are completely different from those of n-type germanium and depend on impurity species. A large annealing stage is observed at 390°–400°K in all kinds of p-type specimens. The kinetics of recovery process of the main annealing stage is second order with an activation energy of about 1.4 eV in the case of gallium doped germanium, whereas the reaction order is less than one with an activation energy of about 1.0 eV in the case of indium doped germanium.

I. INTRODUCTION

Annealing of irradiated n-type germanium has been studied in considerable detail in recent years.[1–5] It has been well established that the behavior of annealing above room temperature depends on the kinds of impurities doped in the crystal.

On the other hand, published results on annealing of irradiated p-type germanium are rather scarce. In the case of annealing below room temperature, however, we still find several papers on p-type germanium.[6–9] But in the case of annealing above room temperature, only a few results have been published; for example, Brown *et al.*[8] described their results on isothermal annealing at about 400°K, and Curtis and Crawford[10] showed the results of their minority carrier lifetime measurements of annealing of an indium doped germanium.

One aim of the present paper is to describe the annealing behaviors of

* Now at the Toshiba Central Research Laboratory, Kawasaki.
† Now at the Department of Nuclear Engineering, University of Tokyo.
‡ Now at the Institute for Solid State Physics, University of Tokyo.

γ-irradiated p-type germanium crystals doped with different kinds of impurities. It is also of interest to see how the position of the Fermi level affects the behaviors of lattice defects. Since it is expected that the charge state of a defect is different in p-type from that in n-type, comparisons of the results of n- and p-type are of great interest.

II. EXPERIMENTAL METHODS

The dopants used in this experiment are indium and gallium which have tetravalent radii of 1.44 and 1.26 Å, respectively.[11] The effect of impurity concentration was also studied. Samples used are listed in Table I together with their carrier concentrations. They were cut in

TABLE I. Samples used in the present experiment.

Series of samples	Dopant	Tetravalent radius of dopant atom (Å[a])	Initial carrier concentration (cm^{-3})
Ga7	Ga	1.26	$3.2\text{–}4.0\times10^{14}$
In2	In	1.44	$6.5\text{–}7.7\times10^{14}$
In7	In	1.44	$2.4\text{–}4.8\times10^{14}$

a The corresponding value for germanium is 1.22 Å.

bridge shape specimens. The electric current is applied along ⟨110⟩ or ⟨112⟩ direction and the magnetic field is applied parallel to ⟨111⟩ direction in measuring the resistivity and the Hall coefficient.

The experimental setup is the same as that described previously.[4] Irradiation was performed at about 20°C with ^{60}Co γ-rays.

III. EXPERIMENTAL RESULTS

1) *Isochronal annealing*

Figure 1 shows the fraction not annealed, $f_p=(p_0-p)/(p_0-p_i)$ and $f_\sigma=(\sigma_0-\sigma)/(\sigma_0-\sigma_i)$, after isochronal annealings of gallium doped germanium crystals, where p_0 and σ_0 are the carrier concentration and the conductivity before irradiation, p_i and σ_i after irradiation, p and σ after each isochronal annealing, respectively. As is seen in the figure, the annealing takes place in an almost single stage centered at 390°K. No remarkable difference is found between annealing curves derived from the Hall coefficient and from the resistivity.

The annealing treatments were performed for 7.5 minutes at an interval

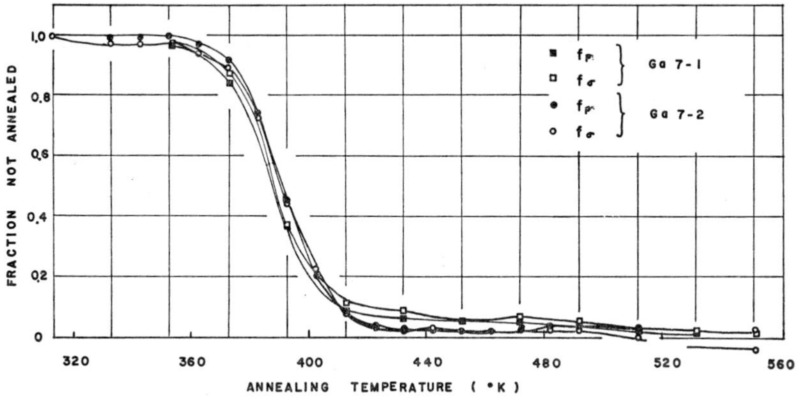

Fig. 1. Isochronal annealing curves for carrier concentration and for conductivity in gallium doped germanium (Ga 7).

of 10°K for specimens Ga 7-2, In 7-2 and In 2-6, and for 15 minutes at an interval of 20°K for specimens Ga 7-1, In 7-1 and In 2-5.

Figures 2 and 3 show the results obtained from two series of indium doped samples. The resistivity of the samples In 7 is slightly higher than that of In 2. The main recovery stage in In7 is again found at about 390°K, while a tailing shoulder remains at around 410°K. The annealing temperature of the main stage of In 2 is found to be about 10°K higher than

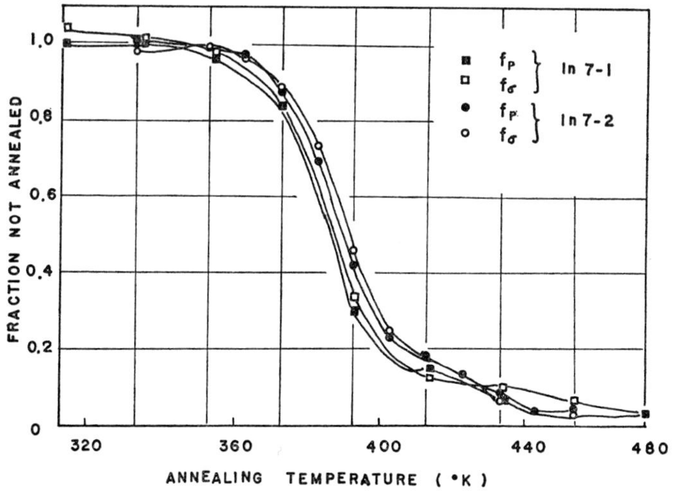

Fig. 2. Isochronal annealing curves for carrier concentration and for conductivity in indium doped germanium (In 7).

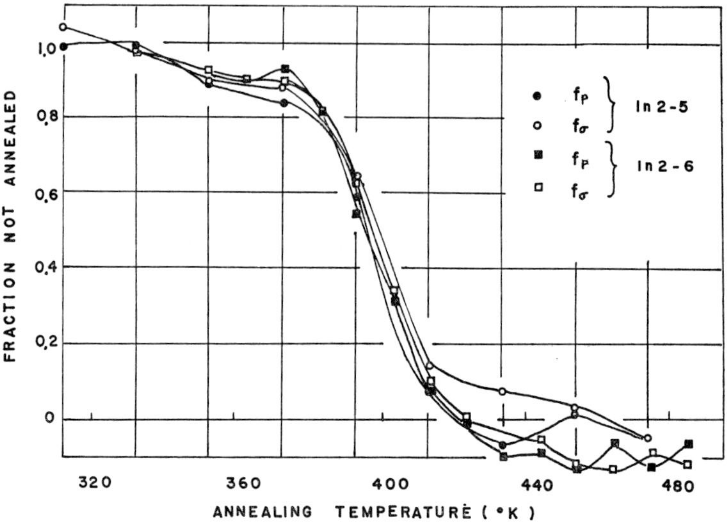

Fig. 3. Isochronal annealing curves for carrier concentration and for conductivity in indium doped germanium (In 2).

that of In 7. In addition, a small stage is observed at about 350°K in In 2 samples.

2) *Isothermal annealing*

The isothermal annealing of gallium doped samples was performed at 373°K, 393°K, and 403°K. The list of the samples, including the initial carrier concentrations and the fractions of removed carrier concentrations, is shown in Table II. The isothermal annealing curves are shown in Fig. 4 together with first and second order kinetics curves. The second

TABLE II. List of samples for isothermal annealing.

Sample	Annealing temperature (°K)	Initial carrier concentration (cm^{-3})	Removed carrier / Initial carrier
Ga 7-4	373	3.4×10^{14}	0.134
Ga 7-5	393	4.0×10^{14}	0.198
Ga 7-6	403	3.9×10^{14}	0.178
In 7-5	393	4.7×10^{14}	0.183
In 7-6	403	2.4×10^{14}	0.350
In 7-7	413	4.5×10^{14}	0.180
In 2-1	403	6.8×10^{14}	0.128
In 2-4	413	6.8×10^{14}	0.119
In 2-2	423	7.4×10^{14}	0.115

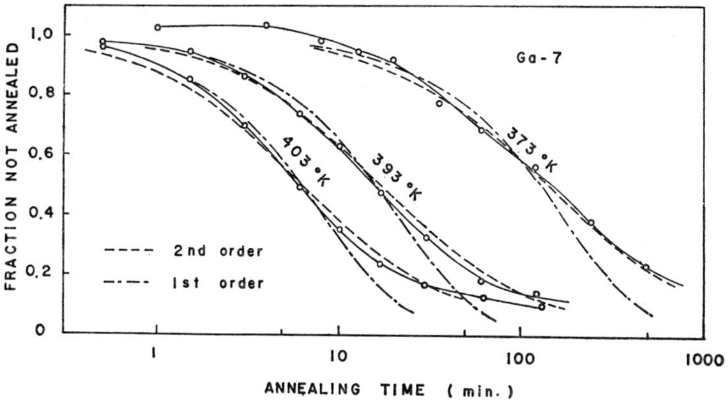

Fig. 4. Isothermal annealing curves for carrier concentration in Ga 7.

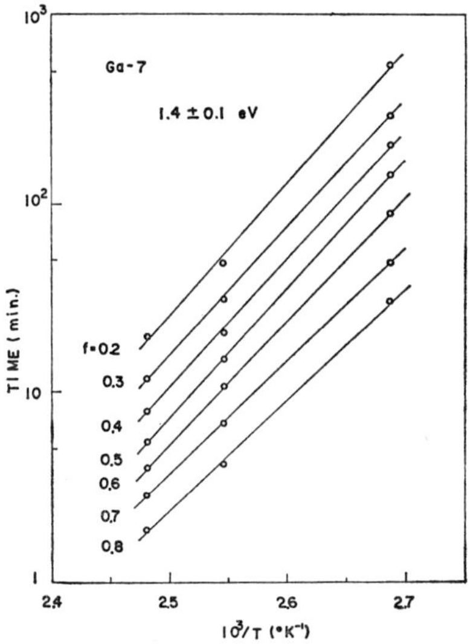

Fig. 5. Determination of the activation energy of Ga 7 from Fig. 4.

order curves are well fitted to the experimental results. Plots of the annealing time vs. temperature in Fig. 5 give the activation energy of 1.4± 0.1 eV.

The isothermal annealing of two sets of indium doped samples was performed as shown in Figs. 6 and 7. The samples used are also listed in Table II. Appropriate asymptotic values for respective annealing curves were assumed to evaluate the activation energies. Plots of the annealing time vs. reciprocal temperature give the activation energies of

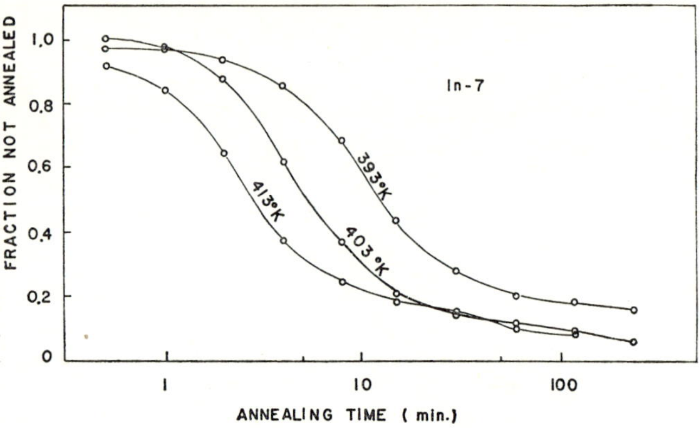

Fig. 6. Isothermal annealing curves for carrier concentration in In 7.

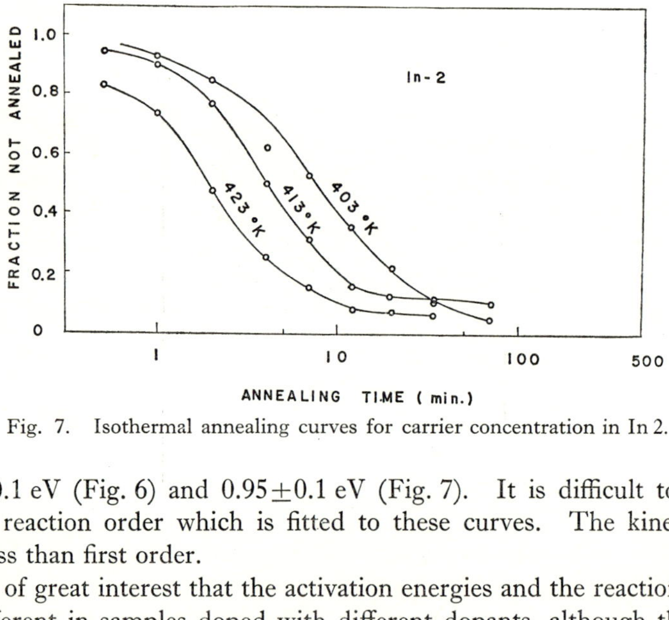

Fig. 7. Isothermal annealing curves for carrier concentration in In 2.

1.05 ± 0.1 eV (Fig. 6) and 0.95 ± 0.1 eV (Fig. 7). It is difficult to find a simple reaction order which is fitted to these curves. The kinetics are even less than first order.

It is of great interest that the activation energies and the reaction orders are different in samples doped with different dopants, although the main annealing occurs at the same temperature range.

IV. DISCUSSION

The experimental results of the main annealing stage or the 390°–400°K stage are summarized in Table III. Here the pre-exponential factor $1/\tau_0$ was obtained from the equation

$$\frac{1}{\tau} = \frac{1}{\tau_0} \exp(-E/kT)$$

where τ is the time required for the fraction not annealed to reach the value $1/e$, E is the observed activation energy, and k and T have the usual meanings.

From the results described in the preceding section and shown in Table III, it is evident firstly that the annealing behaviors of γ-irradiated p-type germanium above room temperature are, as expected, completely different from those of n-type germanium,[1–5] and secondly that the annealing behaviors of p-type germanium depend on impurity species as those of n-type germanium do.[4, 5]

In view of the above mentioned impurity dependence together with the recent knowledge that a single vacancy as well as a single interstitial in germanium migrates at temperatures far below room temperature, it is highly probable that the annealing defects which we observe are some complex defects which include an impurity atom, i.e. perhaps an impurity-vacancy or an impurity-divacancy complex. The difference between the behaviors of gallium doped crystals and those of indium doped ones cannot be understood at present in terms of simple models including the concept of atomic radii of impurities.

As seen in Table III, the frequency factors $1/\tau_0$ are somewhat too large to be understood as a manifestation of a simple atomic jump process. But

TABLE III. Summary of the results of 390°–400°K stage.

Sample	Activation energy (eV)	$1/\tau_0$ (sec^{-1})	Reaction order
Ga7	1.4 ± 0.1	8×10^{14}	2
In7	1.05 ± 0.1	10^{14}	$\lesssim 1$
In2	0.95 ± 0.1	10^{14}	$\lesssim 1$

similarly large frequency factors are often observed, e.g. in the results of p-type germanium by Brown *et al.*[8] and in the results of n-type silicon by Hirata *et al.*[12] This seems to offer an interesting problem to be solved in future.

Acknowledgements

The authors sish to thank Dr. F. Nagatani of Toshiba Central Research Laboratory and Dr. T. Abe of Toshiba Transistor Works for supplying them with various kinds of germanium crystals.

REFERENCES

1. *J. Appl. Phys.*, **30**, 1117–1322 (1959) (Gatlinburg Conference Issue).
2. in *Proc. Symp. on Radiation Damage in Semiconductors, 1964*, Dunod, Paris; Academic Press, New York, 1965.
3. R. R. Hasiguti and S. Ishino, in *Proc. Symp. on Radiation Damage in Semiconductors, 1964*, p. 259, Dunod, Paris; Academic Press, New York, 1965.
4. S. Ishino, F. Nakazawa and R. R. Hasiguti, *J. Phys. Chem. Solids*, **24**, 1033 (1963).
5. F. Nakazawa, S. Ishino, I. Higashinakagawa and R. R. Hasiguti, *J. Phys. Chem. Solids*, **26**, 1895 (1965).
6. G. W. Gobeli, *Phys. Rev.*, **112**, 732 (1958).
7. J. W. MacKay and E. E. Klontz, *J. Appl. Phys.*, **30**, 1269 (1959).
8. W. L. Brown, W. M. Augustyniak and T. R. Waite, *J. Appl. Phys.*, **30**, 1258 (1959).
9. E. E. Klontz, J. W. MacKay and T. M. Flanagan, in *Proc. Symp. on Radiation Damage in Semiconductors, 1964*, p. 41, Dunod, Paris; Academic Press, New York, 1965.
10. O. L. Curtis and J. H. Crawford, Jr., *Phys. Rev.*, **126**, 1342 (1962).
11. L. C. Pauling, in *The Nature of the Chemical Bond*, 3rd ed., p. 246, Cornell University Press, Ithaca, N. Y., 1960.
12. M. Hirata, M. Hirata, H. Saito and J. H. Crawford, Jr., in this book.

Part III

Defects in Compound Semiconductors

Lattice Defects and Electron Mobilities in Gallium Arsenide

ROBERT K. WILLARDSON

Bell & Howell Research Center, Pasadena, California

ABSTRACT

Crystals of gallium arsenide can contain significant concentrations of defects. The actual concentration of point defects depends on the temperature and arsenic vapor pressure environment after growth as well as the conditions of growth. Usually the concentration of defects is related to the concentration of n-type impurities.

At high doping levels, melt grown single crystals have lower defect concentrations than those grown by gaseous epitaxy. It has been possible at carrier concentrations of $2 \times 10^{18}/cm^3$ to obtain in Czochralski grown crystals electron mobilities at 300°K which are twice the values reported in the literature for epitaxial gallium arsenide, i.e., 3,300 cm^2/volt sec. as compared to 1650. Results are consistent with theoretical calculations of electron mobilities.

I. THEORY

Ehrenreich[1] has calculated the electron mobility as a function of carrier concentration in uncompensated GaAs assuming the predominant scattering mechanisms at room temperature to be polar lattice modes and ionized impurities. He found that variational calculations using the semiclassical theory of Brooks and Herring agreed well with experiment in the lower concentration regions where polar scattering is more important. However, there was a discrepancy of approximately 40% at the higher concentrations where ionized impurity scattering was assumed to dominate.

More recently, Moore and Ehrenreich[2] have developed a quantum transport theory for n-type semiconductors. The Kohn and Luttinger transport equations for ionized impurity scatterings were simplified. A simple closed formula for the mobility was derived that contains corrections to the semiclassical Brooks-Herring result, including scattering from pairs of impurities, higher Born corrections for single scattering and dressing effects of the impurities on the electron wave functions and energy levels. From these results, it is straight-forward to derive the quantum mobility for GaAs. The simplest procedure consists of first computing the Brooks-Herring mobility and then dividing by a correction factor of $1+\delta_B+\delta_M+\delta_D$. In GaAs for impurity concentrations over $10^{18}/cm^3$ the largest correction is the multiple scattering term δ_M, which is about 0.35. The normalization correction δ_D and the incoherent Born correction δ_B are about 0.1.

The results of Moore and Ehrenreich's[2] quantum theory calculations at 77°K, as well as Ehrenreich's[1] variational calculation at 300°K, are shown in Fig. 1 along with experimental data on Czochralski-grown GaAs single crystals which are considered to be essentially defect free. The results of Moore and Ehrenreich's quantum mobility calculations at 300°K are identical with the variational calculation for 3×10^{17} n-type impurities/cm³, but are 10% higher at $5\times10^{16}/cm^3$ and 25% lower at $1\times10^{19}/cm^3$. Since the polar mobility is significant at room temperature, the combination of polar and impurity mobilities by means of the variational principal is to

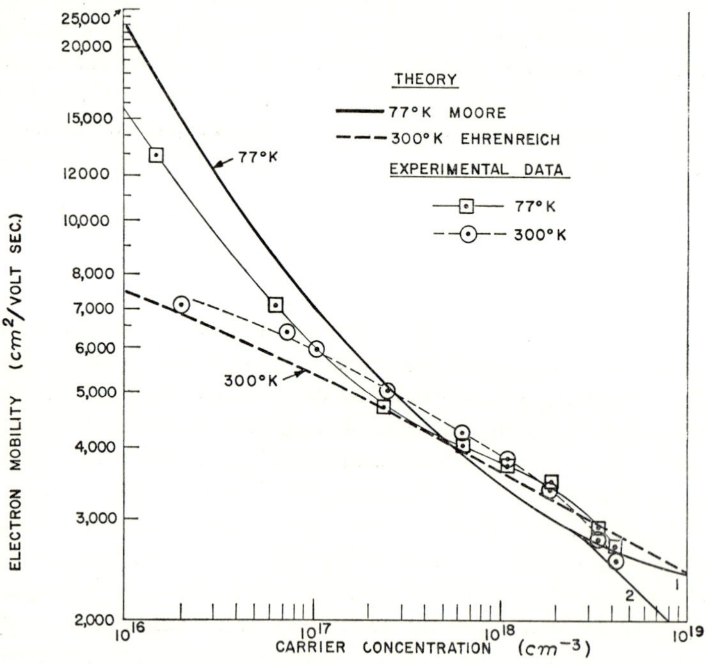

Fig. 1. Electron mobility at 77 and 300°K as a function of carrier concentration in gallium arsenide.

be preferred over the $1/\mu = 1/\mu_I + 1/\mu_L$ approximation used by Moore and Ehrenreich.

The quantum mobility calculations at 77°K are for impurity scattering only. For concentrations above $10^{17}/cm^3$ the effect of polar scattering is negligible. In the range of 10^{16} to $10^{17}/cm^3$ the addition of polar scattering and electron-electron scattering should lower the theoretical curve to a position very near the experimental points.

At high carrier concentrations, the effect of the non-parabolic conduction band in GaAs becomes apparent. The variational calculation and the quantum curve labeled 1 assume a parabolic band, while the calculation of curve 2 includes the appropriate effective masses at the Fermi surface.

The experimental data at 300°K suggest slightly different values for some of the parameters used by Ehrenreich[1] to calculate a polar scattering mobility of 9,300 cm^2/volt•sec that would give a value closer to 11,000 cm^2/volt•sec.

It is interesting to note that the theoretical curves in Fig. 1 predict a region in the $10^{18}/cm^3$ impurity range where the electron mobility at 77°K is actually less than at 300°K. This is experimentally observed at a somewhat lower carrier concentration. Of course, an increase in the 300°K polar mobility used in the calculations would bring the actual positions of the crossover points into better agreement. However, the agreement between theory and experiment is much better than should be expected.

II. DEFECTS

While studies of electron mobilities in GaAs with large carrier concentrations provided experimental values 40% too low as compared to Ehrenreich's variational calculation, more recent data gives values in excellent agreement with theory. The question arises as to why such a difference exists. Mass spectrographic analyses prove that this large change in electron scattering is caused by defect scattering and not by a change in impurity content. This suggests that GaAs single crystals prepared by different techniques could have much different electron mobility versus carrier concentration relationships. This is shown in Fig. 2, where the circles give data of Blum[3] and Willardson on Czochralski grown single crystals. Blum used conventional growth procedures, while Willardson used special techniques to minimize defect concentrations.[4] The data of Weisberg *et al.*[5] was obtained from single crystals prepared

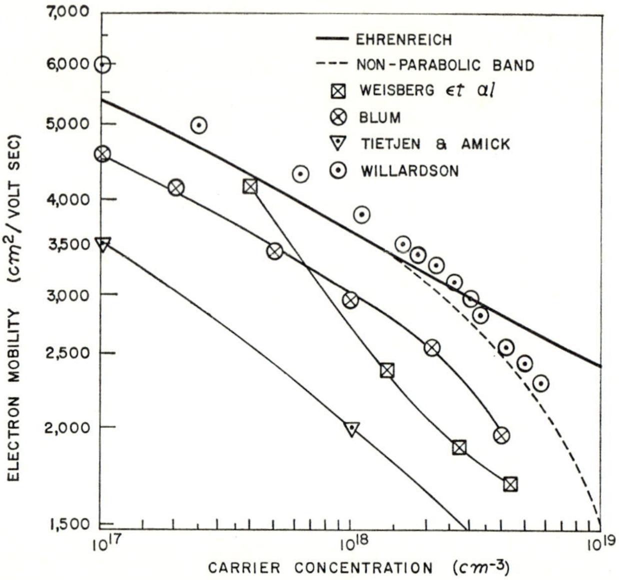

Fig. 2. Electron mobility at 300°K as a function of carrier concentration in heavily doped gallium arsenide.

by the horizontal Bridgeman technique and is the experimental data which Ehrenreich used for comparison with his theoretical calculations.

The epitaxial layers of Tietjen and Amick[5] were prepared with a novel apparatus using gaseous arsine, hydrogen chloride, hydrogen, and gallium. Typical undoped specimens had carrier concentrations in the $2 \times 10^{15}/cm^3$ to $1 \times 10^{16}/cm^3$ range with electron mobilities of 20,000 to 30,000 cm^2/volt·sec at 77°K, but values as high as 58,000 cm^2/volt·sec were reported. Hydrogen selenide was used to provide the n-type dopant.

At carrier concentrations ranging from $10^{17}/cm^3$ to $5 \times 10^{19}/cm^3$ the electron mobility of the epitaxial layers is only about one-half that of the " defect free " GaAs. At $2 \times 10^{18}/cm^3$, the electron mobility in the epitaxial sample was 1,650 cm^2/volt·sec compared to 3,300 cm^2/volt·sec for the better Czochralski grown material. However, using the same epitaxial apparatus, carrier concentrations of $3.5 \times 10^{14}/cm^3$ and electron mobilities of 7,600 cm^2/volt·sec at 300°K were obtained.

It is clear that the defect concentration which reduces the electron mobility is related to the carrier concentration. To obtain further information on this relation let us assume that the defect scattering mobility is inversely proportional to the defect concentration.

$$\mu_{\text{Defect}} = \frac{C}{N_{\text{Defect}}} \tag{1}$$

In the extreme degenerate case where all the carriers have the same energy, Eq. (2) applies.

$$\frac{1}{\mu} = \frac{1}{\mu_{\text{I}}} + \frac{1}{\mu_{\text{L}}} + \frac{1}{\mu_{\text{Defect}}} \tag{2}$$

Assuming this to be an adequate approximation for heavily doped n-type GaAs:

$$N_D = C\left[\frac{1}{\mu} - \left(\frac{1}{\mu_{\text{I}}} + \frac{1}{\mu_{\text{L}}}\right)\right] \tag{3}$$

Using the experimental epitaxial mobility value for μ and the reciprocal of the experimental mobility in "defect free" GaAs for the term in parenthesis, the data on the upper curve in Fig. 3 was calculated. The same procedure was used to calculate the corresponding data points in Fig. 3 for the conventional Czochralski and horizontal Bridgeman grown crystals. The best fit of this data gives a cube root of the carrier con-

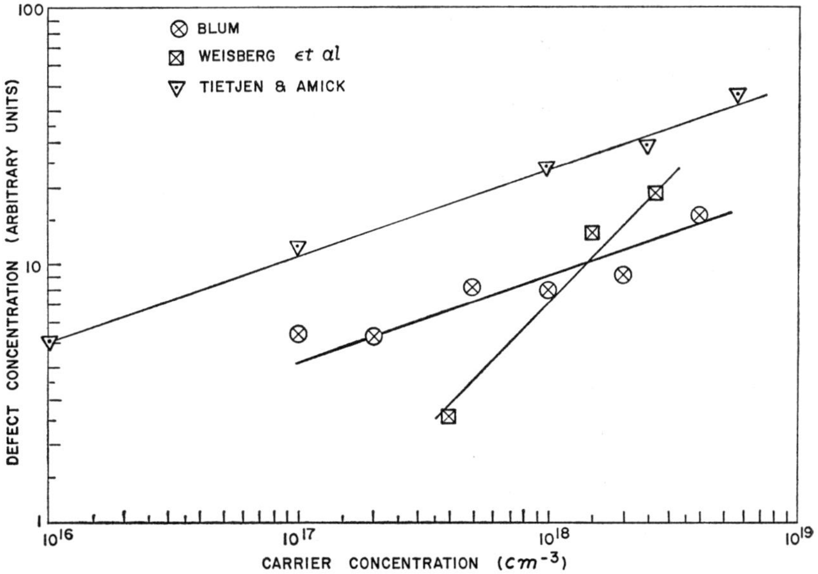

Fig. 3. Defect concentration as a function of carrier concentration in gallium arsenide grown by Czochralski, horizontal Bridgeman and epitaxial techniques.

TABLE I. Summary of studies of defects in GaAs.

Defect formation	Effect of defects	Comments
as grown	lower electron mobility[7]	$N_{\text{defects}} \sim N_{\text{donor}}^{1/3}$
2 MeV electrons at 50°K	lower maximum thermal[7] conductivity	$\frac{1}{K}-\frac{1}{K_0} \sim \Phi$ at 50°K
2 MeV electrons at 50°K+52°C anneal	new low temp. max. at 20°K[7]	localized defect mode scattering
2 MeV electrons at 50°K+300°C anneal	density increase[7]	$K \sim T^{1.0}$ below 30°K sheets of precipitate
2 MeV electrons at 50°K+300°C anneal	boundary scattering[7]	
0.6–40 MeV electrons at 50°C	quenching of luminescence[15,16]	peaks at $\begin{cases}1.4433\ \text{eV}\\1.4415\ \text{eV}\\1.4402\ \text{eV}\end{cases}$
0.6–1.0 MeV electrons at 50°C +220°C anneal	reappearance of luminescence + fine structure[15]	
0.6–1.0 MeV electrons at 50°C +260°C anneal	reappearance of luminescence no fine structure[15]	
1.5 and 2 MeV electrons at 50°C+220°C anneal	reappearance of luminescence no fine structure[15]	
450°C anneal with Cu	large trap densities introduced[8]	traps $\begin{cases}0.2\ \text{eV}\\0.5\ \text{eV}\end{cases}$
450°C lithium diffused	no infrared Li-defect local mode peaks[14]	

TABLE I. Continued.

Defect formation	Effect of defects	Comments
700–1,000°C lithium diffused	8 infrared Li-defect local mode peaks[14]	^{7}Li peaks at 364, 383, 421, 379, 391, 394, 398 cm^{-1}
700°C anneal with KCN	density increase[8]	float zoned—0.01% horizontal Bridgeman—0.0005%
700°C anneal with KCN	decrease in thermal conductivity at 25°C[8]	float zoned—10% horizontal Bridgeman—5%
700°C anneal with KCN	large trap densities introduced[8]	horizontal Bridgeman material 0.2, 0.5, 0.7 eV traps
900–1,200°C anneal	electron paramagnetic resonance[11,12]	appears only in p-type GaAs spin density 2–5 $\times 10^{19}/cm^3$ $g = 2.046 \mid a \mid = 374G$
1,100°C anneal[170 Torr] As pressure and quench	density change[9]	0.11%
1,100°C anneal[170 Torr] As pressure and quench	density change[9]	0.024%
as grown	2 internal friction peaks[13]	80° and 140°C at 13.9 kc/sec defect concentration $4 \times 10^{18}/cm^3$
after 930°C anneal (without As)	no internal friction peak[13]	
after 1,040°C anneal (with As)	one internal friction peak[13]	140°C at 13.9 kc/sec

centration dependence for the defect concentration for the epitaxial and conventional Czochralski crystals. A linear dependence is observed for the horizontal Bridgemen crystals, but this is primarily because of silicon impurities, which at high concentrations prefer gallium sites. The Czochralski and vapor epitaxial GaAs contained selenium or tellurium impurities.

The defect concentration is given in arbitrary units because their character and the scattering mechanism are unknown. The assumption that the defect scattering mobility varies directly as the inverse of the defect concentration may not be correct. In fact, a variation such as $(1/N_{\text{Defect}})^{1/3}$ could be suggested by the above discussion.

If the defect is ionized and the ordinary ionized impurity scattering formula applies, the concentration of defects in the epitaxial material would be $3 \times 10^{18}/\text{cm}^3$ for a carrier concentration of $2 \times 10^{18}/\text{cm}^3$. The scale in Fig. 3 would then be multiplied by about 1×10^{17} and have units of defects/cm^3. Of course, the use of these simple equations is at best only a rough approximation.

III. DISCUSSION

Since high values of electron mobilities are obtained in high purity GaAs, it has been generally assumed that the effects of lattice defects were negligible, particularly for heavily doped materials. During the past few years, evidence of appreciable concentrations of lattice defects in GaAs has been obtained by a variety of experimental techniques. These include measurements of thermal conductivity,[7,8] density,[7–9] photo-electronic analysis,[10] thermally-stimulated currents,[8] electron paramagnetic resonance,[11,12] internal friction,[13] local modes,[14] luminescence,[15,16] electroluminescence,[17] and impurity distribution coefficients.[18]

In most of the studies in which these techniques are discussed, the experiment involved the introduction of defects by electron bombardment or annealing at elevated temperatures followed by a quenching operation. For the latter case the introduction of defects was at least partially related to variations in arsenic pressure which in some cases was known and controlled, but usually not specified. While the measurement techniques are valuable, much work remains to be done to correctly interpret the results in terms of specific lattice defects.

Summarized in Table I are data taken from the literature as indicated. There can be no question as to the existence of large concentrations of defects in ordinary as-grown GaAs, as well as in samples that have been annealed at various temperatures and quenched. A correlation of data on similar samples utilizing several of the techniques described in Table I is needed to obtain hypotheses regarding specific defects which are consistent.

Since our studies of the effects of defects on the electron mobility indicate a concentration dependence on the carrier concentration in n-type samples, the law of mass action predicts that the defects are acceptors. They may be the same acceptors as the " impurities " referred to by Woodall and Woods.[19] The distribution of the acceptors referred to by them does not agree with that required for impurities, but can be explained on the basis of defects.

Of all the data in Table I the most pertinent to describing the nature of mobility reducing defects are the internal friction results. These were on conventional Czochralski grown materials. A bound pair of gallium vacancies is the simplest defect that fits the experimental characteristics of the damping peak.

However, it is not certain that the acceptors created during the formation of the crystal remain unchanged during the cooling cycle to room temperature. Measurements made 4, 16, and 350 hours after growth to conventional Czochralski crystals indicate an increase, (*1*) from 2.00×10^{18} to 2.22×10^{18} and (*2*) from 2.50×10^{18} to 2.85×10^{18} carriers/cm^3 between the 4 and 16 hour measurements. No further change was observed after the 16 hour measurement. This might be related to the first stage of room temperature annealing of high temperature induced defects proposed by Potts. However, his model which proposed vacancy-interstitial recombination presents a few problems.

REFERENCES

1. H. Ehrenreich, *Phys. Rev.*, **120**, 1951 (1963).
2. E. J. Moore and H. Ehrenreich, *Bull. Am. Phys. Soc.*, II **11**, 199 (1966).
3. E. J. Moore, *Tech. Report No. ARPA-24*, March 1966, Harvard Univ., Cambridge, Massachusetts.
4. R. K. Willardson, in *Purification of Materials*, Vol. 137, p. 41, Annals of the New York Academy of Sciences, 1966.
5. L. R. Weisberg, J. R. Woolston and M. Glicksman, *J. Appl. Phys.*, **29**, 1514 (1958).
6. J. J. Tietjen and J. A. Amick, *J. Electrochem. Soc.*, **113**, 724 (1966).
7. F. L. Vook, *Phys. Rev.*, **135**, A1742 (1964); *ibid*, A1750 (1964); *Proc. of the Intern. Conf. on Crystal Lattice Defects, Kyoto, 1962*: *Conf. J. Phys. Soc. Japan*, **18** Suppl. II, 190 (1963); in *Proc. Intern. Conf. Phys. Semicond., Paris, 1964*, p. 713, Dunod, Paris and Academic Press, New York, 1964.
8. J. Blanc, R. H. Bube and L. Weisberg, *J. Phys. Chem. Solids*, **25**, 225 (1964).
9. H. R. Potts, *Ph.D. Thesis*, Stanford Univ., 1964.

10. R. H. Bube, in *Semiconductors and Semimetals* (R. K. Willardson and A. C. Beer, eds.), Vol. 3, Academic Press, New York, 1967.
11. B. Goldstein and N. Almeleh, *Appl. Phys. Letters*, **2**, 130 (1963).
12. B. Goldstein, in *Semiconductors and Semimetals* (R. K. Willardson and A. C. Beer, eds.), Vol. 2, p. 196, Academic Press, New York, 1966.
13. B. K. Chakravery and R. W. Dreyfus, *J. Appl. Phys.*, **37**, 631 (1966).
14. O. G. Lorimor, W. G. Spitzer, R. K. Willardson and W. P. Allred, *Conf. on The Physics of Semiconducting Compounds*, Swansea, Wales, 1966.
15. G W Arnold, *Bull. Am. Phys. Soc.*, II **11**, 189 (1966).
16. D. M. J. Compton and S. J. Black, *Bull. Am. Phys. Soc.*, II **11**, 754 (1966).
17. I. J. Bell, H. Robbins and R. K. Willardson, to be published.
18. R. K. Willardson and W. P. Allred, *Symp. on Gallium Arsenide*, Reading 1966, p. 35, Institute of Physics and The Physical Society, London, 1967.
19. J. M. Woodall and J. F. Woods, *Solid State Communications*, **4**, 33 (1966).

Energies of Formation of Metal-vacancies in II-VI Semiconducting Tellurides (HgTe, CdTe, ZnTe)

F. Bailly

Laboratoire de Magnétisme et de Physique du Solide,
C.N.R.S., Bellevue (S. & O.), France

ABSTRACT

The energy of formation of metal vacancies in tellurides has been calculated with the assumption that covalent bonding occurs between the Te neighbors. The model involves the use of Morse functions to describe the bond strength between nearest neighbors and the approximation of Suchet's crystallochemical model. The mean displacement of the atoms around the vacancy is calculated by minimizing the sum of the energy terms associated with the various interactions and is shown to be important. It is shown that, in the particular case of HgTe, this type of vacancy has a negative energy of formation (neglecting the interactions between vacancies). These defects are thus expected to be naturally found with a high concentration, and they are expected to behave as neutral defects. Previous experimental observations of " complex vacancies " support this conclusion. The bonds between Te neighbors may be broken at high temperatures, which transforms the complex vacancies into normal ones: the calculated energy of transformation is in fair agreement with experiment. In the cases of CdTe and ZnTe, the covalent bonding between Te neighbors affects only slightly the energy of formation of the vacancy.

The same kind of calculation, applied to the group IV elements, leads to an energy of formation of vacancies in good agreement with other theoretical calculations and experimental determinations, which supports the validity of the model.

I. EXPERIMENTAL PROCEDURES

The presence of lattice defects in semiconducting compounds plays an important role for the study of the structural and electronic properties of these materials. Up to now, the thermodynamic equilibria between defects of any kinds have been theoretically studied as quasi-chemical equilibria[1] and these studies enable us to interpret some experimental data about the electronic behavior of the materials. Simultaneously, calculations of the formation energies of defects have been made, particularly in elemental semiconductors, in order to describe the local behavior of the crystal lattice in the neighborhood of the defects.[2–4]

This work is an attempt to calculate the energies of formation of metal vacancies in semiconducting tellurides by an atomic approach, considering the various possibilities of binding between the Te-nearest neighbors of the vacancy.

The first part of the paper is devoted to the description of the proposed model. The second part is an application of the derived theoretical relations to the case of II-VI tellurides, with a discussion of the results obtained. The third part compares the results of the generalization of the model to the elemental semiconductors with other theoretical calculations and experimental determinations.

1. *Description of the model*

The II-VI semiconducting compounds, such as HgTe, CdTe, and ZnTe, crystallize in the zinc-blende structure. Figure 1 represents the disposition of the atoms in such a structure. The nature of the iono-covalent bonds in these materials has been studied in other papers[5,6] and their ionicities

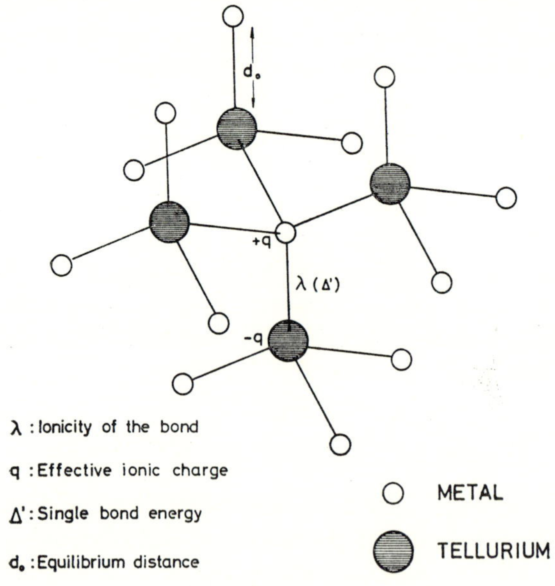

Fig. 1.

λ have been evaluated. We recall only that it has been possible to define and to calculate an effective chemical ionic charge q, which is carried by the metal M (the charge $-q$ being carried by the metalloid Te). We recall also that if we consider a pair of binding electrons, the crystallochemical model shares these electrons between the two considered atoms, $1-\lambda$ electron being " localized " on the atom M and $1+\lambda$ on the Te atom.

To transport an atom M from its normal position in the lattice to the surface, it is necessary to spend an energy ΔH_l. This energy is the difference between the energy spent to break four bonds and to transport the atom into the gas phase, and the energy gained by the transport of the atom from the gas phase onto the surface with the formation of two bonds. We can evaluate the energy ΔH_l from thermochemical data

$$\Delta H_l = \frac{S_{\mathrm{M}} - \Delta H_{\mathrm{f}}^{\circ}}{2} \tag{1}$$

$\Delta H_{\mathrm{f}}^{\circ}$ is the standard energy of formation of the compound from its elements in the standard state, and S_{M} is the sublimation energy of the metal from its standard state.

Figure 2 represents the lattice after this operation. As a first approxima-

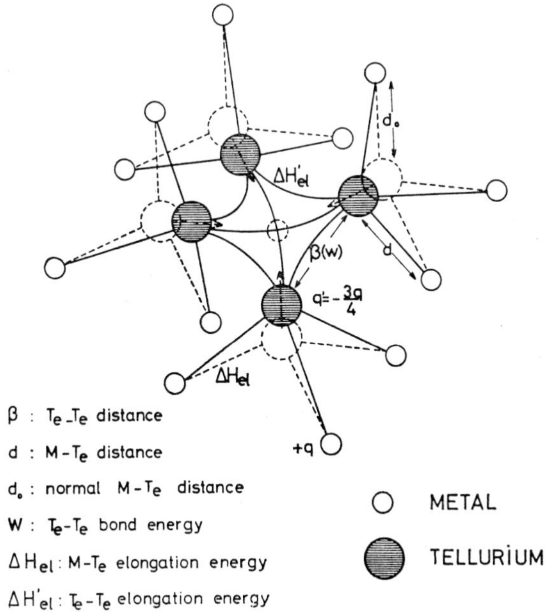

Fig. 2.

tion we suppose, in order to simplify the following calculations that the nature of the remaining bonds between the atoms in the neighborhood of the metal-vacancy is not changed by the formation of this vacancy; that is to say, that the remaining bonds are always characterized by the same ionicity λ. But there are now only three bonds in the place of the four initial ones, and the effective ionic charge carried by the atoms is now $q'=-3q/4$; it remains 3/2 non-utilized electron on each Te atom. These electrons are able to create new bonds between the Te atoms, the formation of these bonds leading to a minimization of the energy of the lattice around the vacancy.

We will consider two types of possible interactions between these remaining electrons: the covalent interaction and the Van der Waals interaction, both existing in solid tellurium. The formation of covalent bonds (with two electrons) between the Te atoms (considering that, statistically, each Te atom is bound to the three others) needs the sharing (from each Te atom) of $(2-1.5)/3=0.5/3$ electron in common with the others. If we assume that the bond-energy is proportional to the number of participating electrons, we obtain for the total energy for six bonds $W=6\times(0.5/3)\times\Delta=\Delta$; Δ being the single bond energy of the tellurium for a covalent bond ($\Delta=2.13$ eV).

However these bonds being fractionary ones, they are no longer stabilized for the distance r_0, corresponding to a single bond ($r_0=2.86$ Å), but for the distance β_0, such that[7]

$$\beta_0 = r_0 - 0.6 \log (0.5/3) \tag{2}$$

The minimizing of the energy of these bonds around the vacancy needs the evaluation of the various elongation energies.

For the M-Te bonds, we have, assuming a representation by a Morse-potential

$$\Delta H_{\text{el}} = 12\Delta' \{1 - \exp[-\alpha(d-d_0)]\}^2 \tag{3}$$

In this relation, ΔH_{el} is the total energy for the elongation of all the M-Te bonds around the vacancy; Δ' is the single bond energy of the M-Te bond in its normal state; α is a constant which can be calculated by the relation $\alpha=(f/2\Delta)^{1/2}$ (where f is the strength-constant of the bond); d_0 is the distance between the M and Te atoms in the normal state, and d is the new distance between these atoms. The factor 12 in the relation (3) arises because there are four Te atoms around the vacancy, each being bound to three M atoms.

For the Te-Te bonds, we have similarly

$$\Delta H_{el}' = W\{1-\exp[-\alpha'(\beta-\beta_0)]\}^2 \qquad (4)$$

where the constant α', calculated for tellurium, plays the same role as α ($\alpha'=2.77$ Å^{-1}) and β is the new distance between the Te atoms.

We must add to these energies the Coulomb interaction energy U, due to the fact that the displaced atoms are electrically charged.

The equilibrium distance of the bonds corresponds to a value of β which minimizes the sum

$$\Sigma = \Delta H_{el} + \Delta H_{el}' + U \qquad (5)$$

The energy of the vacancy, evaluated from the normal state, is then

$$\Delta H_{cov.} = \Delta H_l + (\Sigma)_{min} - W \qquad (6)$$

For the Van der Waals interactions, the same type of calculation has been used, with other values of the constants. In particular we have chosen the value $W'=0.186$ eV for the energy of the Van der Waals interaction in tellurium, equal to the energy of melting in tellurium. This choice can be justified by the fact that the semiconducting properties of tellurium remain almost unchanged across the melting point; this observation can be interpreted by the assumption that the melting destroys the long-range order of the crystal (due to the Van der Waals interactions between the tellurium chains) but preserves the short-range order (due to the covalent interactions).

Obviously all these relations are only approximate. *(1)* We have considered interactions as far as the second nearest neighbors only. *(2)* We have simplified the expressions of the energies, especially those for fractional bonds. *(3)* We have used the values of single bond energies in pure tellurium neglecting the distortions of bond angles. *(4)* We have expressed the variation of the covalent bond strength versus distance by means of Morse functions, which are only approximate in the case of these compounds. *(5)* We have assumed that the nature of the binding orbitals for covalent bond between the metal and the tellurium atoms remained unchanged (sp^3 type) by the formation of the vacancy; we have not taken into account the fact that the true binding orbitals between the tellurium atoms in pure tellurium are of (p^2) type.

With these rough approximations the numerical results obtained can only be indicative. However, we shall see that these results are qualitatively and sometimes quantitatively in agreement with thermodynamical data and can explain some experimental results about the electrical properties of the compounds considered.

Before we study the applications of these calculations to the II-VI tellurides, we must indicate another possibility of relaxation of the atoms around the vacancy; this relaxation consists of the nearing of the electrically charged Te-atoms, to the oppositely charged M-atoms. In this case also it is a problem of minimum, because there are two contributions to the total energy: the Coulomb interaction which lowers the energy of the whole, and the shortening of the M-Te bonds which increases this energy. The calculation for CdTe (the only compound for which $q \neq 0$ in the sequence ZnTe, CdTe, HgTe) shows that the gain of energy is very small ($\sim 4 \times 10^{-3}$ eV for a shortening of 0.025 Å of the CdTe bonds).

2. *Application to the cases of ZnTe, CdTe and HgTe*

Table I shows the results of the calculation for these compounds. We see that in all cases the calculated energy of formation of " Van der Waals " vacancies is in good agreement with the thermochemical results. The energy of formation of " covalent " vacancies is smaller than the former. This type of vacancy with a neutral electrical behavior (due to the fact that all the free electrons on tellurium participate in covalent bonds) are then expected to be present in CdTe and ZnTe in appreciable number. For HgTe we note that their energy is even negative. Evidently this result is only valid for an isolated vacancy (we have not taken into account the possible interactions between vacancies) and does not mean that solid HgTe cannot exist; but this result indicates that such " covalent " vacancies must be present in great number in this compound; this fact is confirmed by the studies of H. Rodot on HgTe.[8] In all these cases, the displacement of the tellurium atoms around the vacancy is important for the covalent bonds (about 0.31 Å, 0.30 Å, 0.22 Å, for HgTe, CdTe and ZnTe respectively).

TABLE I.

Compound	E_f (eV)	$\Delta H_{cov.}$ (eV)	$\Delta H_{vdw.}$ (eV)
HgTe	0.50	−0.13	0.51
CdTe	0.96	0.85	1
ZnTe	1.16	0.41	1.22

3. *Calculation for the elemental semiconductors.*

In order to verify the validity of our model, we have calculated in the same manner the energy of formation of vacancies with covalent bonds in the elemental semiconductors (Ge, Si, C) for which some theoretical calculations

TABLE II. E_f (eV)

Element	SWALIN	SCHOLZ	BENNEMANN	MAYBURG	LOGAN	Our work
C	4.16	—	3.68	—	—	3.56
Si	2.32	2.78	2.13	—	—	2.28
Ge	2.07	1.62–2.54	1.91	2.01	2.0	2.06
	Theoretical			Experimental		Theoretical

or experimental measuremements have been performed by other workers.

Table II compares the results of these previous determinations with our own values; the agreement is good and can be considered as a test for the validity of our model.

II. CONCLUSION

Our calculation shows that in the II-VI tellurides, metal vacancies with covalent interactions between the Te neighbors and subsequent displacement of these Te atoms can appear in order to form electrically neutral defects. The number of this type of defect is expected to be very large in the compound HgTe; this fact is confirmed by the experimental measurements. These defects are also expected to be present to a smaller extent in CdTe and in ZnTe.

Such calculations can also be performed for other compounds, for instance selenides, which crystallize in the zinc-blende structure, and with the help of the crystallochemical model for other structures. This enables us to predict whether certain types of lattice defects can or can not be present in the materials.

REFERENCES

1. F. A. Kröger, in *The Chemistry of Imperfect Crystals,* North-Holland, Amsterdam, 1964.
2. R. A. Swalin, *J. Phys. Chem. Solids,* **18**, 290 (1961).
3. A. Scholz, *Phys. Stat. Sol.,* **3**, 42 (1963).
4. K. H. Bennemann, *Bull. Am. Phys. Soc.,* II **9**, 298 (1964).
5. J. P. Suchet and F. Bailly, *Ann. Chimie,* **10**, 517 (1965).
6. F. Bailly, *J. Phys.,* **27**, 335 (1966).
7. L. Pauling, *The Nature the of Chemical Bond,* Cornell Univ. Press, New York, 1960.
8. H. Rodot, *J. Phys. Chem. Solids,* **25**, 85 (1964).

Conditions of Formation and Transformation of Mercury Vacancies in HgTe and HgTe-CdTe Alloys

H. Rodot, R. Triboulet and A. Hruby*

Laboratoire de Magnétisme et de Physique du Solide, C.N.R.S., 92—Meudon/Bellevue, France

ABSTRACT

Two kinds of mercury vacancies can exist in HgTe. One is acceptor for all annealing temperatures, the other is called a " complex vacancy " and is neutral if $T<350°C$ and acceptor if $T>350°C$. Only short-range movements of atoms are implied in this transformation. The concentration of the neutral vacancies is decreased in crystals formed from a solution at low temperatures, and also in the case of $Hg_{1-x}Cd_xTe$ alloys.

I. INTRODUCTION

Because the semiconducting compounds may dissolve one or several of their constituents in rather large concentrations, they present lattice defects, whose influence on the electronic properties often overwhelms that of foreign impurities. In order to control and increase their metallurgical perfection, it is necessary to know the concentration and nature of these defects and their relation with the concentration of charge carriers. The semiconductors offer some accurate methods of investigation (electrical and optical measurements) but nevertheless the exact determination of the departures from stoichiometry is generally difficult, especially because a large amount of neutral defects may be present. They must be detected by more indirect ways.

We present here a study of lattice defects in mercury telluride and in its solid solutions with cadmium telluride.

II. EXPERIMENTAL PROCEDURES

1. *Summary of previous work*

By varying the conditions of solid-vapor equilibria, we have established

* Detached from Institute of Solid State Physics, Czech. Acad. Sci., Prague.

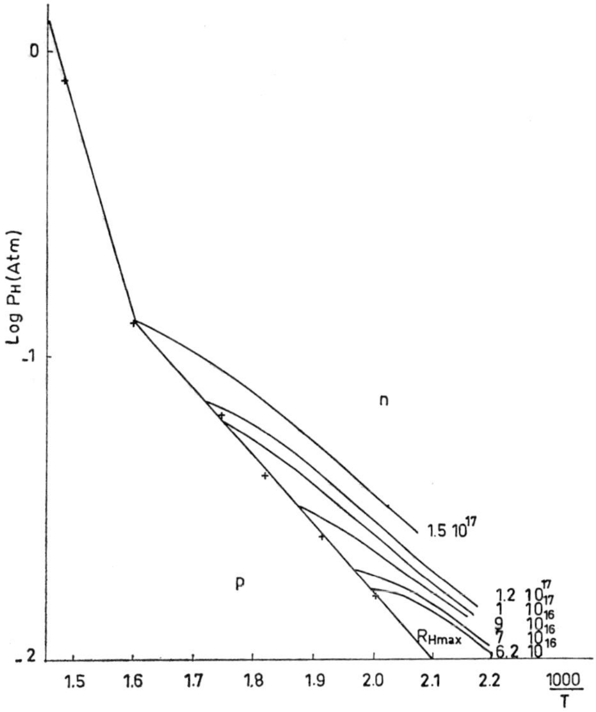

Fig. 1.

the relation between the mercury vapor pressure p_{Hg}, the annealing temperature of HgTe T and the electronic concentration in the compound x (p-T-x diagram).[1] Some parts of these results are shown in Fig. 1 in a ($\log p$) vs. ($1/T$) diagram. The temperature variation of the pressure for which $n=p$ (intrinsic regime) presents a change of slope, which implies that the type of disorder determining electrical properties was different below and above 350°C.

The existence of Frenkel disorder (mercury interstitials and vacancies) was supported by measurements of density and lattice parameter on samples of different annealing conditions having thus different electronic concentrations. However, if this disorder was determining electronic properties below 350°C, it was necessary to find the origin of the electrical behavior of the compound for $T>350$°C. At this point it was noted, through weighing Hg and Te before synthesis, and HgTe and residual Hg after synthesis, that mercury vacancies existed in HgTe melted at 675°C, in a very high concentration (4×10^{19} to 10^{20} cm^{-3}). These vacancies, which are clearly

TABLE I. HgTe: Synthesis at 675°C

Influence of the defects on the carrier-concentration	Annealing at 175°C$<T<$350°C 120 hours			Annealing at 350°C$<T<$650°C 5 minutes to 120 hours		
	Defects	Electronic state	Concentration (cm^{-3})	Defects	Electronic state	Concentration (cm^{-3})
Determining	Interstitial Hg	Donors	10^{15}–10^{19}	Complex vacancies of Hg	Acceptors	4×10^{19}–10^{20}
	Hg vacancies	Acceptors				
Small	Complex vacancies of Hg	Neutral	4×10^{19}–10^{20}	Interstitial Hg	Donors	10^{15}–10^{19}
				Hg vacancies	Acceptors	

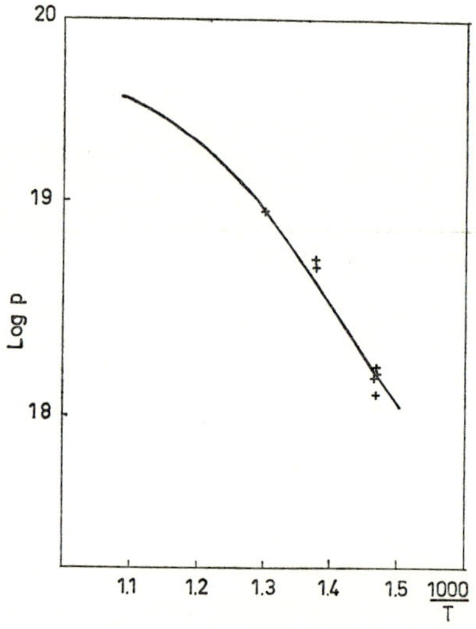

Fig. 2.

neutral electrically, were called complex vacancies, as opposed to simple vacancies which occur in Frenkel disorder. Table I summarizes the nature and concentration of defects in HgTe melted at 675°C, as a function of the annealing conditions. The dominant feature is that the complex vacancies, neutral for $T<350$°C, can be transformed by annealing at $T>350$°C into acceptors which then dominate the electrical properties. From the variation with T of the concentration of these excess acceptors (Fig. 2), the energy of transformation was determined to be 1.0 eV.[1]

In order to get a deeper insight into the nature of complex vacancies in melt-grown HgTe, we will first examine, the kinetics of their transformation from neutral to acceptor state, and then experiments made on HgTe grown from solution at low temperatures.

2. *Kinetics of transformation of complex vacancies*

The principle of our experiment is the following: Samples of HgTe grown from the melt at 675°C are first annealed at a temperature lower than 300°C, so that the complex vacancies are neutral, and under a vapor pressure minimizing the concentration of interstitial mercury. We then anneal these samples at a temperature higher than 350°C, in order both to ionize the complex vacancies and to change the concentration of interstitials according to the vapor pressure chosen for this second annealing. We can follow the variations of defect concentrations by electrical measurements after different times of annealing at various temperatures.

Experiment. Purified HgTe was melt-grown and cut into plates $8\times4\times2$ mm. These plates were annealed at 200°C under 1×10^{-2} atm. These conditions, according to ref. 1, lead to the lowest carrier concentrations.

The second annealing was performed at 300, 350, 400, 430, 470 and 500°C. For each temperature, the vapor pressure was chosen, according to ref. 1, so as to obtain after a time of 120 hr an acceptor concentration of the order of 10^{18} cm^{-3}. This choice allows a direct determination of the hole concentration from Hall effect at 77°K. The annealing was performed in silica ampules, of external and internal diameters 8 mm and 6 mm, which were quickly introduced into the furnace previously set to the annealing temperature. The thermal inertia of the system is only a few minutes. The annealing times were 5, 10, 15, 30, 60 min and 2, 4, 6, 8, 16, 32, 64 and 130 hr (different samples being used for different annealing times and temperatures). After the annealing, the ampules are taken off and water-quenched.

Results. Figure 3 shows the hole concentration p vs. annealing time t for typical temperatures: $T=300$, 350, 400 and 500°C. Two different phenomena appear on several of these curves. For $T=400$°C, t varying from 0 to 30 min, p increases rapidly from 6×10^{16} to about 10^{20} cm^{-3}. After 30 min, p decreases slowly to the equilibrium value according to a diffusion-like law.

For $T=500$°C, the second step sets up faster, so that the maximum hole concentration is lower, due to a compensation between the two types of defects respectively introduced by the two steps.

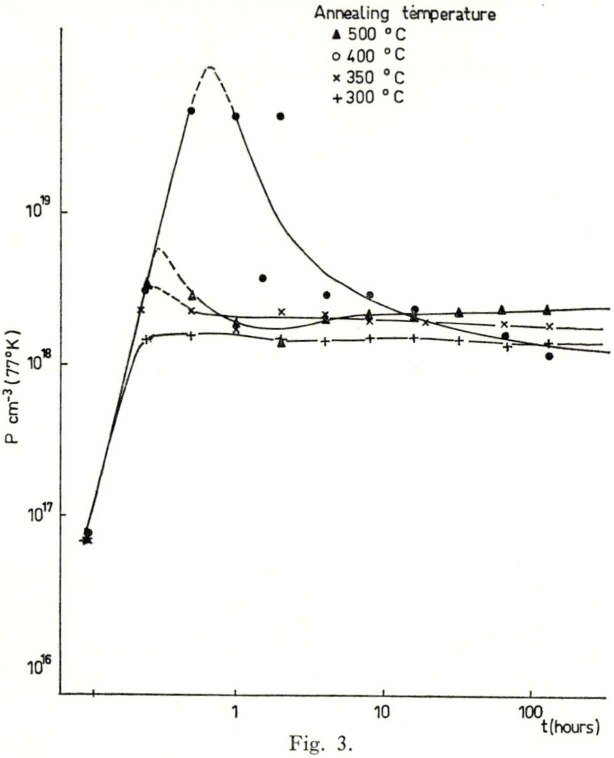

Fig. 3.

For $T=300$ and 350°C, the first step is hardly visible and the material quickly reaches its equilibrium concentration.

The typical curve obtained for 400°C may be interpreted as follows: The rapid initial increase of hole concentration is due to the transformation of neutral complex vacancies into ionizable ones. The speed of this transformation (in 5 min an important variation is already obtained) suggests that only a short-distance movement of atoms is involved. We had previously proposed[1] to consider the complex vacancies as mercury vacancies associated with displacements of the Te neighbors. Our kinetics experiment is in agreement with this assumption.

The second step of the annealing, on the other hand, is interpreted as a diffusion of interstitial Hg from the vapor phase, whose equilibrium concentration is higher after than before the second annealing, for the chosen vapor pressure. When this diffusion begins, the plate has more Hg interstitials near the surface than inside, which causes an unhomogeneity that explains the scattering of the experimental points between 1 and 6 hr.

By contrast, the samples were verified to be homogeneous just after the first step, by cutting them into four parts and measuring the hole concentration separately in the four parts.

3. *Experiments on HgTe grown from solution at 200°C*

The complex vacancies studied in Section 1 and 2 are formed during the synthesis at 675°C. We have also grown HgTe crystals from Hg-rich solutions at 200°C. This materials was first annealed to equilibrium at 200°C, then annealed again for 10 min at 400°C. Table II shows that the acceptor concentration of two different samples remained practically unchanged by this second annealing, though the same conditions would have led to an increase of one order of magnitude in the case of melt-grown crystals. From this we conclude that complex vacancies are absent in crystals grown at 200°C, so that the amount of defects present in as-grown crystals seems to be essentially related to the internal energy of the crystallizing solid.

TABLE II. HgTe : Synthesis at 200°C

Annealing temperature and time	$T = 200$°C 120 hours	$T = 400$°C 10 minutes
Acceptor concentration (cm^{-3})	8.5×10^{16}	8.0×10^{16}
	1.3×10^{17}	1.4×10^{17}

4. *Experiments on $Hg_{1-x}Cd_xTe$ solid solutions*

By a study very similar to that mentioned in Section 1 and 2, we have seen that these alloys, for $0<x<0.4$, contain also interstitial mercury and simple and complex vacancies. The concentration of complex vacancies apparenly decreases when x increases. For instance, for $x=0.34$, the hole concentration is 4×10^{17} cm^{-3} after an annealing of 30 min at 500°C, as compared with 3×10^{18} cm^{-3} in pure HgTe.

III. CONCLUSIONS

The presence, in melt-grown HgTe, of neutral "comples vacancies" in high concentration, which can be transformed to acceptor defects at high temperatures, can be considered as well established. The kinetics of this transformation indicates that only short-distance movements of atoms are involved. Indeed, the possibility that complex vacancies are associations of vacancies is to be ruled out because of the extreme speed of transformation. The diffusion of interstitials involves displacements on the order of 10^3 Å/min at 400°C, as seen from the second step of annealing. The diffusion of vacancies would be at least an order of magnitude less rapid, so that it is hardly possible that the first step involves displacements of 10^3 Å/min in HgTe and even 10^4 Å/min in the alloys, as would be needed for the vacancies to travel a distance equal to the mean distance between vacancies and form aggregates in times of the order of 10 min.

The possibility that the transformation of complex vacancies involves only displacements of Te neighbors a short distance from their normal site, as implied by the first model given for a complex vacancy,[1] is compatible with our results. This explanation is also supported by recent calculations[2] showing that vacancies with covalent bonds between Te neighbors must occur in HgTe, and can be transformed to normal vacancies with an activation energy of the order of the measured value.

The conclusion that has also been drawn, that the concentration of complex vacancies depends on the internal energy of the solid during crystallization, is also valuable for a further study of these defects.

REFERENCES

1. H. Rodot, Thése de Doctorat, Paris (1965); *J. Phys. Chem. Solids*, **25** (1964); *C. R Acad. Sci.*, **258**, 6386 (1964); C. Verie and M. Rodot, in *Proc. Int. Conf. Semicond. Physics, Paris 1964*, p. 1237.
2. F. Bailly, this volume.

Optical Effects in InSb Irradiated with Slow Neutrons

L. K. Vodopyanov, V. S. Vavilov and N. I. Koordiani

P. N. Levedev Institute of Physics, Academy of Sciences of the USSR Moscow, U.S.S.R.

Fast neutrons and various types of charged particles produce radiation defects which are regarded as being unstable. They are easily annealed at high temperatures. In the case of thermal-neutron irradiation there occur nuclear reactions (thermal neutron captures) which result in transmutations of one chemical element into another. The most probable reaction when they interact with solids is the (n, γ) reactions.

This paper reports on investigations into the possibility of using the method of nuclear doping for the experimental verification of some theoretical aspects of the physics of semiconductors.

We chose to investigate the compound semiconductor InSb, which has a large neutron absorption cross-section (140 barns). The isotope ^{115}In (which makes up 95.7% of the natural mixture) transmutes as a result of the (n, γ) reaction to a stable isotope, ^{116}Sn. The latter while replacing the In atoms in the InSb lattice, behaves like a donor impurity.

In our experiments, n-type specimens of InSb were exposed to large integrated fluxes of slow neutrons in the nuclear reactor. To decrease the background of fast neutrons, experimental channels were chosen which were remote from the uranium rods. The ambient temperature for the pile irradiation was always less than 70°C. The specimens after irradiation showed large induced radioactivity. But since the half-life of ^{116}In is only 54 min, after 25 to 30 days the activity of the sample falls.

The concentration of induced impurities was evaluated from the known neutron absorption cross-section of In and from the fixed flux of thermal neutrons. Then the concentration of the induced impurities was measured experimentally using the Hall coefficient and the resistivity measurements. It can be seen from Table I that the calculated data are smaller than the experimental data. In order to decrease this discrepancy we took into account in our calculations the resonance absorption by In atoms of neutrons with an energy of 1.46 eV. The neutron absorption cross-section for these neutrons is 2.4×10^4 barns. The concentration of the impurities after taking into account resonance absorption is listed in Table I. We see that

TABLE I.

sample N	Thermal neutron flux n/cm²	Resistivity	Carrier mobilities cm²/V·sec	Measured concentration of carriers	Calculated concentration (without resonance absorption)	Calculated concentration (with resonance absorption)
1	—	7.4×10^{-2}	3.5×10^{5}	2.4×10^{14}	—	—
2	5×10^{16}	$2\ \times10^{-3}$	1.2×10^{4}	1.3×10^{17}	5.2×10^{16}	1.1×10^{17}
3	1.5×10^{17}	9.8×10^{-4}	6.7×10^{3}	6.5×10^{17}	1.6×10^{17}	$5\ \times10^{17}$
4	4×10^{17}	$8\ \times10^{-4}$	$5\ \times10^{3}$	1.5×10^{18}	4.1×10^{17}	1.2×10^{18}
5	2×10^{18}	$3\ \times10^{-4}$	2.5×10^{3}	8.1×10^{18}	2.1×10^{18}	$6\ \times10^{18}$
6	—	1.4×10^{-3}	$7\ \times10^{5}$	$6\ \times10^{13}$	—	—
7	—	7.7×10^{-3}	$4\ \times10^{2}$	$2\ \times10^{18}$	—	—
8	6×10^{17}	7	$3\ \times10^{3}$	$3\ \times10^{4}$	—	$2\ \times10^{18}$

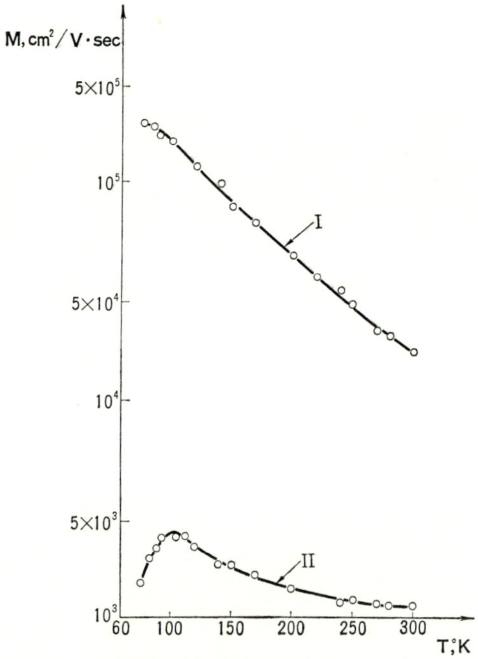

Fig. 1. Temperature dependence of the carrier mobilities.
I: before irradiation,
II: after irradiation with a flux of 2×10^{18} n/cm².

the agreement is much better now.

First of all we investigated the effect of neutron irradiation on the electrical properties of InSb. As one can easily predict, irradiation with slow neutrons causes the resistivity of the samples to decrease. This can be explained by the increase in charge-carrier concentration caused by nuclear doping. But the carrier mobility after irradiation noticeably decreased (see Table I). To obtain more information about the change in mobility after irradiation, we measured the temperature dependence of the Hall coefficient and of the resistivity before and after exposure (Fig. 1). From these experiments we found out that after irradiation mobility over a wide range of temperatures is less than before irradiation. In the temperature range from 77° to 100°K one can observe an increase in μ_e with temperature, which is probably due mainly to scattering of the ionized impurities created by irradiation.

Our main interest was in investigating the effect of thermal-neutron irradiation on the width of the fundamental absorption band. Therefore

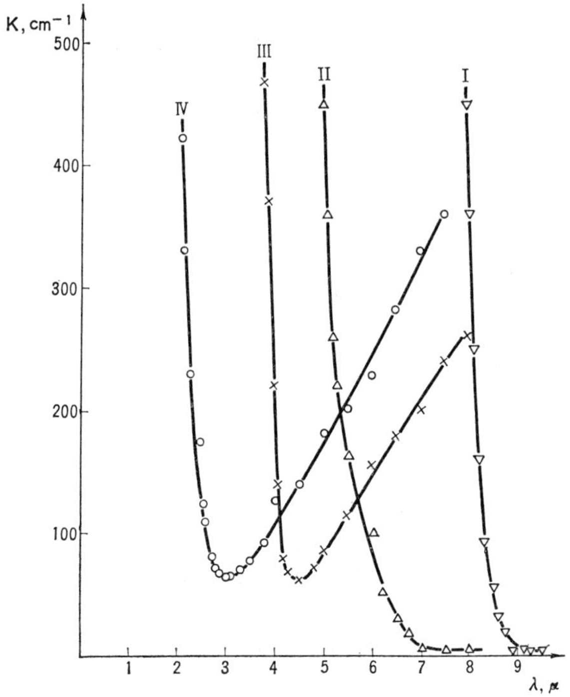

Fig. 2. Absorption coefficient spectrum for InSb before and after irradiation (300°K).

I: sample N 2 before irradiation

II: sample N 2 irradiated with a flux of 5×10^{16} n/cm^2

III: sample N 4 irradiated with a flux of 4×10^{17} n/cm^2

IV: sample N 5 irradiated with a flux of 2×10^{18} n/cm^2

we studied the optical properties of InSb. Measurements of optical absorption by samples of InSb in the infrared part of the spectrum showed that after irradiation the edge of the fundamental absorption band shifted towards the lower wavelengths, and the extent of this shift increased with the dosage (Fig. 2). These shifts were, respectively, for the sample N2 irradiated with a flux of 5×10^{16} n/cm^2—0.08 eV, for the sample N4 irradiated with 4×10^{17} n/cm^2—0.16 eV, and for the sample N5 which took the maximum dose of 2×10^{18} n/cm^2—0.36 eV. This phenomenon may be explained by the Burstein effect.[1]

By measuring the concentration of carriers in the irradiated specimens, we were able to estimate the position of the Fermi level in the conduction band. The value of the effective mass for the given impurity concentration was taken from refs. 2 and 3, where the non-parabolicalness of the conduction bond of InSb was taken into account. The shifts of the Fermi level from the bottom of the conduction band after thermal-neutron exposure were as follows: after irradiation with a flux of 5×10^{16} n/cm^2—0.05 eV, with a flux of 4×10^{17} n/cm^2—0.32 eV. It was also possible to determine the position of the Fermi level from the shift of the edge of the absorption band after irradiation by using the method described in ref. 1.

However, these two methods yield results differing by 30%.

As was shown by Keldysh[4] and Bonch-Bruevich,[5] besides the effect mentioned above in the heavily doped specimens of InSb, there should occur a narrowing of the forbidden zone. In fact, the excited impurities distort the periodic potential of the crystal lattice. This leads to the appearance of a " tail " of the density of states in the forbidden zone, and optical transitions with photon energies of $h\nu < E_g$ now become possible. The overlapping of the wave functions of the current carriers in degenerate semiconductors should also result in a reduction of the energy gap.[6] Therefore, there are three effects which have to be separated.

For observing the total narrowing caused by the " tail " of states and the interaction between the charge-carriers we studied optical transitions beyond the edge of the foundamental absorption band of InSb. The energy dependence of the absorption coefficient was calculated from the reflection spectrum for InSb. In order to detect differences in photon energies corresponding to optical transitions at a fixed value of the wave vector **K**, there must occur some deviations from monotonicity in the absorption coefficient spectrum. Such deviations from monotonicity for InSb might correspond to optical transitions between so-called combined densities of states in the valence band and in the conduction band (transitions $L_{3'}\rightarrow L_1$ in $\langle 111\rangle$ direction). When these transitions take place we

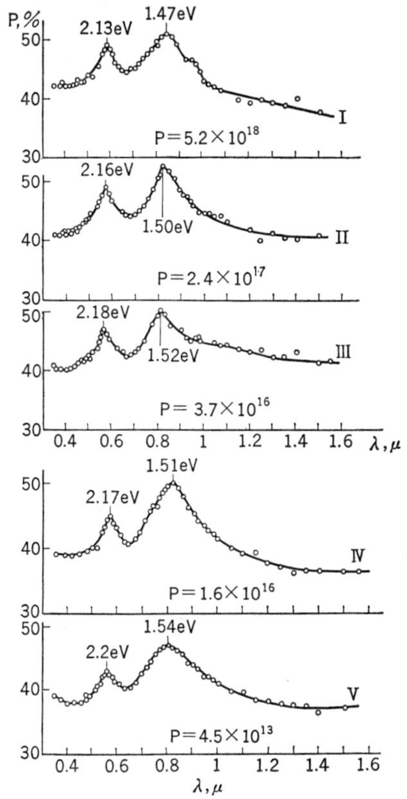

Fig. 3. Reflection spectrum for InSb (pure sample, $t=77°K$).

begin to observe a peak in the monotone spectrum of the absorption coefficient. In fact we observed a double peak due to the splitting of the state $L_{3'}$ by a spin-orbit interaction. We used these peaks as base points for observing the shift caused by narrowing of the forbidden zone (see Fig. 3). It can be seen from Fig. 3 that for heavily doped specimens, these peaks shifted into the longer wavelengths by comparison with their positions in pure samples. For p-type samples of InSb, having a doping impurity concentration of 5×10^{18} cm^{-3}, this shift was 0.07 eV. But this shift was a result of two effects. In order to remove the second effect (that is, interaction between the carriers) and eliminate the Burstein effect we tried to remove the charge carriers from the conduction band of the heavily doped specimen. To accomplish this we employed the method of nuclear doping to compensate for the presence of impurities. This method enables us, by choosing the initial acceptor concentration and thermal neutron

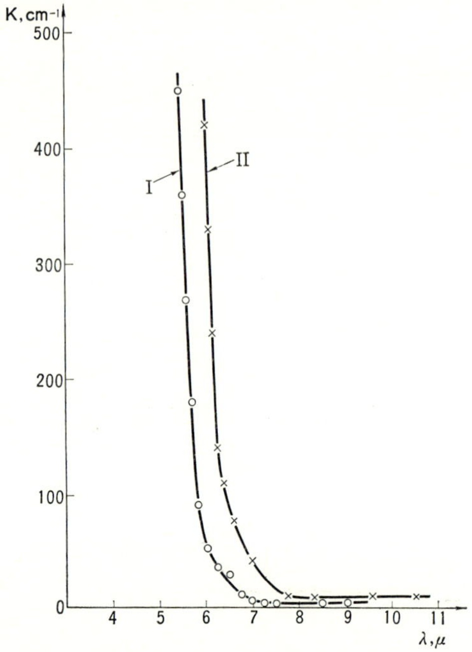

Fig. 4. Absorption coefficient spectrum for pure (I) and compensated (II) samples of InSb. ($t=83°$K).

exposure, to predetermine the final degree of compensation between acceptors and donors to within very narrow limits. We irradiated a number of p-type samples with initial hole concentration of 2.2×10^{18} cm^{-3}. After irradiation all samples showed n-type conductivity. From all irradiated specimens we chose the ones with a maximum degree of compensation. They had an electron concentration of 3×10^{14} cm^{-3}. The Hall coefficient mesurements showed that the Fermi level lies at 0.026 eV below the lower edge of the conduction band. The absorption coefficient spectrum for specimens with maximum compensation (curve II), and for the pure specimens with electron concentration at 83°K of 6×10^{13} cm^{-3} (curve I) is shown in Fig. 4. It is evident that for the compensated specimens the edge of the fundamental absorption band has shifted to lower energies; this shift was equal to 0.02 eV, which is in good agreement with the value calculated by Stern and Dixon.[7] But, as was mentioned above, the total effect (including carrier interaction) caused a shift of 0.07 eV. Therefore the narrowing of the forbidden zone due to the interaction between holes is in our case 0.05 eV.

REFERENCES

1. E. Burstein, *Phys. Rev.*, **98**, 632 (1954).
2. S. Smith and T. Moss, *Phys. Chem. Solid.*, **11**, 131 (1959).
3. W. Spitzer, and H. Fan, *Phys. Rev.*, **106**, 882 (1957).
4. L. Keldych and G. Prochko, *Soviet Physics-Solid State*, **5**, 3378 (1963).
5. L. Bonch-Bruevich, *Soviet Physics-Solid State*, **5**, 1852 (1963).
6. L. Bonch-Bruevich and P. Rozman, *Soviet Physics-Solid State*, **6**, 2535 (1964).
7. F. Stern and R. Dixon, *J. Appl. Phys.*, **30**, 268 (1959).

GaSb Tunnel Diode Degradation and Lattice Defects

R. P. Nanavati

Syracuse University, Syracuse, New York, U.S.A.

ABSTRACT

When GaSb tunnel diodes are biased at a fixed junction voltage past the valley region, their characteristics undergo irreversible changes. Both the peak and the valley currents rise exponentially towards well defined steady state values. The useful negative resistance region usually disappears altogether. It is believed that the rise of valley current is due to the creation of lattice defects and the consequent increase in the density of band-gap states.

The rise of the peak current is thought to be due to the change in the tunneling probability due to the creation and motion of charged lattice defects. Mathematical formulation of the above model presented in this paper does indeed satisfactorily explain the major experimental findings. The overall model explains all known experimental observations on the degradation of GaSb tunnel diodes.

I. INTRODUCTION

Some of the early high hopes for the use of tunnel diodes made from the wider band gap compound semiconductors have unfortunately not been fulfilled. One of the important reasons can be found in the rather rapid degradation of tunnel diodes. The first promising compound semiconductor was GaAs. It quickly developed, however, that when GaAs tunnel diodes were biased for a period of time in the forward conduction region past the valley, permanent changes in the tunnel diode characteristics occurred. For this reason these tunnel diodes are not used except under certain carefully defined circumstances.

In this paper we wish to discuss the degradation of GaSb tunnel diodes and suggest a quantitative analytical model involving lattice defects to explain the experimentally observed modes of failure. Some qualitative models have previously been put forward[1,2] to explain the changes in the i-v characteristics of GaSb tunnel junctions. Some reversible changes in the characteristics of GaSb tunnel diodes were reported by Carr. The model to be presented below explains both reversible and irreversible changes.

II. THEORY

Three conditions are being suggested as sufficient to cause permanent changes in the characteristics in GaSb tunnel diodes. These are:

(1) A sufficient source of energy must be available within the device to create lattice defects by removing host and/or impurity atoms from their normal lattice sites.

(2) The electric field intensity in the transition region must be high enough to move charged defects across the transition region. A sufficiently high gradient of lattice defects will also result in their motion.

(3) Conditions must exist so that the device is subjected to conditions beyond *deterioration threshold.* Below such a threshold self-healing of defects proceeds faster than defect generation.

That a sufficient source of energy is available in GaSb p-n junctions is quite clear from the work of Calawa.[3] He reported that high intensity radiation at 1.6μ is observed in forward-biased GaSb p-n junctions. Furthermore all of this radiation is emitted at the p-n junction with little or no spreading as a function of increasing current. He also concluded from his experimental data that no significant absorption re-emission processes are taking place away from the junction region. Furthermore Calawa's measurements indicate a quite high efficiency (i.e., greater than one) of photons of energy 0.775 eV produced per injected carrier. Only those photons which are absorbed in the tunnel junction can produce lattice defects. From the above considerations we would expect such defects to be produced primarily in the junction region. These lattice defects, if charged, can move across the transition region due to both diffusion and high electric field normally present in the junction region of a p-n tunnel junction.

The fact that the radiation is an increasing function of voltage certainly indicates that conditions will exist beyond deterioration threshold for a sufficiently high voltage in the forward direction.

These lattice defects produced as described above will influence both the valley and peak currents in p-n tunnel junctions. Consider first the valley current. The valley current usually consists of a very large component of excess current.[4] A model that seems to work well for excess current in GaSb tunnel diodes[5] is that of rate controlling tunneling into band gap states from n-side. Lattice defects are known to cause band gap states. These defects could be substitutional atoms going into interstitial sites, or the defects could be more complex in nature, such as clusters.

We may say that the valley current I_v is given by

$$I_v = AN_D N_{bg} \tag{1}$$

where A is a constant. The rate of change of band gap states N_{bg} is expected to be proportional to the net density of substitutional atoms $(N_s - N_{bg})$. Thus,

$$\frac{dN_{bg}}{dt} = B(N_s - N_{bg}) \tag{2}$$

The coefficient B is probably simply related to the injection current and may possibly be simply proportional to it. N_s is total density of substitutional states.

The solution to Eq. (2) substituted into Eq. (1) leads to the following expression for the valley current:

$$I_v = EN_D \{N_s - [N_s - N_{bg}(0)] \exp(-t/\tau_v)\} \tag{3}$$

where $N_{bg}(0)$ is simply the initial value of band gap states. From Eq. (3) an exponentially rising valley current is expected, characterized by a time constant $\tau_v = 1/B$.

In considering the peak current variation with time the following hypothesis is suggested. The lattice defects produced are charged and hence are subject to motion in the transition region due both to the electric field always present there, and to diffusion. The motion of ions (charged lattice defects) can change the width of tunneling barrier w. It is postulated that the rate of change of this barrier width is proportional to barrier width as follows:

$$\frac{dw}{dt} = -k_1 w \tag{4}$$

where k_1 is a constant. Equation (4) leads to

$$w = w_0 \exp(-t/\tau_p) \tag{5}$$

where w_0 is the original barrier width and $\tau_p = 1/k_1$.

The peak current I_p may be roughly stated as

$$I_p = GN_D N_A T_t + I_{p0} \tag{6}$$

where G is a coefficient, N_A is the acceptor density, N_D is the donor density, T_t is the tunneling transmission coefficient, and I_{p0} is the value of peak current before degradation.

Using Kane's[6] expression for T_t, writing $k_4(v_b - v)/w$ for the electric field, and using a two term Taylor series expansion for the exponential term, the following approximation is obtained for T_t:

$$T_t = k_2[1 - k_3 w_0 \exp(-t/\tau_p)] \tag{7}$$

where k_2, k_3 are coefficients. With the help of Eq. (7), Eq. (6) can be written as

$$I_p = G N_D N_A k_2 [1 - k_3 w_0 \exp(-t/\tau_p)] + I_{p0} \tag{8}$$

where G is a coefficient.

We therefore expect an exponentially rising peak current as a function of time characterized by a different time constant τ_p when the junction voltage is held fixed.

III. EXPERIMENTAL RESULTS

Figure 1 shows some data on how a typical GaSb tunnel diode degrades. It may be noted that both peak and valley currents increase with time.

In order to obtain a fairly continuous monitoring of the tunnel diode characteristics a system was set up in which the following currents were monitored in the given sequence (repeated many times)

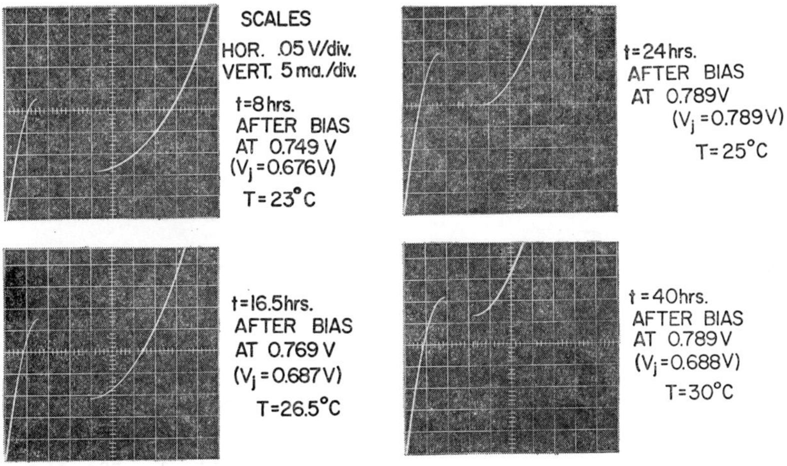

Fig. 1. GaSb S-5 life test forward bias.

peak current: 10 sec
valley current: 10 sec
forward current at a fixed voltage: 20 sec
zero applied voltage: 20 sec

In an effort to avoid instability the peak current was always measured at a constant voltage slightly below the initial peak voltage, and the valley current was monitored at a constant voltage slightly above the initial valley voltage. For room temperature and higher the diodes were immersed in a circulating oil bath in order to ensure constant temperature.

The peak and valley currents were recorded on a strip chart recorder. Figure 2 shows a typical record. It is clearly seen that a definite final value existed for both peak and valley currents. Figure 3 shows the plot of in I vs. time for peak and valley currents for three different voltage settings corresponding to the three different regions. It is clearly seen that both the peak and valley do indeed increase toward their steady state values exponentially, which is consistent with the theory just presented. Note that the peak and valley time constants are indeed quite different with

$$\tau_p < \tau_v \tag{9}$$

The peak current always reached its steady state value before the valley reached its steady state value.

The rate of degradation, i.e., the τ's are only a function of voltage and are relatively insensitive to current. Empirically

$$\tau = A \exp(-v/B) \tag{10}$$

Interestingly the coefficient B was the same for both the valley and peak current data.

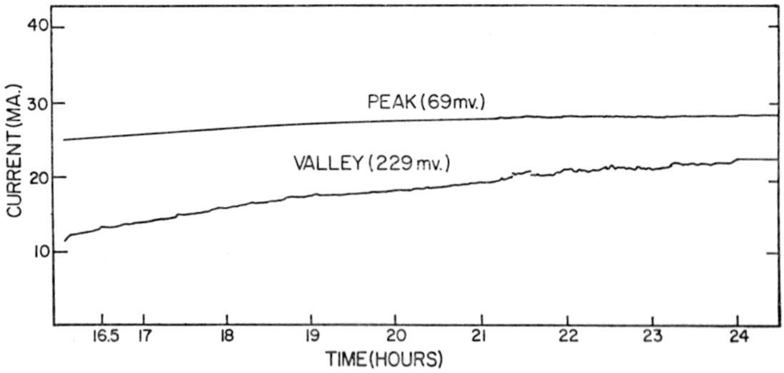

Fig. 2. Strip chart record of GaSb tunnel diode biased at 789 mV ($V_a - IR = 688$ mV) during degradation.

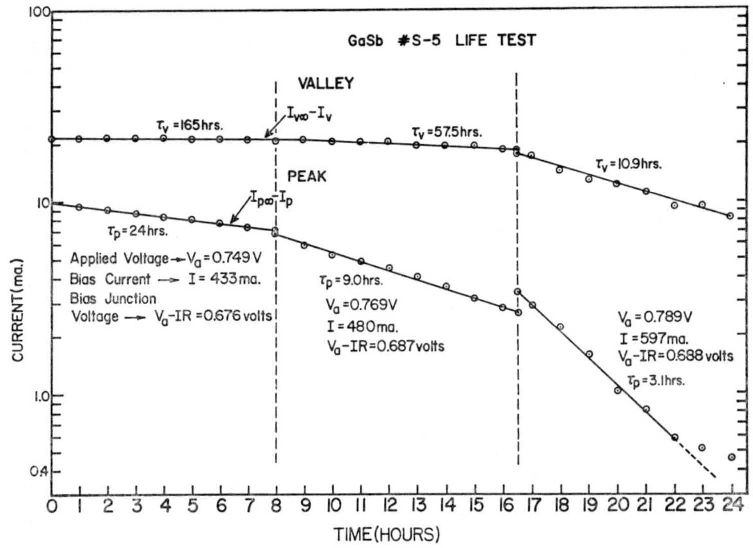

Fig. 3. GaSb #S-5 life test.

Some additional experimental observations of interest are the following:

(*1*) The total change in valley current ΔI_v was always greater than the total change in peak current ΔI_p. Often the negative resistance region disappears.

(*2*) Changes in peak and valley currents started immediately after application of bias sufficient to initiate degradation.

(*3*) In some devices the peak current decreased after reaching its steady state value.

(*4*) The bulk resistance of the diode was always higher after degradation.

(*5*) Figure 4 shows the extraordinary data on a diode which showed significant degradation *after* forward bias was removed. Figure 4(f) shows that some temporary recovery is possible. Finally Fig. 4(g) shows that the elegant Esaki diode has been reduced to a mundane resistance.

(*6*) Capacitance increases at all voltages with failure.

IV. DISCUSSION OF RESULTS

The experimental data seem to be consistent with the model proposed in the theory section. Excess current increases rapidly due to the rapid

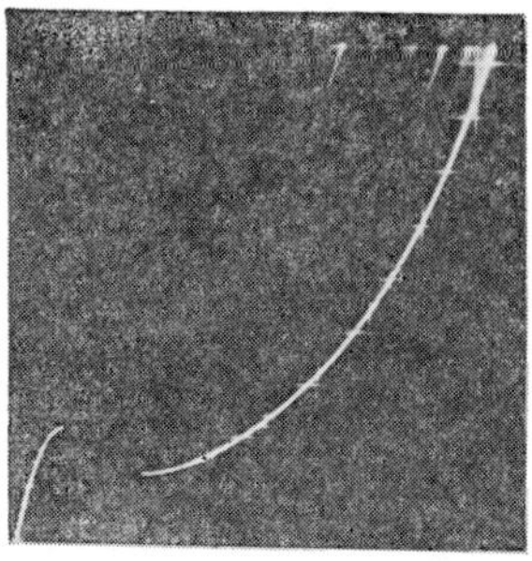

Vert. 5 mA/div ; Horiz. 0.05 V/div.
(a) Forward i-v characteristics before bias.

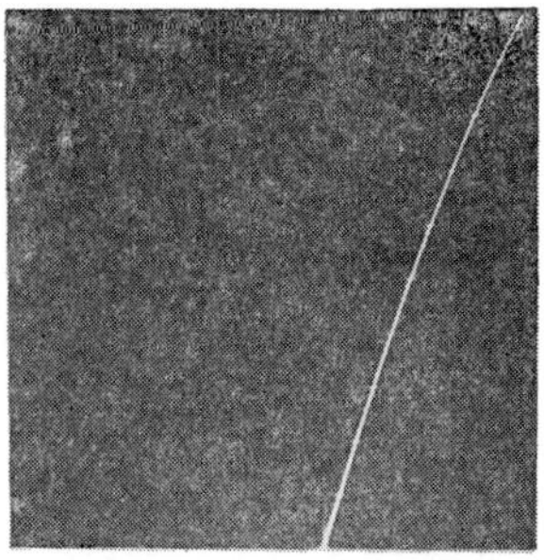

Vert. 5 mA/div ; Horiz. 0.02 V/div.
(a′) Reverse i-v characteristics before bias.

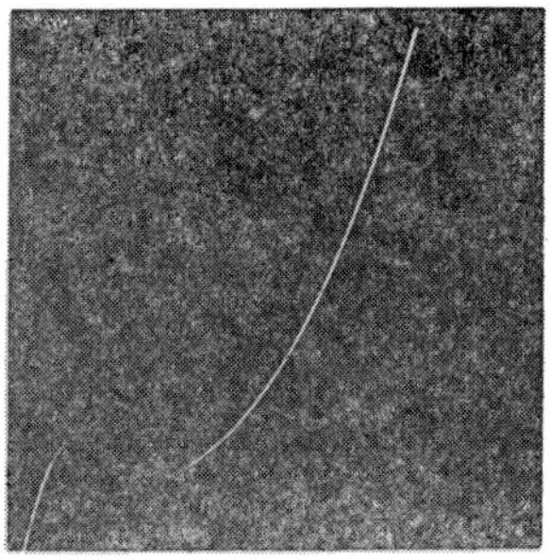

Vart. 5 mA/div, Horiz. 0.5 V/div.

(b) Forward i-v characteristics after 45 hours of bias well into the " injection " region.

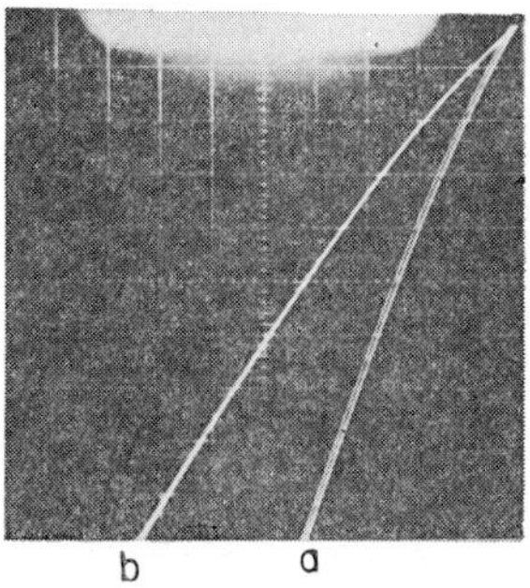

Vert. a : 5 mA/div, b : 10 mA/div ; Horiz. 0.02 V/div.
(b′) Reverse i-v characteristics after 45 hours of bias well into the " injection " region.

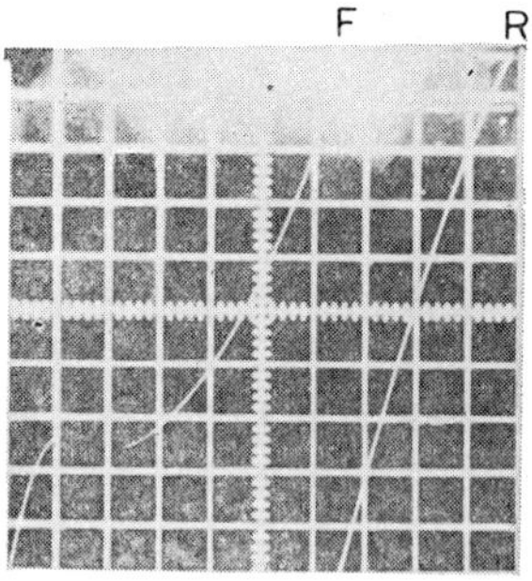

Forward (F) : 5 mA/div, 0.05 V/div.
Reverse (R) : 5 mA/div, 0.02 V/div.
(c) Forward and reverse i-v characteristics 60 hours after removal of bias.

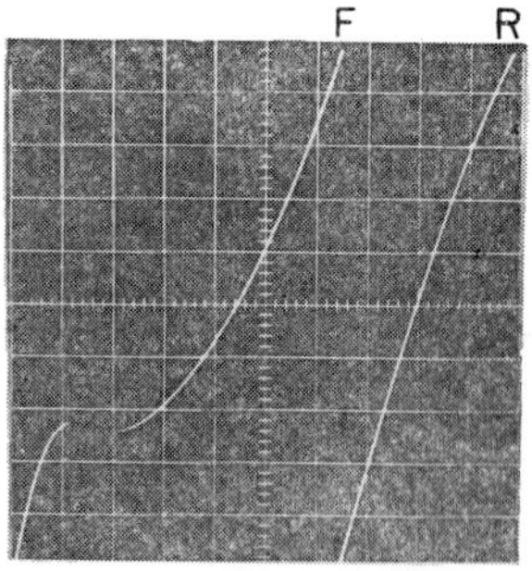

F : 5 mA/div, 0.05 V/div.
R : 5 mA/div, 0.02 V/div.
(d) Forward and reverse i-v characteristics 80 hours after removal of bias.

Fig. 4. Failure of GaSb tunnel diode after bias was removed after initially being biased in the " injection " region.

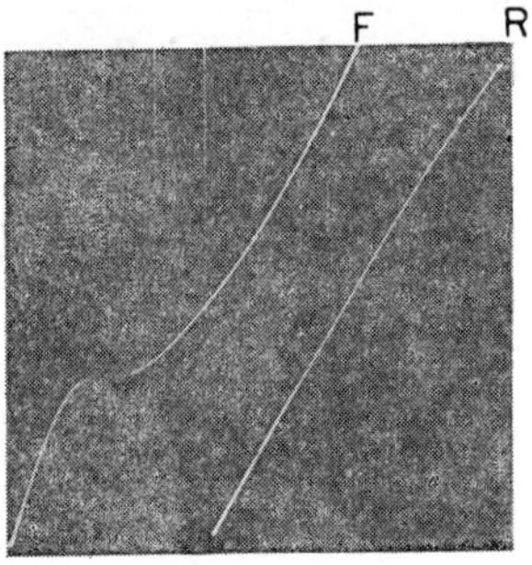

F : 5 mA/div, 0.05 V/div.
R : 5 mA/div, 0.02 V/div.

(e) Forward and reverse i-v characteristics 9 days after removal of bias.

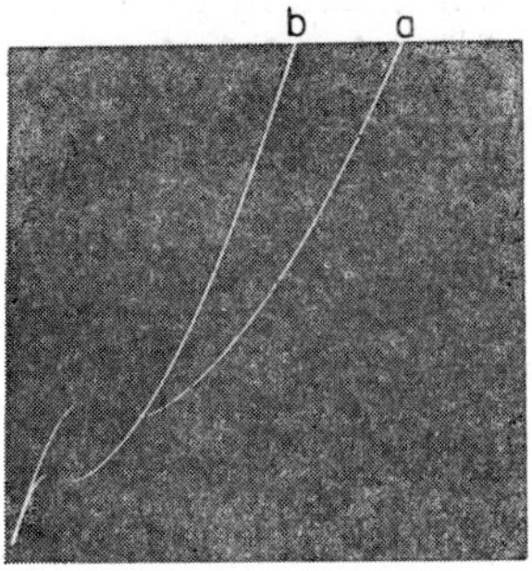

F(a) : 5 mA/div, 0.05 V/div.
F(b) : 10 mA/div, 0.1 V/div.

(f) Forward i-v characteristics after passing 200 mA forward current through device for a brief period of time.

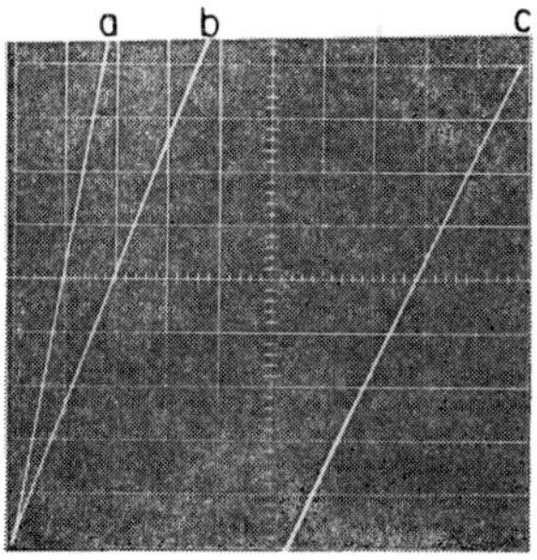

F(a) : 5 mA/div, 0.05 V/div.
F(b) : 20 mA/div, 0.1 V/div.
R(c) : 5 mA/div, 0.02 V/div.

(g) Foward and reverse i-v characteristics after applying forward bias when the diode failed completely

Fig. 4. Continued.

creation of lattice defects which in turn increase the density of band gap states and hence the consequent rise in the valley current.

The fact that the peak reaches its final value before the valley reaches its value and the fact that $\tau_p < \tau_v$ can be explained by the model, by the following reasoning. The valley current according to the proposed model is primarily a function of band gap states. It should be remembered that even after a defect has left the transition region of the p-n junction it can still contribute to valley current. The peak current however is primarily influenced in GaSb tunnel diodes by the transition region width, which in turn is influenced by the motion of charged lattice dcfects. Once a charged lattice defect has left the transition region it can no longer influence w or I_p.

The data of Fig. 4, i.e., the significant changes in i-v characteristics after bias has been removed is convincing proof that the motion of *charged* lattice defects is an important contributing factor in the mechanism of degradation. The validity of Eq. (5) is confirmed by the fact that the capacitqnce increases with increasing degradation. Figure 5 gives an easy way to understand this. The negative charges within circles, for example, are the additional charges presumably due to new lattice defects leading to a narrowing of the transition region, and a consequent increase in capacitance.

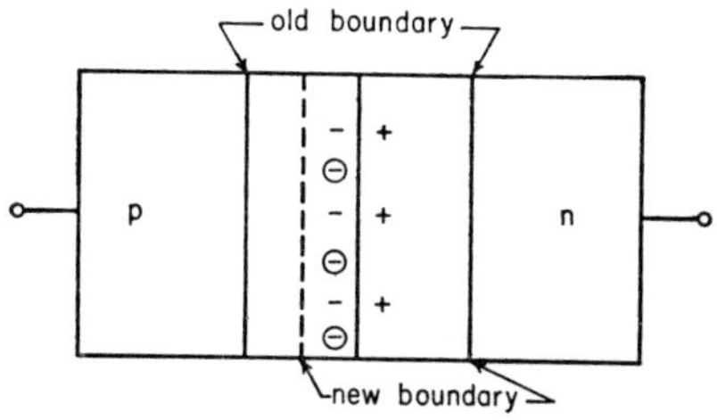

Fig. 5. Transition region narrowing with degradation in GaSb tunnel diode.

The increase in series resistance with failure is another confirmation of the proposed model. A more poorly organized lattice structure should offer higher resistance.

The report of some reversible changes in GaSb tunnel diodes reported by Carr can also be explained by saying that his experiments were conducted below degradation threshold. That is, atoms which left their normal lattice sites were able to reoccupy them on reversal of bias conditions.

As indicated by Eq. (10) the degradation is a function of voltage. Only when the diode is biased at a fixed voltage can reproducible and understandable data be obtained. Hence the more prevalent practice of observing tunnel diode degradation at fixed currents is bound to lead to incomprehensible results.

In conclusion all the known experimental data on GaSb tunnel diode degradation are consistent with the model proposed in this paper.

REFERENCES

1. R. P. Nanavati and Marcos Eisencraft, " On the Degradation of GaSb Tunnel Diodes," presented at *the 1964 Electron Devices Meeting*, Washington, D. C.

2. R. P. Nanavati, in *Physics of Failure in Electronics* (M. F. Goldberg and Joseph Vaccaro, eds.), Vol. 2, pp. 550–559, RADC Series in Reliability, 1964.
3. A. R. Calawa, *J. Appl. Phys.*, **34**, 1660 (1963).
4. A. G. Chynoweth *et al.*, *Phys. Rev.* **121**, 684 (1961).
5. R. P. Nanavati and M. Eisencraft, " On Thermal and Excess Current in GaSb Tunnel Diodes" (to be published).
6. E. O. Kane, *J. Appl. Phys.*, **32**, 83 (1961).

Part IV

Radiation Damage

Defect Complexes Induced in Semiconductors by High-energy Radiation*

J. C. Corelli

Division of Nuclear Engineering and Science, Rensselaer Polytechnic Institute, Troy, New York

ABSTRACT

Radiation damage induced in silicon and germanium by high energy electrons (45–50 MeV) was studied using infrared spectroscopy as the primary probe to study specific defects. In silicon, the annealing behavior and uniaxial stress response of the defects causing the 1.8, 3.3 and 3.9 μ infrared absorption bands indicate that these three bands all arise from the same defect, the divacancy[2] in different charge states. The effect of oxygen and phosphorous impurities on the production and recovery of the divacancy are found to be significant. Infrared active modes of vibration of oxygen defect complexes are observed in the 600 to 1200 cm^{-1} ($\approx$8 to 17 μ) region of the spectrum for irradiated silicon and germanium containing $\gtrsim 10^{17}$ oxygen atoms/cm^3. These oxygen-associated defect complexes exhibit some very striking similarities in both their production and annealing processes. One specific defect complex consisting of a single vacancy plus an oxygen atom (A-center) is observed in both silicon (12 μ) and germanium (16.2 μ). The electrical properties (conductivity and carrier concentration) were also used as probes to study the defects. In the case of n-type oxygen-doped germanium we do not observe a single defect energy level to be associated with the oxygen-vacancy complex as is found for the silicon A-center ($E_c - 0.17$ eV). The carrier concentration recovery in low ($<10^{16}$ cm^{-3}) and high ($>10^{17}$ cm^{-3}) oxygen containing n-type germanium is strongly affected by the presence of oxygen.

I. INTRODUCTION

Radiation damage induced in silicon and germanium by high-energy electrons (45–60 MeV) has been studied using infrared spectroscopy as

* Work supported by the National Aeronautics and Space Administration under Grant NsG-290.

the primary probe to study the microscopic properties and characteristics of specific radiation-induced defects. Additionally, standard DC electrical properties (conductivity and carrier concentration) have been utilized in some instances in an attempt to obtain further information on the correlation of defects.

The primary purpose of this paper will be to present results on studies of the divacancy defect in silicon and a comparison of defect complexes in silicon and germanium associated with oxygen impurity atoms. In the case of the divacancy studies[1] we have used infrared spectroscopy with polarized light and we have compared the uniaxial stress response of the defects causing the 1.8, 3.3 and 3.9 μ infrared absorption bands to the EPR divacancy studies of Watkins and Corbett.[2] In an attempt to obtain a correlation between defect complexes in silicon and germanium consisting of vacancies plus oxygen atoms we have studied the electrical properties of electron-irradiated germanium containing low oxygen concentration $<10^{16}$ cm^{-3} and high oxygen concentration $>10^{17}$ cm^{-3}. The experimental methods used are the same as those given previously for infrared[3] and electrical property[4] studies and will not be repeated here. Apart from one cold temperature (125°K) irradiation of oxygen-doped germanium all infrared studies were performed on samples irradiated at 300°K, and the electrical property studies were performed on samples of either silicon or germanium irradiated at temperatures of $\approx$80°K.

II. RESULTS AND DISCUSSION

A) *Infrared studies of the silicon divacancy*

In these experiments the silicon samples were oriented single crystals so that the incident electron beam direction, the direction of application of stress, and the direction of the incident polarized infrared beam were known. Investigation of stress-induced ordering of defects in oriented silicon single crystals has been used successfully by several workers in spin resonance[2, 5, 6] and infrared absorption[7, 8] studies to ascertain on a microscopic scale the defect symmetries in silicon, and, as will be seen below the technique has been applied in this work with fruitful results.

The defect infrared active absorption bands at 1.8, 3.3 and 3.9 μ were observed in reactor neutron, deuteron (9.6 MeV), and electron (4.5 MeV) irradiated silicon by Fan and Ramdas[9] and were observed in silicon irradiated with 45 MeV electrons by Corelli *et al.*[10] Fan and Ramdas[9] first suggested that the same defect was responsible for both the1.8 and 3.3 μ bands and in another paper[8] suggested a model for the defect center responsible for the 1.8 μ band as being one consisting of two vacancies or two interstitials on the next nearest sites with an axis of symmetry in the $\langle 110 \rangle$ direction. In addition it was known[1,8–10] that the three defect bands in question were dependent upon the position of the Fermi level for their observation in a given sample. Cheng *et al.*[1] found that the bands were independent of the doping impurity in the sample which would suggest an intrinsic defect, e.g. a divacancy. The observations[1,10] that the annealing properties of the 1.8, 3.3 and 3.9 μ bands in low oxygen-containing floating zone and in high oxygen-containing pulled silicon were very similar to the annealing of the divacancy defect from EPR studies[2] suggested that further correlative measurements be made on the infrared bands to attain additional evidence as to whether or not the divacancy was in fact the defect which gave rise to the three bands. The experimental method used to achieve a correlation was to study the stress-induced dichroism in the three infrared bands and compare this with the symmetry and orientation found from EPR studies[2] on both the electronic and atomic reorientation of the divacancy. The infrared dichroism is measured by taking the difference in absorption coefficient with light passing through the sample polarized perpendicular and parallel to the stress direction.

One particular set of samples was stressed at $\approx 160°C$ for 15 minutes with pressures which were typically in the range 2000 to 3000 kg/cm^2. The stress was applied in a $\langle 110 \rangle$ direction and the polarized infrared beam passed through the sample in the $\langle 001 \rangle$ direction. The dichroism remaining after 15 min isochronal anneals at temperatures in the range 20°C to 200°C was compared to similar isochronal annealing where the loss of divacancies was measured from the alignment as determined by EPR.[2] The comparison is shown in Fig. 1, where we have plotted the dichroic ratio (IR) (ratio of absorption coefficient for light polarized $\perp$ to stress direction to absorption coefficient for light polarized $\|$ to stress direction) and alignment (EPR) as a function of the annealing temperature. The results in Fig. 1 show that the loss of defect alignment for the three infrared bands correlates closely with what has been observed for the divacancy from EPR. Further correlations with the EPR divacancy data were obtained from a comparison of isothermal annealing experiments in which both IR and EPR yielded first order annealing kinetics and nearly the same value of activation energy 1.25 eV (IR) and $\sim$1.3 eV (EPR). Additional evidence bearing on the identification of the 1.8, 3.3 and 3.9 μ bands with the divacancy can be found in ref. 1 and will not be repeated here.

Although the evidence given above and in ref. 1 constitutes a fairly strong argument that the three IR bands are due to the divacancy defect,

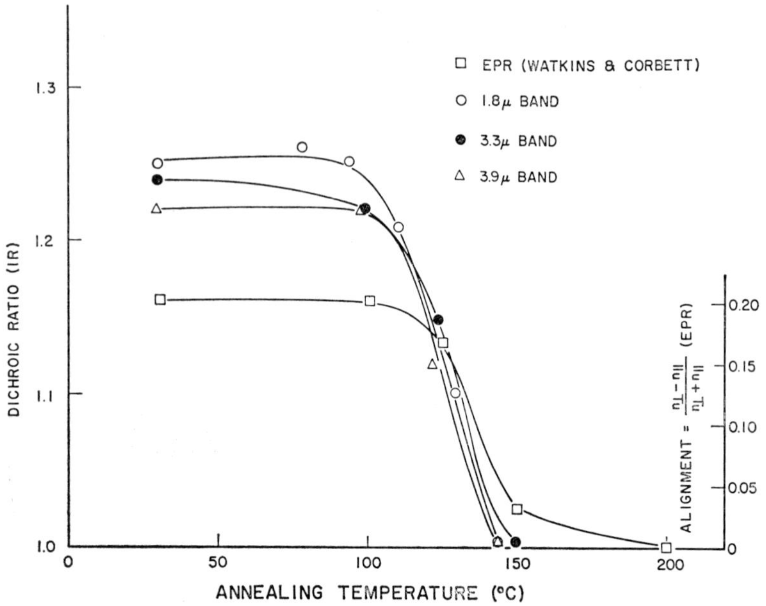

Fig. 1. Isochronal (15 min) anneal of the quenched-in dichroism in ⟨110⟩ stressed crystals for the 1.8, 3.3 and 3.9 μ bands. Also shown is the anneal of the quenched-in divacancy alignment as studied by EPR (Fig. 11 of ref. 2).

we do not conclude that the results given here are to be taken as positive proof. It can be seen that one important experiment not yet performed which can yield additional correlative evidence would be to do the stress-induced dichroism studies *in situ* at low temperatures and compare to EPR.

We have found two additional characteristics of the defect responsible for the 1.8 μ band which seem worthy of note here. The first characteristic is the fact that the presence of oxygen in the sample enhances the production rate of the 1.8 μ band, an effect also observed by Fan and Ramdas.[8] Unless one argues that the oxygen atoms have a much larger (factor of ~100) collision cross section than silicon atoms, which is unlikely, we conclude that the role of the oxygen atom in the production process is not understood and requires further study.

A second characteristic is the effect of donor concentration on the annealing of the 1.8 μ band. We have observed that the isochronal annealing of the band occurs in two stages in floating zone material but only in one stage in pulled crystals. Figure 2 shows the isochronal annealing of the 1.8 μ band for several n-type silicon samples doped either with phosphorus or arsenic to the resistivities shown on the figure (1 and 100 ohm·cm). It can be seen that the lower temperature stage accounts for a larger percentage of the annealing in lower resistivity floating zone material than it does in the higher resistivity floating zone material. The temperature of this lower stage ($\approx$170°C) is about the same temperature at which the Si-G8 center (formerly called E-center)[11] comprising a phosphorus plus vacancy complex has been observed to anneal.[12] Since E-

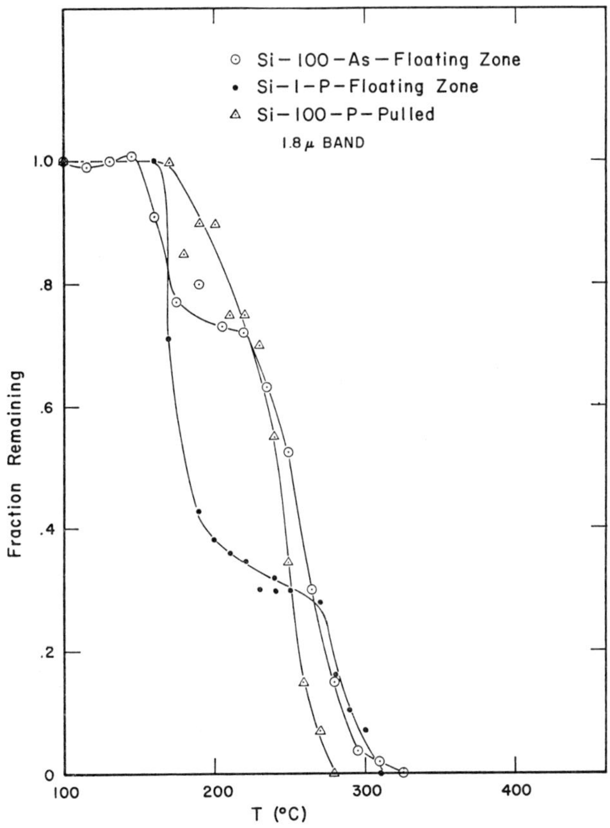

Fig. 2. Isochronal annealing of the 1.8 μ band for two floating zone silicon samples (1 and 100 ohm·cm before irradiation) and one pulled silicon (100 ohm·cm before irradiation). The incident electron energy was 45 MeV and total integrated flux was 2.5×10^{18} e/cm^2 (100 ohm·cm samples) and 1.3×10^{18} e/cm^2 (1 ohm·cm sample) at a sample temperature during irradiation of 300°K.

centers are a dominant radiation-induced defect in floating zone but not in pulled crystals, this suggests that the annealing of the E-center has an effect on the 1.8 μ band. The higher the E-center concentration as results from irradiation of the lower resistivity material (1 ohm·cm, see Fig. 2), the larger the effect on the annealing of the 1.8 μ band (divacancy). Not enough work has been done to pinpoint the nature of this effect. Two possible explanations are given here: either the mobile entity from the annealing E-center becomes trapped at the divacancy (which causes the 1.8 μ band) changing it to a new type of center or, since the 1.8 μ band is Fermi level dependent, the annealing of the E-center changes the Fermi level in such a way as to decrease the absorption of the 1.8 μ band.

B) *Radiation-induced oxygen-associated complexes in silicon and germanium*

1) *Infrared Studies.* Irradiation of silicon containing oxygen with concentrations $>10^{17}$ cm^{-3} results in the production of a number of infrared active bands in the 9–12 μ range. The first observations of these bands were reported several years ago by Fan and Ramdas[8,13] and by Corbett, Watkins, Chrenko and McDonald.[7] Of all the bands observed one found at 12 μ appears irrespective of bombarding energy (e.g. electrons 1–60 MeV) and particle type (neutrons and electrons). A detailed study by Watkins and Corbett combining EPR[5] and infrared[7] measurements resulted in a definitive identification of the defect causing the 12 μ band as one composed of a single vacancy coupled to an oxygen atom (Si-B1 center formerly called the A-center). In the intervening time since the work of Watkins and Corbett none of the other oxygen associated bands in silicon have been identified with defects whose specific configurations are known. However, in some very recent experiments Whan[14,15] and Vook[15] have made a detailed study of the oxygen-associated defect bands induced in silicon as a function of the sample temperature during 2 MeV electron irradiation and have accumulated empirical evidence which separates the defect bands (9–12 μ range) into two groups: one group in which the complexes are " composed of a single vacancy or a single interstitial trapped at an oxygen impurity atom or group of atoms " (this group would include the Si-B1 center at 12 μ), and a second group " composed of two intrinsic defects trapped at an oxygen impurity center." Whan and Vook[15] find in addition to the 12 μ band several new oxygen associated bands in the 10–12 μ range produced in low temperature irradiations. It is clear that much work still remains to be done on the identification of specific defects with oxygen-associated bands in silicon.

Detailed annealing studies on the defect infrared bands due to oxygen-

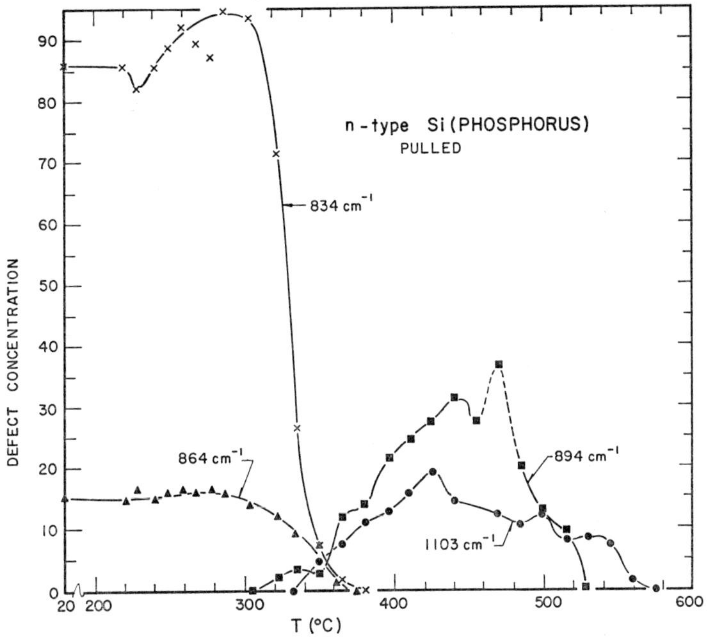

Fig. 3. Isochronal annealing of radiation-induced (60 MeV electrons) defect lattice vibrational bands associated with oxygen-defect complexes for n-type silicon (100 ohm·cm). In addition to the anneal of the 834 cm^{-1} (12 μ) and 864 (11.6 μ) bands, the growth and decay of two other bands at 894 cm^{-1} (11.2 μ) and 1103 cm^{-1} (9.07 μ), are also included. Infrared measurements made with sample at 80°K.

associated complexes have been made recently by Ramdas and Rao[16] for silicon irradiated by reactor neutrons and 4.5 MeV electrons. Similar annealing experiments using 45 MeV electrons to produce the oxygen defect complexes have been reported by Corelli *et al.*[10] Figure 3 shows the recovery and growth and decay of various oxygen associated bands in an *n*-type silicon sample (100 ohm·cm before irradiation) that has been irradiated with 60 MeV electrons at 300°K to 2.5×10^{18} e/cm^2. The annealing results given in Fig. 3 are exactly the same as those we obtained in *p*-type silicon irradiated to the same total integrated flux of 60 MeV electrons. The band at 834 cm^{-1} (12 μ) has been identified as the A-center, and we have only observed the bands at 864 cm^{-1}, 894 cm^{-1} and at 1103 cm^{-1} in oxygen-containing silicon and conclude that the bands are all manifestations of oxygen defect complexes. The results given in Fig.

3 indicate a growth of the 864 and 1103 cm^{-1} bands as the 834 and 864 cm^{-1} bands anneal out. We have previously reported[10] exactly similar annealing behavior in these oxygen bands induced in silicon by 40 MeV electrons, and since Corbett, Watkins and McDonald[17] find similar annealing behavior for the 834 and 894 cm^{-1} band in oxygen-containing silicon irradiated by 1.5 MeV electrons the production of the defect complexes and their configuration are independent of the bombarding energy.

In all samples studied we observe an increase in the silicon A-center concentration in the 225 to 300°C annealing temperature range (see Fig. 3). From Fig. 2 we see that this temperature range (225–300°C) corresponds to the recovery of the divacancy as monitored by the 1.8 μ band and this suggests that as the divacancy " breaks up " during annealing additional A-centers are produced. The Si-A center at 834 cm^{-1} is known to be associated with a net acceptor level at $E_c - 0.17$ eV found from the electrical properties of irradiated oxygen-containing silicon. The 0.17 eV level has been observed by many workers[12] in low energy electron (≤ 2 MeV) and ^{60}Co gamma (1.25 MeV) irradiation of silicon, and was recently studied in our laboratory in 1.5 MeV[12] and 45 MeV[4] electron irradiations. We find the detailed isochronal annealing results on the $E_c - 0.17$ eV level are the same as for the 12 μ band (see Fig. 3).

The first reported results on the use of infrared spectroscopy techniques in the study of irradiated germanium was recently given by Whan[18] who used 2 MeV electrons and reactor neutrons to produce defect infrared active bands due to oxygen-defect complexes. Defect absorption bands were only observed in germanium containing high oxygen concentration ($>10^{17}$ cm^{-3}). In room temperature 45 MeV electron irradiations many[3] defect bands are observed in the 11 to 16 μ region of wavelength. In subsequent 2 MeV electron irradiations performed at low temperature (25°K) Whan[19] found that only upon warm-up to temperatures between 58 and 120°K did complexes with bands at 719, 736 and 618 cm^{-1} develop, thus implying that the complexes giving rise to these bands were relatively simple. In comparison with oxygen complexes formed in silicon the annealing of the oxygen associated defect bands in germanium were found[3,19] to have some marked similarities, e.g. the 618 cm^{-1} band anneals with subsequent growth of bands at higher frequencies (700 to 850 cm^{-1} range) in a manner which closely resembles what occurs in silicon (see Fig. 3). Baldwin[20] succeeded in making EPR measurements on the oxygen complexes in germanium which when combined with the IR results strongly suggested that the band at 618 cm^{-1} was a germanium A-center consisting of a single vacancy coupled to an oxygen atom.

In order to further check the nature of the complexes in germanium a cold temperature (125°K) irradiation by 48 MeV electrons was performed on high oxygen-containing germanium (1.2×10^{17} cm^{-3}) cut from the same ingot as was used in the room temperature irradiation experiments.[3] Figure 4 shows the spectra measured in the 600 cm^{-1} to 880 cm^{-1} region after irradiation and after 20 min anneals at each of the indicated temperatures. The results shown in Fig. 4 are in good agreement with Whan's[19] results using 2 MeV electrons, in that we observe the 715 cm^{-1} band, which after a twenty minute anneal at 154°K disappears, and the strong growth of the band at 618 cm^{-1}. Presumably both these bands are due to relatively simple complexes. It should be noted that the band at 618 cm^{-1} (16.2 μ)

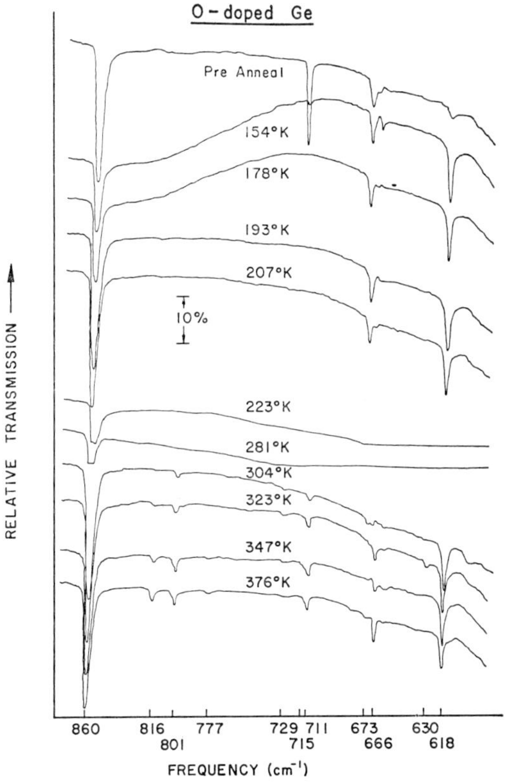

Fig. 4. Infrared spectra for irradiated oxygen-doped germanium showing in detail the wavelength region 11.4 μ (880 cm^{-1}) to 16.7 μ (600 cm^{-1}) after 48 MeV electron irradiation at 125°K before anneal and after 20 minute anneals at the indicated temperatures. The numbers on the abscissa indicate the frequency (cm^{-1}) where bands are present. All spectra were measured at 80°K.

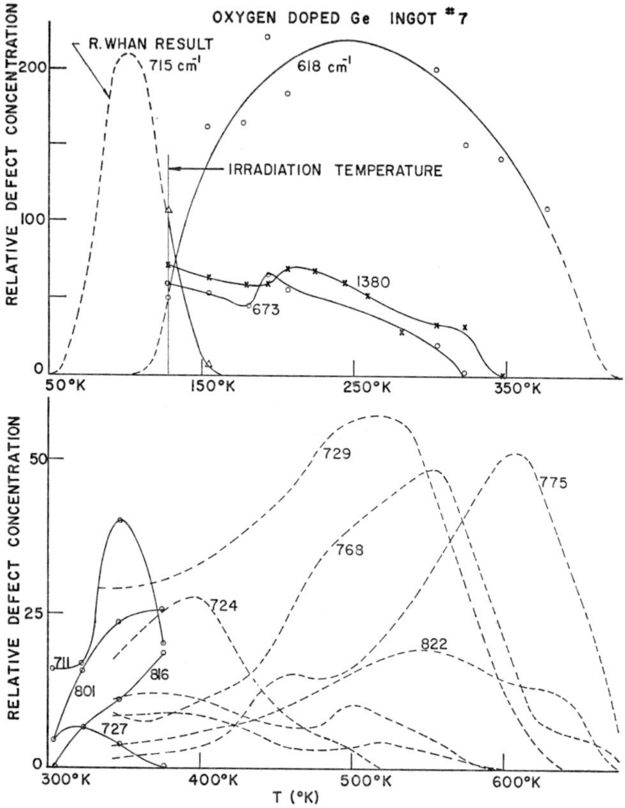

Fig. 5. Isochronal annealing of oxygen-associated defect absorption bands in irradiated oxygen-doped germanium. The result by Whan 715 cm^{-1} band is shown schematically for purposes of completeness and is inferred from Fig. 2 of ref. 19.

was the longest wavelength band observed in measurements made to 40 μ. Although not included in Fig. 4, the pre-anneal spectrum had an absorption band at 1380 cm^{-1} in addition to those at 715, 673 and 618 cm^{-1}.

After the first anneal at 154°K, the 715 cm^{-1} band is nearly completely annealed out and the 618 cm^{-1} band exhibits considerable growth, a result similar to that found by Whan.[19] The annealing results are shown in Fig. 5, where the defect concentration (area under the absorption peak) is plotted as a function of annealing temperature. Whan[21] has suggested that the 715 cm^{-1} band results possibly from a different charge or configuration state of the 618 cm^{-1} band (Ge A-center). The annealing results in Fig. 5 indicate very similar annealing behavior for a band at 673 cm^{-1} and for a band of nearly twice the frequency at 1380 cm^{-1}. These bands occur at frequencies above and below the normal oxygen band in germanium at 860 cm^{-1} and have not been observed by Whan.[19] The frequency relation between these two bands suggests they may be different vibrational modes of the same defect which, since the bands are not seen by Whan[19] in 2 MeV liquid helium temperature irradiation, may require a more energetic primary atom (as would occur in our 48 MeV irradiation) for their production, thereby being characteristic of a high energy primary defect and oxygen complex since the temperature at which the 673 and 1380 cm^{-1} anneal out is markedly different from the annealing temperatures of the 618 and 715 cm^{-1} bands.

After the 303 and 323°K anneals, weak bands at 630, 711, 729, 778, 801 and 816 cm^{-1} appear for the first time. All of these have been observed after room temperature 2 MeV[18] and 40–60 MeV[3] electron irradiations. The low temperature irradiation reported here for 48 MeV electron irradiation and by Whan[19] have served to clarify which of the defect bands arise from primary simple defects, e.g. single vacancy or interstitial, and because of the relatively high temperature required to produce the bands after low temperature irradiation we can conclude that the oxygen-defect complexes do not originate from primary defects but from defects formed by migrating vacancies (or interstitials).

One striking and unexpected feature of the annealing experiments are the large changes in background transmission. A very wide absorption band centered about the 860 cm^{-1} oxygen interstitial band appeared after the 153°K anneal and disappeared after the 193°K anneal, as can be seen in Fig. 4. The wide absorption could be some broadening effect of the 860 cm^{-1} band. A similar but more widespread absorption occurred after the 223°K anneal and lasted until the 304°K anneal (see Fig. 4). This absorption resulted in no transmission below 650–700 cm^{-1} and consequently no spectra could be measured in this temperature region. It is possible that vapors may have condensed on the sample, or alternatively by some mechanism which we cannot explain electrons were freed and later recaptured. Such an effect has been observed in electrical measurements performed on heavily irradiated *n*-type germanium and was recently reported by Gerasimov and Konovalenko,[22] but there is not adequate data on the experiments of Gerasimov and Konovalenko[22] to draw definitive conclusions.

2) *Electrical Property Studies.* In order to study the possible correlation of defect energy levels as deduced from electrical properties with specific defect infrared bands in germanium an irradiation of an oxygen-

doped germanium sample cut from the same ingot used in the infrared studies (see Figs. 4 and 5, and ref. 3) was irradiated with an oxygen-free antimony-doped germanium sample. In the case of oxygen-containing silicon one finds a level at $E_c - 0.17$ eV associated with the 12 μ band. This led us to perform the experiment to search for analogous behavior in germanium. The samples were cut in the form of 8-arm bridges used in typical DC measurements of carrier concentration and conductivity, and were irradiated at a temperature of 82°K with 48 MeV electrons using a cryostat and methods described elsewhere.[4] The antimony-doped sample (2.1×10^{15} Sb/cm^3) served as a check to observe correlations and differences between oxygen-free and oxygen-doped germanium. After irradiation

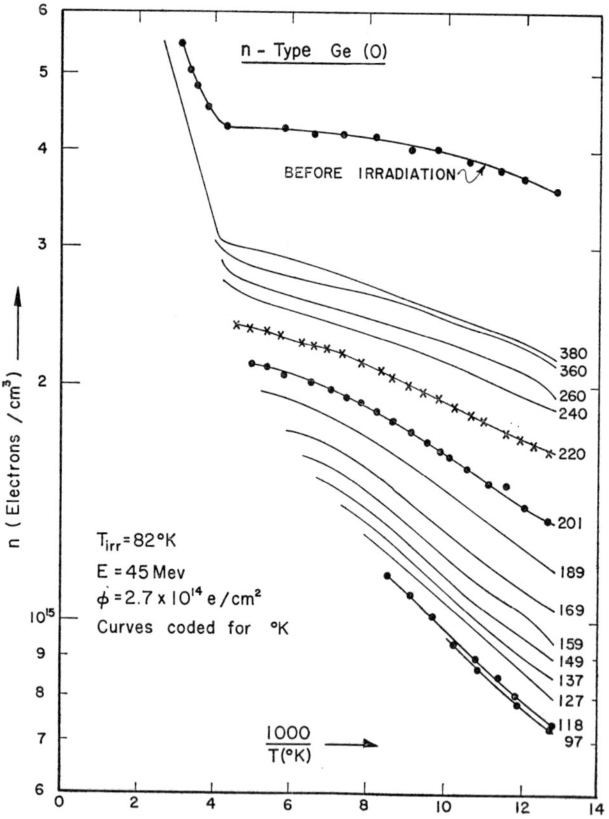

Fig. 6. Temperature dependence of carrier concentration of oxygen-doped germanium before and after irradiation at 82°K and annealing for twenty minutes at each of the temperatures indicated on the figure.

both samples were annealed isochronally using 20 min pulses at various temperatures between $\approx 80°$K up to 380°K, and in between anneals measurements were made of the temperature dependence of conductivity and carrier concentration.

The temperature dependence of the carrier concentration for the oxygen-doped sample after each anneal is shown in Fig. 6; comparable data for the antimony-doped sample is not shown because we have previously reported similar measurements on ≈ 45 MeV electron-irradiated germanium.[23] It is significant to note that the carrier removal rate for both samples was the same ≈ 12 cm^{-1}.

In the oxygen-doped germanium sample we find a donor level at $E_c - 0.20$ eV in unirradiated material. Fuller and Doleiden[24] have made detailed studies on the electrical properties of oxygen-doped germanium and have found good evidence for attributing the donor level at $E_c - 0.20$ eV as composed of two oxygen atoms in a Ge O_2 complex. From Fig. 6 we see that a defect energy level is introduced by irradiation of the oxygen-doped sample which acts as an acceptor and is located at $\geqslant 0.20$ eV from the conduction band edge which makes the determination of the radiation-induced defect difficult since it is masked by the oxygen donor level. Examining Fig. 6 and taking into account the position of the Fermi level we find that any radiation-induced defect level must be deeper 0.14 eV below the conduction band. No single dominant energy level is resolvable for the oxygen-doped sample (Fig. 6), and no shallow defect energy levels are observed to be located 0.08 to 0.14 eV of the conduction band.

In the antimony-doped sample the dominant radiation-induced defect energy level is located at $E_c - 0.20 \pm 0.03$ eV with about 75% of the removed carriers trapped on this level. We also find, as is usually the case for irradiated *n*-type germanium, a deep energy $\gtrsim E_c - 0.30$ eV level which traps carriers (about 25%).

The isochronal annealing of the carrier concentration for the germanium samples is shown in Fig. 7, where we have plotted the fraction of carrier concentration damage remaining as a function of the annealing temperature. In Fig. 7 we have referenced our carrier concentration measurements after each anneal at 85°K. For the oxygen-doped sample where we cannot resolve any one specific energy level we include all defects which remove carriers, while for the antimony-doped sample we have shown separately the annealing for all defects and the $E_c - 0.20$ eV level. The marked differences in the annealing temperatures for the oxygen-doped and antimony-doped germanium, coupled with infrared annealing results of Fig. 5 on the growth of the 618 cm^{-1} band (in the 120–250°K temperature region), give

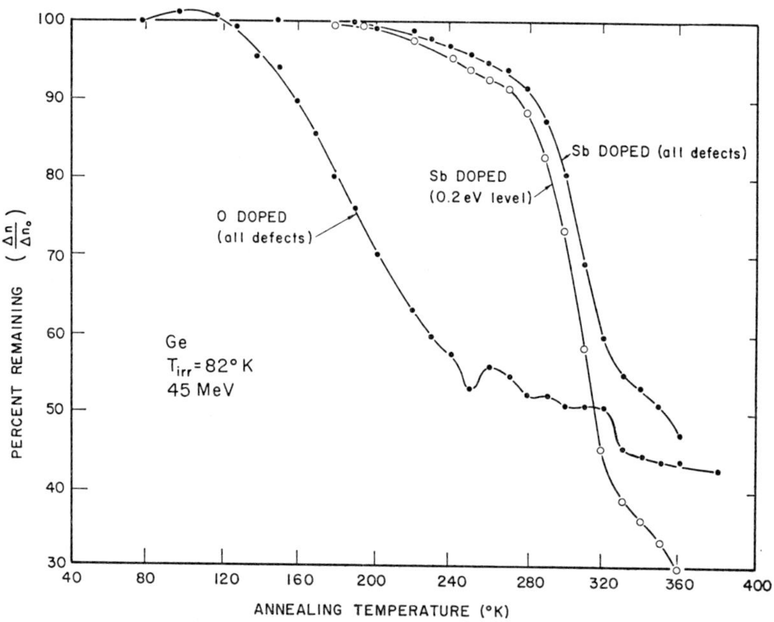

Fig. 7. The percent of carrier concentration damage remaining as a function of annealing temperature (20 min pulses) for irradiated oxygen- and antimony-doped germanium which had carrier concentrations of 3.4×10^{15} e/cm^3 and 2.1×10^{15} e/cm^3 respectively, total integrated flux was 2.7×10^{14} e/cm^2. Note: for the annealing of the antimony-doped sample an isochronal curve is given for all defects and the $E_c-0.20$ eV level separately for comparison.

strong evidence for the conclusion that the much larger concentration of oxygen atoms may be acting as a strong sink for migrating vacancies and interstitials which are apparently in a charge state such that they are electrically inactive in trapping carriers; and because of this effect we observe the recovery of carrier concentration at substantially lower temperatures. It can be recalled here that the action of oxygen atoms acting as sinks for vacancies has been observed in EPR[2] and IR[15] studies of irradiated silicon. We can discuss the annealing for the antimony-doped sample of Fig. 7 in somewhat the same light insofar as the strong annealing peak which occurs at 280–320°K is known[23] to be dependent upon impurities both in concentration and in type. A detailed study of the impurity-dependent annealing behavior for the case of ^{60}Co gamma irradiated germanium has been made by Saito, Pigg and Crawford[25] who find that the annealing in the vicinity of 300°K depends on both the type and concentration of impurity. If the results shown in Fig. 7 are compared with 45 MeV electron irradiation of purer material obtained by Fischer and Corelli[23] one concludes that the annealing near 300°K is dominated by an impurity-defect complex whose only known characteristic is that it has associated with it an energy level 0.20 eV from the conduction band.

From the results on the oxygen- and antimony-doped germanium sample we see that both the induced energy levels and annealing behavior of the radiation induced defects are different, implying the strong dependence of defects on the dopant in the sample. Although we have measured the same carrier removal rates for both samples, the conductivity changes we observed during irradiation lead us to conclude that the defects formed in the Sb-doped sample lower the mobility to a larger extent than the defects in the oxygen-doped sample, thereby implying that the " antimony defects " are more efficient scatterers.

III. SUMMARY

Radiation-induced defect infrared absorption bands in silicon at 1.8, 3.3 and 3.9 μ have been found to exhibit a very close correlation to the divacancy[2] and are concluded to be caused by different charge states of the divacancy defect. The production and annealing of the divacancy as detected by the 1.8 μ band is affected by the presence of oxygen and donor impurities, e.g. phosphorus and arsenic, in a manner which is not understood at present.

Both silicon and germanium containing high oxygen concentrations exhibit oxygen-defect complexes which are formed after primary defects migrate to oxygen atoms, and give rise to a series of vibrational bands in the 9–12 μ range for silicon and 11–17 μ range for germanium. Combining EPR[20] and IR[3,19] studies of the oxygen complexes we conclude that a defect which may be a germanium A-center is formed in the irradiation of oxygen-containing germanium. Both germanium and silicon containing high oxygen concentration show similar annealing behavior as observed by infrared properties of radiation-induced defect infrared bands. Detailed studies of the oxygen-associated defect bands in silicon show that the intensity of the A-center band at 834 cm^{-1} is saturated after a total integrated flux of $\sim 8 \times 10^{18}$ e/cm^2 (45 MeV electrons). Weak bands induced in oxygen-containing silicon at 933, 937, 1022 cm^{-1} in addition to the more intense band at 864 cm^{-1} increase with radiation flux, initially reach a maximum intensity and decrease upon further irradiation; the reason for the flux dependence of these bands is not known at present and is to be studied further in future experiments.

Electrical property studies on oxygen-doped germanium have indicated that no single defect energy level responsible for removing carriers can be associated with the germanium A-center in a manner which would be similar to what occurs in silicon (A-center and the $E_c - 0.17$ eV energy level). However, we conclude that as vacancies migrate to form the germanium A-center at 618 cm^{-1} (in the 100–250°K temperature range), the vacancies forming the 618 cm^{-1} band are trapped by oxygen atoms and are no longer available to act as carrier traps (see Figs. 5 and 7, and compare the annealing of oxygen and antimony doped germanium). Finally, we conclude that further experiments on optical and electrical properties of irradiated silicon designed to make further correlations with EPR are needed, and would prove quite fruitful in helping to unravel defects; in the case of germanium, where EPR techniques are much more difficult to apply, infrared and electrical properties correlations appear to be most promising for future work.

Acknowledgment

The experiments reported here have been performed by Dr. Arne H. Kalma, Mr. M. R. Gaerttner and Dr. L. J. Cheng, assisted in most cases by Mr. J. W. Westhead, and their work is gratefully acknowledged. We also thank Drs. J. W. Corbett and G. D. Watkins of the General Electric Research and Development Center for some very informative discussions on the divacancy defect in silicon, and their willingness to help us in our research at all times.

REFERENCES

1. A detailed account of this work appears in a paper by L. J. Cheng, J. C. Corelli, J. W. Corbett and G. D. Watkins, is published in *Phys. Rev.*, **152**, 761 (1966); and for the present paper we shall mainly refer to some of the important results as they relate to our more recent experiments.
2. G. D. Watkins and J. W. Corbett, *Phys. Rev.*, **138A**, 543 (1965).
3. J. F. Becker and J. C. Corelli, *J. Appl. Phys.*, **36**, 3606 (1965).
4. L. J. Cheng and J. C. Corelli, *Phys. Rev.*, **140A**, 2130 (1965).
5. G. D. Watkins and J. W. Corbett, *Phys. Rev.*, **121**, 1001 (1961).
6. G. D. Watkins and J. W. Corbett, *Phys. Rev.*, **134A**, 1359 (1964).
7. J. W. Corbett, G. D. Watkins, R. M. Chrenko and R. S. McDonald, *Phys. Rev.*, **121**, 1015 (1961).
8. H. Y. Fan and A. K. Ramdas, *Proc. Int'l Conf. on Semicond. Phys.*, Prague, 1960, p. 309.

9. H. Y. Fan and A. K. Ramdas, *J. Appl. Phys.*, **30**, 1127 (1959).
10. J. C. Corelli, G. Oehler, J. F. Becker and K. J. Eisentraut, *J. Appl. Phys.*, **36**, 1787 (1965).
11. G. D. Watkins and J. W. Corbett, *Phys. Rev.*, **134A**, 1359 (1964).
12. O. H. Merrill, *Masters Thesis*, Rensselaer Polytechnic Institute, June 1965 (unpublished). See also H. Saito, M. Hirata, and T. Horiuchi, *J. Phys. Soc. Japan*, **18**, Suppl. III, 246 (1963). Earlier references to papers on electrical properties of electron ($\leqslant 2$ MeV) and ^{60}Co gamma irradiated silicon are cited in the paper by Saito *et al.*
13. A. K. Ramdas and H. Y. Fan, *J. Phys. Soc. Japan*, **18**, Supplement II, 33 (1963).
14. R. E. Whan, *Appl. Phys. Letters*, **8**, No. 6, 13 (1966).
15. R. E. Whan and F. L. Vook, *Phys. Rev.*, **153**, 814 (1967).
16. A. K. Ramdas and M. G. Rao, *Phys. Rev.*, **142**, 451 (1966).
17. J. W. Corbett, G. D. Watkins and R. S. McDonald, *Phys. Rev.*, **135A**, 1381 (1964).
18. R. E. Whan and H. J. Stein, *Appl. Phys. Letters*, **3**, 187 (1963).
19. R. E. Whan, *Phys. Rev.*, **140A**, 690 (1965).
20. J. A. Baldwin, Jr., *J. Appl. Phys.*, **36** (Part 1), 793 (1965); *J. Appl. Phys.*, **36**, 2079 (1965).
21. R. E. Whan, *Appl. Phys. Letters*, **6**, 221 (1965).
22. A. B. Gerasimov and B. M. Konovalenko, *Soviet Phys.-Solid State*, **6**, 2544 (1965).
23. J. E. Fischer and J. C. Corelli, *J. Appl. Phys.*, **37**, 3287 (1966).
24. C. S. Fuller and F. H. Doleiden, *J. Phys. Chem. Solids*, **19**, 251 (1961).
25. H. Saito, J. C. Pigg and J. H. Crawford, Jr., *Phys. Rev.*, **144**, 725 (1966).

Photoconductivity of Electron-irradiated Silicon at 20°K

K. MATSUI* AND P. BARUCH

Laboratoire de Physique, École Supérieure, Paris

ABSTRACT

The defects introduced into silicon crystals by 1.5 MeV electron irradiation were studied by infrared photoconductivity measurement. Analysis of p-type crystals doped with boron, gallium or aluminium irradiated at about 20°K and 90°K indicates the presence of the following types of defects: unstable interstitial-boron complex with its energy level at $E_v+0.525\pm0.005$ eV, stable interstitial-boron complex at $E_v+0.445\pm0.005$ eV and stable interstitial-oxygen complex at $E_v+0.29\pm0.01$ eV. A fourth type with its level at $E_v+1.0\pm0.03$ eV emerging in p-type crystals irradiated at 20°K followed by subsequent annealing at around 180°K for 20 minutes is probably the vacancy-oxygen pair with its level at $E_c-0.16$ eV. The same level was introduced by 90°K irradiation without annealing, which might be attributed to an ionization effect of electron beams on vacancy migration.

In n-type crystals irradiated at 23°K, the most remarkable event is the appearance of defect levels at $E_c-0.16$ eV, $E_c-0.43$ eV (P-doped) and $E_c-0.48$ eV (As-doped). They seem to be close to the levels of vacancy-impurity pairs.

I. INTRODUCTION

In the course of the investigation of radiation damage effects in silicon, analysis by means of infrared photoconductivity techniques has played an important rôle. The technique is most suited to an exact determination of the energy level scheme of defects. We may expect that the technique is still useful in many phases of the study as a supplement of e.p.r., a typical microscopic means.[1]

---

* Present adress: Department of Nuclear Engineering, University of Tohoku, Sendai, Japan.

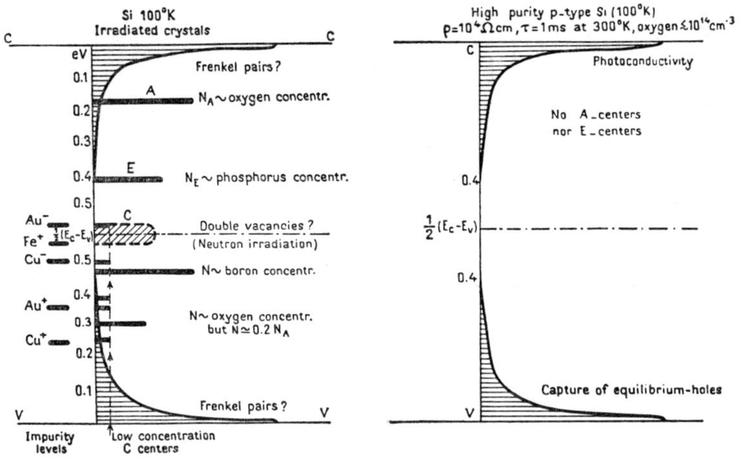

(a) (b)

Fig. 1. Energy levels in irradiated silicon (after Vavilov, ref. 2).

An extensive effort has been made in this regard by Vavilov and his associates. Figure 1 shows their results presented at the Royaumont conference.[2] They found a number of impurity-defect complexes in crystals bombarded at liquid nitrogen temperature and above. Now it is widely recognized that liquid nitrogen temperature is not low enough to keep the elemental defects namely I and V, in their isolated states.[1, 3, 4] We intended therefore to investigate the liquid hydrogen range by photoconductivity measurement. The results to be reported here cover the first-step data of the study.

The main difficulty inherent in such a study is the formation of non-equilibrium charge states of defects. The effect will be pronounced in high resistivity materials, as is the case in non-degenerated silicon bombarded at low temperatures. Therefore we limited ourselves to lightly bombarded crystals, in which the photoconductivity increment at an excitation threshold is roughly proportional to the defect density.[5] The relaxation time of the photoconduction is then governed by fast recombination on chemical dopants, which allows the use of a lock-in amplifier with chopped light.[10]

II. EXPERIMENTAL METHODS

The experimental set-up is shown in Fig. 2. The optical path was evacuated to remove the errors due to atmospheric absorption. The

Abbreviations, I: Interstitial, V (V^+, V^0, V^-): vacancy (charge +, 0, −), V_2: divacancy.

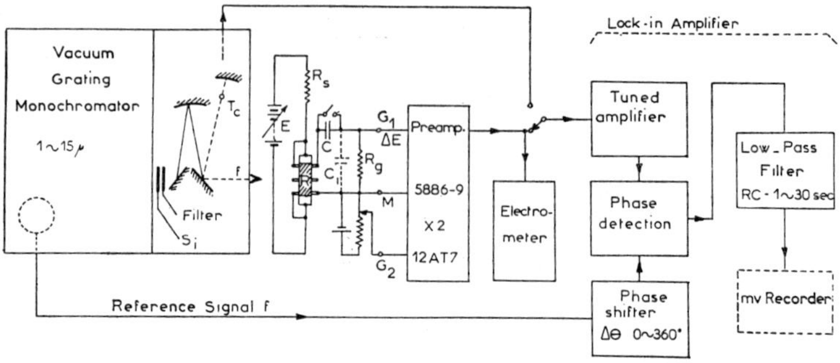

Fig. 2. Block diagram of the photoconductivity equipment.

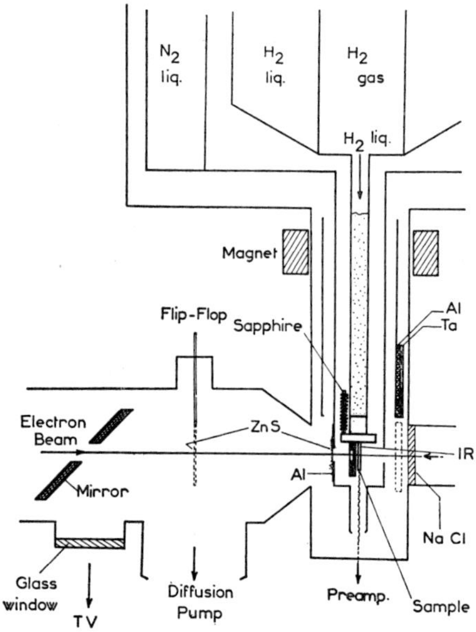

Fig. 3. Detail of the sample arrangement.

modulation frequency was chosen to 3 or 10 cps. As shown in Fig. 3 the specimen was soldered to a copper block in good thermal contact with coolant liquids and the temperature of the specimen was monitored during the course of electron bombardments or annealings. The electrical or thermal contacts were protected from the electron beam. The surface of the specimen was also protected from gas adsorption as far as possible by using a mercury pump and an activated charcoal trap.

Some photoconductivity spectra, normalized with respect to unit photon flux, are shown in Fig. 4, where some anomalies can be seen in the longer

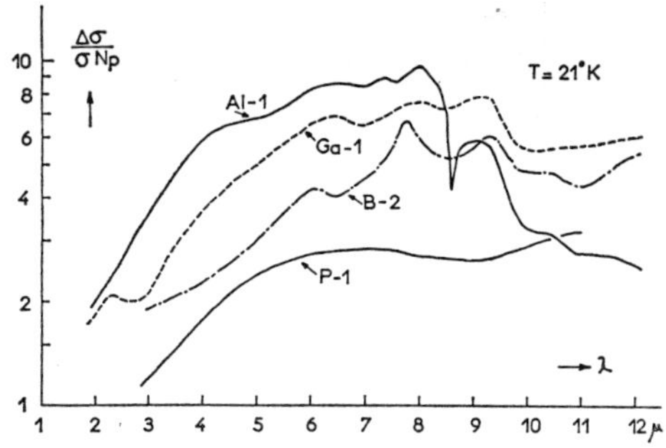

Fig. 4. Photoconductivity spectra at 21°K before irradiation.

B-2:	boron doped,	4.2×10^{12} B/cm^3;	floating zone crystal
Ga-1:	gallium ,,	8.0×10^{13} Ga/cm^3;	pulled ,,
Al-1:	aluminium ,,	1.0×10^{15} Al/cm^3;	pulled ,,
P-1:	phosphorus ,,	2.4×10^{13} P/cm^3;	floating zone ,,

wavelength sides of p-type specimens: the over-all feature of photoconductivity spectra shown in the figure is to be explained as neutral chemical dopant absorption, roughly with a dependence of $(h\nu)^{-3.5}$.[6] Such anomalous regions will be excluded hereafter from the discussion on bombarded crystals.

III. RESULTS AND DICUSSION

Now we shall enter into the discussion on electron-irradiated specimens. The lower curve in Fig. 5 indicates that all defect complex levels are absent in a pure p-type crystal. A floating-zone crystal with 1.0×10^{12} boron/cc was irradiated at around 50°K. We assign the photoconductivity rise at about 5 μ to the lowest energy level of *divacancy*, the position of which has been definitely located at $E_v+0.25$ eV by an e.p.r. study of Corbett and Watkins.[3,7]

On the other hand, in crystals containing a large amount of boron and oxygen (pulled crystals), impurity effects are prominent. The upper curve of Fig. 5 shows the photoconductivity spectrum obtained in a pulled crystal with 4.0×10^{15} boron/cc at 21°K after electron irradiation at about 25°K: three dominant levels were observed at $E_v+0.53$ eV, $E_v+0.45$ eV and $E_v+0.3$ eV, along with a weaker level of V_2 (a primary defect) at $E_v+0.25$ eV.

If we recall that long range motion of interstitials is possible even at 4°K in p-type crystals (Watkins, Royaumont symposium),[1,8] we may be safe in assigning the $E_v+0.3$ eV level to *interstitial-oxygen complex* in agree-

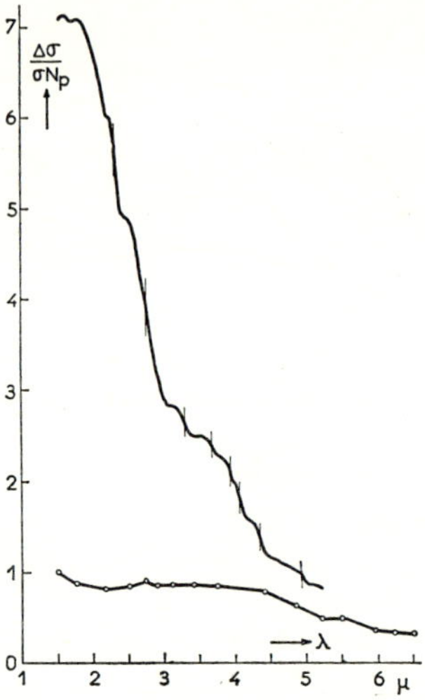

Fig. 5. Photoconductivity spectra of boron doped crystals after irradiation.

Lower curve: sample: B-1, 1.0×10^{12} B/cm^3, floating zone;
dose: 10^{15} e/cm^2, temperature 50°K.

Upper curve: sample: B-3, 4.2×10^{15} B/cm^3, pulled:
dose: 7×10^{15} e/cm^2, temperature 25°K.

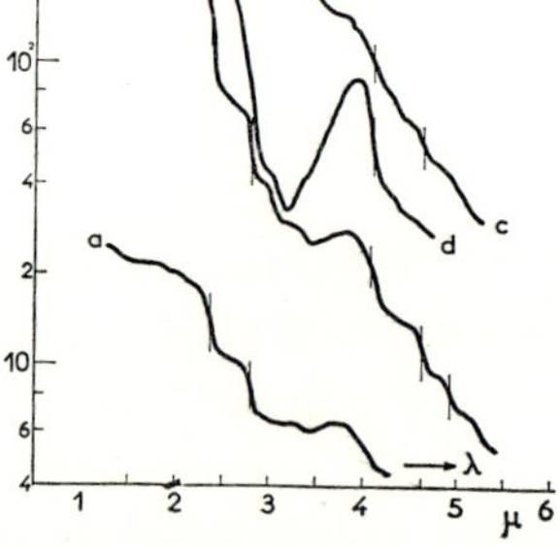

Fig. 6. Effect of annealing on photoconductivity at 21°K.

Sample B-4: 1.3×10^{16} B/cm^3, floating zone.

Curve *a*, first irradiation, $T = 21$°K, dose 2×10^{15} e/cm^2.
,, *b*, second irradiation, ,, total dose 3×10^{15} e/cm^2.
,, *c*, annealed at 80°K.
,, *d*, annealed at 280°K.

ment with Malovetskaya, Galkin and Vavilov[9] based on a different reasoning. We may conclude also that the partner of a *defect-boron complex* (Vavilov, Royaumont symposium)[2] with its level at $E_v+0.45$ eV is interstitial.

The intense level at $E_v+0.53$ eV has not been reported in literature. That it is stable only at the liquid hydrogen temperature range is demonstrated in Fig. 6. In this specimen, with 1.3×10^{16} borons/cc and floating zoned, the intensity of the $E_v+0.525$ eV level increased with electron dose (curves *a* and *b*) but decayed by annealing at 80°K (5 min, *c*). This is not due to a recharging effect (thermal bleaching) but to an annealing effect. We are inclined to an interpretation that the $E_v+0.53$ eV level belongs to a second, unstable configuration of *interstitial-boron complex* from the impurity dependence.

We then tried to compare the situation in p-type materials doped with dopant other than boron. As we have seen, it is expected that interstitials introduced by bombardment are likely to be distributed among " sinks " such as oxygen, boron, and other group III dopants.[1] Unfortunately, in our case, floating zone material doped with Al, Ga, etc. was not available.

Figure 7 shows photoconductivity spectra taken in a Ga-doped pulled crystal, with 8.0×10^{13} Ga/cc. The most intense effect of 25°K bombardment was the introduction of the $E_v+0.3$ eV level which was assigned to an interstitial-oxygen complex (curve *b*). Appearance of slight amounts of

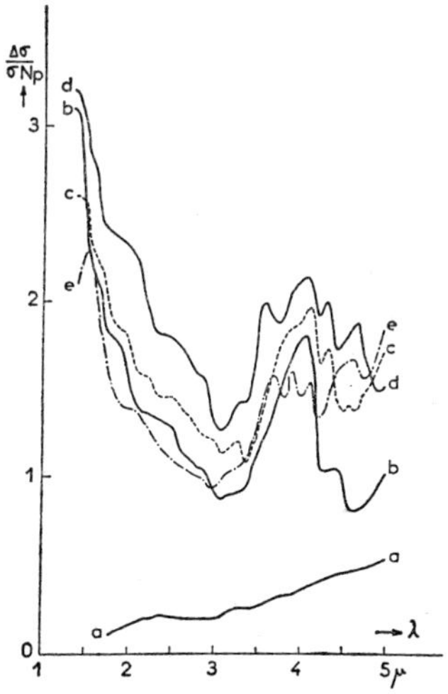

Fig. 7. Gallium doped, pulled crystal Ga-1, 8.0×10^{13} Ga/cm^3. Curve *a*, before irradiation; Curve *b*, irradiation 3×10^{14} e/cm^2 at 25°K; Curve *c*, annealed at 60°K; Curve *d*, annealed at 100°K; Curve *e*, annealed at 195°K.

the E_v+0.53 eV and E_v+0.45 eV levels may be attributed to the presence of " residual " boron impurity: the distribution coefficients of Ga and Al are in fact two orders of magnitude smaller than that of boron. No new levels to be associated with Ga impurity were found. Quite similar results were obtained in an Al-doped specimen bombarded at 25°K.

We then compared the effect of 90°K bombardments on three p-type crystals doped with B, Ga and Al, respectively. In the case of 90°K bombardment and 82°K photoconductivity measurement, one is able to give a high dose such that almost all carriers are removed from the valence band. The results indicated no levels to be associated with gallium or aluminium atoms. As *interstitial aluminium* (Al_i^{++}) has been reported in floating-zone material irradiated at 4°K and 20°K,[1,8] it is highly desirable to carry out photoconductivity measurements on such materials.

So far we have interpreted our results in terms of interstitial-impurity type defects. What about the vacancy? We are sure that *single vacancies* were present in the p-type specimens irradiated at 25°K, thanks to the e.p.r. studies of Watkins.[1,4] The single vacancy in p-type material is known in two charge states, V^+ and V^0. The first level V^+/V^0 has been roughly located at E_v+0.05 eV but no definite conclusion has been obtained for the second level V^0/V^-.[1] At present, we believe that the single vacancy took its positive form V^+ in our specimens irradiated at liquid hydrogen temperatures, which is not appropriate for observation through photoconductivity measurement. More careful studies are needed in this regard.

We studied to a certain extent the long range migration of a single vacancy by following the formation of vacancy-impurity pairs at higher temperatures.

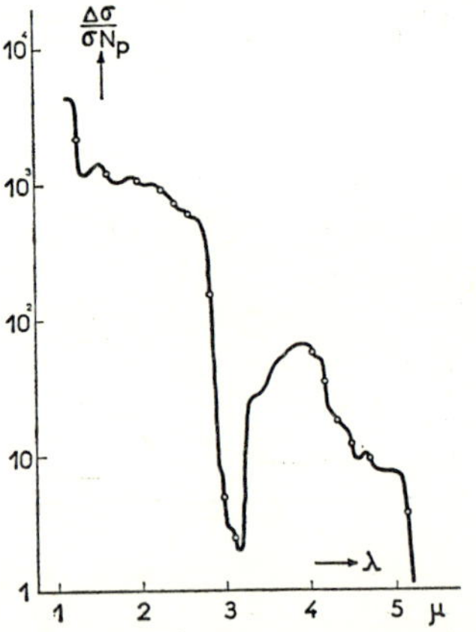

Fig. 8. Boron doped, zone floated crystal B-5, 5.0×10^{14}B/cm^3. Photoconductivity spectrum at 82°K, after 3×10^5 e/cm^2 at 90°K.

In p-type crystals irradiated at 25°K and subsequently annealed at higher temperatures, we observed a new level at $E_v+1.0\pm0.03$ eV. (*cf.* for example the curve *d* in Fig. 6 which shows the result of 280°K annealing). We observed that the level appears by annealing higher than 180°K. From the migration energy of a single vacancy 0.33 ± 0.03 eV reported by e.p.r. study[4] and from the closeness of the level to $E_c-0.16/0.17$ eV of the vacancy-oxygen pair,[3, 11, 12] we assigned the level to the *vacancy-oxygen pair.*

An interesting fact is that the level was introduced directly by 90°K *under irradiation* as a result of change of its charge state. A recent experiment of Gregory's on the effect of intrinsic illumination on solar battery elements seems to support the hypothesis.[13]

A second point of interest is the introduction of two predominant levels $E_c-0.16$ eV and $E_c-0.43$ eV in n-type crystals irradiated at 23°K, as shown in Fig. 9 (curve *b*). The behavior of the single vacancy in n-type material

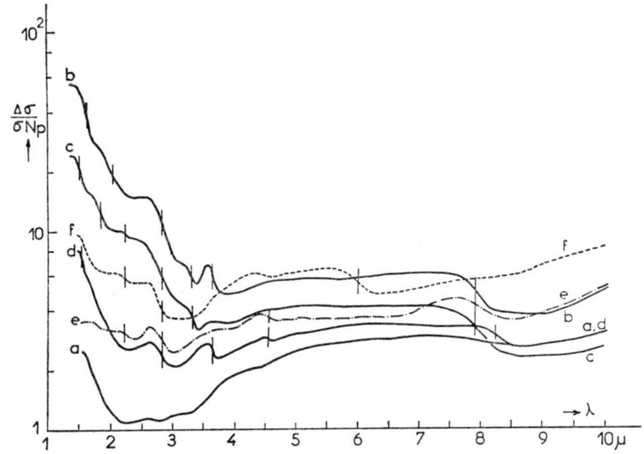

Fig. 9. Phosphorus doped (see Fig. 4); photoconductivity spectra at 21°K.

Curve *a*, before irradiation; Curve *b*, irradiated 2×10^{13} e/cm² at 23°K; Curve *c*, annealed at 60°K; Curve *d*, annealed at 100°K; Curve *e*, annealed at 180°K; Curve *f*, annealed at 280°K.

is far from settled. A recovery of V^- e.p.r. signal was reported at around 60°K in n-type crystals irradiated at liquid helium temperatures.[1] There is an indication suggesting that the recovery stage may be due to the annihilation of $V^=$.[14] There is another indication of vacancy movement even at 11°K.[15]

The two photoconductivity levels observed in our experiment are indeed very close to the levels of a *vacancy-oxygen pair* and *vacancy-phosphorus pair*, respectively.[3, 12] Another level was found at $E_c - 0.48$ eV in an arsenic doped crystal. Evidently more experiments are required to clarify " the apparent vacancy motion " at 23°K.

IV. CONCLUSION

In conclusion, we have investigated the radiation damage effects in silicon at low temperature by means of the photoconductivity technique. The results obtained are found to be very useful in interpreting the nature of elemental defects created by irradiation. Note that most of the defect models presented here are those which have escaped the scope of e.p.r. studies. As it is clear that the main problem of radiation damage study in semiconductors consists in the nature of vacancy and interstitial moving with very low activation energies, we hope to proceed to more refined experiments in the second step, such as photoconductivity measurements with polarized light,[16] effects of optical bleaching, applications to germanium, etc.

REFERENCES

1. G. D. Watkins, in *Effets des rayonnements sur les semiconducteurs*, p. 97, Dunod, Paris 1965.
2. V. S. Vavilov, *ibid.*, p. 115.
3. G. D. Watkins, J. W. Corbett and R. M. Walker, *J. Appl. Phys.*, **30**, 1198 (1959).
4. G. D. Watkins, *J. Phys. Soc. Japan*, **18**, Suppl. II, 22 (1963).
5. K. Matsui, *Thèse de Doctorat d'Etat*, Chap. II, Paris, 1965.
6. H. T. Fan, *Repts. Prog. Phys.*, **XIX**, 107 (1956).
7. G. D. Watkins and J. W. Corbett, *Phys. Rev.*, **138**, A543 (1965).
8. G. D. Watkins, *Disc. Faraday Soc.*, **31**, 86 (1961).
9. V. M. Malovetskaya, G. N. Galkin and V. S. Vavilov, *Sov. Physics Solid State*, **4**, 1008 (1962).
10. A. F. Plotnikov, V. S. Vavilov and B. D. Kopylovskii, *Instr. Exper. Techniques*, n° 3, p. 590 (1962).
11. G. Bemski, *J; Appl. Phys.*, **30**, 1195 (1959).
12. V. S. Vavilov, E. N. Lotkova and A. F. Plotnikov, *J, Phys. Chem. Solids*, **22**, 31 (1961).
13. B. L. Gregory, *J. Appl. Phys.*, **36**, 3765 (1965).
14. G. D. Watkins, private communication.
15. G. D. Watkins, in *Effets des rayonnements sur les semiconducteurs*, p. 112, Dunod, Paris, 1965.
16. Private communications from J. W. Corbett and J. Friedel

Recombination Lifetimes in Gamma-irradiated P-type Float Zone Silicon*†

RALPH A. HEWES‡ and W. DALE COMPTON

Department of Physics, University of Illinois, Urbana, Illinois

ABSTRACT

Photoconductivity decay measurements on boron doped floating zone grown silicon irradiated with ^{60}Co gamma rays indicate that the lifetime in this material is controlled by two recombination levels, one located 0.3 eV from the conduction band edge, and one located tentatively 0.17 eV from the valence band edge. The former level, not previously reported, was seen in all materials ($\rho=25$ to 1,150 ohm·cm) and has a larger capture cross section for holes than for electrons. Its rate of introduction appears to be inhibited by the presence of oxygen. Although the shallow level might possibly arise from the Si-A center (substitutional oxygen), the results of this experiment and previous Hall effect and lifetime measurements suggest that it is nearer the valence band.

I. INTRODUCTION

Much research has been done in recent years on the defects introduced into silicon by irradiation. Only now, however, can one say that a general picture of the nature of the damage is emerging, since the earlier experiments were usually of an insufficiently broad nature, as well as being too few in number, to permit much correlation between their results. Hall effect,[1–6] photoconductivity and infrared absorption,[7–11] and spin resonance measurements,[12–14] along with lifetime studies,[15–20] have now been performed on electron and gamma-irradiated materials, *n*- and *p*-type, crucible grown and floating zone grown. Electron spin resonance studies have

* This work was supported in part by a grant from the National Aeronautics and Space Administration, Grant No. NsG 228-62.

† Based on a thesis submitted to the Department of Physics of the University of Illinois in partial fulfillment of requirements for the degree of Doctor of Philosophy in Physics.

‡ Present address: General Electric Company, Nela Park, Cleveland, Ohio.

been particularly successful in providing detailed models for some of these defects.

Knowledge of the defects controlling recombination is of particular interest, since many important processes in semiconductor devices are controlled by the minority carrier lifetime. When the recombination levels coincide with the major defect levels, as appears to be the case in n-type silicon, lifetime measurements yield the values of the capture cross sections. When the recombination is controlled by defects which are not easily, if at all, detected by other techniques, lifetime measurements may be the only means of obtaining information about these defects. This appears to be the case for at least some of the recombination defects in p-type floating zone grown silicon.

II. THEORY

Recombination of electrons and holes through a single set of monovalent defects was first treated by Hall,[21] and Shockley and Read,[22] who obtained the following solution for the lifetime, valid for steady-state conditions and small excess carrier densities:

$$\tau = \tau_{p0}\frac{(n_0+n_1)}{n_0+p_0}+\tau_{n0}\frac{(p_0+p_1)}{n_0+p_0} \qquad (1)$$

where $\tau_{p0}^{-1}=N\sigma_p v_p$=lifetime of holes in highly n-type material, $\tau_{n0}^{-1}=N\sigma_n v_n$=lifetime of electrons in highly p-type material, N is the concentration of recombination centers, v_p and v_n are the hole and electron thermal velocities, respectively, σ_p and σ_n are the capture cross sections of the recombination center for holes and electrons, n_0 and p_0 are the thermal equilibrium concentrations of free electrons and holes, and n_1 and p_1 are usually defined as the thermal equilibrium concentrations of electrons and holes when the Fermi level lies at the recombination center energy level ($n_1=N_C \exp[(E_R-E_C)/kT]$, $p_1=N_V \exp[(E_V-E_R)/kT]$, N_C and N_V are the effective densities of states in the conduction and valence bands, and E_R, E_V, and E_C are the energy levels of the recombination center, the energy levels of the valence and conductor band edges, respectively). n_0 will be negligible in p-type material except at high temperatures and if no deep-lying acceptors are present, p_0 will be given by N_a, the acceptor concentration.

It should be pointed out that the foregoing definition of n_1 and p_1 does not take into account the fact that the (electron) occupied state of the recombination center contains only one more electron than the unoccupied state.

Because of spin degeneracy, the electron that is added to fill the level is assumed to go into an orbital which may or may not already contain another electron. If the orbital contains another electron, then the level can be filled in only one way (spin opposite to that of the other electron), but can be emptied in two ways (losing either a spin-up or spin-down electron) so that the ratio of the degeneracies of the full to the empty level is 2. Likewise, if the orbital does not contain an electron, it can be filled in two ways but emptied only in one, so that the degeneracy ratio, ω, is now 1/2. Thus, a factor of ω should be included in the definition of n_1. If the hole formalism is employed in the foregoing argument, it can be shown that a factor of $1/\omega$ should be used in the definition of p_1.

Sandiford[23] and Wertheim[24] showed that under conditions of small recombination center concentrations the Shockley-Read result will be valid for transient conditions such as are found in photoconductivity decay experiments. Wertheim also showed that for small enough defect concentrations the net lifetime, τ, in the presence of 2 sets of monovalent defects will be given by

$$\frac{1}{\tau} = \sum_{i=1}^{j} \frac{1}{\tau_i} \tag{2}$$

where the lifetime τ_i is the Shockley-Read lifetime for the ith set of defects. This equation most probaly applies for $j>2$, provided none of the centers act as traps for an appreciable fraction of the minority carriers. Nomura and Blakemore[25] have given a criterion for the concentration of recombination centers for which the Shockley-Read equations are valid. Shockley has shown that the surface recombination is describable as a term in Eq. (2), provided only the fundamental surface decay mode is present.

III. PROCEDURE

A) *Measurement*

Lifetimes (actually half-decay times) of excess carriers were measured by the photoconductivity decay method. A spark discharge lamp, similar to the one used by Swank,[27] with a half decay time of less than 0.6 microseconds was used to create electron-hole pairs. The light was filtered by silicon filters varying in thickness from 1.5 to 7.5 mm, to insure uniform penetration of the exciting light. The 1.5 mm filter was used only when the lifetimes obtained with its use did not differ from those obtained with a much thicker filter, indicating that injection level effects were not present. This was usually the case for the lower resistivity samples.

The samples were connected in series with a current-limiting resistor and a battery, and the transient portion of the sample voltage was amplified by a Tektronix type 1121 amplifier before being passed into a Tektronix type 533A oscilloscope, using Tektronix type L or 1A1 vertical plug-in units. The sweep-delay feature of the type 535A was used to prevent display of at least the first half-decay of the signal, in order to minimize the effects of higher order modes of surface decay than the fundamental and the effect of the finite decay time of the light pulse. The oscilloscope trace was photographed with a Polaroid camera and the photographs of the traces were measured and plotted on semilog paper to obtain the lifetime. Several different sweep rates were used on each photograph so that the zero signal level could be determined.

Discs 5 mm in thickness were cut from cylindrical boules. Rectangular parallelopipeds with dimensions typically $5 \times 5 \times 20$ mm were cut from the discs. At least two of the larger faces were polished to reduce reflection losses and the other faces were ground. The effective surface recombination lifetimes (defined as the reciprocal of the carrier decay rate due to diffusion into surface recombination centers) were found to be very large. The electrical contacts were made by ultrasonically soldering indium to the samples, and the contacts were found to be ohmic in the temperature region where trapping did not prevent measurement of the lifetime.

Each sample was given a code letter and number, and its preirradiation characteristics were measured. The letter designated the boule the sample was cut from and the number was used to distinguish between samples cut from the same boule. The lifetime was measured as a function of position along the sample length, and in those samples where significant variations were found, a region of nearly constant lifetime was selected for measurement. Voltages were chosen such that the excess carrier pulse did not drift out of this region.

The samples were glued onto copper sample mounts with G.E. #7031 insulating varnish, with layers of 1/2 mil mylar between the sample and the mount, and the sample and the copper-constantan thermocouple. The sample mounts were fixed to the bottom of a vacuum dewar, but separated from it by several resistors imbedded in a copper block. The temperature was varied by changing the power dissipated in the heater. For below room temperature measurements a cooling mixture was placed in the dewar.

The samples were irradiated at the U. S. Naval Research Laboratory in Washington, D. C., at approximately room temperature. The exposure rate was approximately 2×10^6 roentgens per hour.

The post-irradiation lifetime measurements were done from low to high temperature. On the first sample of each resistivity to be measured, note was taken of the temperature at which annealing first occurred, and measurement of the other crystals(s) of that material was restricted to lower temperatures.

Once the lifetime had been determined for a sample as a function of inverse temperature, the data were plotted on semilog paper, and a smoothed curve drawn through the data points. Each set of points $(10^3/T, \tau_m)$, where τ_m is the measured lifetime, was identified by a code number affixed after the sample number, indicating the total number of measurements taken after irradiation, including that set. Thus E5-0 indicates a pre-irradiation measurement, and E5-2 indicates the second post-irradiation measurement. The value of $\tau_{1/2}$ of the smoothed curve was read from the curve for equally spaced intervals of inverse temperature. If the lifetime was seen to be a strongly varying function of inverse temperature, the interval chosen was $\Delta(10^3/T)=0.05$, otherwise an interval of $\Delta(10^3/T)=0.10$ was chosen. A table was then constructed listing the half-decay time for these values of inverse temperature. This set of points was also given the same code as the original set for which the smoothed curve was drawn. After additional defects had been introduced into the material by irradiation, the lifetime was again measured as a function of inverse temperature, the data plotted, the smoothed curve drawn, and the values of τ_m entered into the table mentioned above.

The multilevel lifetime formula Eq. (2) can be rewritten as

$$1/\tau_m = 1/\tau_{pre} + 1/\tau_\gamma \tag{3}$$

where τ_γ is the lifetime which would be measured if the radiation induced defects were the only recombination centers present, τ_{pre} is the pre-irradiation lifetime, and τ_m is the lifetime measured after irradiation. The set of points $(10^3/T, \tau_\gamma)$ is defined as a data set, and is identified by the sample code followed by the two code numbers of the smoothed curves. Thus E5-2-0 denotes the data set describing the lifetime change caused by defects introduced between the pre-irradiation measurement (the —0) and the second post-irradiation measurement (the —2).

It was noted earlier that the surface states contribute to the lifetime, and hence there are more terms than indicated above in the formula relating pre-irradiation and post-irradiation lifetimes. The pre-irradiation lifetime will be given by

$$1/\tau_{pre} = 1/\tau_{bo} + 1/\tau_{so} \tag{4}$$

where τ_{bo} is the lifetime due to bulk defects present before irradiation, and τ_{so} is the effective pre-irradiation surface lifetime. After irradiation the lifetime will be given by

$$1/\tau_m = 1/\tau_{bo} + 1/\tau_\gamma + 1/\tau_{sf} \qquad (5)$$

where τ_{sf} is the effective surface lifetime for the post-irradiation measurement. If the two surface lifetimes are greatly different, and are comparable with the other lifetimes involved, large errors can arise in the measurement of the radiation induced lifetime. In this experiment, the surface lifetimes were much longer than any of the other lifetimes. The errors introduced by considering the two surface lifetimes to be equal is small. The error introduced by the difference in surface decay times is relatively insensitive to moderate changes in the post-irradiation lifetime, if that lifetime is much less than the pre-irradiation lifetime. The error in the radiation induced lifetime due to surface lifetime differences was calculated from the data given by Blakemore and Nomura[28] and was usually found to be 1% or less.

B) *Analysis of lifetime-temperature data*

The changes in the lifetime due to radiation induced defects were analyzed on an IBM 7094 computer for those cases where measurements could be made over a temperature range large enough to allow meaningful interpretation of computer fits to the S-R theory.

The computer program can be described as a "lesser squares fitting program," in that it adjusts the parameters A_j of the lifetime formula supplied to it, so that the quantity

$$\chi^2 = \sum_{j=1}^{N} \left(\frac{\tau_j - \tau_{\gamma j}}{W_j} \right)^2 \qquad (6)$$

is constantly reduced. In the above expression the sum runs over the N data points $(10^3/T_j, \tau_{\gamma j})$, W_j is the estimated probable error of the jth value of $\tau_{\gamma j}$, and τ_j is the lifetime calculated for the temperature of $10^3/T_j$. The W_j were chosen to be equal to $\tau_{\gamma j}/20$, since the error in the lifetime measurement is a percentage error. It was also possible to weight a given temperature region by increasing the density of data points in that region, so that that region contributed more to χ^2 than the other region(s). With the weights as chosen, χ^2 should be equal to the number of points less the number of parameters if the random error is 5%.

For the p-type material used in this experiment, the Shockley-Read equation becomes

$$\tau = \tau_{n0}(1+p_1/N_a)+(n_1/N_a)\tau_{p0} \tag{7}$$

Unless the defect level is very close to the center of the gap, one of the terms $p_1\tau_{n0}$ and $n_1\tau_{p0}$ will be much smaller than the other. Thus, if the defect level in p-type silicon is in the lower half of the gap, the lifetime will be given by

$$\tau = \tau_{n0}(1+p_1/N_A) \tag{8}$$

while if it is in the upper half of the gap, it will be given by

$$\tau = \tau_{n0}\left(1+\frac{\tau_{p0}}{\tau_{n0}}n_1/N_A\right) \tag{9}$$

The simplest approximation to Eqs. (8) and (9) is given by

$$\tau = A_1\{1+(10^3/T)^{-1.5}\exp[A_2-11.6(10^3/T)A_3]\} \tag{10}$$

where A_1 will be τ_{n0}, and A_3 the energy separation of the recombination level from the nearest band edge in electron volts. The factor of 11.6 arises from the conversion 1 eV=11,600°K. If the level is nearest the valence band, the constant A_2 will be equal to $\ln[N_V(10^3/T)^{3/2}/(N_A\omega)]-\beta$ with $N_V(10^3/T)^{3/2}$ equal to 7.1×10^{19}°K$^{-1.5}$ cm^{-3}. The quantity β arises from the assumption that the energy level has a linear temperature dependence

$$E_r = E_{r0}+\beta kT \tag{11}$$

If β is greater than zero, the level will approach the conduction band as the temperature increases.

If the level is nearer the conduction band, A_2 will be given by

$$A_2 = \ln[\omega\tau_{p0}N_c(10^3/T)^{1.5}/(N_A\tau_{n0})]+\beta \tag{12}$$

where $N_c=1.73\times10^{20}(10^3/T)^{-1.5}$ cm^{-3}. The above fit was given the code PFZ-1, since it appeared it would fit p-type float zoned material.

In most cases, it will be possible to determine which energy band the level is nearest, if all three constants can be determined. Previous studies enable us to determine reasonable limits for these quantities. The ratio τ_{n0}/τ_{p0} or its reciprocal has been determined to be about 20.[15,17] The factor e^β, from Hall effect measurements on the Si-A center,[1,3] is about 1/2 so that the product ωe^β can range from 4 to 1/4. Thus, if A_2 differs by more than ln 4 from $\ln[N_V(10^3/T)^{1.5}/N_A]$, the level most probably lies in the upper half of the gap. If it is within ln 4, identification is more difficult. Since N_V is not known very accurately, and N_A is calculated from the conductivity using published mobilities, there exists an uncertainty of as much as a factor of two in the ratio of N_V/N_A. If $\tau_{p0}/\tau_{n0}=20$ and $\omega e^{\beta}=4$ it will be impossible on the basis of the foregoing discussion to distinguish this level from one in the lower half of the gap with $\omega e^{\beta}=1/4$. Most previous accounts have tended to neglect these considerations, using $\omega=1$ and $\beta=0$.

If the nature of the defect is known from other investigations, then β may be known, and the energy level position known, making it possible to determine β if the experimental errors are small. Alternatively, if the factor ωe^{β} for a given level is known, it may then be possible to determine whether a level indicated by lifetime measurements is indeed the previously discovered level.

In several experiments, it has been determined that the quantities τ_{n0} and/or τ_{p0} are not constant. Bemski[29] reported temperature dependences of T^n, with $n=1/2$, 2 and 7/2 for several of the levels of gold in silicon. By investigating the injection level dependence of the silicon A-center, Galkin *et al.*[15] concluded that the electron capture cross-section was strongly temperature dependent above 250°K. Baicker's[17] injection level curves also indicate a temperature dependence for $\tau_{n0}+\tau_{p0}$. In view of these findings, a fit designated PFZ-2 was used when it was suspected that τ_{n0} and/or τ_{p0} were temperature dependent.

In the case of sample E, it seems that either 2 separate levels or a multivalent level control the lifetime. The data sets for this material were fit by

$$\tau = \frac{P^*Q}{P+Q} \tag{13}$$

TABLE I. List of computer fits.

Code	Formula
PFZ-1	$\tau = P_0(A_1, A_2, A_3) = A_1\{1+(10^3/T)^{-1.5}\exp[A_2-11.6(10^3/T)A_3]\}$
PFZ-2	$\tau = P_0(A_1, A_2, A_3)(10^3/T)^{A_4}$ $= P_1(A_1, A_2, A_3, A_4)$

Code	Description
PFZ-1	single S-R level in extrinsic temperature range, constant τ_{n0}, τ_{p0}
PFZ-2	same as PFZ-1 except that A_1 now has a temperature dependence of $T-A_4$

where P and Q are PFZ-1 fits. The computer fit for Eq. (13) was denoted by PFZ-10.

IV. EXPERIMENTAL RESULTS

The pre-irradiation characteristics of the 9 samples used are listed in Table II, and their radiation histories are given in Table III. Calculations

TABLE II. Pre-irradiation sample characteristics.*

Sample code	Resistivity (ohm·cm)	Pre-irradiation lifetime (μsec)	Supplier
E 3	24	200	D
E 4	25	270	D
E 5	25	260	D
M 3	71	105	M
M 4	73	98	M
F 3	222	90	D
F 4	238	133	D
U 1	1160	107	S
U 2	1160	125	S

D : Dupont M : Monsanto S : semi-elements, Inc. * At room temperature.

TABLE III. Radiation history of samples used.

Data set*	Total dose (10^6 roentgens)
E 3-1	4
E 4-1	2
E 4-2	4
E 5-1	2
E 5-2	20
F 3-1	2
F 3-2	4
F 3-3	9
F 3-4	18
F 4-1	2.5
M 3-1	2
M 4-1	1
U 1-1	2.5
U 2-1	5
U 2-2	10

* Sample is denoted by first letter and number. Second number denotes number of irradiation.

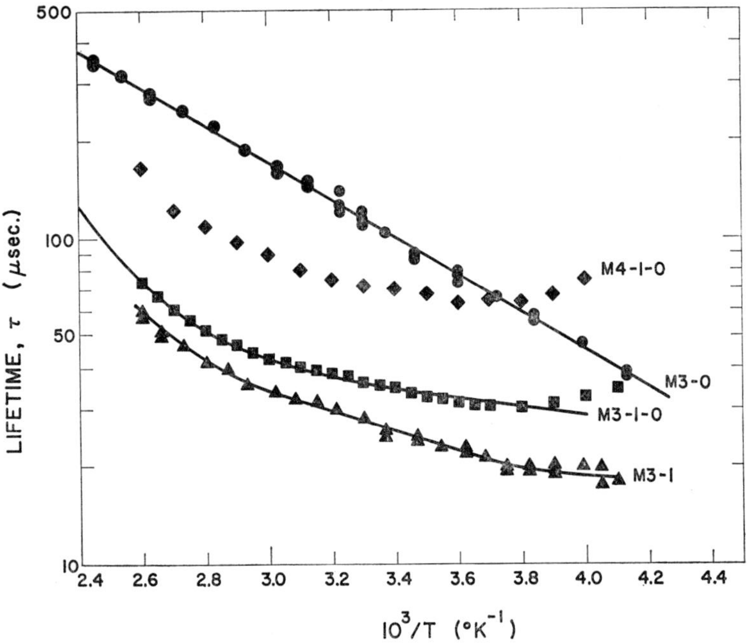

Fig. 1. Variation of radiation induced and measured lifetime with reciprocal temperature in 70 ohm·cm p-type float zoned silicon. The parameters for the theoretical curve fit to the data set M3-1-0 are given in Table IV.

showed that except possibly for the data set U2-2, the Nomura-Blakemore center density criterion was met. Figure 1 shows the variation of the lifetime in sample M3 before (data set M3-0) and after (data set M3-1) an exposure dose of 2×10^6 roentgens. The radiation induced lifetime in sample M3 (data set M3-1-0), obtained from the data sets M3-0 and M3-1 by use of Eq. (3), and for sample M4, is also shown. The parameters used to calculate the theoretical curve shown for the M3-1-0 data set are given in Table IV, and indicate, according to the Shockley-Read theory, a recombination level in the upper half of the forbidden gap, a temperature dependence of $T^{-3/2}$ for the electron capture cross section (or of T for τ_{n0}), and a value of 14.6 for the quantity (σ_p/σ_n) $(e^{-\beta}/\omega)$ indicating that if the capture cross section ratio σ_p/σ_n is about 20, as has been determined previously, that $e^{-\beta}/\omega$ is less than unity. A reasonable consistency would exist if the level is assumed to be in the lower half of the gap with a value of 5.4 for ωe^{β} ($\omega = 2$ and a temperature dependence of 10^{-4}eV/deg

TABLE IV. Computer results and radiation history.

Sample	A_1	A_2	A_3[a]	A_4	Fit	χ^2	Points	A_{20}[b]
M 3-1-0	11.52	10.97	.33	−1.0	PFZ-2	1.20	21	12.65
F 3-1-0	35.7	11.07	.311		PFZ-1	.383	12	13.6
F 3-2-0	21.1	10.57	.302		PFZ-1	1.09	15	13.6
F 3-1-2	50.8	8.10	.236		PFZ-1	1.68	15	13.6
F 3-3-0	9.94	11.92	.339		PFZ-1	1.23	14	13.6
F 3-4-0	5.38	9.06	.250		PFZ-1	.164	12	13.6
F 4-1-0	65.7	13.70	.40		PFZ-1	.40	11	13.6
U 1-1-0	38.6	13.20	.355	− .655	PFZ-2	1.33	13	15.4
U 2-1-0	15.0	12.57	.328	− .5*	PFZ-2	1.94	14	15.4
U 2-2-0	5.66	12.44	.318		PFZ-1	.37	10	15.4
E 4-2-0 E 3-1-0	P183. Q 9.71	10.8* 10.18	.46* .1784		PFZ-10	2.8	17	11.40
E 5-2-0	P 32.8 Q 2.18	7.0 9.79	.30 .170		PFZ-10	.59	13	11.40

* Parameters fixed.

a Energy separation from nearest band edge in eV.

b $A_{20} = \ln [N_v(10^3/T)^{1.5}/N_a]$.

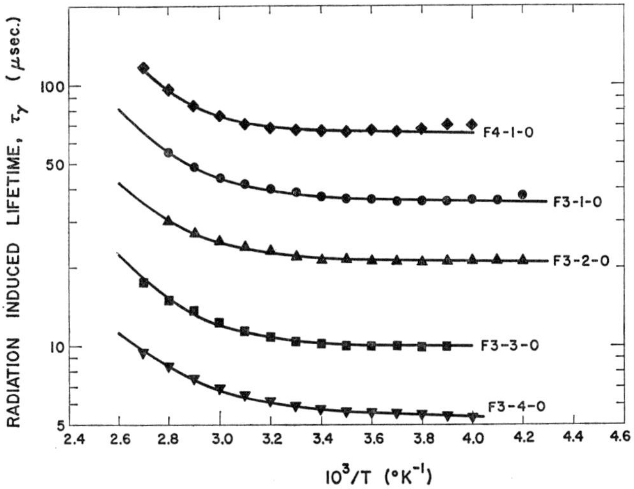

Fig. 2. Variation of radiation induced lifetime with reciprocal temperature in 200 ohm·cm p-type float zoned silicon. The parameters for the theoretical curves are given in Table IV.

for the energy level), but in view of the results on the other samples, the position near the conduction band is much more probable.

Figure 2 shows the radiation induced lifetime in samples F3 and F4,

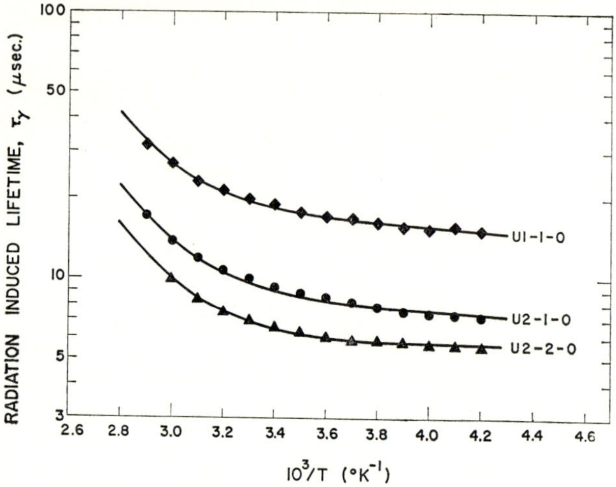

Fig. 3. Variation of radiation induced lifetime with reciprocal temperature in 1200 ohm·cm p-type float zoned silicon. The parameters for the theoretical curves are given in Table IV.

and the computer results for F4 were anomalous (compared to the F3 data sets), probably as a result of annealing at high temperatures during the post-irradiation measurement. Analysis of the F3 computer-determined parameters indicates that the energy level is definitely nearer the conduction band, being between $E_c-0.23$ eV and $E_c-0.34$ eV, with τ_{n0} and τ_{p0} apparently constant. The factor (σ_p/σ_n) $(e^{-\beta}/\omega)$ has values between 13 and 650, with a mean value of 80. Thus σ_n/σ_p is much less than one, indicating that hole capture is coulomb aided (defect negatively charged before a hole is captured).

The variation of radiation induced lifetime in samples U1 and U2 is shown in Fig. 3. The annealing of the defects at lower temperatures in this material than in the others prevented a better determination of the location of the energy level from the nearest band edge. Again the energy level was definitely located in the upper half ot the gap, at about the same separation, with the quantity (σ_p/σ_n) $(e^{-\beta}/\omega)$ ranging between 24 and 49, and a temperature dependence of T^n for τ_{n0}, with n ranging from 0 to 0.66.

In the last boule investigated, the variation of radiation induced lifetime with reciprocal temperature differed greatly from that seen in the other three. Figure 4 shows the data sets for the E samples fit by two Shockley-Read levels according to Eq. (13). One level was presumed to be the

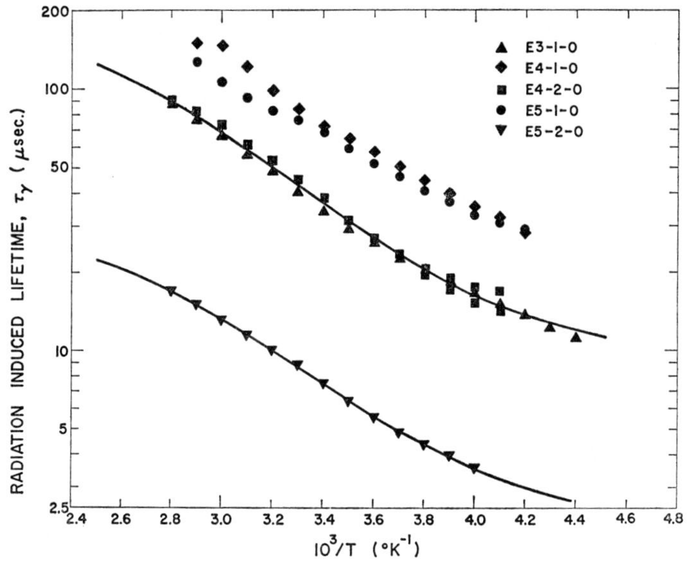

Fig. 4. Variation of radiation induced lifetime with reciprocal temperature in 20 ohm•cm p-type float zoned silicon. Parameters for the theoretical curves are given in Table IV.

level seen in the other samples, and the second level appeared to be located 0.17 eV from a band edge. The shallow (0.17 eV) level controlled the lifetime. Analysis of the computer parameters showed that this level could be located near either band edge, with $\omega e^{\beta}=3.4$ to 5 ($\omega=2$, temperature dependence of 5 to 8×10^{-5} eV/deg, level moving toward midgap with increasing temperature) for a valence band level, or with $\sigma_p/\sigma_n\ e^{-\beta}/\omega$ ranging between 9 and 14 for a conduction band level. If it is assumed that the defect is the A center, Baicker's[17] results on n-type material indicate that for the Si-A center the capture cross section ratio σ_p/σ_n is between 16 and 23, so that ωe^{β} must be greater than unity. However, the spin resonance results of Watkins and Corbett[12] on the A center indicate that $\omega=1/2$, which is in agreement with the previously mentioned Hall effect results of Wertheim,[1] and of Sonder and Templeton,[3] who suggest that ωe^{β} is about 1/4 for the A center. This indicates that if the level lies closer to the conduction band, it is not the A center.

If the introduction rate for the 0.17 eV level observed in the E samples is assumed to hold for the other materials, Eq. (8) predicts that this level will be unobservable in all but the M samples because of their higher resistivities. In the M samples, the 0.3 eV level is expected to control the lifetime except at very low temperatures. If it is assumed that trapping affects the lifetime in the M samples enough so that this behavior is masked, and that a suitable correction for an 0.17 eV level is made according to Eq. (2), the temperature dependence of the M samples appears to be more like that of the U samples.

For a single type of defect, the lifetime-flux product is proportional to the concentration of recombination centers. A calculation of this parameter indicated that, within experimental error, the introduction rates of the recombination centers are constant for material from each boule, except for the U samples. The lifetime-flux product (which for one type of recombination centers is inversely proportional to the center concentration) for the $E_c - 0.3$ eV level was 700 sec-roentgens for the E samples, 70 sec-R for the M samples, 109 sec-R for the F samples, and 47 sec-R for the U samples at fluxes below 10^7 roentgens and 66 sec-R at 10^7 R. The preceding values were calculated at $10^3/T = 3.4$. The lifetime-flux product for the 0.17 eV level in the E samples, using the computer calculated values of τ_{n0}, was about 40 sec-R.

This variation of the lifetime-flux product for the $E_c - 0.3$ eV level among the sample types indicates that any possible dependence of the introduction rate of the defects upon the boron concentration is masked by other factors. The infrared transmission of these samples, in the 9 μ band associated with interstitial oxygen,[30] was measured at room temperature. A correlation between the absorption coefficient of the 9 μ band and the lifetime-flux product indicated that material with a high-lifetime flux product for the $E_c - 0.3$ eV level contained more oxygen.

V. DISCUSSION

The deeper level seen in this experiment, at $E_c - 0.3$ eV, has not been previously reported. Hirata, Hirata and Saito[19] indicate that the slope of the post-irradiated lifetime in 150 ohm·cm ^{60}Co irradiated material is 0.17 eV. Correction of their data for pre-irradiation lifetime indicates a radiation induced level at approximately 0.3 eV from a band edge, with the ratio σ_n/σ_p greater than one. Since the resistivity of this material is close to that of the F samples, a comparison can be made, and it indicates that the two levels, and hence the defects, are not the same, since the lifetime in sample F is nearly constant for $10^3/T$ greater than 2.9 while the lifetime of the sample of Hirata *et al.* is strongly temperature dependent for $10^3/T$ less than 4.0. This difference in temperature dependences of the lifetime in the two cases is due to the different values of σ_n/σ_p.

Nakano, Nakasima and Inuishi[18] report a level at $E_V+0.27$ eV in similar material, but correction of their data for the pre-irradiation lifetime indicates a level less than 0.13 eV from a band edge.

Baicker[17] reported a level 0.18 eV from the valence band edge in 2 ohm•cm p-type float zoned silicon, with injection level measurements confirming the level position. The Hall effect measurements of Vavilov *et al.*[9] also indicate a level near the position $(E_V+0.19)$ eV. The level seen at 0.17 eV in sample E in this experiment appears to be the same as the one seen by Baicker. As mentioned above, the level could be nearer the conduction band, and might be the A center, but the valence band assignment seems more probable on the basis of these measurements.

The luminescence data of Spry[31] indicate that the short wavelength recombination luminescence spectra are the same in n- and p-type floating zone grown silicon, indicating that recombination occurs through the same defects in both materials at low temperatures (77°K and 4°K). Since A centers are known to exist in n-type float zoned silicon, this would argue that either one cannot use the parameters obtained from Hall measurements for lifetime analysis, or that both A centers and defects having a level near the valence band are present, and that at low temperatures the A centers control the recombination, while at higher temperatures the valence band levels dominate.

VI. SUMMARY

Two monovalent recombination levels, one at about $E_c-0.3$ eV and one 0.17 eV from a band edge have been seen in ^{60}Co irradiated boron doped floating zone grown silicon. The introduction rates of the defects giving rise to these levels are constant up to doses of 18×10^6 roentgens in material containing more than 6×10^{13} boron atoms/cm^3. The introduction rate of the $E_c-0.3$ eV level defects is inhibited by the presence of oxygen in the lattice. These defects have a larger capture cross section for holes than for electrons. A conclusive assignment of the 0.17 eV level to the A center or to a level near the valence band cannot be made, although a level near the valence band seems somewhat more probable.

Acknowledgments

Thanks are due to Mr. J. Pfaff and Mr. E. J. West of the U. S. Naval Research Laboratory (Washington, D. C.) for performing the Cobalt 60 irradiations.

REFERENCES

1. G. K. Wertheim, *Phys. Rev.*, **110**, 1272 (1958).
2. V. M. Malovetskaya, G. N. Galkin and V. S. Vavilov, *Soviet Physics-Solid State*, **4**. 1008 (1962).
3. E. Sonder and L. C. Templeton, *J. Appl. Phys.*, **31**, 1279 (1960).
4. E. Sonder and L. C. Templeton, *J. Appl. Phys.*, **34**, 3295 (1963).
5. E. Sonder and L. C. Templeton, *J. Appl. Phys.*, **36**, 1811 (1965).
6. H. Saito, M. Hirata and T. Horuichi, *J. Phys. Soc. Japan*, Suppl. III, 264 (1963).
7. V. S. Vavilov and A. F. Plotnikov, *Proc. of Int. Conf. on Crystal Lattice Defects*, *J. Phys. Soc. Japan*, **18**, Suppl. III, 230 (1963).
8. H. Y. Fan and A. K. Ramdas, *J. Appl. Phys.*, **30**, 1127 (1959).
9. V. S. Vavilov, G. N. Galkin, V. M. Malovetskaya and A. F. Plotnikov, *Soviet Physics-Solid State*, **4**, 1442 (1963).
10. V. S. Vavilov, S. I. Vintovkin, A. S. Lyutovich, A. F. Plotnikov and A. A. Sakolova, *Soviet Physics-Solid State*, **7**, 399 (1965).
11. J. W. Corbett, G. D. Watkins, R. M. Chrenko and R. S. MacDonald, *Phys. Rev.*, **121**, 1015 (1961).
12. G. D. Watkins and J. W. Corbett, *Phys. Rev.*, **121**, 1001 (1961).
13. G. D. Watkins and J. W. Corbett, *Phys. Rev.*, **134**, A1359 (1964).
14. G. D. Watkins and J. W. Corbett, *Phys. Rev.*, **138**, A543 (1965).
15. G. N. Galkin, N. S. Rytova and V. S. Vavilov, *Soviet Physics-Solid State*, **2**, 1819 (1961).
16. R. H. Glaenzer and C. J. Wolf, *J. Appl. Phys.*, **36**, 2197 (1965).
17. J. A. Baicker, Phys. Rev., **129**, 1174 (1963).
18. T. Nakano, K. Nakasima and Y. Inuishi, *J. Phys. Soc. Japan*, **20**, 2140 (1965).
19. M. Hirata, M. Hirata and H. Saito, *J. Appl. phys.*, **37**, 1867 (1966).
20. R. A. Hewes, "Recombination Lifetimes in Gamma-Irradiated Silicon," Thesis, University of Illinois, 1966 (unpublished).
21. R. N. Hall, *Phys. Rev.*, **83**, 228 (1951).
22. W. Shockley and W. T. Read, *Phys. Rev.*, **87**, 835 (1952).
23. D. J. Sandiford, *Phys. Rev.*, **105**, 524 (1957).
24. G. K. Wertheim, *Phys. Rev.*, **109**, 1086 (1958).
25. K. C. Nomura and J. S. Blakemore, *Phys. Rev.*, **112**, 1067 (1958).
26. W. Shockley, in *Electrons and Holes in Semiconductors* (D. Van Nostrand Company, Inc., New York, 1950), Chap. 12, p. 319.
27. R. K. Swank, " Lifetimes of Excited States in Alkali Halide Crystals Containing F Centers," Thesis, University of Illinois, 1962 (unpublished).
28. J. S. Blakemore and K. C. Nomura, *J. Appl. Phys.*, **31**, 753 (1960).
29. G. Bemski, *Phys. Rev.*, **111**, 1515 (1958).
30. W. Kaiser, P. H. Keck and C. F. Lange, *Phys. Rev.*, **101**, 1264 (1956).
31. R. J. Spry, *Bull. Am. Phys. Soc.*, **11**, 193 (1966); R. J. Spry (to be Published).

Recombination Luminescence in N-type Czochralski Silicon*

Robert J. Spry and W. Dale Compton

Department of Physics, University of Illinois, Urbana, Illinois

ABSTRACT

Radiative recombination of optically injected electron-hole pairs has been observed in ^{60}Co gamma ray- and neutron-irradiated n-type Czochralski silicon. At liquid nitrogen temperature, two bands, peaking at 1.31 μ and 1.66 μ, are seen in gamma ray irradiated material. Neutron irradiated material differs in that the long wavelength band in the 1.6 μ region is simply a long wavelength tail. At liquid helium temperature, two complex luminescence patterns are seen in both neutron and gamma irradiated material. Intense bands appear at 0.969 eV and 0.789 eV with much of the structure replicated with a separation of 0.180 eV. It is suggested that the luminescence is due to recombination between free holes and electrons in the ground states of two different defects, 0.196 eV and 0.376 eV below the conduction band minimum, the weaker bands of the two patterns arising from recombination accompanied by phonon emission.

I. INTRODUCTION

In recent years, there has been intensive research on the problem of determining the nature of radiation induced defects in silicon. Electron spin resonance has been particularly successful in providing detailed microscopic models of many of the prominent defects. Measurements of the transport properties of free carriers, either optically or thermally excited, and optical absorption have been successfully used to locate the energy level in the forbidden gap associated with the defect. Although recombination luminescence studies have been made on impurity defects introduced

* This work was supported in part by a grant from the National Aeronautics and Space Administration, Grant No. NsG 228-62.

during crystal growth,[1–4] this technique has been used very little to study the properties of radiation induced defects.[5–8]

Ivanov and Yukhnevich[5,6] studied the recombination luminescence resulting from electron-hole recombination in ^{60}Co irradiated silicon. The carriers were electrically injected into p-n junctions. Yukhnevich suggested that the luminescence resulted from the recombination of an electron-hole pair bound at the Si-A center, a defect consisting of a substitutional oxygen impurity having an energy level about 0.17 eV below the conduction band. It was suggested that phonon assisted transitions also occurred.

The present work was undertaken with the view of comparing the nature of the luminescence introduced by both ^{60}Co and fast neutron irradiations, and to determine the influence of the nature and amount of dopant, both intentionally added impurities and accidental impurities, such as oxygen, upon the luminescence spectra, and to extend the temperature range of measurements to lower temperatures where thermal effects would become less important. The results of a portion of these measurements, specifically those on Czochralski grown n-type material, are presented below.

II. EXPERIMENTAL PROCEDURE

All samples were phosphorus doped and grown by the Czochralski method, thereby containing about $10^{18}/cm^3$ interstitial oxygen atoms.[9] The samples were cut in the form of plates whose thickness ranged between 0.4 mm and 0.9 mm after etching in CP4A ($5HNO_3$, 3HF). About 0.08 mm of material was removed from each side after about one minute in the etch, leaving a surface which had an over-all " orange peel " appearance but which was smooth over distances comparable to the wavelength of the emitted light. After the etching action was quenched by flushing with distilled water, the samples were dried on filter paper and immediately placed in a vacuum dewar. Although the absolute luminescent efficiency was found to depend rather critically upon the etching procedure, no change in the luminescent spectrum was found to result from various surface treatments.

Electron-hole pairs were created by focusing a 500 watt high pressure mercury lamp on the sample. Photons with wavelengths greater than 7,500 Å were removed by passing the incident light beam through a 5 cm water cell whose ends were made of Jena KG3 optical glass 3 mm thick. The incident exciting light was interrupted by a mechanical chopper. This excitation technique is similar to that used by Haynes.[10] The emitted light was collected from the side opposite the incident beam by a CaF_2 lens, analyzed by a monochromator, and detected by a PbS cell cooled to dry ice temperature. The voltage signal from the detector was amplified by a preamplifier and a Princeton Applied Research amplifier whose reference signal was generated by the mechanical chopper. The final signal was displayed on a chart recorder. Data were obtained continuously as a function of wavelength at liquid helium and liquid nitrogen temperature with a Jarrell-Ash model 82-092 grating monochromator. Many of the liquid nitrogen temperature measurements and higher temperature experiments were made with a Perkin Elmer Model 83 prism monochromator. The relative response of the detector and monochromators was determined by a measurement of a blackbody source operating at a known temperature. All data are presented in terms of the relative number of quanta emitted in a given wavelength interval.

The samples were irradiated at room temperature by gamma rays from ^{60}Co and by fast neutrons from a General Dynamics Triga reactor. Thermal neutrons were removed by shielding the samples with a cadmium foil 0.030 inches thick. The samples were re-etched for about 10 sec after each irradiation.

III. RESULTS

A) *Liquid nitrogen temperature*

The luminescence spectrum at liquid nitrogen temperature of ^{60}Co gamma ray bombarded silicon of 100 ohm resistivity is shown in Fig. 1. The band-to-band luminescence at 1.13 μ was first reported by Haynes and Westphal.[11] The lower resolution used in the present measurement produced the greater half width in this band than that reported by these workers. As the integrated gamma flux increases, the band gap luminescence disappears, leaving two new bands peaking at 1.31 μ and 1.66 μ corresponding to 0.947 eV and 0.747 eV respectively. The spectrum for neutron bombarded material shown in Fig. 2 differs in that the distinct second band now appears simply as a long wavelength tail. The slight variation in the location of the peak of the intense band is due to a difference in the instrumental resolution used in the two measurements. Neither the long wavelength band nor tail were observed in similar studies of float zone grown silicon. All the spectra disappeared at dry ice temperature.

These results are in general agreement with those of Ivanov and Yukhnevich,[5] who found two distinct bands at liquid nitrogen temperature which appear to be identical to those in Fig. 1. They found that the luminescence

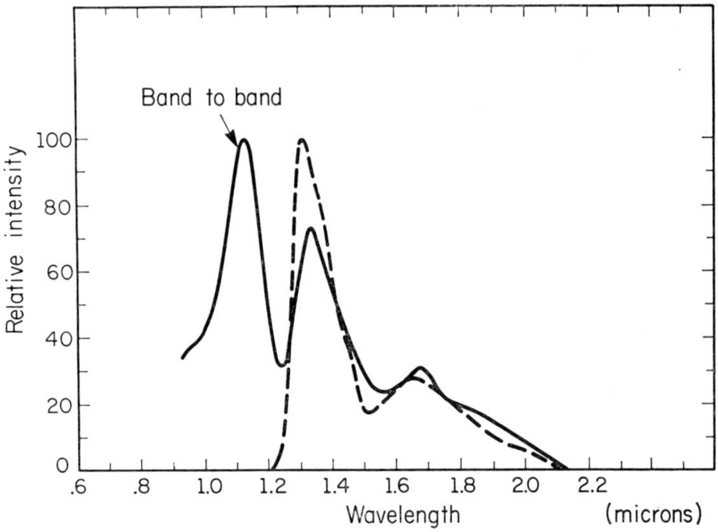

Fig. 1. Luminescence spectrum of 100 ohm·cm n-type Czochralski silicon at liquid nitrogen temperature. Irradiated with ^{60}Co γ-rays. Full line: 1×10^6 roentgens resolution 0.04 μ. Broken line: 40×10^6 roentgens resolution 0.018 μ.

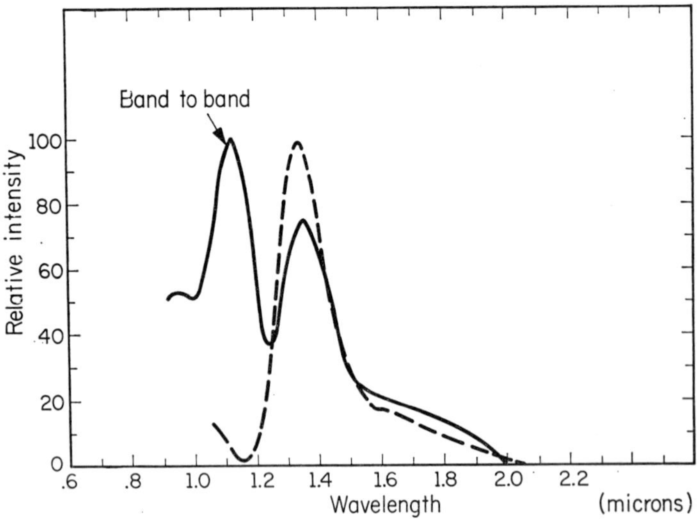

Fig. 2. Luminescence spectrum of 100 ohm·cm n-type Czochralski silicon at liquid nitrogen temperature. Irradiated with fast neutrons. Full line: $10^{12}/cm^2$. Broken line: $10^{14}/cm^2$. Resolution 0.04 μ

disappears at 120°K. In a subsequent study,[6] Yukhnevich was able to observe structure on the 1.31 μ band; in particular he reports an intense band at 0.967 eV that has a measured half width of less than 0.002 eV. By going to an instrumental resolution of 0.0024 eV, we also observed this line but found the half width to be 0.0050 eV. There remains an unexplained discrepancy in the measured half width of this band as reported by Yukhnevich and as found in the present work.

Yukhnevich has also observed a complex emission spectrum[8] in the 2.4 μ to 3.0 μ range which we have not observed.

B) *Liquid helium temperature*

The results for gamma irradiated silicon at liquid helium temperature are shown in Fig. 3. It is particularly striking that two similar complex patterns appear which are displaced by 0.180 eV. The bands located at 0.717 eV, 0.773 eV, and 0.789 eV appear to be uniformly replicated at 0.898 eV, 0.953 eV, and 0.969 eV respectively. In addition to the clear replication of these three luminescent bands, there is an indication that the small peaks at 0.755 eV and 0.739 eV are also replicated at 0.935 eV and 0.919 eV respectively. There are dissimilar complex bands centered about 0.728 eV, 0.752 eV, and 0.923 eV. The spectrum between 0.64 eV and

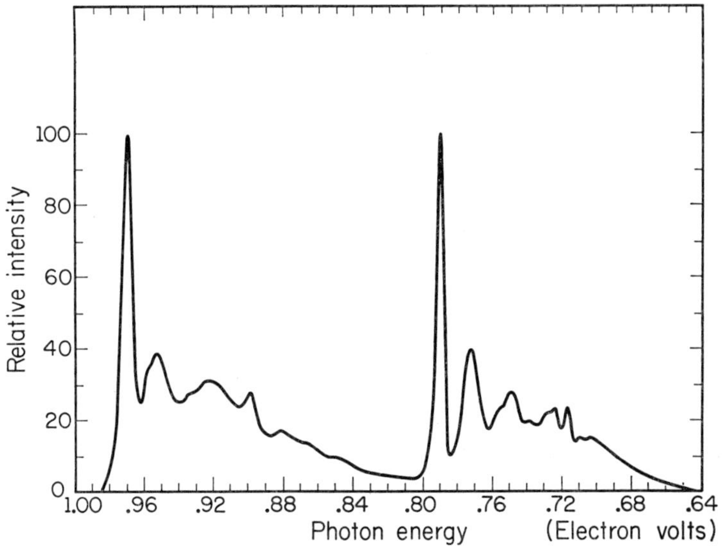

Fig. 3. Luminescence spectrum of 100 ohm·cm n-type Czochralski silicon at liquid helium temperature. Irradiated with ^{60}Co γ flux to 10^8 roentgens. Resolution 0.0064 μ.

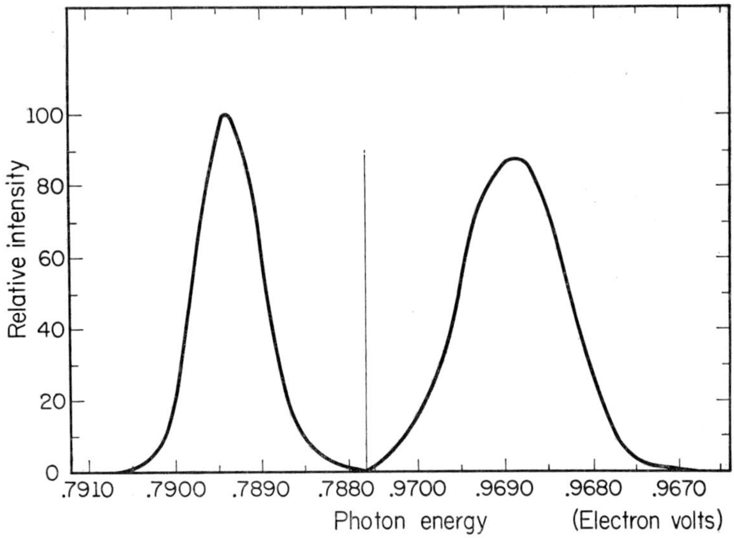

Fig. 4. The intense luminescence peaks of Fig. 3, analyzed with a resolution of 0.0008 eV for the 0.7894 eV line and a resolution of 0.0012 eV for the 0.9688 eV line.

0.80 eV was never observed in float zone grown silicon for which the oxygen content is about two orders of magnitude below that of the Czochralski grown crystals.

The two most intense bands of Fig. 3 were examined under the maximum resolution instrumentally available. The results shown in Fig. 4 indicate that the measured halfwidth is still instrumentally limited. It is clear, however, that the halfwidth of the 0.969 eV line decreases in going from liquid nitrogen to liquid helium temperature.

A comparison of the results obtained on the neutron and gamma ray irradiated silicon suggests that the principal features of the spectra are essentially the same. Thus, it can be concluded that the defects which are active as recombination centers in the silicon are essentially the same in both neutron and gamma ray irradiated material. It must be emphasized, of course, that the defects seen in the luminescence measurement may represent only a single type of a much larger group of defects produced by the radiation.

The fact that the halfwidth of the 0.967 eV line is of the order of kT at liquid nitrogen temperature and the fact that the halfwidth decreases when going to liquid helium temperature strongly suggest that the transition is due to recombination of a trapped carrier with a free carrier having

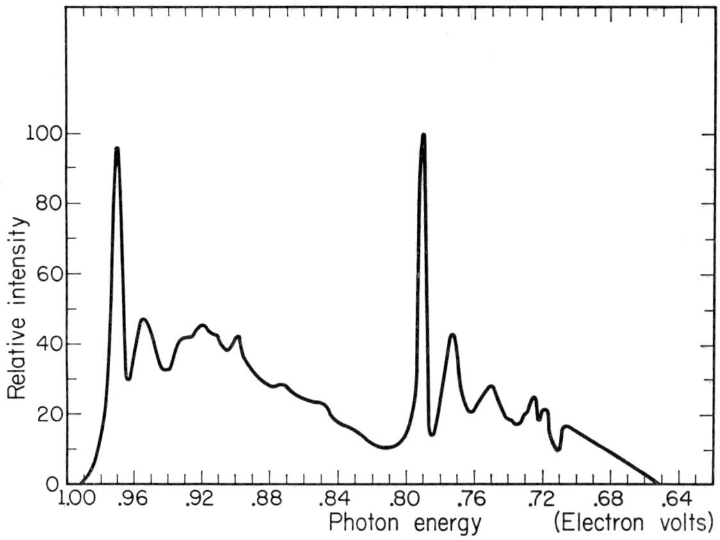

Fig. 5. Luminescence spectrum of 100 ohm·cm n-type Czochralski silicon at liquid helium temperature. Irradiated with fast neutrons to flux of $10^{14}/cm^2$. Resolution 0.0064 μ.

a thermally broadened density of states. This is in disagreement with the conclusion of Yukhnevich[11] who attributed the transition to recombination of an electron and a hole both bound to an A center, after the model of Kurskii.[12,13] The great intensity of this band indicates that the transition proceeds without phonon cooperation.[3,4] It is suggested, therefore, that the recombination occurs between a *free* hole and an electron bound to a defect, the bound electron having an energy level near the conduction band. If the sharp line transition at 0.969 eV involves no phonons, the energy level would, therefore, be 0.196 eV from the conduction band edge. The other alternative would be for the defect itself to take part in the momentum transfer and to dissipate the momentum in a local mode vibration.

The value of 0.196 eV is larger than the depth of the A-center ground state,[14] as determined from transport measurements. Although it is possible that the transport experiments measure an ionization energy which may be less than the value seen here because of the cooperation of lattice vibrations in ejecting the electron, it is also possible that the defect responsible for the luminescence is not the A center.

It is suggested that the 0.953 eV band results from recombination with the accompanying emission of a transverse acoustic phonon. The 0.898 eV band corresponds to recombination with the emission of both a transverse acoustic and a transverse optical phonon.[15] The small bands present at 0.935 eV and 0.919 eV would correspond to emission with two and three transverse acoustical phonons, respectively. The origin of the 0.923 eV band is not known.

By similar arguments the 0.789 eV band is due to phononless recombination between a free hole and an electron trapped at a defect level 0.376 eV below the conduction band minimum. The 0.773 eV band is due to recombination with the emission of a transverse acoustic phonon and the 0.717 eV band is due to recombination with emission of both a transverse optical and a transverse acoustic phonon. By analogy with the above arguments, the bands at 0.755 eV and 0.739 eV may result from recombination with the accompanying emission of two and three transverse acoustic phonons, respectively. The origin of the 0.728 eV and 0.752 eV bands is unknown. Since the lower energy spectrum is found in only Czochralski material, the corresponding defect must depend upon oxygen for its formation but its exact nature is unknown. This model is supported by the data of Fan and Ramdas,[16] who measured the optical absorption and photoconductivity at 4.2°K of neutron and gamma irradiated n-type silicon. They found a strong absorption line at 0.343 eV and a series of weaker absorption bands located at higher energies, the two nearest bands being separated by 0.015 eV and 0.031 eV. The Fermi level had to be close to the conduction band to see the absorption and there was no photoconductivity associated with the center. If this center is the same as the one lying 0.376 eV below the conduction band, then the 0.343 eV absorption band is due to excitation of an electron into a state 0.033 eV below the conduction band and the two nearest absorption bands are due to single and double emission of transverse optical phonons.

In general, the luminescence spectrum appears to depend mainly on the local environment of the traps and upon the phonon spectrum of the crystal. It appears to be less sensitive to the exact structure of the defect. It seems clear that the techniques described in this paper are capable of yielding important information about the energy levels associated with the radiation induced defects and the interaction of these defects with the lattice phonons.

REFERENCES

1. J. R. Haynes, *Phys. Rev. Letters*, **4**, 361 (1960).
2. Ya E. Pokrovsky and K. I. Svistunova, *J. Phys. Chem. Solids*, **22**, 39 (1961); *Soviet Physics-Solid State*, **3**, 2058 (1962); **5**, 1373 (1964); **6**, 13 (1964); **7**, 1275 (1965); **7**, 1478 (1965).

3. V. L. Bonch-Bruevich, *Soviet Physics-Solid State*, **4**, 215 (1961).
4. V. L. Bonch-Bruevich and A. A. Drugova, *Soviet Physics-Solid State*, **7**, 2491 (1966).
5. Yu L. Ivanov and A. V. Yukhnevich, *Soviet Physics-Solid State*, **6**, 2965 (1965).
6. A. V. Yukhnevich, *Soviet Physics-Solid State*, **7**, 259 (1965).
7. R. J. Spry, *Bull. Am. Phys. Soc.*, **11**, 193 (1966).
8. A. V. Yukhnevich and V. D. Tkachev, *Soviet Physics-Solid State*, **7**, 2746 (1966).
9. W. Kaiser, P. H. Keck and C. F. Lange, *Phys. Rev.*, **101**, 1264 (1956).
10. J. R. Haynes, in *Methods of Experimental Physics*, Vol. 6B, p. 321, Academic Press, New York, 1959.
11. J. R. Haynes and W. C. Westphal, *Phys. Rev.*, **101**, 1676 (1956).
12. Yu A. Kurskii, *Soviet Physics-Solid State*, **6**, 1162 (1964).
13. Yu A. Kurskii, *Soviet Physics-Solid State*, **6**, 1795 (1965).
14. See, for example, V. S. Vavilov, in *Effects of Radiation on Semiconductors*, pp. 169–189, Consultants Bureau, New York, 1965.
15. B. N. Brockhouse, *Phys. Rev. Letters*, **2**, 256 (1959).
16. H. Y. Fan and A. K. Ramdas, *J. Appl. Phys.*, **30**, 1127 (1959).

Radiation Damage in Silicon

Kenshiro Nakashima and Yoshio Inuishi

Faculty of Engineering, Osaka University, Osaka, Japan

ABSTRACT

Radiation damages mainly in p-type silicon single crystals irradiated with γ-ray from ^{60}Co were investigated by the carrier lifetime and Hall effect measurement. The negative annealing was ascribed to the temporary formation of a deeper level which has a larger recombination cross section for minority carrier. The energy level 0.3 eV from the band edge was obtained from the temperature dependence of carrier lifetime in a high resistivity floating zoned p-type sample in accordance with the Hall effect measurement. In the case of neutron irradiation, different results from that of γ-ray irradiation were obtained, i.e. the introduction and annealing of the defect are rather insensitive to impurities. In p-type silicon an energy level at about 0.3 eV above the valence band is mainly introduced, and the rate does not dpend on the kinds of impurities, free hole numbers, or crystal growth method.

I. INTRODUCTION

It has been found by us that the introduction rate of a recombination center due to γ-irradiation in silicon measured from inverse lifetime is sensitive to impurities such as oxygen, gold, and iron.[1–3] For example the introduction rate of recombination centers for n-type floating zoned silicon is appreciably higher than for the n-type pulled silicon containing a larger amount of oxygen.[1] This difference is ascribed to the predominant formation of an A-center in the case of pulled silicon, which has a smaller effective recombination cross section for holes than that of the E-center predominant in floating zoned silicon. The author intend to describe first the features of the radiation damage in silicon introduced by γ-ray irradiation at room temperature. The methods for measuring the lifetime of minority carrier and the theoretical grounds for analyzing the experimental results were published by one of the authors previously.[4]

Neutron irradiation gives remarkably different results from γ-irradiation, as was reported previously.[4] The rate of lifetime degradation and annealing behavior does not depend on the dopant impurities or the crystal growth method. This fact was confirmed by Curtis recently.[5] In this paper the carrier removal in p-type samples with a wide range of resistivity was used as a measure of defect concentration. The results show that the fact stated above holds for p-type silicon.

II. EXPERIMENTALS

The silicon samples used in lifetime measurements were floating zoned (FZ) and pulled silicons (PU) of p-type mainly doped with boron, the resistivity of which was about 100 Ω·cm. High resistivity samples of about 1,200 Ω·cm were especially used to investigate the energy level position of the recombination center from the temperature dependence of carrier lifetime. Characteristics of the samples are listed in Table I. In the case of neutron irradiation, floating zoned and pulled samples doped with boron and aluminum were used to study the influence of dopant impurities and crystal growth method. The resistivity range of these samples was from 16 to 600 Ω·cm. Oxygen content was obtained from the intensity of infrared absorption at 9.1 μ.[6] Gamma irradiation from ^{60}Co with the dose rate of 1×10^6 rö/hr was carried out at room temperature. Neutron irradiation was done at room temperature in a KUR reactor whose dose rate was 2×10^{11} fast neutrons/cm²·sec.

Carrier lifetime and Hall coefficient were measured in the temperature range between liquid nitrogen and room temperature. Inverse lifetime $1/\tau$ can be taken as a measure of defect concentration, when the Fermi level does not move appreciably owing to the irradiation.

TABLE I.

Sample	Resistivity (ohm·cm)	Dopant	Type	Oxygen concentration (atoms/cm³)	Crystal growth method
P_l-1-5	60	B	P	1.5×10^{17}	Pulled
F_l-1	80	B	P	oxygen free	F.Z.
F_l-H1,2	1200	B	P	”	”
P_l-6	80	P	N	6×10^{17}	Pulled
P-1-3	100	B	P	—	”
F-1,2	100	B	P	oxygen free	F.Z.
Al-1	16	Al	P	”	”
NP-1	40	P	N	7.2×10^{17}	Pulled
NF-1	20	P	N	oxygen free	F.Z.

III. RESULTS AND DISCUSSION

(a) *Gramma irradiation*

The A-center, which is introduced predominantly in n-type pulled silicon, is not so effective as the E-center for recombination. Accordingly, the apparent introduction rate of recombination center in n-type pulled silicon is much less than that of n-type floating zoned silicon[1] where E-centers are produced predominantly. In p-type silicon, however, the situation differs. Figure 1 shows the introduction rate of inverse lifetime or the recombination center as a function of total γ-ray dose. It should be noted that the introduction rate of inverse lifetime in pulled and floating zoned silicon having the same order of hole density is nearly the same.

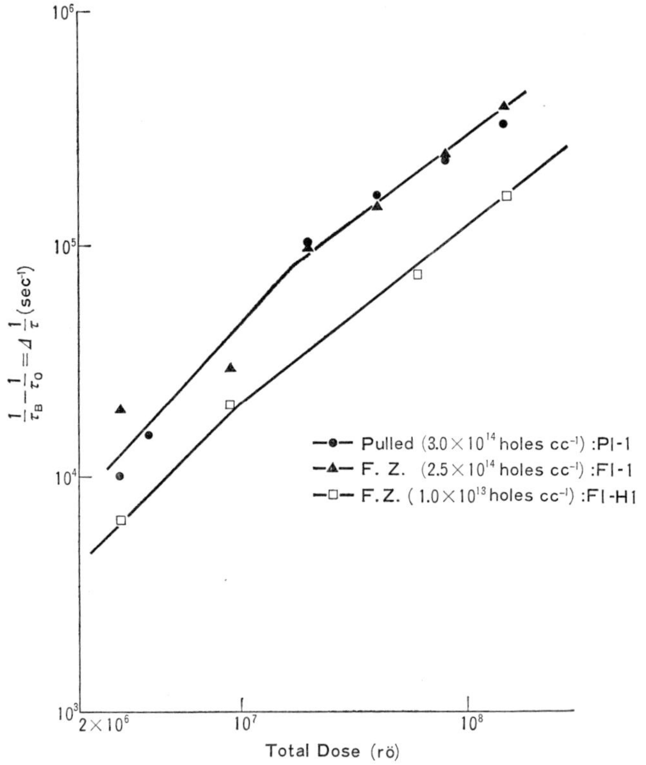

Fig. 1. Introduction rate of inverse lifetime of P-type silicon vs. γ-ray dose τ_0 and τ_B represent the carrier lifetime at room temperature before and after γ irradiation.

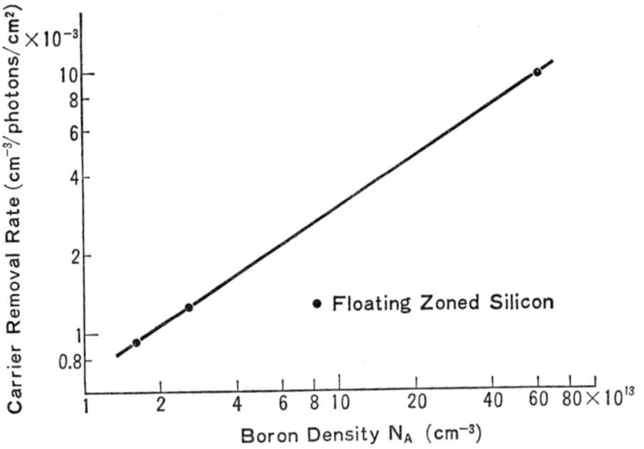

Fig. 2. Carrier removal rate due to J′-center for various boron densities.

In other words oxygen does not affect the apparent introduction of recombination center in p-type silicon as remarkably as it does n-type silicon. Figure 1 also shows the introduction rate in high resistivity samples, which is smaller than that of medium resistivity samples.

In p-type silicon two kinds of centers are introduced. One is the A-center which lies at $E_c-0.17$ eV and has the nature of an acceptor. This center is introduced in pulled silicon predominantly; the evidence for this will be discussed later. The other is a donor-like center that is favorably introduced at $E_v+0.27$ eV in p-type floating zoned silicon. The carrier removal rate due to this donor-like center measured from the Hall effect is shown as a function of acceptor or boron density in Fig. 2.[9] It should be noted that the carrier removel rate in boron doped floating zoned silicon is nearly proportional to the boron density. Accordingly the defect introduced may be a complex defect of vacancy-boron or divacancy-boron. This center was named J′ by one of the authors[9] to distinguish it from the Watkins divacancy model or J-center.[7,8] Both A-center and J′-center may act as a recombination center in the case of p-type silicon having nearly the same recombination cross section.

The position of energy level of the recombination center in p-type floating zoned samples can be decided from the lifetime temperature dependence using the Shockley-Read formula

$$\tau = \tau_{n0}\left(1+\frac{p_1}{p_0}\right)$$

for the lower half of the energy gap and

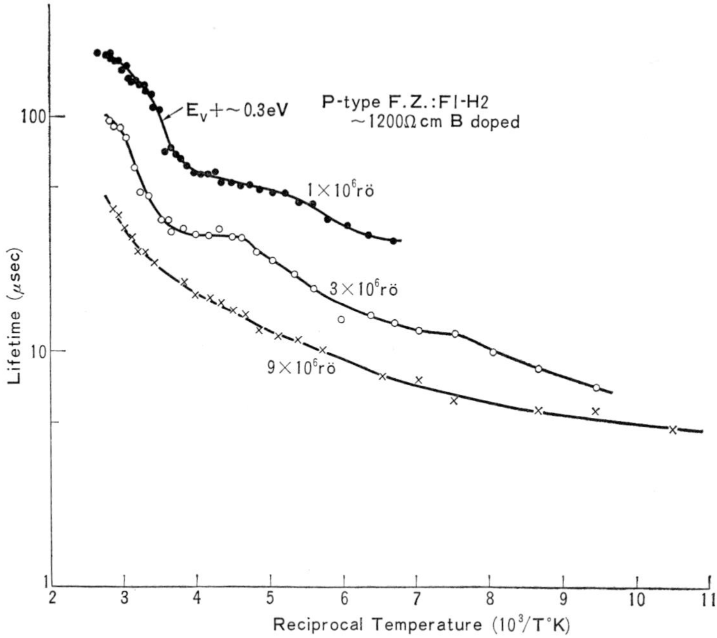

Fig. 3. Temperature dependence of carrier lifetime for vaious γ-ray doses.

$$\tau = \tau_{n0} + \tau_{p0}\frac{n_1}{p_0}$$

for the upper half of the energy gap if the number of recombination centers and injection level are small, where conventional notations are used. From the slope of the high temperature part in Fig. 3 and the above-mentioned formula, the position of the recombination center was estimated to be at 0.3 eV from the band edge in accordance with the Hall effect measurement of the J′-center[9] (E_V+0.27 eV). It should be noted that the V-shape curve of the temperature dependence of minority carrier lifetime due to trapping effect[2] which is remarkably apparent in n-type silicon is not so remarkable in p-type silicon.

Ten minute isochronal annealings of p-type pulled silicon irradiated with γ-rays from 2×10^7 rö to 2×10^8 rö were performed at various temperatures. The unannealed fractions of inverse lifetime $1/\tau$ are shown in Fig. 4, where the negative annealing peak becomes more remarkable with increasing γ-ray dose. These negative annealings were previously ascribed to A-center dissociation by the authors.[1] The peak tends to move

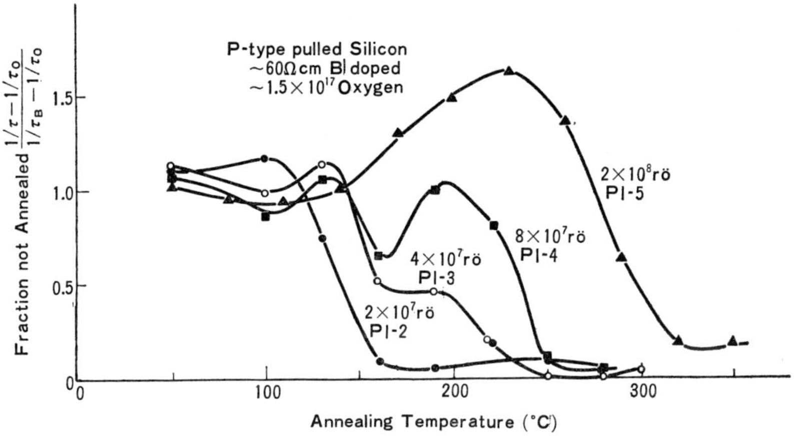

Fig. 4. Ten minute isochronal annealing in p-type pulled silicon for vaious γ-ray does. τ_0 τ_B, and τ represent carrier lifetime at room temperaure before and after irradiation, and after annealing at each temperature, respectively.

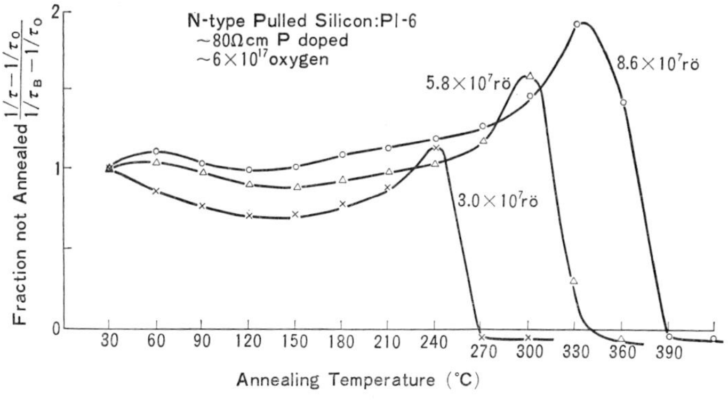

Fig. 5. Ten minute isochronal annealing in n-type pulled silicon taking γ-ray dose as a parameter.

towards a higher temperature with increasing γ-ray dose or increasing A-center density. Figure 5 shows a similar effect in n-type pulled silicon, as previously published by one of the authors.[2] Figure 6 shows the minority carrier lifetime of p-type pulled silicon irradiated with 2×10^8 rö γ-rays as a function of reciprocal temperature for various annealing stages. It is

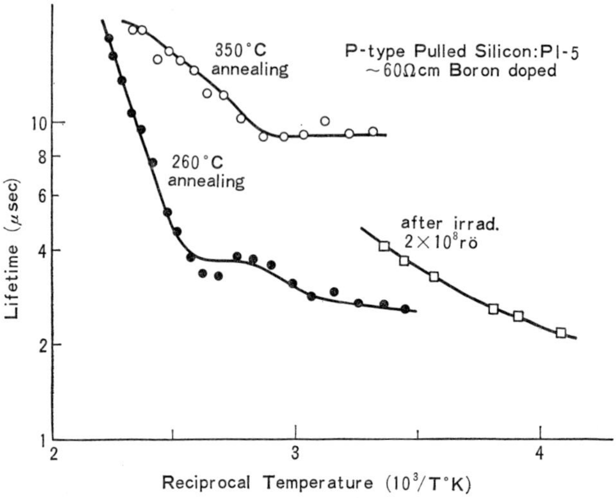

Fig. 6. Temperature dependence of carrier lifetime at various annealing stages.

noted that the deep recombination centers at about 0.6 eV from either band edge appear temporarily when the sample is annealed at a negative annealing temperature 260°C, and are annihilated by annealing at 350°C. Accordingly the negative annealing may be related to the appearance of this deep level which should have a larger recombination cross section. The reason why the negative annealing peak shifts to a higher temperature with increasing dosage is not clear at this stage of the experiments. The authors, however, suspect *(1)* the effect of the surface or dislocations which might have smaller activation energy for annealing at smaller dosages, *(2)* the more favorable formation of divacancies due to interaction of two dissociated A-centers with increasing A-center density.

(b) *Neutron irradiation*

Neutron irradiation of samples was performed in a KUR reactor at room temperature. The reactor spectrum consists of 2×10^{11} neutrons/cm²·sec fast neutrons, 4×10^{12} neutrons/cm²·sec thermal neutrons, and 10^8 rö/hr γ-ray. The apparent neutron irradiation effect can be determined by measurement of the Hall coefficient as a function of temperature. Both floating zoned and pulled p-type samples in a resistivity range from 16 to 600Ω·cm, doped with boron or aluminum were investigated. Figure 7 shows the curves of hole concentration thus obtained as a function of reciprocal temperature. It was found that at least two donor-like centers were

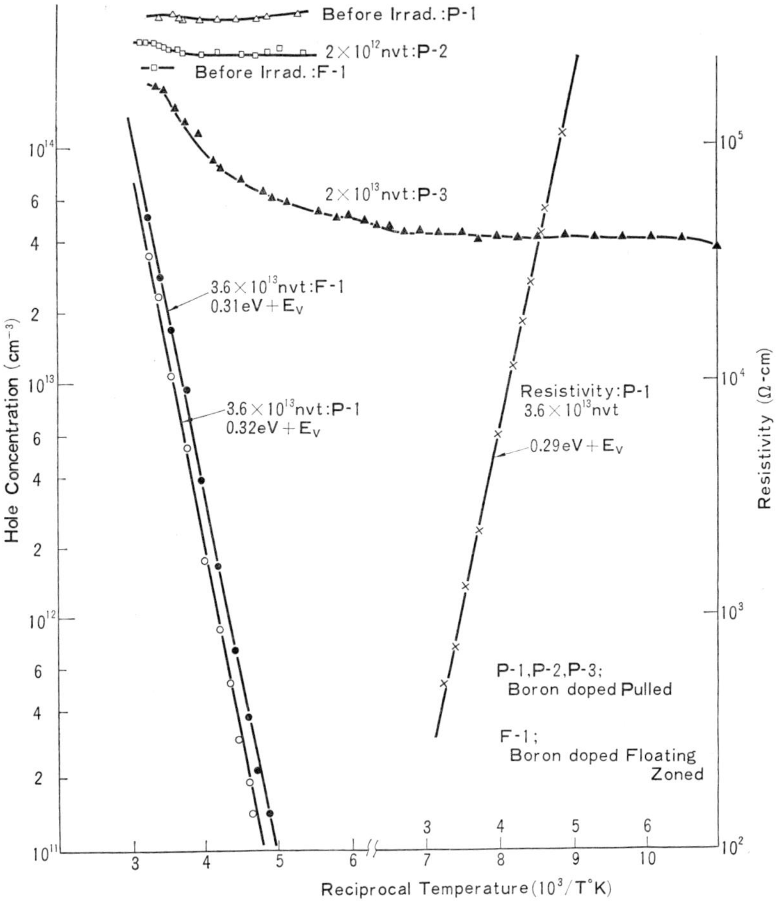

Fig. 7. Carrier concentration vs. reciprocal temperature for the samples irradiated with various total neutron flux.

introduced, one of which lies at about 0.3 eV above the valence band as follows. Fermi levels are plotted as a function of temperature in Fig. 8 in order to determine the level depth. In heavily irradiated samples Fermi level is pinned at about 0.3 eV above the valence band showing the existence of an $E_V+0.3$ eV level. Carrier removal rates of various samples with different kinds and amounts of dopant are shown in Fig. 9. It was found that the carrier removal rate of neutron irradiated samples is independent of acceptor density, types of dopant impurities and the crystal growth method. The isochronal annealing of inverse lifetime was carried out by one of the

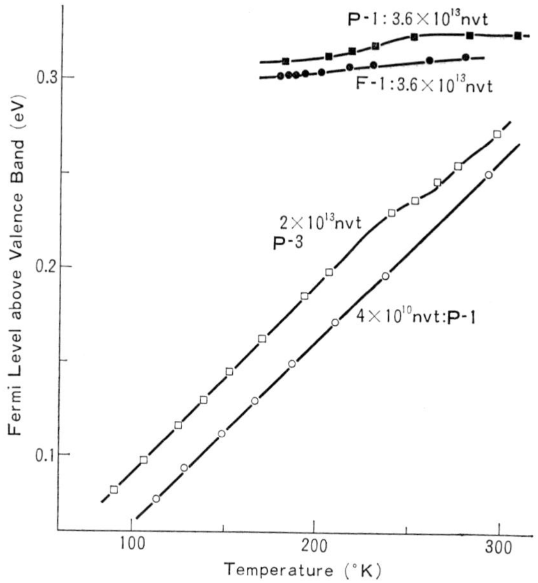

Fig. 8. Fermi level plots of the samples shown in Fig. 7. vs. temperature.

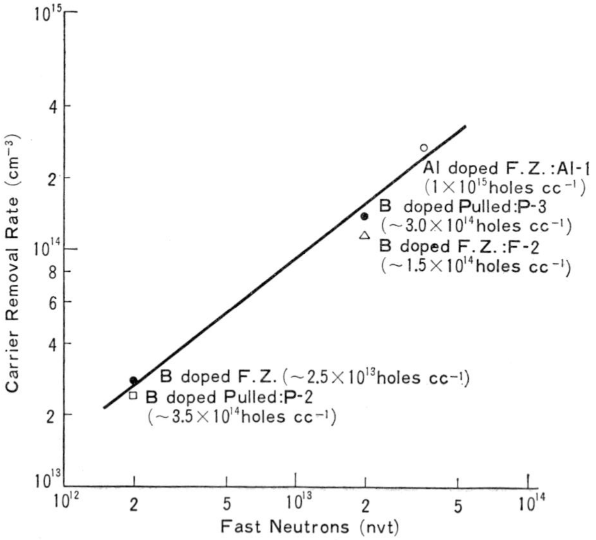

Fig. 9. Carrier removal rate of different kinds of smples irradiated with neutrons. Carrier density and crystal growth method are shown in the figure.

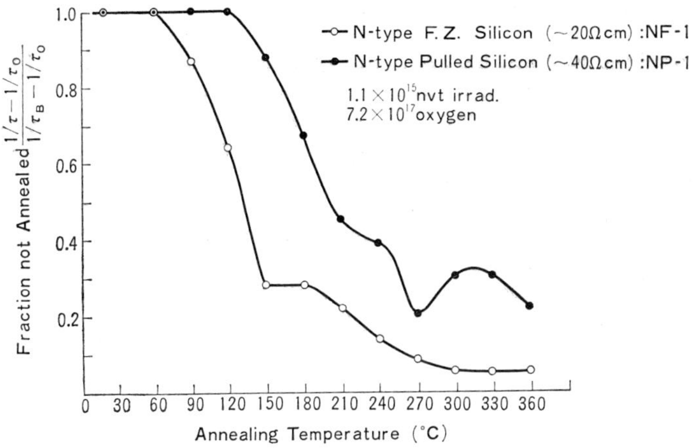

Fig. 10. Ten minture isochronal annealing of neutron irradiated n-type samples.

authors as shown in Fig. 10.[1] Appearance of negative annealing and distinction between floating zoned and pulled samples are not found in this case. These facts were reconfirmed by other investigators.[5,10] The $E_V+0.3$ eV donor-like center in the case of neutron irradiation may be a divacancy or a vacancy cluster due to high concentration of localized Frenkel-pairs.

IV. CONCLUSION

1) The appearance of negative annealing in p-type pulled silicon is related to the temporary formation of a deep recombination center with a larger recombination cross section after the dissociation of the A-center. The negative annealing peak shifts to a higher temperature with increasing dosage as in the case of n-type silicon.

2) The recombination center at 0.3 eV from the valence band is introduced by γ-ray irradiation in p-type floating zoned silicon, which may be in accordance with the J′-center measured by the Hall effect.

3) In p-type silicon irradiated with neutrons, a donor-like center is introduced at $E_V+0.3$ eV predominantly. The carrier removal rate due to this center does not depend on the dopant impurities or the crystal growth method. The $E_V+0.3$ eV center in this case may be a divacancy or a vacancy cluster.

4) The introduction and annealing of radiation damage in the case of neutron irradiation is insensitive to impurities such as oxygen, etc.

REFERENCES

1. Y. Inuishi and K. Matsuura, *Proc. Int. Cryst. Latt. Def.*, **18**, Supp. III, 240 (1963).
2. T. Tanaka and Y. Inuishi, *J. Phys. Soc. Japan*, **19**, 851 (1964).
3. T. Nakano, K. Nakashima and Y. Inuishi, *J. Phys. Soc. Japan*, **20**, 2140 (1965).
4. K. Matsuura, U. Kubo and Y. Inuishi, *Technol. Repts. Osaka Univ.*, **12**, 39 (1962).
5. O. L. Curtis, Jr., this volume.
6. W. Kaiser, P. H. Keck and C. F. Lange, *Phys. Rev.*, **101**, 1264 (1956).
7. J. W. Corbett and G. D. Watkins, *Phys. Rev. Letters*. **7**, 314 (1961).
8. G. D. Watkins and J. W. Corbett, *Phys. Rev.*, **138**, A543 (1965).
9. T. Tanaka and Y. Inuishi, *Japan. J. Appl. Phys.*, **4**, 725 (1965).
10. H. J. Stein, *J. Appl. Phys.*, **37**, 3382 (1966).

Defects Induced in Silicon by Irradiation in the Reactor at 77°K

R. F. KONOPLEVA, S. R. NOVIKOV AND E. E. RUBINOVA

A. F. Ioffe Physico-Technical Institute, Academy of Sciences of the USSR, Leningrad K-21, U.S.S.R.

1. *Energy spectrum of defects induced in silicon by fast neutrons at 77°K*

In the present paper electro-conductivity and spectral distribution of impurity photo-conductivity in n- and p-Si irradiated by fast neutron fluxes (from 5×10^{14} to 10^{15} n/cm^2) in the reactor at 77°K and then annealed at temperature intervals of 77–400°K were investigated.

Specimens of n- and p-type with specific resistivity 3,10 and 20 ohm·cm were provided by palladium contacts formed electrolytically. The removal of specimens and the measurement of their characteristics on irradiation were carried out without warming them. Later on, the specimens were subjected to annealing for 10 min at each temperature. The annealing was followed by measurements of electro-conductivity and photo-conductivity at 77°K.

Figure 1 illustrates the curves of the change of electro-conductivity by

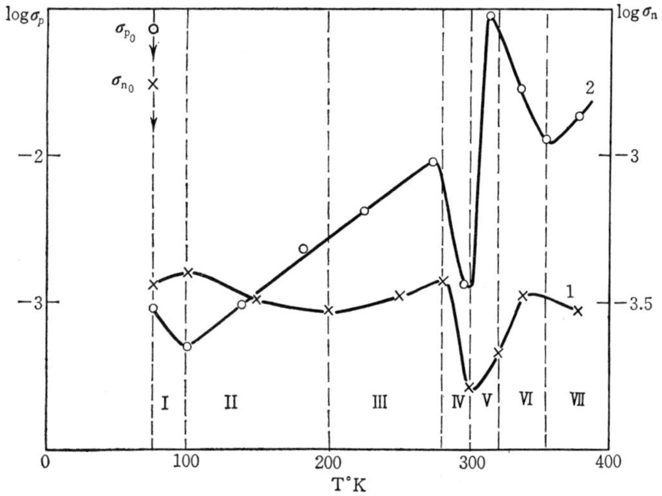

Fig. 1. Electro-conductivity change in n-Si (1) and p-Si (2) during isochronal annealing (10 min at each temperature).

annealing for n- and p-Si. As seen from the curves, seven annealing stages are clearly observed in the process of isochronal annealing within a temperature range of 77–400°K.

Such curves of isochronal annealing allow us to determine the character of appearing and disappearing defects at each annealing stage. If defects formed in n- and p-material are supposed to be equal, the corresponding correlation must be observed during conductivity change of n- and p-Si. So the formation of donor states must have resulted in the decrease of conductivity in p-material and in the increase of conductivity in n-material. (Acceptor formation results in a contrary effect.)

Figure 1 illustrates that irradiation at 77°K results in a compensation of both n- and p-Si. In the process of irradiation, therefore, both donor and acceptor levels are formed. Such compensation for p-Si specimens lasts up to 100°K while an insignificant increase of conductivity is observed in n-material at this annealing stage. Such change of conductivity at the first annealing stage can be explained by the formation of " deep " donor levels which will decrease conductivity in p-material and will influence a little the conductivity of n-material, since measurements are performed at 77°K when conductivity is mainly defined by the levels near the valence band and the conductivity band. (A similar analysis can be done for each annealing stage.) Together with the study of the electric properties

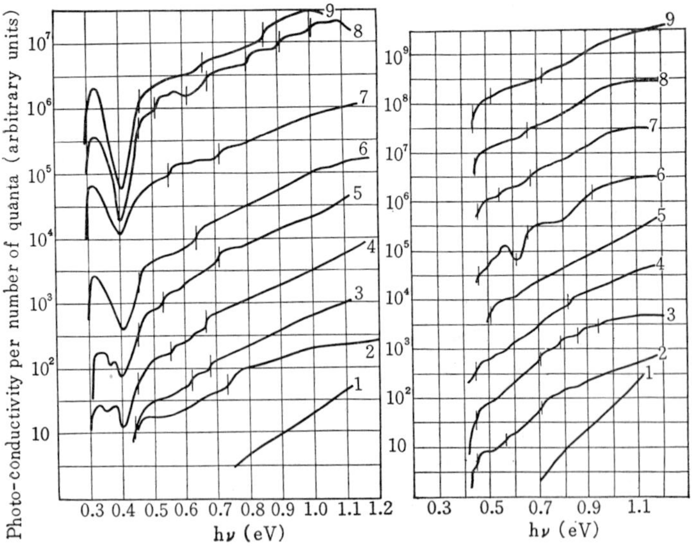

Fig. 2. Spectral distribution change of photo-conductivity in p-Si (a) and n-Si (b) specimens at different isochronal annealing stages.

the specimens at each annealing stage, the measurements of spectral distribution of impurity photo-conductivity were performed. This allowed us to determine the energy state of defect levels. Measurements were carried out at 77°K with a IKS-12 spectrometer.

In Fig. 2 the photo-conductivity spectra for p- and n-Si at different annealing stages are shown. Irradiation is immediately followed by only two transitions with activation energy $E_c-0.43$ eV and $E_v+0.45$ eV. During further specimen heatings at corresponding annealing stages a number of transitions appear or disappear in the spectra of impurity photo-conductivity. Comparing the process of electro-conductivity change with photoconductivity curves we attempted to state the character of the observed levels. The main difficulty was that some levels appeared or disappeared at each annealing stage and, therefore, there was no certainty about the character of each of them separately at some annealing stages. We managed, however, to determine the character of a number of levels. Levels $E_c-0.45$ eV, $E_v+0.53$ eV, $E_v+0.37$ eV and $E_v+0.31$ eV, for instance,

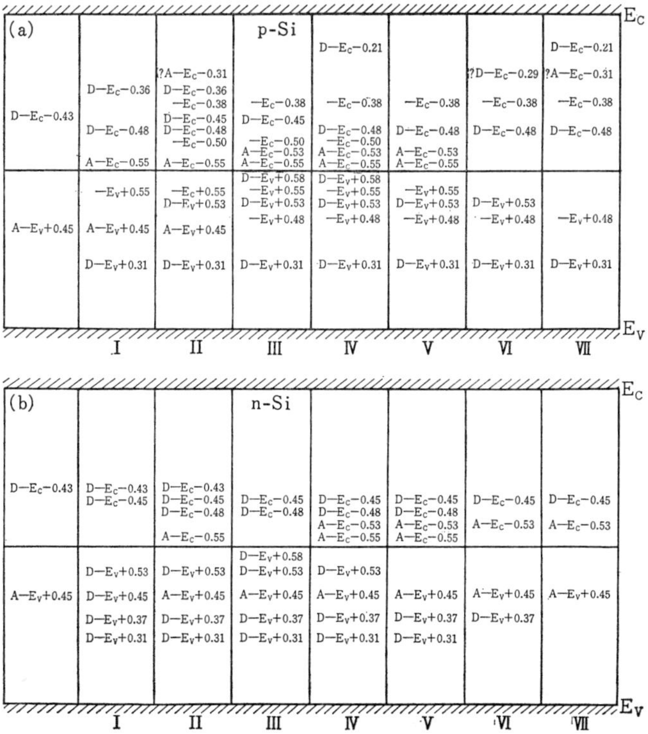

Fig. 3. Energy level spectra for p- and n-Si at different annealing stages.

appear in n-Si at the first annealing stage (Fig. 3). It is seen from the electro-conductivity process (Fig. 1) that conductivity in n-Si at this annealing stage insufficiently increases and all the levels, therefore, can be referred to donors since the appearance of "deep" donors in the measurements in liquid nitrogen has no practical effect on the n-material conductivity change. Levels $E_c-0.36$ eV, $E_c-0.48$ eV, $E_c-0.55$ eV, $E_v+0.55$ eV and $E_v+0.31$ eV (Fig. 3) appear at the same stage. Conductivity in p-Si at this stage decreases and all levels appearing together therefore behave as effective donors. Some levels can be treated as acceptors since "deep" acceptors have no effect on p-Si conductivity. Therefore, we could not come to any conclusions about the character of each level appearing, from the results of this annealing stage in p-Si. Identical analyses for all annealing stages were carried out. Figure 4 reveals a total level spectrum. We managed to determine the character of only 13 levels out of 19 from photo-conductivity spectra. Ten levels are donor levels and 3 levels are acceptor levels.

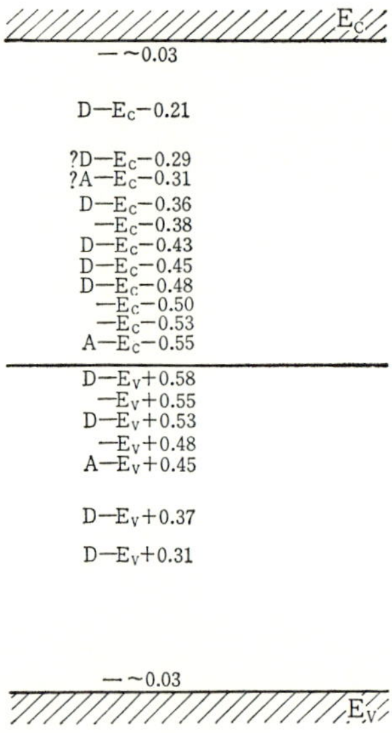

Fig. 4. Total level scheme observed in n- and p-type silicon on irradiation by fast neutrons at 77°K and due to some "ripening" processes.

2. *Defect "ripening" process at low-temperature annealing*

A "rearrangement" process of defects on annealing which we shall call the "ripening" of defects should be determined by such factors as the kind of bombarding particles and the atomic bond energy in the lattice of an irradiated material. Fast neutrons, indeed, supplying originally knocked-out atoms with sufficiently high energy, will ripen, in general, a kind of aggregate of defects, while electron or γ-radiation will create single defects. In addition, lattice atom bond energy must influence various rearrangement and migration phenomena of defects. It was interesting, therefore, to compare the defect " ripening " process in material irradiated with different particles with the process of irradiating different materials with the same kind of particles. Figure 5 gives three spectral curves of photo-conductivity. Curves 1 and 2 correspond to germanium and silicon irradiated by fast neutrons at 77°K and removed after 40 min irradiation. Curve 3 corresponds to Si irradiated by fast neutrons at 100°K and removed also after 40 min irradiation. (Taken from the paper by V. S. Vavilov, A. F. Plotnikov and V. D. Tkatchev.[1])

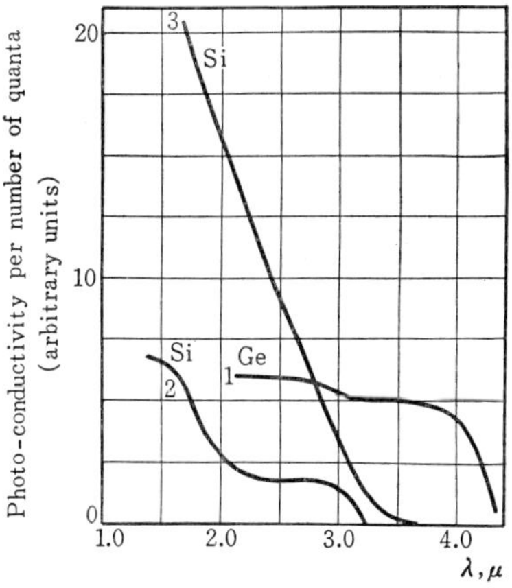

Fig. 5. Si and Ge photo-conductivity spectra taken after 40 min on irradiation at low temperature.

Curve 1: Ge—p-type on irradiation by fast neutrons at 77°K.
Curve 2: Si—p-type on irradiation by fast neutrons at 77°K
Curve 3: Si—p-type on irradiation by fast electrons at 100°K

As seen from the curves, defect levels are clearly observed in Ge and in Si immediately after irradiation in the case of neutron bombardment. At the same time, the curves corresponding to Si irradiated by electrons represent a number of activation energies.

V. S. Vavilov and his colleagues explained this phenomenon as the result of the presence (in the crystal immediately on irradiation at 100°K) of a number of Frenkel pairs with different distances between vacancies and interstitial atoms. The whole number of energy levels in the forbidden band corresponds to such pair distribution. In the case of neutron irradiation there is obviously no such random distribution of Frenkel pairs. Defects appear as aggregations. In this case the concentration of single defects and their quantity is small. Discrete levels observed at neutron irradiation can be supposed to belong to Frankel pairs, since due to random distribution of those pairs even at their small concentrations, the whole activation energy number should be expected. Such differences in the specimen states irradiated by different types of radiation should have a further influence on rearrangement processes and on defect migration. " Ripening " of defects in specimens irradiated with electrons can be explained by simple diffusion of defects to impurity atoms and by the formation of new, more stable, complexes having their defect levels in the forbidden band. The " ripening " process in specimens irradiated by fast neutrons may be attributed, for instance, to simple defects or some defect aggregations splitting off from the clusters and their migration in the crystal. Corresponding energy levels appear in the forbidden band. It is interesting to compare the " ripening " process in Si and in Ge irradiated by fast neutrons since the atomic energy bond in the lattice of Si is considerably larger than in Ge. As seen from such a comparison, " ripening " processes resulting in the appearance of a maximum number of levels in the forbidden band and the compensation formation of both n- and p-material in the case of Ge are complete up to 150°K while in Si these processes are complete up to 300°K.

REFERENCES

1. V. S. Vavilov, A. F. Plotnicov and V. D. Tkatchev, *FTT.* **4**, 3447 (1962).

The Role of Disordered Regions in Recombination of Carriers in Neutron-irradiated Silicon and Germanium*

O. L. CURTIS, JR.

Nortronics, A Division of Northrop Corporation,
Applied Research Department,
Newbury Park, California 91320, U.S.A.

ABSTRACT

There are several indications that carrier recombination in neutron-irradiated silicon and germanium occurs primarily at disordered regions. First, the amount of lifetime change in both materials is much greater for neutron than for electron or gamma irradiation, when comparable changes in carrier concentration occur. Second, the rate of introduction of recombination centers by neutrons is very nearly independent of oxygen and dopant impurities. Third, the annealing above room temperature of neutron-induced recombination centers is relatively insensitive to impurities. Fourth, short term annealing is too complicated to be accounted for by simple defects. In spite of the fact that the recombination level lies inside the disordered region, it seems to display an energy position (measured from the band edges) very similar to that observed for defects generated by electrons or gamma rays. In accordance with the model of Gossick, the difference in the recombination process presumably lies in the potential wells associated with defect clusters which greatly increase minority carrier capture probabilities. This fact can be used to explain departures from a simple theory of recombination behavior in neutron-irradiated material.

I. INTRODUCTION

The existence of disordered regions in neutron-irradiated semiconductors is well established and in fact has been directly observed.[1,2] Much remains to be learned about their role in determining various electrical

* Work supported by Harry Diamond Laboratories and the Defense Atomic Support Agency.

properties, however. An important question is whether defects introduced inside disordered regions are sufficiently dilute to maintain well-defined crystal properties. Furthermore, do defect energy levels remain well-defined or does a complete spectrum of defect states emerge?

Analysis of the effect of disordered regions on mobility, conductivity, and Hall coefficient was performed by Gossick,[3] and experimental observation has been in basic agreement.[4–6] However, only brief consideration has been given to the problem of recombination. We hope to provide qualitative answers to several questions which arise concerning the nature of the recombination process in material containing disordered regions. Experimental results reveal that, in both silicon and germanium, recombination centers inside disordered regions are more effective than an equal concentration uniformly distributed.

II. EXPERIMENTAL PROCEDURE

1. *Rate of lifetime degradation*

Consider first the relative effectiveness of recombination centers, compared to the total rate of introduction of defects. It is not possible to obtain the total concentration of defects by measurement of a single parameter, but carrier removal is a quantity which should be nearly proportional to total defect introduction.[7] In contrast, other parameters, such as lifetime, ordinarily reflect only a single level.

Table I summarizes results of gamma-ray and neutron irradiation of germanium. As a measure of the total rate of introduction of defect centers,

TABLE I. Comparison of carrier removal and lifetime changes in ~2 ohm·cm Germanium.

Type of irradiation	$\frac{\Delta n}{\Delta\phi}$ (cm^{-1})	$\frac{\Delta(1/\tau)}{\Delta n}$ (sec$^{-1}\cdot$cm^3)	
		n-type Ge	p-type Ge
^{60}Co	6×10^{-4} [a]	1.5–15×10^{-9} [b]	$\sim6\times10^{-10}$ [f]
Fission	2.7 [a]	$\sim4\times10^{-8}$ [c, d]	$\sim4\times10^{-9}$ [f]
Fusion (14 MeV)	6 [a]	$\sim1.1\times10^{-7}$ [e, d]	$\sim1.6\times10^{-8}$ [e]

a See ref. 7.
b See refs. 8 and 9.
c See refs. 8 through 11.
d Corrections in dosimetry have been applied.
e See ref. 12.
f See ref. 8.

we have used the carrier removal rates in n-type germanium from Cleland, Bass and Crawford.[7] A wide range of lifetime degradation is seen in gamma-irradiated n-type germanium, a fact which will be considered in detail below. However, the greater effectiveness of fission neutrons in producing lifetime changes is apparent, and even greater effectiveness is shown by 14-MeV (fusion) neutrons. A similar effect is seen in p-type material; however, the data in this column were taken from only one sample for each type of irradiation and may not be too accurate. Nonetheless, the effects are large enough to be well outside observational error.

All the lifetime data of Table I were obtained by the same observer. Since a similar situation does not exist for silicon, a comparison of this material is more difficult. However, Smits and Stein[13] have found 14-MeV neutrons to be ~3 times more effective than fission neutrons in removing carriers from silicon and ~3.5 times more effective in producing lifetime changes. This smaller difference for silicon is probably due to the different mechanism of defect production caused by the dissimilarity in atomic weights.

2. *Dependence of lifetime damage on impurities*

Under ^{60}Co gamma-irradiation, both germanium and silicon display a considerable range of lifetime degradation rates depending upon impurities. Degradation rates in silicon vary in order of magnitude, depending upon whether vacuum-float-zone ("oxygen free") or Czochralski (oxygen containing) material is used.[14, 15] In germanium a similar dependence on dopant has been observed. This is illustrated in Table II in which arsenic and antimony-doped material are compared. The data shown are from refs. 8 and 9. The sample designation used here and throughout this

TABLE II. Relative lifetime changes in Germanium divided by the product of carrier concentration and gamma dose.

Sample	$\left(\frac{\Delta \frac{1}{\tau}}{n\phi}\right)_{\text{Initial}}$ $(\times 10^{-27})$	$\left(\frac{\Delta \frac{1}{\tau}}{n\phi}\right)_{\text{Final}}$ $(\times 10^{-27})$
NCAs 1.3	4.8	1.3
NCAs 2.5	4.6	1.3
NCAs 5.4	5.2	2.3
NCSb 1.4	5.8	6.2
EPSb 2.3	—	14
NCSb 3.7	—	5.7

paper is made up of symbols for the manufacturer, the chemical dopant, and the resistivity. The temperature dependence shows that for small excitation

$$\tau \approx \frac{p_1}{c_n N n} \tag{1}$$

where standard Hall-Shockley-Read notation is used, so that the lifetime should be inversely proportional to n, the carrier concentration. The quantity $\Delta(1/\tau)/n\phi$ is thus a measure of the recombination center concentration. The first four samples listed in Table II were irradiated at 0°C and the last two samples at room temperature. Data taken shortly after irradiation at 0°C are shown in the first column, and data taken several days after irradiation appear in the second column. Actually the " initial " damage seemed similar in the specimens, the difference apparently occurring as defects moved about and interacted with impurities.

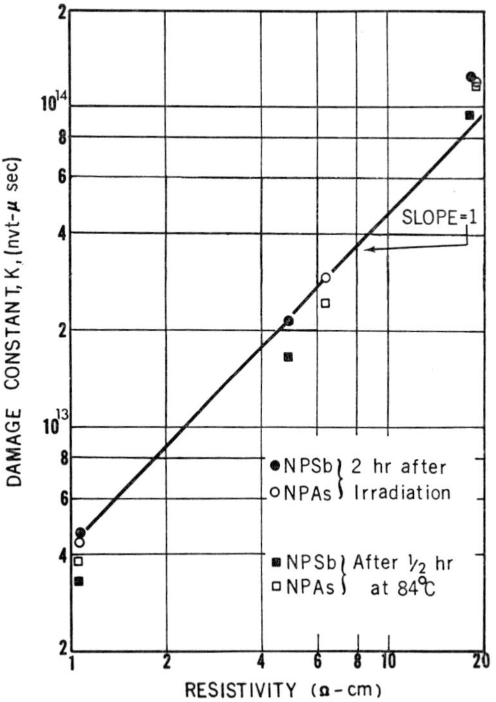

Fig. 1. Damage constant, $\Delta(1/\tau)/\phi$, for antimony- and arsenic-doped germanium. Expressed in these units, K is the neutron dose required to reduce the lifetime of an initially perfect sample to 1 μsec.

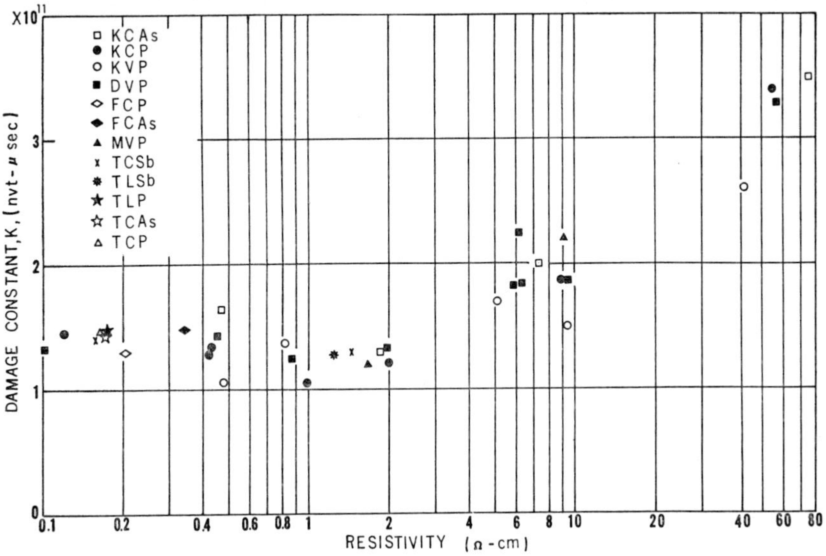

Fig. 2. Damage constant for n-type silicon.

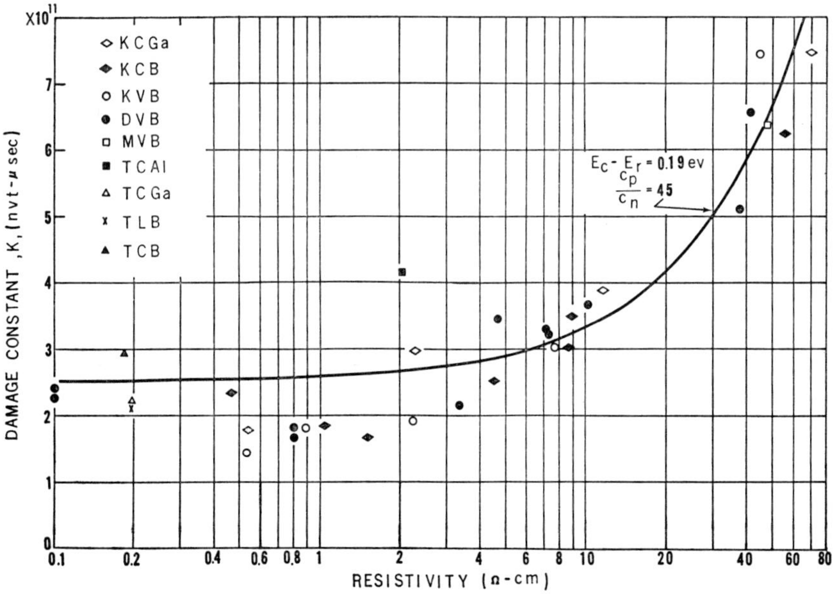

Fig. 3. Damage constant for p-type silicon.

On the other hand, neutron-irradiated germanium shows no such effect. Figure 1 is a plot of damage constant, $\Delta(1/\tau)/\phi$, versus resistivity for neutron-irradiated material shortly after irradiation, and again after a 1/2-hour anneal at 84°C. The pre-anneal points, especially, display a linear dependence on carrier concentration, and no difference between antimony and arsenic-doped material is seen. After the anneal, which should be roughly comparable to several days at room temperature, the difference between antimony- and arsenic-doped material remains small.

Figures 2 and 3 are similar plots for neutron-irradiated silicon, except that many points, representing a wide variety of samples, are shown. Here the sample designation contains an additional symbol following the manufacturer, C, V, or L corresponding to Czochralski, vacuum-float-zone, or LOPEX material. Preliminary data were presented earlier,[16] and a complete discussion will be published soon.[17] Some conclusions reached were that, in contrast to the case for gamma-irradiated silicon, no effect of oxygen or dopant is observed in n-type silicon, while only moderate dopant effects are seen in p-type material, and then only for low resistivities.

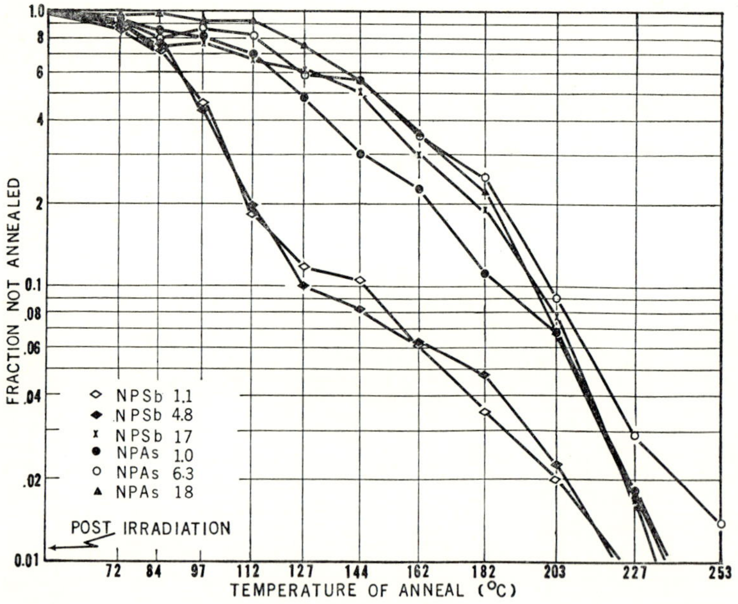

Fig. 4. Isochronal annealing of neutron-irradiated n-type germanium. The anneals were 1/2 hour each.

3. *Annealing data*

Eventually, as defect clusters begin to break up, impurities can play a role. In n-type germanium this apparently happens shortly above room temperature and, as shown in Fig. 4, the apparent number of recombination centers becomes smaller in the lower resistivity antimony-doped samples. Since the arsenic-doped and high resistivity antimony-doped samples behave similarly, evidently only the antimony atoms participate in impurity defect complex formation. This seems to support an earlier model[18] which postulated that the recombination centers were vacancies or divacancies that lost their effectiveness when they combined with antimony atoms. Presumably, antimony is favored over arsenic because of its larger size. However, this is almost certainly an oversimplification of the situation.

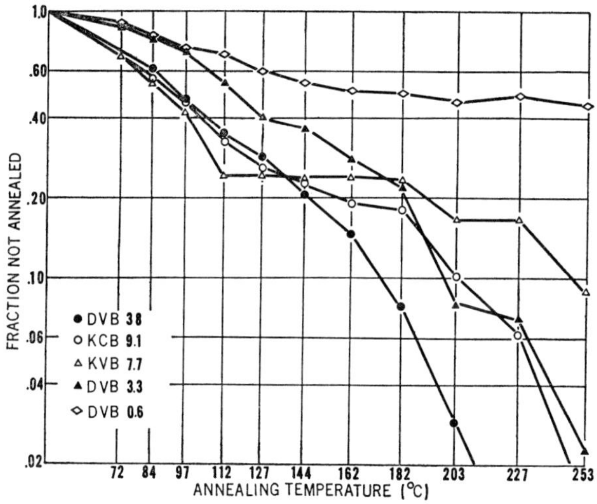

Fig. 5. Isochronal annealing of neutron-irradiated p-type silicon.

Data for p-type silicon are shown in Fig. 5. Except for lower resistivity samples, the annealing is similar up to 144°C. Evidently, a sufficient concentration of dopant tends to reduce the annealing rate. This behavior is consistent with the dependence of damage rate on dopant concentration at room temperature.[17]

The most striking demonstration of an absence of dopant effect is shown in Fig. 6, which shows the annealing of n-type silicon. (In Figs. 4 to 6 the annealing time at each temperature was 1/2 hour.) Here the annealing is virtually identical for a variety of samples up to 200°C. Data for

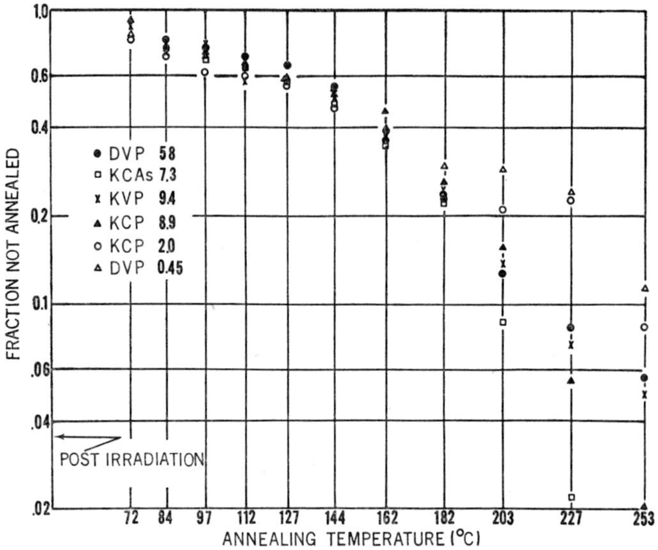

Fig. 6. Isochronal annealing of neutron-irradiated n-type silicon.

an additional sample, DVP 9.5, have been omitted to avoid confusion, but they agreed well with the samples shown. Of course, the oxygen concentration in these samples varied widely. Both arsenic and phosphorus dopants were used, and the phosphorus concentration ranged from $\sim 8\times 10^{13}$ to $\sim 1.3\times 10^{16}$ cm^{-3}.

Annealing occurs for times much shorter than those conventionally

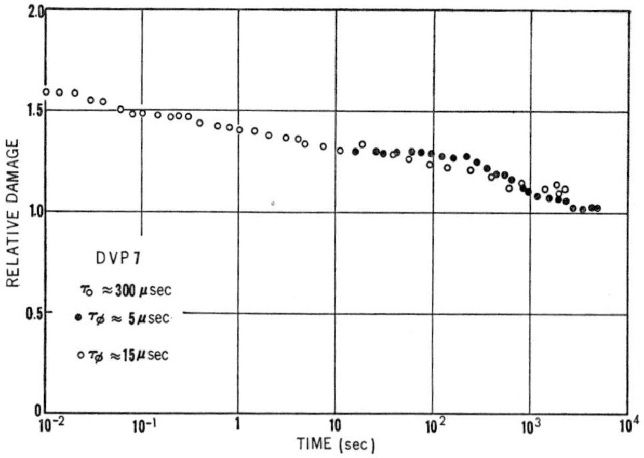

Fig. 7. Short term annealing in neutron-irradiated n-type silicon (DVP 7).

used for " before and after " measurements. Figure 7 is a plot of relative lifetime for two 7 ohm•cm n-type silicon specimens in the time interval following a short burst of neutrons. The relative lifetime was determined from steady-state photoconductivity using the relationship

$$\Delta n = gI\tau \tag{2}$$

Annealing shown in Fig. 7 is also an indication of the dominant role that disordered regions play in the recombination process in neutron-irradiated silicon. Certainly, any annealing process which occurs over so many decades of time must be very complicated, and such complications would be expected for the reorganization which must occur inside a disordered region.

4. *Comparisons with carrier concentration changes*

R. F. Bass[11] has performed measurements of carrier removal induced by neutrons in n- and p-type silicon with resistivities of about 2 ohm•cm. Table III gives his values of carrier removal rates at room temperature. In contrast to the results for recombination, there seems to be a clear indication of an effect of oxygen on removal rate and possibly some dopant effect. In p-type material any impurity effect is small. Similar results are indicated by mobility changes.

Annealing studies confirm these results, showing differences between samples grown by different techniques or containing various dopants. Also, the annealing pattern is much different for carrier concentration than for lifetime, indicating that the recombination centers have little effect on total carrier removal. In general, carrier concentration anneals more slowly than lifetime, especially in p-type material. In n-type material only about 50 per cent of the carrier concentration is recovered at 253°C while about 90 per cent of the carrier concentration change still remains in p-type material. (Virtually all lifetime change has recovered at these temperatures.)

Since impurity effects do occur in n-type silicon, a large fraction of defects evidently exists outside the disordered regions. From the range of carrier removal rates observed, a reasonable estimate is that these represent about one-half the total defect concentration. The fact that simple defects are present in n-type material, but do not affect recombination, further indicates the greater effectiveness of recombination centers within disordered regions.

TABLE III. Carrier removal rates in neutron irradiated 2-ohm·cm silicon measured at 26°C (after Bass[11]).

Sample	Initial carrier concentration ($cm^{-3} \times 10^{15}$)	Carrier concentration change ($cm^{-3} \times 10^{14}$)	φ $\begin{pmatrix} \text{nvt} \times 10^{14} \\ E > 10 \text{ keV} \end{pmatrix}$	Carrier removal rate ($\text{neutron}^{-1} \cdot cm^{-1}$)
KCAs 1.82	2.22	8.32	5.59	1.49
KCAs 1.70	2.28	7.76	4.86	1.60
TCSb 1.41	2.56	7.65	4.86	1.57
TCSb 1.43	2.56	7.60	4.86	1.56
KCP 1.74	2.13	9.12	6.06	1.50
KCP 1.89	2.17	8.27	4.86	1.70
KVP 1.59	2.59	10.09	3.89	2.59
KVP 1.69	2.49	10.31	3.86	2.68
DVP 1.69	2.43	2.73	1.20	2.28
DVP 1.52	2.29	8.07	2.69	3.00
DVP 1.56	2.21	8.24	2.69	3.06
TLSb 1.21	2.91	9.29	4.86	1.91
TLSb 1.20	2.97	9.01	4.86	1.85
TCAl 1.87	6.33	15.52	4.86	3.19
TCAl 1.70	6.43	13.84	4.86	2.85
KCGa 1.41	7.85	17.22	5.84	2.95
KCGa 1.30	9.62	20.90	7.45	2.80
KCB 1.22	9.53	28.47	9.47	3.01
KCB 1.35	8.36	28.22	9.47	2.98
KVB 1.84	5.72	13.90	4.60	3.02
KVB 1.86	6.01	19.25	5.84	3.29
DVB 3.21	3.71	11.79	3.70	3.19
DVB 2.61	3.95	12.30	3.86	3.19

5. *Behavior of recombination centers within a disordered region*

One method for investigating recombination center parameters is to measure carrier lifetime as a function of temperature. In Figs. 8 and 9 plots of lifetime following neutron irradiation and subsequent anneals are given for samples DVP 58 and DVB 38. The behavior exhibited in the figures is typical of that observed for a number of specimens measured in the resistivity range down to 2 ohm·cm.

Because of the lack of similarity between pre- and post-irradiation temperature behavior for the n-type samples, it was useful to plot the quantity $1/\tau - 1/\tau_0$ versus temperature to obtain apparent energy level positions. This is illustrated for DVP 58 in Fig. 10. Correcting for the temperature

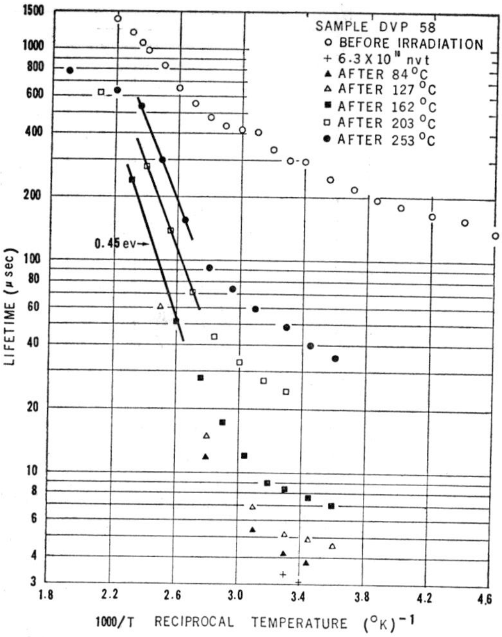

Fig. 8. Temperature dependence of lifetime before and after neutron irradiation and following anneal (DVP 58).

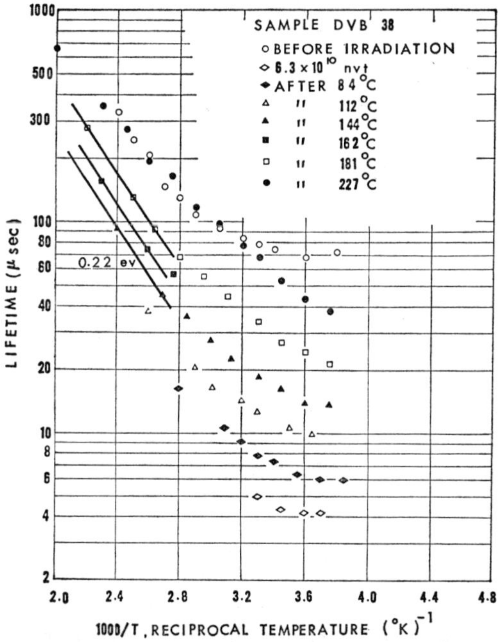

Fig. 9. Temperature dependence of lifetime before and after neutron irradiation and following anneal (DVB 38).

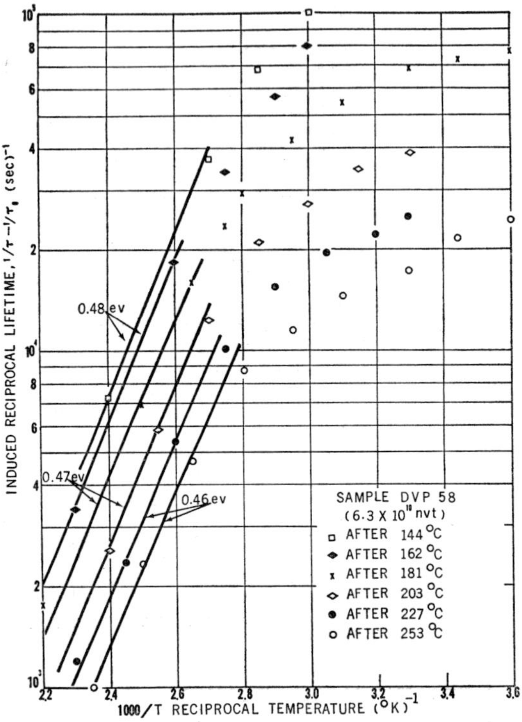

Fig. 10 Induced reciprocal lifetime vs. reciprocal temperature after neutron irradiation and subsequent anneals (DVP 58).

dependence of the density-of-states function, and averaging for several samples, the indicated positions of the recombination center as measured from a band edge are ≈0.42 eV in n-type silicon and ≈0.18 eV in p-type material. These values are consistent with those obtained by others for gamma and electron-irradiated material.[14, 15, 19, 20]

The occurrence of the ≈0.18 eV level is particularly interesting because it has a value similar to that observed for the oxygen-vacancy complex ("A center") identified from electron resonance. It seems unlikely that this center is an oxygen-related defect because of the lack of dependence of its introduction rate on oxygen concentration, and the evidence that recombination centers within disordered regions dominate the recombination process.

Another lifetime property, the observation of which can lead to a determination of energy level position, is the variation of lifetime with density of injected carriers.[21] The technique is illustrated in Fig. 11 where relative

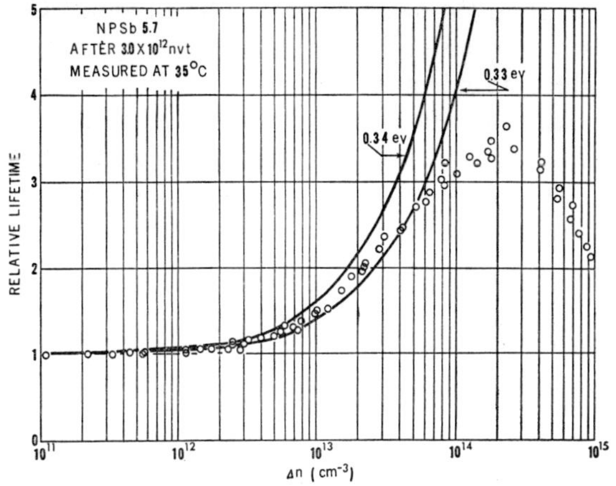

Fig. 11. Relative lifetime vs. concentration of injected pairs, Δn, for NPSb 5.7, following neutron irradiation.

lifetime is plotted versus injected carrier concentration for a germanium sample, NPSb 5.7. The solid curves show behavior expected from simple one-level Hall-Shockley-Read theory. Experimental points obey the predicted behavior at low injection levels, and the departure at higher levels is well explained in ref. 21. Figure 12 shows similar results for a

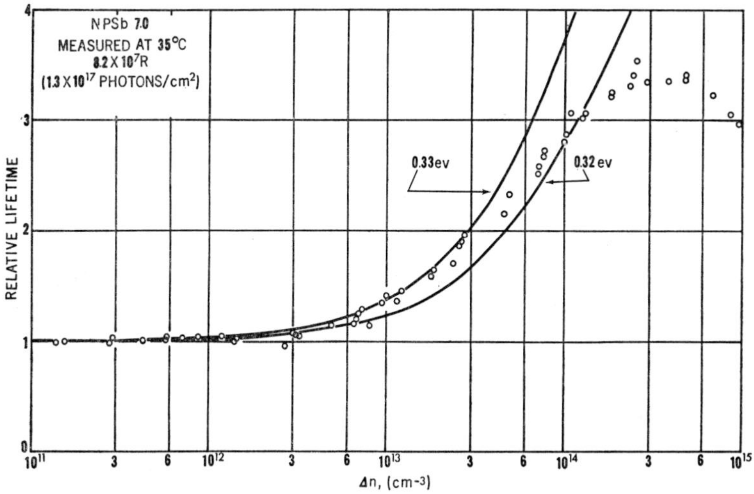

Fig. 12. Relative lifetime vs. concentration of injected pairs, Δn, for NPSb 7.0, following ^{60}Co irradiation.

sample NPSb 7.0, from the same ingot, irradiated with ^{60}Co gamma rays. The higher resistivity is an indication of the carrier concentration change required to produce an adequate degradation of lifetime; the resistivity change in the neutron-irradiated sample was negligible. The results are strikingly similar, especially at low injection levels where the data are used to determine the recombination level position. These measurements have been performed on samples having a variety of resistivities and dopants with very consistent results.[21]

An important conclusion drawn from the data of Figs. 8 to 12 is that the effective position of the recombination level is evidently independent of whether or not it lies inside a disordered region.

6. *A qualitative model*

Since the recombination level position is not affected by being inside a disordered region, the difference in recombination level effectiveness must be caused by a change in capture probability. A consideration of Gossick's model[3] shows that this should occur. To illustrate this model we have plotted the energy diagram for ~10 ohm•cm n-type silicon in Fig. 13.

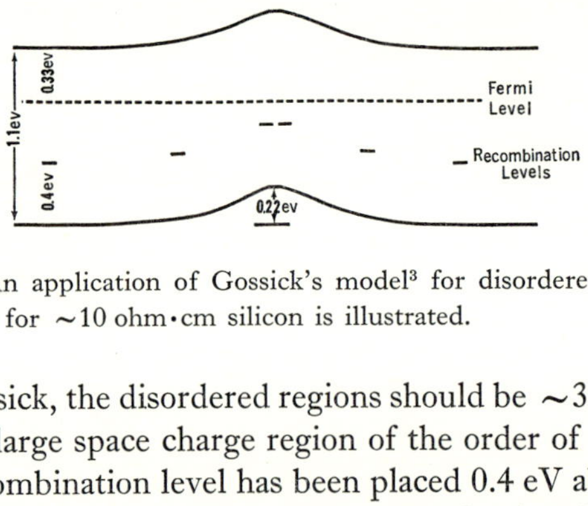

Fig. 13. An application of Gossick's model[3] for disordered regions. The case for ~10 ohm•cm silicon is illustrated.

According to Gossick, the disordered regions should be ~350 Å in diameter surrounded by a large space charge region of the order of 1 μ in diameter. Note that the recombination level has been placed 0.4 eV above the valence band. Others have placed it nearer the conduction band[15, 19, 20] primarily because of the desire to associate it with an EPR center. However, for our data the fit to the Hall-Shockley-Read theory is satisfactory only if the level lies in the lower half of the forbidden gap.[10] Wysocki[22] has used thermally-stimulated conductivity of neutron-irradiated silicon to deduce the presence of a level at $E_v+0.41$ eV, which is probably the same level we have observed.

One possibility suggested by Gossick is that the large spacecharge regions behave like spheres with an infinite surface recombination velocity.

When a minority carrier encounters such a sphere, it disappears. Gossick discounted this possibility because of the excessive effects such regions would have. Another objection which is apparent is that if this behavior did exist, the individual properties of the defects would be completely obscured, contrary to evidence. Apparently then, one must consider the potential well to be primarily a means of enhancing the probability of a minority carrier being in the region of a recombination center, thus increasing the capture probability. However, it is also clear that the carrier must have a high probability of escaping the disordered region without recombination. Otherwise, the individual defect properties could not manifest themselves. That is, the potential well for minority carriers must be sufficient to enhance the probability of capture by a recombination center, but not great enough to insure recombination once the carrier has come within its influence. Thus, the depth of potential well associated with disordered regions may be much smaller than used by Gossick to explain other phenomena. His model and Fig. 13 illustrate the largest effect one would expect. Two factors could make the potential change smaller. (*1*) If the total number of defects inside the disordered region were too small, such a large space charge region could not be maintained. This factor was neglected by Gossick in his assumption that the field inside the region containing the defects is zero. (*2*) The defect concentration may be sufficiently dilute that the Fermi level within the region is not the same as that observed throughout the bulk material for very large irradiations, contrary to another assumption of Gossick's. Other experimental evidence, including thermal conductivity measurements,[23] indicate that the disordered regions may be smaller than Gossick estimated. Other regions with deeper potential wells probably exist, but are not favored for recombination because of their smaller concentration or the statistics of the recombination process.

Observations for n-type germanium, where Gossick treats the disordered region as p-type, demonstrate very effectively that, if recombination does occur within a disordered region, the potential well must be shallower than predicted. The recombination center is well established at $E_r - E_v =$ 0.34 eV, very near the center of the gap.[21] The observed behavior can exist only if the equilibrium Fermi level remains above this position; thus, the interior of the disordered region must appear at least slightly n-type. Probably some regions do convert to p-type but are ineffective as recombination centers. Since any hole which enters such a region becomes a majority carrier, it seems that another hole would reappear elsewhere to maintain charge neutrality. The possibility that such a mechanism would prevent recombination at disordered regions in n-type germanium was suggested by Gossick.

Perhaps the most difficult problem posed by this model is the fact that in p-type germanium the rate of lifetime change also points to enhancement of capture probability by the disordered region, an unexpected occurrence for moderate resistivity p-type material. There are two possible explanations for this behavior. (*1*) The interior of the disordered region actually tends toward intrinsic as it does for silicon, a surprising occurrence, but not inconceivable. Cleland, Bass, and Crawford[24] showed that the ultimate Fermi level is a complicated function of relative amounts of various types of disorder. Possibly the disordered region, when present in small concentrations, does have the required balance of defects to make it tend towards intrinsic. Such a possibility would also eliminate the problem mentioned above for n-type germanium. (*2*) In p-type germanium the relative rate of introduction of recombination centers by more energetic radiation rather than the existence of disordered regions might be the important factor in enhancing recombination center effectiveness.

This introduces the problem that, for both n- and p-type germanium, the relatively fast degradation of lifetime under neutron bombardment could be explained by different ratios of defect production rates. However, the injection level data give a measure of the rate of introduction of recombination level centers relative to that for the 0.2 eV level,[21] a center largely responsible for carrier removal. We have found that the 0.2 eV level is relatively *less* effective in gamma-irradiated material.* This point seems to rule out the possibility mentioned for n-type germanium. Furthermore, the difference in defect stability and the lack of an impurity dependence are strong indications that the recombination levels are related to disordered regions.

As the carrier concentration is changed, both the depth and extent of the potential well change. Thus, the dependence of lifetime on carrier concentration is not expected to follow the simple Hall-Shockley-Read theory, in agreement with observation.[17] Perhaps even more interesting is the fact that at high carrier concentrations, where the potential well is deeper, the properties of individual defects should begin to vanish. We have observed such an effect in both n- and p-type silicon. The lifetime versus reciprocal temperature curves lost the characteristic shape of the

* On the other hand, if one assumes that the ratio of shallow to deep levels remains constant, the difference between the case for neutron and gamma-ray irradiation can be explained well by the decrease in electron concentration within the disordered region, an effect ignored in the above comparison.

Hall-Shockley-Read theory, and behaved much as would be expected from the modulation of potential well depth with temperature.[11] In n-type silicon, the behavior can also be explained by the combined effects of two recombination levels. A similar explanation does not seem possible in the case of p-type material.

III. SUMMARY AND CONCLUSIONS

There are a number of facts indicating the dominance of recombination centers within disordered regions in germanium and silicon. The evidence is more complete in n-type material on which more extensive studies have been performed. The following observations are noted:

1. In germanium the ratio of lifetime degradation to carrier concentration change is much greater for neutron than for gamma irradiation, and an enhancement of this effect occurs when 14-MeV neutrons are used rather than reactor neutrons.
2. A comparative lack of dependence of lifetime changes on impurities exists under neutron irradiation when compared to gamma irradiation.
3. The annealing of recombination centers shows little dependence on impurities, especially in n-type silicon.
4. Short-term annealing at room temperature is too complicated to be accounted for by simple defects.
5. In n-type silicon both the carrier removal rate and the recovery of carriers upon annealing are dependent upon impurities. No such dependence is observed for lifetime changes.

These facts are consistent with Gossick's model except that the depth of potential well for the effective regions must not be as great as he pre-This can easily dicted. come about if the total number of defects per cluster or the defect concentration is smaller than expected. Deeper potential wells probably exist but are ineffective as recombination centers because of changes in occupancy of the recombination level or, in the case of n-type germanium, because of the fact that inside the disordered region the hole is a *majority* carrier and charge neutrality requires an expulsion of another hole. This model then explains qualitatively two other observations: (*1*) the failure of lifetime versus carrier concentration to follow the Hall-Shockley-Read theory and (*2*) the temperature behavior of lifetime in low resistivity n- and p-type silicon.

As annealing occurs, defect clusters break up, and impurity effects are observed. Evidently the disordered region becomes less dominant.

However, no impurity effects occur in silicon up to ~200°C, indicating disordered region dominance throughout this range.

Acknowledgement

The author thanks R. F. Bass and C. A. Germano for their help in various phases of this work.

REFERENCES

1. J. R. Parsons, R. W. Ballufi and J. S. Koehler, *Appl. Phys. Letters*, **1**, 57 (1962).
2. Bertolotti, Papa, Sette, Grasso and Vatali, *Nuovo Cimento*, **29**, 1200 (1963).
3. B. R. Gossick, *J. Appl. Phys.*, **30**, 1204 (1959).
4. H. J. Stein, *J, Appl. Phys.*, **31**, 309 (1960).
5. N. Van Dong, L. Koch and N. D. Thong, *Trans. Faraday Soc.*, **57**, 1968 (1961).
6. W. H. Closser, *J. Appl. Phys.*, **31**, 1693 (1960).
7. J. W. Cleland, R. F. Bass and J. H. Crawford, Jr., in *Radiation Damage in Semiconductors*, p. 401, Dunod, Paris, 1965.
8. O. L. Curtis, Jr. and J. H. Crawford, Jr., *Phys. Rev.*, **124**, 1931 (1961).
9. O. L. Curtis, Jr. and J. H. Crawford, Jr., in *Radiation Damage in Semiconductors*, p. 143, Dunod Paris, 1965.
10. O. L. Curtis, Jr., R. F. Bass and C. A. Germano, *Report* 235-1, Harry Diamond Laboratories (1965).
11. O. L. Curtis, Jr., R. F. Bass and C. A. Germano, *Report* 235-2, Harry Diamond Laboratories (1966).
12. O. L. Curtis, Jr. and J. W. Cleland, *J. Appl. Phys.*, **31**, 423 (1960).
13. F. M. Smits and H. J. Stein, *Bull. Am. Phys. Soc.*, **9**, 289 (1964).
14. T. Nakano and Y. Inuishi, *J. Phys. Soc. Japan*, **19**, 851 (1964).
15. R. A. Hewes, *Thesis*, University of Illinois.
16. O. L. Curtis, Jr., in *Lattice Defects and Their Interactions* (R. R. Hasiguti, ed.), to be published.
17. O. L. Curtis, Jr., *IEE Trans. Nuclear Sci.*, to be published.
18. O. L. Curtis, Jr. and J. H. Crawford, Jr., *Phys. Rev.*, **126**, 1342 (1962).
19. R. H. Glaenzer and C. J. Wolf, *J. Appl. Phys.*, **36**, 2197 (1965).
20. M. Hirata, M. Hirata and H. Saito, *J. Appl. Phys.*, **37**, 1867 (1966).
21. C. A. Germano and O. L. Curtis, Jr., *IEE Trans Nuclear Sci.*, to be published.
22. J. J. Wysocki, *J. Appl. Phys.*, **36**, 2968 (1965).
23. N. Van Dong, *J. Appl. Phys.*, **36**, 3450 (1965).
24. J. W. Cleland, R. F. Bass and J. H. Crawford, Jr., *Appl. Phys. Letters*, **2**, 113 (1963).

On Damage Regions Due to High Energy Neutron Irradiation in Si, InSb, GaAs Semiconductors

M. Bertolotti, T. Papa, D. Sette and G. Vitali

Istituto di Fisica della Facoltà di Ingegneria dell' Università di Roma, Roma, Italy

Fast particle irradiation in various materials may produce regions of localized damage which manifest themselves in different ways according to the nature of the material.[1–15].

The existence of these regions in semiconductors was first theoretically treated in detail for n-type germanium by Gossick[16] and by Crawford and Cleland.[17] Experimental evidence of these regions has also been obtained by several authors in germanium[6–12] and GaSb.[13–15]

We report here on an investigation concerning damage regions in Si, GaAs and InSb semiconductors. We have observed the damage with an electron microscope by making a carbon replica of the etched surface of the specimen. This method has been succesfully applied in the case of n-type germanium,[9,10] and with a few modifications it is applicable also to other semiconductors.

The semiconductors were irradiated at ambient temperature with an integrated flux of about 6×10^{12} 14 MeV monoenergetic neutrons/cm^2 by a Van de Groaf.

The crystals of GaAs and InSb have been treated with the following etch:

50%	in volume	HF
25%	,, ,,	HNO_3
25%	,, ,,	acetic acid

The GaAs samples have been etched in the [111] direction for a time of 6 seconds. Regions were observed having dimensions on the order of 2,000–2,500 Å, and density about 2×10^8 cm^{-2}. The regions generaly have an irregular shape, as shown in Fig. 1 (70,000× at the microscope). The theoretically expected density of regions is about 0.7×10^8 cm^{-2} which is in fairly good agreement with the figure found, and the dimensions are also comparable with the expected ones.

In the case of Si the chemical etch used was a CP-4A on the [111]

Fig. 1.

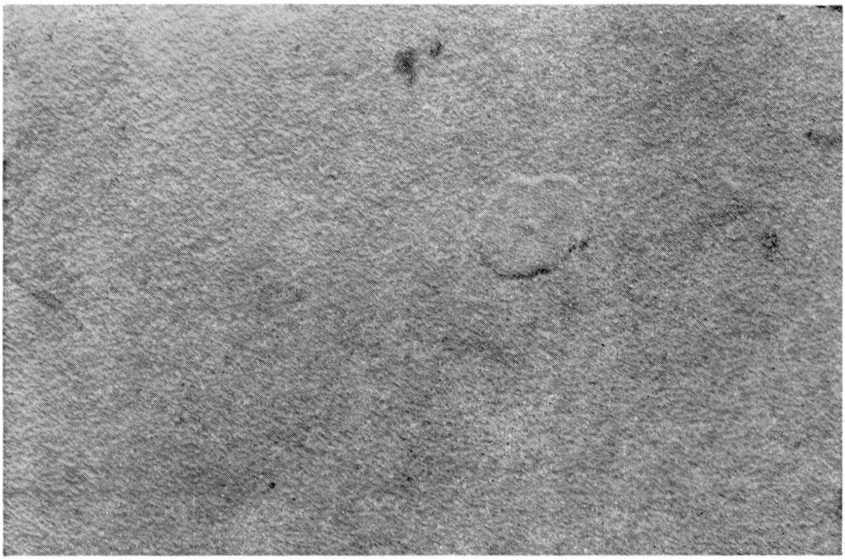

Fig. 2.

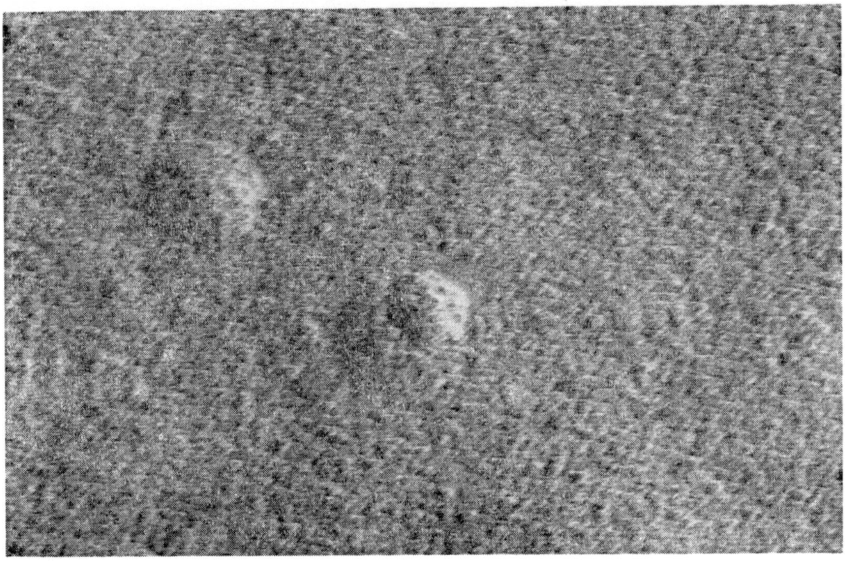

Fig. 3.

surface, and regions have been detected (see Fig. 2: 60,000 × at the microscope) which clearly show a central crater.

Due to the lower atomic number of silicon than germanium it is to be expected that the dimension of the damage region (the crater in Fig. 2) be larger than in germanium. A theoretical value is 600 Å. The crater dimension is about 500 Å. To this corresponds an effective radius of the space-charge region[10,16,17] of the order of 2,000–2,500 Å (resistivity of Si sample is 10 $\Omega \cdot$cm) which is to be compared with the dimension of the region in Fig. 2 which is about 2,000 Å.

The situation in the case of InSb is not equally clear. In this case regions are visible (see Fig. 3: 60,000 × at the microscope) which seem to be damage regions, with the correct expected dimensions (2,000–2,500 Å). The density of such regions, however, is nearly two orders of magnitude lower than the theoretically expected value.

The experiments indicate therefore that damage regions are present also in Si and III-V semiconductors. In the case of III-V semiconductors, however, it is not possible at this stage of the research to decide whether the regions are due to space-charge effects, i.e. are crystalline in nature, or whether they are due to the precipitation of one of the constituents of the semiconductors.

REFERENCE

1. E. W. Mitchell and P. T. Wedepohl, *Phil. Mag.*, **3**, 1280 (1958).
2. H. H. Atkinson, R. E. Smallman and K. H. Westmacott, *J, Appl. Phys.*, **30**, 646 (1959).
3. T. S. Noggle, *Bull. Am. Phys. Soc.*, **4**, 137 (1959).
4. R. E. Smallman and K. H. Westmacott, *J. Appl. Phys.*, **30**, 603 (1959).
5. M. W. Thompson and R. S. Nelson, *Phil. Mag.*, **7**, 2015 (1962).
6. W. H. Closser, *J. Appl. Phys.*, **31**, 1963 (1960).
7. H. J. Stein, *J. Appl. Rhys.*, **31**, 1309 (1960).
8. J. R. Parsons, R. W. Balluffi and J. S. Koehler, *Appl. Phys. Letters*, **1**, 57 (1962).
9. M. Bertolotti, V. Grasso, T. Papa, D. Sette and G. Vitali, *Nuovo Cimento*, **29**, 1200 (1963).
10. M. Bertolotti, T. Papa, D. Sette and G. Vitali, *J. Appl. Phys.*, **36**, 3506 (1965).
11. J. R. Parsons and R. W. Balluffi, *J. Phys. Chem. Solids*, **25**, 263 (1964).
12. J. R. Parsons, *Phil. Mag.*, **12**, 1159 (1965).
13. U. Gonser and B. Okkerse, *Phys. Rev.*, **105**, 757 (1957).
14. U. Gonser and B. Okkerse, *Phys. Rev.*, **109**, 664 (1958).
15. F. E. Fujita and U. Gonser, *J. Phys. Soc. Japan*, **33**, 1068 (1958).
16. B. R. Gossick, *J. Appl. Phys.*, **30**, 1214 (1959).
17. J. H. Crawford, Jr. and J. W. Cleland, *J. Appl. Phys.*, **30**, 1024 (1959).

Electron Bombardment Induced Recombination Centers and Radiation Damage in GaAs*

J. J. Loferski, H. Flicker,† R. M. Esposto and M. H. Wu‡

Division of Engineering, Brown University, Providence, Rhode Island, U.S.A.

ABSTRACT

The recombination radiation spectra of n- and p-type gallium arsenide produced by electron-hole pairs injected by excitation with electrons of energies between 150 and 200 keV were used to monitor the introduction of radiation defects into material containing various concentrations of impurities. Irradiations were performed over a temperature range between 100° and 300°K. For p-GaAs, it was found that the peak intensity of light emitted by specimens containing between 10^{16} and 10^{19} acceptors per cc was approximately proportional to the acceptor concentration provided that the beam current I_B and sample temperature T was kept constant. The intensity was proportional to I_B for the range used in the irradiations, namely $2.0 \leqslant I_B \leqslant 20\ \mu A/cm^2$, and it was extremely sensitive to temperature; it decreased by a factor of ten when T increased from 100 to 300°K. When the bombarding electron energy E_B was 300 keV or higher, the intensity of the emitted line at 1.49 eV (at 120°K) began to decrease, the amount of decrease being greater as the electron energy increased. For $E_B = 475$ keV, about 5×10^{15} electrons/cm^2 caused a reduction of peak intensity by a factor of two. The threshold for damage was found to lie between 250 and 300 keV which is higher than the value deduced by Bauerlein[32] on the basis of changes in electron voltaic properties of a gallium arsenide p-n junction. Rates of degradation in material doped with Cd and Zn are compared. Recombination radiation spectra of n-GaAs are also discussed. It is shown that the spectra emitted by n-GaAs cannot be easily correlated with donor concentrations; that the intensity

* This work was supported by the Aerospace Research Laboratories, Office of Aerospace Research, United States Air Force.

† Present address: TRW Space Technology Lab., Redondo Beach, California.

‡ Present address: Semiconductors Products Division, General Electric Company, Syracuse, New York.

ELECTRO

by	Excess carrier generation		Samples GaAs	Impurity density (cm^{-3})	Detector
		Excitation level			
INJECTION ACROSS POTENTIAL BARRIER	point or silver paint contact	up to 1 A/cm^2 50% pulse	n non-degenerate	10^{17}	PbS Dry ice cooled
	p-n junction Zn diffused	82 mA	n degenerate	4.5×10^{18}	
		200 mA			
		80 mA			
	p-n junction Zn diffused	0.2, 0.5 and 0.8 A per 2×10^{-2} cm^2	n non-degenerate	10^{17}	
	p-n junction		p on n photo cell		
	p-n junction Zn or Cd diffused	Up to 10 A/cm^2	n	2×10^{17}	Pbs PM
	p-n junction Zn diffused	above 0.1 A/cm^2	n epitaxial	10^{16}	
			substrate Te doped	3×10^{18}	
	p-n junction Zn diffused		n	1.5×10^{17} Sn doped	
	p-n junction		n Te, Se	Zn diffused	
			p Zn	Sn-Te Pellet alloy	
	p-n junction		n Zn diffused		
			n on Zn doped substrate Zn diffused		
	p-n junction Be diffused	1.2 μsec 50 cps pulse: 2.5×10^3 A/cm^2	n	5×10^{17}–10^{18}	PM
		3000 A/cm^2			
		6000 A/cm^2			
	p-n junction Zn diffused	9 mA	n Te	10^{18}	
		20 mA			
		30 mA			
			n Te	10^{17}	
	p-n junction Zn diffused		n		

LUMINESCENCE

Spectral composition				Purpose of observation	Reference
Low energy emission	Near edge emission	Exciton emission	Temp. (°K)		
1.10			room temp.	Studies of injection luminescence	Braunstein[5] 1955
1.19			77		
0.9	1.45		4.2	Studies of injection luminescence	Pankov and Massoulie[6] 1962
0.95	1.33–1.42		78		
0.95	1.35		300		
1	1.47		78	Studies of injection luminescence	Black *et al.*[7] 1963
and red light					
	1.45		77	Studies of injection luminescence	Nasledov *et al.*[8] 1962
1.41	1.43		200		
1.36	1.39		300		
1.02–1.28	1.48		77	Compare the correspondence between photo and electro luminescence	Nathan *et al.*[9] 1963
1.35	1.48	1.505	77	Studies of exciton recombination radiation	Wilson[10] 1963
0.95	1.37		300	Effect of Cu impurities on low energy emission	Larsen[11] 1963
1–1.25	~1.47		77	Effect of doping on near edge emission	Lucovsky *et al.*[12] 1963
	1.472–1.48		78	Effects of doping on near edge emission	Dousmanis *et al.*[13] 1963
	1.482–1.488		27		
	1.433–1.469		78		
	1.443–1.478		27		
	~1.472		77	To compare the luminescence with the Zn diffused diode	Yunovich *et al.*[14] 1964
	~1.468		77		
	1.468		77		
	1.435		4.2	To study dependence of near edge emisson on diode current	Bagaev *et al.*[15] 1964
	1.445		4.2		
	1.45		4.2		
	1.495		4.2		
	1.485		77		
	1.38		290		
0.95, 1.27	1.44		78	Studies of injection luminescence	Keyes and Quist[38] 1962
~0.95	1.33		298		

RADIATION

by	Excess carrier generation	Excitation level	Samples GaAs	Impurity density (cm^{-3})	Detector
PHOTO-IONIZATION	mercury arc filtered by Cu_2SO_4 solution		n	5.5×10^{17} Te doped	PbS PM
			p	1.6×10^{18} Zn doped	PbS PM
	mercury arc filtered by Cu_2SO_4 solution		n	3×10^{17} undoped	Dumont 6911 PM PbS
			p		
	mercury arc filtered by Cu_2SO_4 solution		high purity crystal Czochralski	10^{15}–10^{16}	Dumont 6911 PM
			high purity crystal Hori. Bridgman	10^{16}–10^{18}	PbS
	mercury arc filtered by Cu_2SO_4 solution		n Hori. Bridgman		PM, PbS PbSe
FAST PARTICLES-IONIZATION	linear Acc. electrons	0.6 MeV up to 1 μA 2 μsec. Pulse 400 cps			FÉU PM
	Van de Graaff electrons	pulsed	n	non-degenerate	R.C.A. 7102 PM
			p	non-degenerate	
	linear Acc. electrons	30 MeV $>$ 400 mA	n		Dumont K 2276 R.C.A. 7102
			p	3×10^{18}	
	linear Acc. electrons	30 MeV $>$ 400 mA	n	3×10^{16}	Dumont K 2276 R.C.A. 7102
			n	2×10^{16} Se	
			p	4×10^{16}–2×10^{18}	
			p	4×10^{18}	
	electrons	600 keV 0.1 μA pulse	n	$> 2\times10^{17}$	PM
			n	$< 2\times10^{17}$	
	electrons	15 keV, 2–10×10^{-2}	n	10^{17}	PM
		35 keV, 1–8×10^{-2} A/cm^2	n	10^{17}	

LUMINESCENCE

Spectral composition				Purpose of observation	Reference
Low energy emission	Near edge emission	Exciton emission	Temp. (°K)		
0.98 eV	1.515 eV		77	Compare the correspondence between photo and electro luminescence	Nathan *et al.*[9] 1963
1.25	1.485		77		
1.5	1.52		4.2	Studies of optical injection luminescence	Nathan and Burns[16] 1963
	1.47		4.2		
1.418* 1.455	1.491	1.514	4.2	Studies of exciton emission line	Nathan and Burns[16] 1963 * Longitudinal optical mode phonon energy
1.414* 1.451	1.487	1.487	4.2		
▲ 0.65, ■ 1	1.49		6	Effects of O_2 pressure during crystal growth on low energy emission (▲↑, ■↓ as O_2 ↑)	Turner *et al.*[17] 1963
-1.29			77	Observe luminescence by electron beams	Basov *et al.*[18]
	-1.42		300		
1.1	1.4			Emission time constant studies	Flicker *et al.*[19] 1964
	1.4				
	-1.31		80	Dependence of emission intensity on the dose rate	Compton Bryant *et al.*[20] 1964
	-1.31				
	1.395		104	Dependence of emission intensity on the dose rate	Compton Bryant *et al.*[20] 1964
	1.3		295		
-1.22	1.5		80	Studies of activation energy of trap levels by temp. dependence of emission relax. time	Van de Does de Bye[21] 1964
1.22-1.35	1.47		80		
1.215	1.45		100	Studies of spectral variation versus electron beam energy	Bisson and Mayer[22] 1964

of the " band gap " emission is even more temperature dependent than in p-GaAs and that the intensity of emission begins to decrease for beam energies $\geqslant$290 keV.

I. INTRODUCTION

When a semiconductor crystal is irradiated by high energy particles, the particles dissipate their energy in two ways. Most of the energy is expended in ionization within the solid and this leads to an excess density of both holes and electrons. This excess density reaches a steady state value because the excess holes and electrons recombine either directly or through intermediate states in the forbidden energy gap. The recombination can be either radiative or non-radiative. If it is radiative, light will be emitted by the semiconductor while it is being bombarded by the high energy particles. By studying the form of this particle bombardment induced luminescence spectrum, it is possible to acquire information about localized impurity states through which hole-electron recombination is occurring.

The second way that high energy particles lose energy in the solid is by displacing atoms from their equilibrium sites, thus producing structural defects in the crystals. Such structural defects have associated with them localized levels in the forbidden energy gap. It is evident, therefore, that if one were to bombard a semiconductor by high energy electrons or protons, one could study the introduction of defects into the forbidden energy gap by following changes in the recombination radiation spectrum. This concept is the basis of the study of radiation damage in gallium arsenide described in this paper.

The feasibility of using recombination radiation to study radiation damage was first demonstrated by Braunstein.[1] In his experiment, he injected carriers into a germanium diode by applying a forward bias to the p-n junction. He then observed the intensity of emitted light before and after bombardment by β-particles emitted by a ^{90}Sr radioactive source. No attempt was made to observe changes in the spectral composotion of the emitted light during the course of irradiation. Later, Kulp and Kelley had studied changes in the cathodo-luminescence spectrum of CdS to determine displacement energies of the sulphur atom[2] and cadmium atom[3] in CdS crystals, and again Kulp and Detweler,[4] by a similar technique, studied the damage threshold in ZnSe due to electron bomdardment.

Recently the investigation of radiation induced defects in n-GaAs based on the above technique has been reported by us.[24, 25] The purpose of this paper is to review those results and to present further data on radiation luminescence studies in p-GaAs crystals, with emphasis on thresholds for producing defects by electron bombardment.

A brief summary of the references pertinent to this study is given in the Table. It is concerned primarily with electro-luminescence studies on gallium arsenide, although it also includes photo-ionization and fast particle ionization studies.

II. EXPERIMENTAL TECHNIQUES AND PROCEDURES

(1) *Instrumentation*

The experimental apparatus has three basic components, i.e. (i) beam source, (ii) sample housing and (iii) signal detector.

The equipment is shown schematically in Fig. 1.

(i) *Beam source.* The beam source assembly includes a horizontal Van de Graaff Electrostatic Accelerator, capable of developing up to 500 keV and of delivering a 100 μA DC current of either electrons or protons, and associated apparatus which includes a generating voltmeter, voltage stabilizing circuits, a focus magnet (for electron beams), electrostatic deflection plates, solenoid operated shutter, and vacuum pumping systems. All controls could be operated from outside the room in which irradiation was occurring.

Since the accelerator is not only being used to introduce permanent defects into the crystals but also to inject carriers into the solid, it was extremely important to have a particle beam in which both the energy and current of the incident particles were stable and known to a precision of a few per cent. A feedback circuit which kept the output of the generating voltmeter constant by controlling the charge deposited on the Van de Graaff belt assured constancy of voltage to within a few per cent. The beam energy was measured by a voltmeter, whose output had been adjusted to yield correct readings after calibration of the Van de Graaff voltages by (p, γ) reactions in a water cooled LiF target.[25]

A second low energy ($\leqslant$20 keV) electron gun was mounted at an angle perpendicular to the main beam of the Van de Graaff. This second gun allows one to examine spectra produced by low energy electrons and to examine changes produced when the Van de Graaff is converted to serve as a proton source.

The area of the sample covered by the beam was controlled by the focusing magnet. Positioning of the beam on the sample was accomplished by applying DC voltages of several kilovolts to two pairs of electrostatic

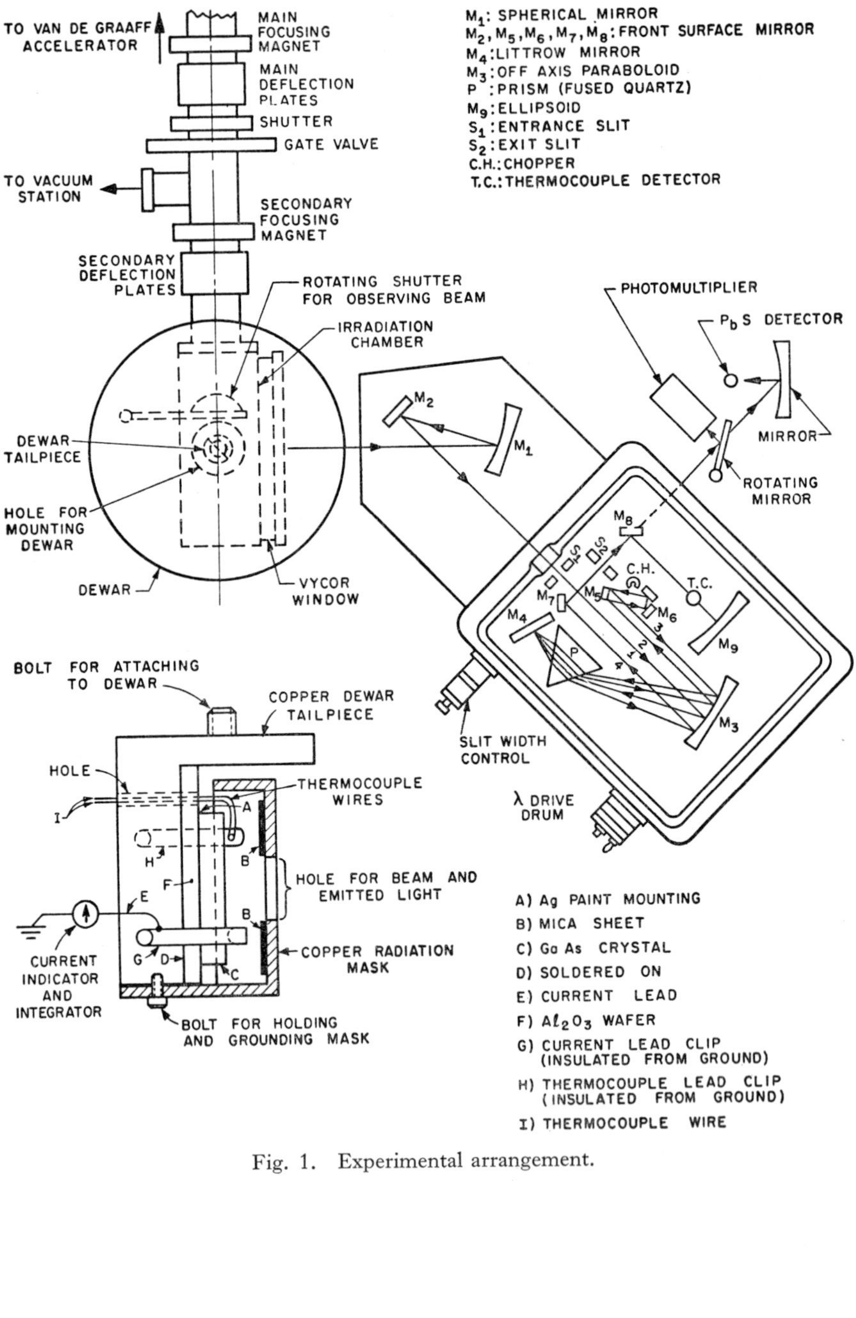

Fig. 1. Experimental arrangement.

deflection plates (one horizontal, the other vertical). In some of the experiments, the electron beam intensity on the sample was modulated at 13 cps in synchronism with the phase sensitive amplifiers associated with the optical system. Such modulation of the electron beam increased the overall sensitivity of the light detection system and in addition reduced sample heating and eliminated spurious thermal radiation reflected from the sample or its mounting.

The solenoid operated shutter which can be set to close after a prescribed integrated flux had been received by the sample prevents the electron beam from impinging on the sample when no irradiation is intended.

The vacuum pumping system consists of two pairs of standard cascade combinations of mechanical fore pump and oil diffusion pump. One of them pumps the main column of the Van de Graaff machine from the gun to the gate valve; the other pumps the irradiation chamber after the sample is changed. During irradiation and scanning both systems are pumping simultaneously to maintain vacuum of better than 10^{-5} mmHg as monitored by the Philips Gauges.

(ii) *Sample housing.* The samples were mounted on the cold finger of a dewar which was evacuated by the Van de Graaff pumping system, i.e. there was no metallic " window " through which electrons had to pass before striking the sample. The dewar was fitted with a Vycor window so that the light resulting from recombination of excess carriers could be directed via a series of mirrors onto the entrance port of the monochromator. The sample holder was designed so that the electron beam was incident at 45° to the sample normal and the light emitted by the sample could also be viewed at 45° to the same normal.

The sample was mounted on a sapphire slab which provided thermal conduction to and electrical isolation from the cold copper block. The sample-sapphire block assembly was covered by a copper plate which had a hole in it over the sample position (see Fig. 1) but which was electrically isolated from the sample. Thus the beam current could be determined by measuring the current received by the sample. A sample cover plate could be driven into place over the sample by means of a motor. This plate was covered by ZnS phosphor so that one could observe the beam distribution, position, homogeneity, etc. and adjust these parameters to their optimum values before exposing the sample to the beam.

A copper-constantan thermocouple was placed in good thermal contact with the top of the specimen and a second thermocouple was placed in contact with the copper block. Thus any differences in temperature between the sample and the block were easily detected.

(iii) *Photo-Detectors.* The radiation emitted by the sample during electron bombardment was first collimated by adjusting the relative positions of a plane and a spherical surface mirror. The collimated signal was then fed through a Perkin Elmer, Model 112 single beam double pass prism monochromator and detected by the R.C.A. 7102 photomultiplier or Kodak lead sulphide photo conductor at the exit port of the monochromator. A quartz prism was used for these experiments.

Some experiments were performed with DC excitation of the sample, in which case the emitted light was modulated at 13 cps by the chopper inside the monochromator. Other experiments were performed with the electron beam modulated at 13 cps. In this case, the 13 cps signal was provided by the motor driving the switches for the phase sensitive amplifier of the monochromator. In either case, the output signal from the detector was fed to the phase sensitive 13 cps chopper amplifier whose output was in turn recorded by an electronic recorder. The spectral response of both the photomultiplier and lead sulphide photodetector were established by comparison to the response of a thermocouple detector.

(2) *Sample preparation*

The samples were purchased from the Monsanto Chemical Company. They were usually pulled crystals, although some of them were boat grown. The carrier concentrations were those measured by the manufacturer. No attempt was made to analyze Hall effect or conductivity data to establish acceptor and donor concentration more accurately so that compensation was likely in the samples which had low carrier concentrations. However, the mobility gives some indication of the amount of compensation. The samples were cut to a size $4\times4\times0.5$ mm^3 and subjected to the following treatment prior to irradiation. The samples were first lapped carefully and then etched in a HNO_3: HF: H_2O (2: 1: 4)[26] mixture for about 45 minutes. After this process they were rinsed with distilled water several times to assure the removal of any deposit and again cleaned with acetone and alcohol before mounting for irradiation.

(3) *Irradiation procedure*

Specimens prepared as indicated above were first irradiated at the reference beam current and energy. The energy chosen as reference was well below the threshold for bulk damage production. The spectra of emitted light due to this bombardment was scanned and recorded. The reference beam was chosen to be 15 μA/cm^2 at 200 keV for the first set of experiments. In later experiments, this reference was changed to 20 $\mu A/cm^2$ at 125 keV.

The beam energy was then increased to a higher value (usually 25 keV to 50 keV higher than the previous value) and the sample was irradiated at much lower beam current for some specific time or dose (usually a current integrator was used). Then the voltage and the current were set back to the reference level for spectrum recording. Thus after each bombardment at successively higher beam energies spectra were recorded at the reference energy and examined for changes in intensity and position.

As mentioned before, a similar technique has been used by Kulp and Kelley to determine displacement energies of the sulphur atom[2] and cadmium atom[3] in CdS crystals and by Kulp and Detweiler to measure damage threshold in ZnSe.[4]

III. RESULTS AND DISCUSSION

(1) *n-GaAs—Summary*

We have previously[24] reported our results on n-GaAs. In summary of those observations we found that the n-GaAs spectrum consisted of a broad band which has a maximum at about 1.1 eV and which covers the

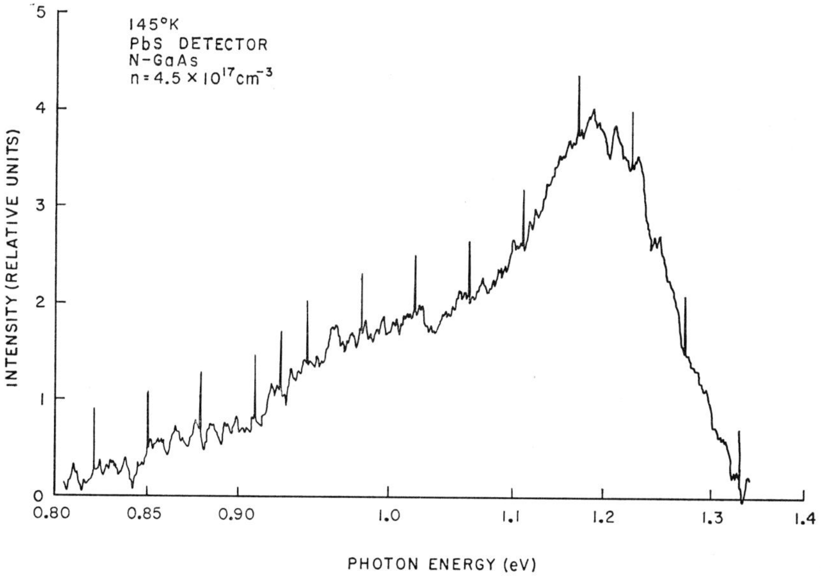

Fig. 2. Emission from n-GaAs ($N_{Te}=4.5\times10^{17}$/cc) as seen by a PbS detector; copy of record on *x-y* plotter.

region from 0.9 to 1.4 eV. The spectrum also contains a second narrow " line " centered at about 1.49 eV (115°K). Figure 2 is a reproduction of the trace of the spectrum emitted by n-GaAs ($n=4.5\times10^{17}$/cc) as seen by a PbS detector, the detector temperature being about 300°K. For this particular sample the amplifiers had to be set at maximum possible gain and the noise level was therefore appreciable. Note that the spectrum as viewed by the PbS detector shows no evidence of a peak at 1.49 eV. The magnitude of this peak differed in a random fashion from piece to piece of the n-GaAs. It was more readily detectable by the R.C.A. 7102 photomultiplier whose sensitivity is very high at 1.49 eV but whose long wavelength cutoff caused extreme distortion in the form of the 1.1 eV band. Our earlier experiments showed that two distinct phenomena occur as the energy of the bombarding electrons increases. When the energy of the electrons exceeds 300 keV, the 1.49 eV peak begins to decrease in intensity, but the 1.1 eV peak remains the same. When the energy of the electrons exceeds 350 keV, the 1.1 eV band begins to decrease.[24]

We have, however, recently examined the temperature dependence of the n-GaAs spectrum. The form of this dependence is seen in Fig. 3. Here it is evident that the intensity of the 1.49 eV peak remains virtually constant as the temperature is raised from 180° to 243°K. However, the 1.1 eV peak undergoes a considerable change in amplitude (a factor of about 10) over this same temperature range. (Note that the spectrum shown in Fig. 3 is the emission from n-GaAs as seen by the photomultiplier so that the shape of this main band is considerably distorted). Now we also found that the temperature of the sample could be increased because of absorption of energy from the bombarding electrons. Therefore changes in relative magnitudes of the 1.1 eV and 1.4 eV bands must be examined very carefully before assertions are made about the disappearance of one of these bands, which could be caused by a substantial rise in temperature during the experiment. In spite of this new information concerning the strong temperature dependence of the n-GaAs spectrum, it still appears to us that there are two thresholds in n-GaAs.

We have not been able to correlate the relative magnitude of the 1.1 and 1.49 eV luminescence or the absolute magnitudes of either with impurity content or any other parameter of the n-GaAs. We found that some specimens emit an intense 1.49 eV line while it is almost absent in other essentially identical specimens. The possiblity of orientation dependence was examined at least to the extent that we irradiated both the $(\bar{1}\bar{1}\bar{1})$ and $(\bar{1}\bar{1}\bar{1})$ faces of a given slice of n-GaAs, but no obvious difference could be found between the spectra from the two different faces of the slice.

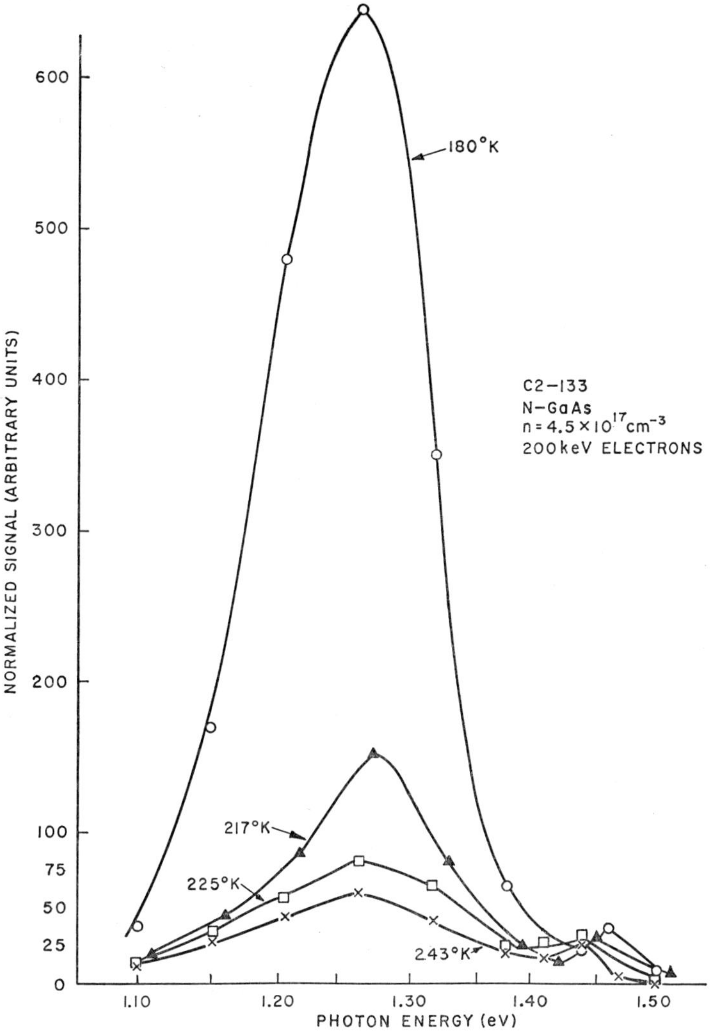

Fig. 3. Temperature dependence of n-GaAs spectrum.

Figure 4 shows the emission (as recorded by the R.C.A. 7102 photomultiplier) from 5 specimens of n-GaAs doped by various concentrations of n-type impurities. (Note the changes in scale needed to plot all these curves in a single drawing). The most intense emission was that from the most heavily doped specimens (Se doped; $N_{Se} \sim 7.1 \times 10^{17}$/cc) while the least intense line was that emitted by the specimen which was not intentionally doped. However, in-between concentrations do not come in

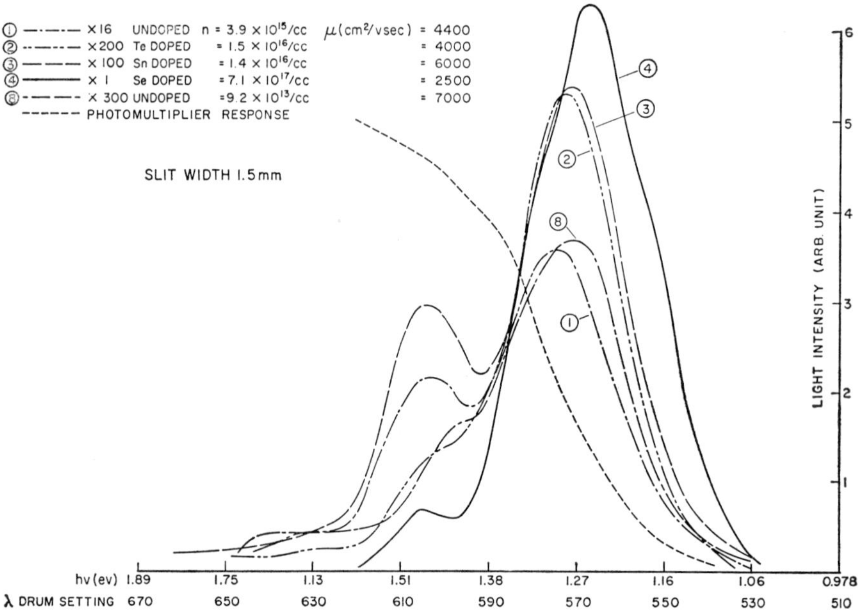

Fig. 4. Emission from n-GaAs specimens containing various concentrations of various donor impurities.

proper sequence. For example, the Sn doped specimen (with $N_{Sn} \sim 1.4 \times 10^{16}$/cc) exhibited an emission which was about a factor of two more intense than the emission from Te doped specimens with $N_{Te} \sim 1.5 \times 10^{16}$/cc. As we pointed out above, these samples may be compensated and therefore the carrier concentrations may not reflect donor concentrations correctly.

One other observation should be made with respect to the position of the 1.49 eV line as a function of temperature. It was found that this line shifted with temperature at about the same rate as does the energy gap of GaAs. To be specific, its position moved at the rate of -3.1×10^{-4} eV/°K over the temperature range 180° to 250°K.

(2) *p-GaAs—Description of the spectrum*

Figure 5 shows the emission (as seen by the R.C.A. 7102 photomultiplier whose spectral response is also plotted) from three p-type specimens containing different concentrations of impurities. Figure 6 shows the spectrum for a sample with a very high Zn concentration (9.4×10^{18}/cc) whose emission was so intense that the entrance slits of the monochromator could be narrowed to permit a resolution of about 20 Å. Figures 7 and 8 show

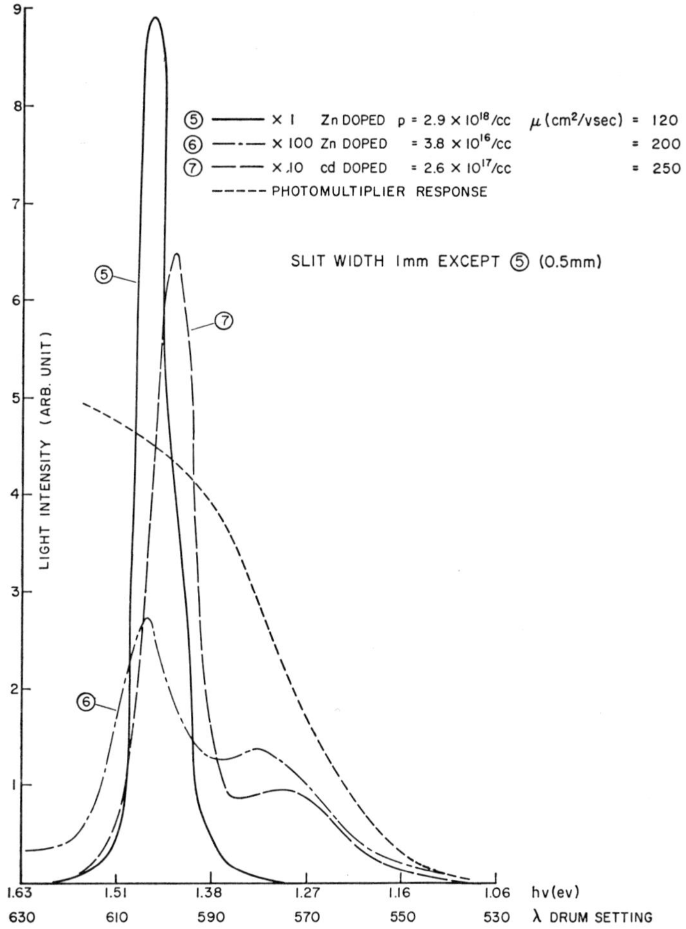

Fig. 5. Emission from p-GaAs specimens containing various concentrations of acceptor impurities.

spectra for Zn doped ($p=1.8\times10^{17}$/cc) and Cd doped ($p=2.3\times10^{17}$/cc) specimens in which the carrier concentrations were close to each other.

The first striking difference in the emission from the n-GaAs is the intensity of emission at an energy close to the band gap energy of GaAs. The 1.1 eV broad band is not seen in the p-GaAs spectrum. However, a second lower energy band can be seen in the less heavily doped specimens (Figs. 7 and 8). Figure 9 shows the temperature dependence of the position of the " band gap " line between 120° and 280°K. From this plot, we calculate that the peak shifts about -6×10^{-4} eV/°K compared to -3.1

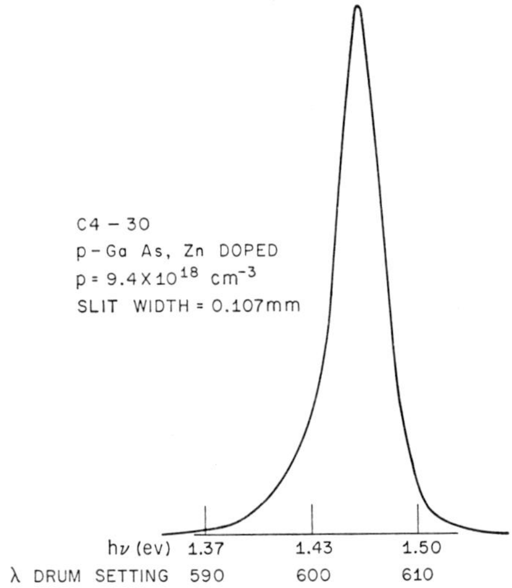

Fig. 6. Emission spectrum of p-GaAs Zn-doped ($p = 9.4 \times 10^{18}$/cc) caused by 200 keV electrons, 8.6 $\mu A/cm^2$ at 99°K.

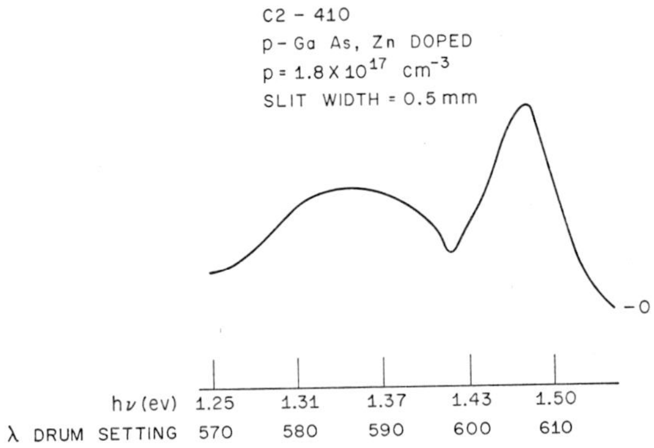

Fig. 7. Emission spectrum of p-GaAs Zn-doped ($p = 1.8 \times 10^{17}$/cc) caused by 200 keV electrons, 23 $\mu A/cm^2$ at 130°K.

$\times 10^{-4}$ eV/°K in n-GaAs. Figure 10 shows how the intensity of this line changes with temperature. Here it is seen that the intensity decreases by a factor of two when the temperature is increased from 180° to 240°K.

Having measured the temperature dependence of the position of the peak, it became possible to reduce the data for specimens doped with roughly equal concentrations of different impurities or with widely differ-

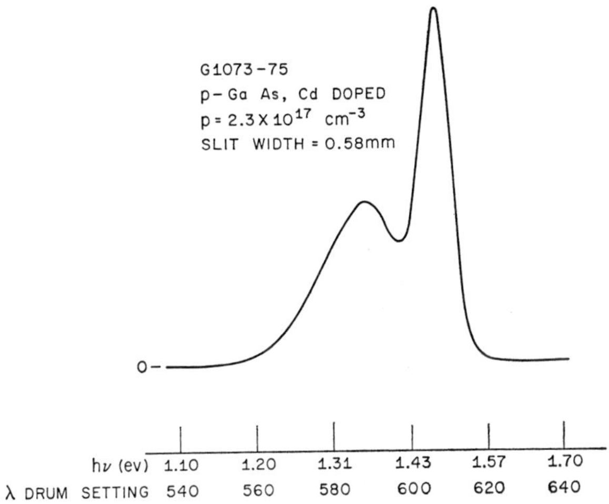

Fig. 8. Emission spectrum of p-GaAs Cd-doped] ($p=2.3\times10^{17}$/cc) caused by 200 keV electrons, 7.7 μA/cm^2 at 98.5°K.

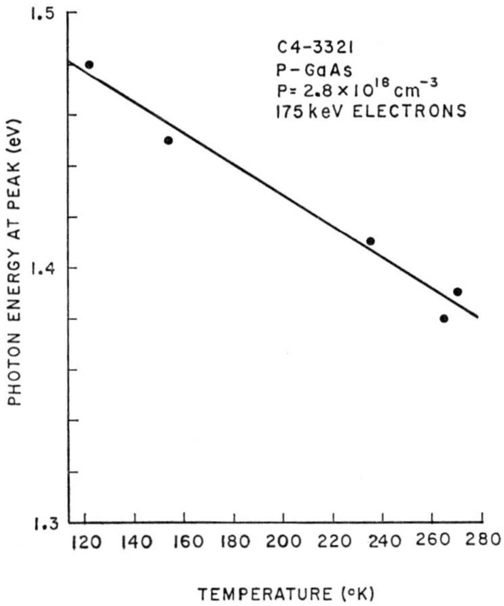

Fig. 9. Temperature dependence of 1.48 eV peak position in p-GaAs ($N_{Zn}=2.8\times10^{18}$/cc).

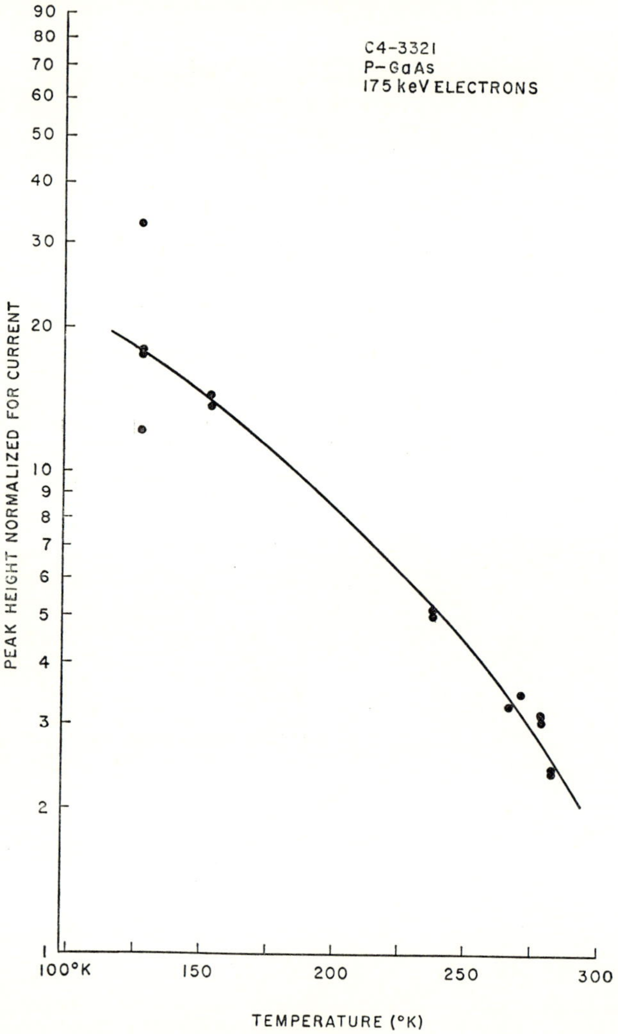

Fig. 10. Temperature dependence of 1.48 eV peak intensity in p-GaAs ($N_{Zn}=2.8\times10^{18}$/cc).

ent concentrations of the same impurity to the same temperature in order to compare the effects of these parameters on peak positions. Thus, for the Cd doped specimen (2.3×10^{17}/cc) (Fig. 8), the peak lies at 1.477 eV at 99°K or at 1.479 eV at 96°K, which will be the temperature at which comparisons are made. For the Zn doped specimen of comparable carrier concentration (1.8×10^{17}/cc) the peak lies at 1.473 eV at 130°K. Extrapolation to 96°K yields a peak position of 1.493 eV. The difference between these is 0.014 eV which can be construed as resulting from the difference in activation energies of Zn and Cd.

As for the effect of different carrier concentrations with the same impurity, compare Figs. 6 and 7. For the sample strongly doped with Zn ($p=9.4\times 10^{18}$/cc), the line position is at 1.465 eV at 96°K. The data of Fig. 7 extrapolated to 96°K yielded a line position of 1.493 eV. The difference between these values is 0.028 eV. If it is assumed that this line is caused by recombination of an electron in the conduction band and a Zn acceptor, the difference between the energy of this line and the band gap would be equal to the activation energy of Zn. According to Ermanis and Wolfstirn,[29] the activation energy of Zn in the case of $N_A=1.5\times 10^{17}$ is 0.018 eV. Thus the energy gap of GaAs would be equal to 1.51 eV at 96°K. This should be compared to the value 1.510 eV at 77°K deduced from absorption edge data by Holden and Sturge.[30] If their value is extrapolated to 96°K, the energy gap at that temperature whould be 1.500 eV. It appears therefore that the transition must be to an energy level somewhat farther removed from the top of the conduction band than the position of the Zn level deduced from temperature dependence of the Hall coefficient. This view is supported by the lower energy of the transition in the more heavily doped specimen (Fig. 6). In this case the activation energy of the Zn is zero,[29] yet the transition energy is lower than for the less heavily doped specimen. As the zinc concentration increases, impurity banding occurs. The impurity band broadens as zinc concentration is increased even further and the observed radiative transition probably involve levels close to the top of the impurity band.[27]

(3) *p-GaAs—Radiation damage*

Three p-GaAs specimens, cut from the same Zn doped ($p=2.8\times 10^{18}$/cc; $\mu_p=117$ cm^2/V·sec) crystal were irradiated by electrons of sufficient energy to cause damage. Figures 11 and 12 show the result of irradition of the first piece of this crystal at 400 and 475 keV, respectively. The crystal had previously been irradiated at 300 keV to a high integrated dose and no change was observed. The emission in Figs. 11 and 12 was produced by the same electrons which produced the damage, i.e. 400 and 475 keV, respectively. It is evident that substantial damage is occurring at 400 keV. Data from the second piece of this crystal are shown in Fig. 13. The emission was produced by 200 keV electrons, i.e. the specimen was irradiated by the " damage producing electrons " and the spectrum produced by the base (200 keV) electrons was observed. Here it is obvious that

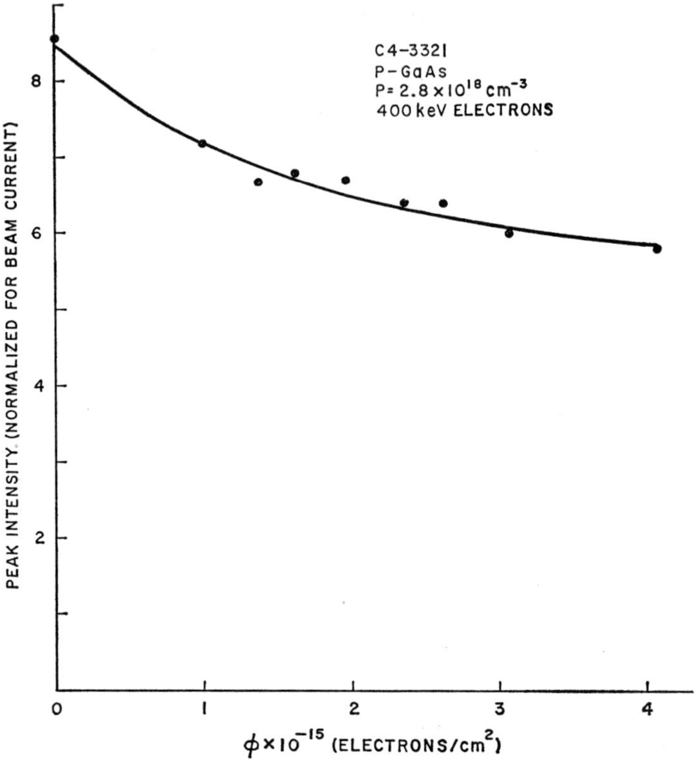

Fig. 11. Intensity of emission peak caused by 400 keV electrons at 1.48 eV as a function of integrated flux for p-GaAs ($N_{Zn}=2.8 \times 10^{18}$/cc). First sample.

damage has occurred at 350 keV. The rate of degradation of intensity of the peak was 3.7% for 10^{15} electrons/cm^2 at 350 keV. This figure also shows the degradation of emission cuased by 475 keV electrons for this same specimen. The rate of degradation of peak intensity caused by 475 keV electrons was 18% for 10^{15} electrons/cm^2. The third sample was irradiated by 400 and 475 keV electrons. The rate of degradation at 400 keV was 12% for 10^{15} electrons/cm^2 and 20% for 10^{15} eletcrons/cm^2 at 475 keV.

The degradation rates at these three energies are plotted in Fig. 14. The data look reasonable and imply that the threshold for damage in p-GaAs lies between 300 and 350 keV.

In another series of experiments, the samples whose pre-irradiation spectra are shown in Figs. 6, 7 and 8 were irradiated at 300 keV to a much higher integrated flux (2×10^{17} electrons/cm^2) than those shown in Figs.

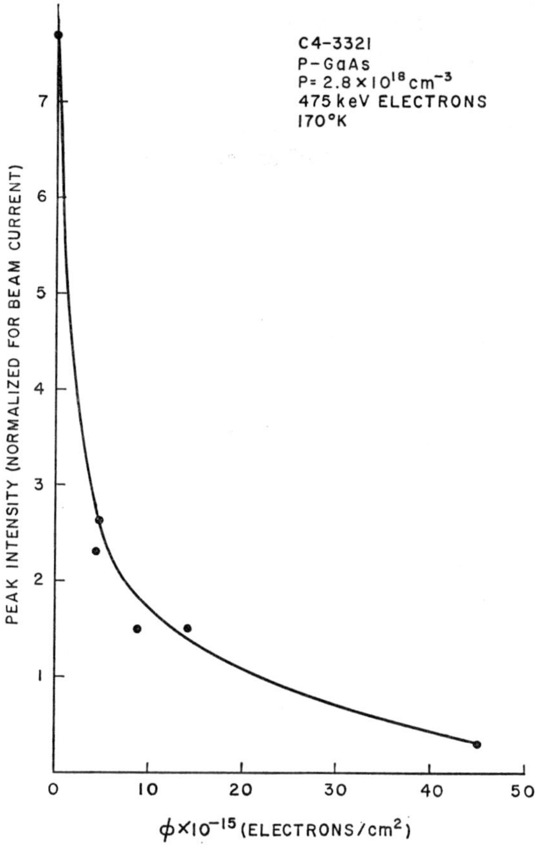

Fig. 12. Same as 11 except that the energy of the bombarding electrons is 475 keV.

11 through 13. Their spectra under excitation by 200 KeV electrons were then re-examined and compared to the pre-irradiation spectra. They were then subjected to about 1×10^{17} 450 keV electrons/cm^2 and the spectra excited by 200 keV electrons were re-examined.

First of all, according to those experiments, the intensity of the spectrum decreased but no shift in line position or appearance of new lines was observed. Thus after the Zn doped sample corresponding to Fig. 7 ($p=1.8 \times 10^{17}$/cc) was exposed to 2×10^{17} 300 keV electrons, the intensity of the 1.47 eV peak decreased by 25%. If it is assumed that the decrease is linear down to fluxes of 10^{15}/cm^2, this means that the decrease in peak intensity would have been only 1% at 300 keV which is consistent with the data of Fig. 14.

We also compared the degradation rates of the main (~1.47 eV) peaks of Cd doped and Zn doped specimens. No substantial difference was observed in the amount of degradation after a given flux of electrons. For example, 1×10^{17} 450 keV electrons/cm^2 caused a reduction of 62% in Cd doped material (p=2.3×10^{17}/cc) and 67% in Zn doped (p=1.8×10^{17}/cc) material.

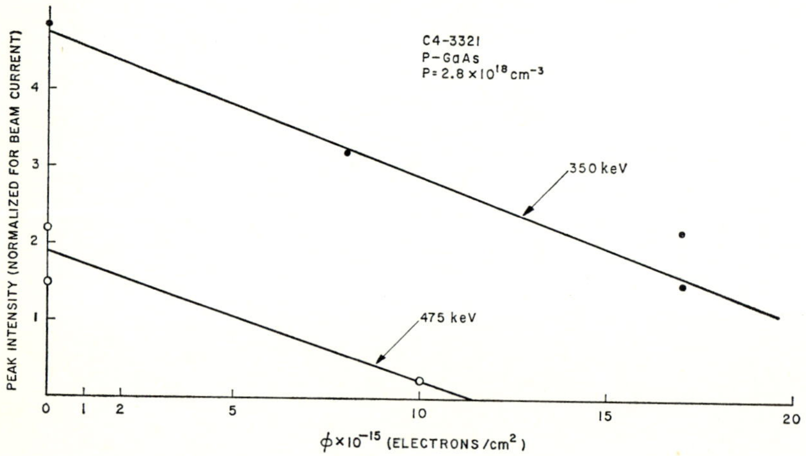

Fig. 13. Intensity of 1.48 eV emission peak (injection by 200 keV electrons) as a function of integrated flux of 350 keV and 475 keV electrons. Second p-GaAs sample ($N_{Zn}=2.8 \times 10^{18}$/cc).

Furthermore, we compared the relative degradation rates of the 1.47 and 1.34 eV peaks in the Zn doped specimen whose pre-irradiation spectrum is shown in Fig. 7. After exposure to 2×10^{17} 300 keV electrons/cm^2, the 1.47 eV peak decreased 25% while the 1.34 eV peak decreased 17%. After an additional 1×10^{17} 400 keV electrons/cm^2, the 1.47 eV peak was down to 33% of its initial value while the 1.34 eV peak was down to 60% of its initial value. These results should be compared to the relative degradation rates reported by Millea and Auckerman[31] for various bands of electroluminescent GaAs diodes. They reported that at 80°K the 1.47 eV line degrades by a factor of 2.4 while the 1.28 eV emission decreases by a factor of five after the diodes are irradiated by 2.8×10^{15} 2 MeV electrons/cm^2.

The rates of degradation of the (111) and ($\bar{1}\bar{1}\bar{1}$) faces of the Zn doped sample of Fig. 6 (p=9.4×10^{18}/cc) were also measured in the hope of detecting differences in damage rates or even of damage thresholds.

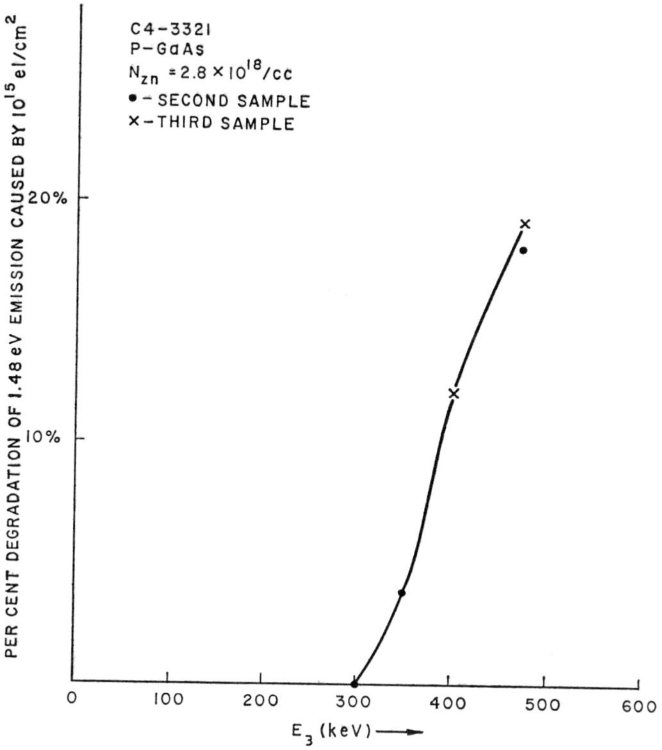

Fig. 14. Percentage degradation of 1.48 eV peak response after 10^{15} electrons per cm^2 as a function of electron bombardment energy. p-GaAs, $N_{Zn}=2.8\times 10^{18}/cc$.

However, the small differences which were observed were within our experimental error and we therefore concluded that no such difference could be detected by this technique.

IV. DISCUSSION OF THE RESULTS

In the case of p-GaAs, no changes in position of the lines in the radiative recombination spectrum were observed. For that matter, for those spectra which contained more than one line (e.g. Figs. 7 and 8), the degradation of intensity differed somewhat for the two lines although the difference was not as striking as that which we reported in n-GaAs.[24] However, in this case just as in n-GaAs, the peak close to the band gap value degraded more rapidly than the lower energy band which is probably caused by impurity centers in the forbidden energy gap.

There are certain questions concerning the data for which we are presently seeking answers. First of all substantial changes in the spectrum have occurred after exposure to about 10^{16} electrons/cm^2. If the 1.47 eV line is associated with transitions between the conduction band and Zn or Cd acceptor levels whose concentration is between 10^{17} and 10^{19}/cc, it is difficult to see how such a small number of electrons/cm^2 can produce such large changes in spectral intensity. To complicate matters further, there wasn't any appreciable difference between the number of electrons needed to produce given degradation in a sample doped with 10^{19} Zn acceptors and one doped with 10^{17} such entities.

As for the threshold for radiation damage, we conclude from these experiments that in p-GaAs, it lies between 250 and 300 keV. In our experiments on n-GaAs we concluded that the threshold in that material lay between 275 and 290 keV so that these results are consistent with the idea of equal threshold values for both conductivity types.

It has been disappointing to find that no new radiative paths between the conduction and valence bands of GaAs had been established as a result of the incorporation of radiation defects. This was the striking feature of the work of Kulp and Kelley[2,3] on CdS and Kulp and Detweiler[4] on ZnSe.

Finally our values of the threshold are higher than those reported by Bauerlein[32] on the basis of electron-voltaic effect measurements in GaAs diodes. He reported two thresholds, one at 233, the other at 256 keV and attributed them to the displacement of the galliun and arsenic atom, respectively. His measurements were made at room temperature.

V. SUMMARY AND CONCLUSIONS

(1) Changes in the recombination radiation spectra of GaAs have been used to determine radiation damage thresholds. In both n- and p-GaAs, it was found that the line whose energy corresponds closely to the forbidden energy gap begins to decrease in intensity when the electron energy lies between 250 and 300 keV for $T \sim 100°$K.

(2) No new lines or bands were detectable in the spectrum during the course of irradiation.

(3) The number of electrons/cm^2 (about 10^{16}) needed to cause substantial decay of line intensity was much lower than the density of acceptors in the p-GaAs ($>10^{17}$/cc).

(4) No difference in degradation rates or thresholds was observed between Cd or Zn doped specimens.

(5) If the Zn concentration was of the order of a few times 10^{17}/cc, the spectrum exhibited two peaks which decayed at different rates.

Acknowledgements

The authors wish to thank Mr. William Patterson, III for devising control circuits for stabilizing the energy of the Van de Graaff and for discussions of various phases of this work. They also wish to thank Mr. Richard Sutton for help in conducting the experiments.

REFERENCES

1. R. Braunstein, *Bull. Am. Phys. Soc.*, Ser. II, **2**, 157 (1957).
2. B. A. Kulp and R. H. Kelley, *J. Appl. Phys.* **31**, 1057 (1960).
3. B. A. Kulp, *Phys. Rev.*, **125**, 1865 (1962).
4. B. A. Kulp and R. M. Detweiler, *Phys. Rev.*, **129**, 2422 (1963).
5. R. Braunstein, *Phys. Rev.*, **99**, 1892 (1955).
6. J. I. Peunkove and M. Massoulie, *Elect. Chem. Soc. Electronics Division Abst.*, Vol. 11, No. 1, pp. 71–75 (1962). Also *Bull. Am. Phys. Soc.*, **7**, 88 (1962).
7. J. Black, H. Lockwood and S. Mayburg, *J. Appl. Phys.*, **34**, 178 (1963).
8. D. N. Nasledov, A. A. Rogachev, S. M. Ryvkin, V. E. Kartsiev, and B. V. Tsarenkov, *Fizika Tverdogo Tela*, **4**, 3346 (1962).
9. M. I. Nathan, G. Burns, S. E. Blum and J. C. Marinace, *Phys. Rev.*, **132**, 1482 (1963).
10. D. K. Wilson, *Appl. Phys. Lett.*, **8**, 127 (1963).
11. T. L. Larsen, *Appl. Phys. Lett.*, **7**, 113 (1963).
12. G. Lucovsky and C. J. Repper, *Appl. Phys. Lett.*, **5**, 71 (1963).
13. G. C. Dousmanis, C. W. Mueller and H. Nelson, *Appl. Phys. Lett.*, **8**, 133 (1963).
14. A. E. Yunovich, P. G. Elisejev, I. A. Nakchodnova, A. B. Ormont, L. A. Osadchaja and V. M. Stuchebnikov, in *Radiative Recombination in Semiconductors, 7th International Conference on the Physics of Semiconductors*, Vol. 4, p. 143 (1964).
15. V. S. Bagaev, Y. N. Berozashvili, L. V. Keldysh, A. P. Shotov, B. M. Vul and E. I. Zavaristskaya, in *Radiative Recombination in Semiconductors, 7th International Conference on the Physics of Semiconductors*, Vol. 4, p. 150 (1964).
16. M. I. Nathan and G. Burns, *Phys. Rev.*, **129**, 125 (1963).
17. W. J. Turner, G. D. Pettit and N. G. Ainslie, *J. Appl. Phys.*, **34**, 3274, (1963).
18. N. G. Basov and O. V. Bogdankevich, *J. Exptl. Theoret. Phys.* (U.S.S.R.), **44**, 1115 (1963).
19. H. Ficker and J. A. Baicker, *Bull. Am. Phys. Soc.*, **9**, 446 (1964).
20. D. M. J. Comton, G. T. Cheney and J. F. Bryant, in *Radiative Recombination in Semiconductors*, 7*th International Conference on the Physics of Semiconductors*, Vol.4, p. 235 (1964).; also J F. Bryant, G. T. Cheney and D. M. J. Compton, *Bull. Am. Phys. Soc.* (1964).
21. J. A. W. Van de Does de Bye, in *Radiative Recombination in Semiconductors, 7th International Conference on the Physics of Semiconductors*, Vol. 4, p. 243 (1964).

22. G. Bisson and G. Mayer, in *Radiative Recombination in Semiconductors, 7th International Conference on the Physics of Semiconductors*, Vol. 4, p. 251 (19634)
23. M. I. Nathan, *Solid State Electronics*, **6**, 425 (1963).
24. J. J. Loferski and M. H. Wu, in *Radiation Damage in Semiconductors, 7th International Conference on the Physics of Semiconductors*, Vol. 3, p. 213 (1964).
25. J. J. Loferski, *Technical Report*, AF 33(657)/1, Division of Engineering, Brown University, Providence, R. I. (June 1964).
26. J. G. Harper and M. S. Astor, *Electrochem. Soc. Electronics Div., Extended Abstracts*, **8**, 61 (1959).
27. D. A. Cusano, *Appl. Phys. Lett.*, **7**, 151 (1965).
28. N. G. Basov, O. V. Bogdankevich and B. M. Larvashin, *Fiz. Tverd. Tela*, **8**, 21 (1966)
29. F. Ermanis and K. Wolfstirn, *J. Appl. Phys.* **37**, 1963 (1966).
30. M. V. Hobden and M. D. Sturge, *Proc. Phys. Soc.*, **78**, 615 (1961).
31. M. F. Auckerman, *J, Appl. Phys.* **37**, 1788 (1966).
32. R. Bauerlein, *Zeits. F. Naturfor.*, **14a**, 1069 (1959).
33. R. J. Keyes and T. M. Quist, *Proc. IRE*, **50**, 1822 (1962).

On the Nature of Damages in Indium Antimonide Produced by γ- and X-irradiation

R. Ju. Khansevarov, T. V. Mashovetz, N. A. Vitovskii and B. Chelustka*

A. F. Ioffe Physical Technical Institute, Leningrad, USSR

ABSTRACT

As was convincingly proved in Eisen's papers,[1,2] the damages produced in InSb by electron irradiation at low temperatures are mainly two kinds of Frenkel pairs: In-interstitial (i-In)-In-vacancy (v-In) and Sb-interstitial (i-Sb)-Sb-vacancy (v-Sb), both acting as acceptors. The minimum electron energy required to displace In atom was found to be 268 keV that for Sb atom is equal to 398 keV.

Vook[3] observed, however, the creation of damages in InSb under X-irradiation at peak voltage, equal to only 100 kV. This means that some other mechanism of defect production must exist.

This paper is concerned with the investigation of defects induced by ^{60}Co γ-rays and 55 kV peak X-rays. The experiments included carrier concentration measurements at 78°K after different doses of γ-irradiation or X-irradiation as well as after isochronal and isothermal annealing.

I. EXPERIMENTAL RESULTS

X-irradiation

The carrier concentrations in n- and p-InSb versus irradiation time are plotted in Fig. 1 (see also ref. 4). No changes are visible in n-InSb, though the hole concentration in p-InSb increases under irradiation. The initial hole production rate is equal to $(4 \pm 2) \times 10^{-4}$ cm^{-1}.

The isochronal recovery data are shown in Fig 2. The whole recovery of hole concentration occurs at one stage at 95°K.†

The value of activation energy of this recovery process obtained from isothermal data is equal to 0.27 ± 0.03 eV.

* Institute " Ruder Boshkovich," Zagreb, Yugoslavia.

† The temperature of every stage will be indicated for 50% recovery. The heating rate was 1°K/min in all cases.

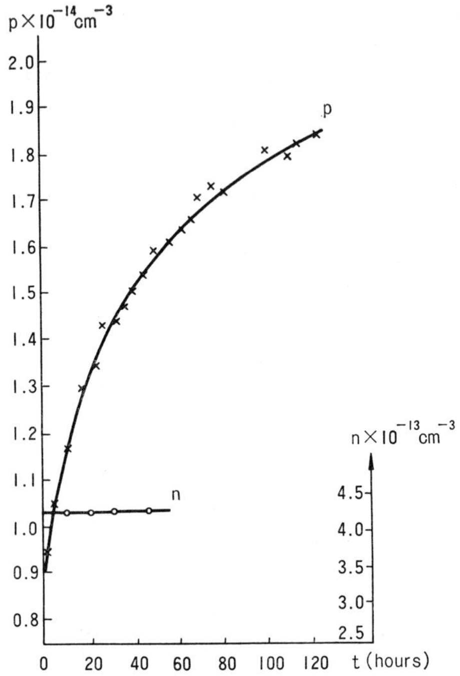

Fig. 1. The carrier concentration dependence on X-irradiation time for n- and p-InSb.

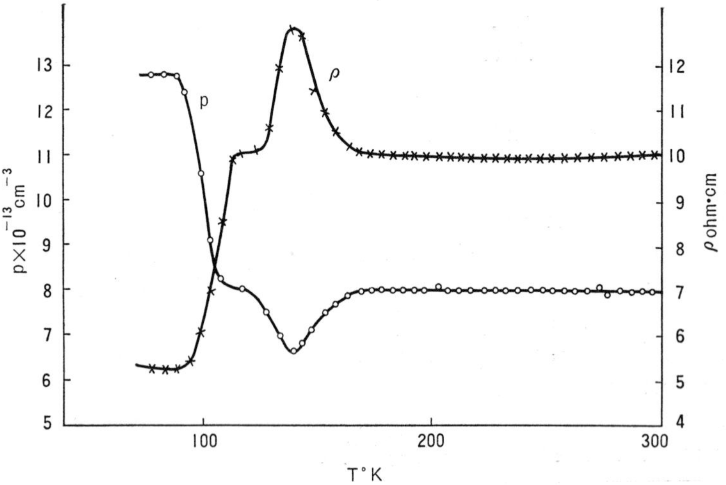

Fig. 2. The isochronal recovery of p-InSb after X-irradiation. p-hole concentration, ρ-conductivity. The measurements were made at 78°K after 5 min, annealing at each temperature.

After the whole recovery occurs the hole concentration decreases to values lower than the initial and then increases again to the initial value.

γ-irradiation

Figure 3 shows the typical dose dependences of carrier concentration in n- and p-InSb (see also refs. 5 and 6). The irradiation of n-InSb leads to a decrease in electron concentration, then to n-p conversion, and after that to an increase in hole concentration. The irradiation of p-InSb leads to an increase in hole concentration.

The initial rates of concentration changes in n- and p-InSb are approximately equal to each other and are $(6\pm2)\times10^{-3}\ cm^{-1}$. The rate of hole production becomes lower as the dose increases. This accounts for the annealing at 78°K in p-InSb. Such annealing was not observed in n-InSb. The recovery curves for n- and p-InSb are presented in Fig. 4,

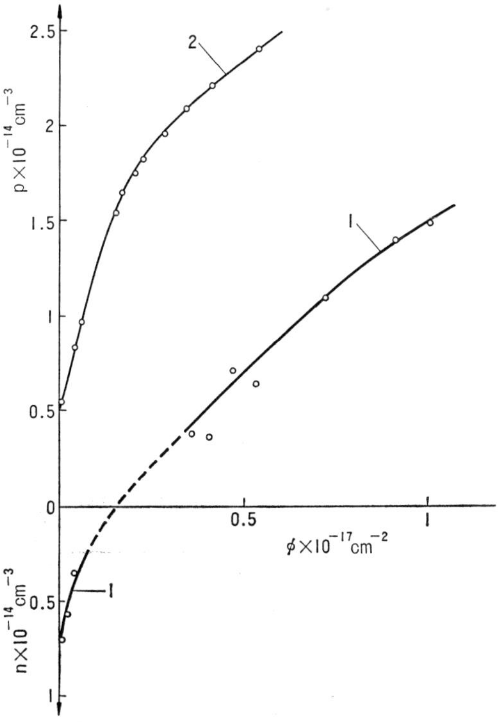

Fig. 3. The carrier concentration dose dependence for γ-irradiated n- and p-InSb.

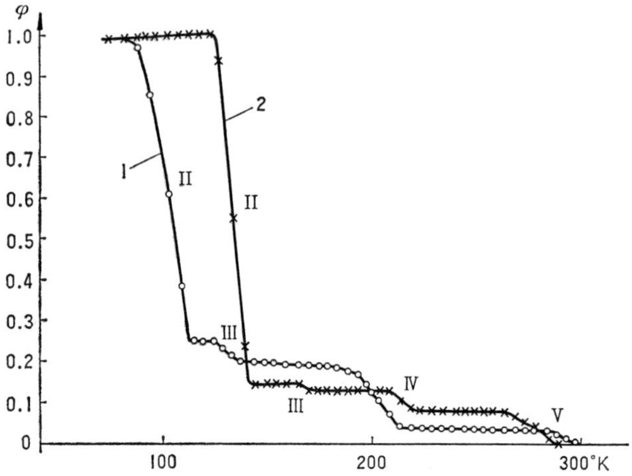

Fig. 4. The isochronal recovery of γ-irradiated p-type (curve I) and n-type (curve II) samples, φ-fraction of non-recovered damage determined from the changes of carrier concentration. The measurements were made at 78°K after 5 min, annealing at each temperature.

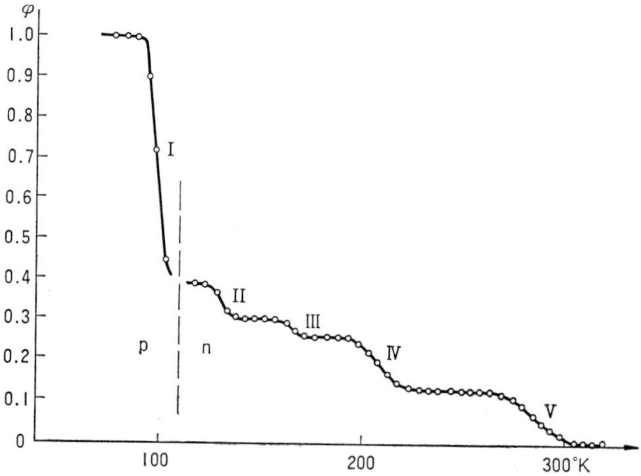

Fig. 5. The isochronal recovery of γ-irradiated " converted " sample.

and for n-InSb converted to p-type under irradiation in Fig. 5.

The p-part of the latter curve coincides with the corresponding part of Curve I in Fig. 4 and the n-part of the curve with that of Curve II in Fig. 4.

The temperatures and fractions of different stages φ, as they were ob-

TABLE I. Recovery stages (temperatures and percentage)

N	Type of conductivity and concentration (10^{14} cm^{-3}) before irradiation	after irradiation	Stages I (100)	II (150)	III	IV (200)	V (250)	Radiation	Ref.
1	n 380	n 120	23	51			26	e 4.5 MeV	10
2	n 150	n >75	23	62	3,4	2,7	9,2	e 1.0 MeV	1
3	n 150	n >75	19	77	1	1	2	e 0.7 MeV	1
4	n 150	n >75	1	99				e 0.4 MeV	1
5	n 34	n 4	15	59			26	e 4.5 MeV	10
6	n 6	n 5	6	60	8		28	e 4.5 MeV	10
7	n 6	n >0.5	5	95				e 0.5 MeV	2
8	n 1	n 0.3		84	3	6	7	γ	Present paper
9	n ⩾0.7	—		92	1	2	5	γ	12
10	n 0.9	p 1.5	63 →	13	4	10	10	γ	Present paper
11	p 0.8	p 1.8	100					X-ray	Present paper
12	p 1	p 6	68		9	14	9	γ	Present paper
13	p 4.6	p 6.3	91		3		6	γ	11
14	p 140	p <215	16 84					e 1.0 MeV	8

served by some authors, are shown in Table I.

The value φ is usually referred to as a fraction of unrecovered defects. This is true, if all the effective acceptors, due to the defects, are occupied by electrons in the whole recovery process. In our experiments this condition is fulfilled, as aboslute initial rates of carrier concentration alterations

TABLE II. The acceptor production rates for different types of defects.

Stage	$\frac{dM}{d\phi} \times 10^{-3}\ cm^{-1}$
II	5
III	0.2
IV	0.4
V	0.4

are the same in n- and p-In-Sb under irradiation. This means that all the acceptor levels which recover at stages II-V lie no further from the valence band then the Fermi level does in the whole irradiation and recovery process, i.e. $E_V+0.05$ eV is the upper possible position of the levels*.

The rates of acceptor level production under irradiation for the different types of defects are listed in Table II.

II. DISCUSSION

The results obtained can be explained on the basis of a model, proposed by Eisen for defects in electron irradiated InSb.[1,2,7,8]

According to the model, the close pairs i-Sb-v-Sb recombine at recovery stage I, and those of In at stage II.

One can see from Table I that both stages I and II take place at a lower temperature the closer to the valence band the Fermi level is. This effect was calculated by Eisen for stage II, supposing that recovery activation energy for i-In is lower when i-In levels are empty. The fact that the position of both stages depends in the same way on the Fermi level position shows that the recovery activation energy of i-Sb is also less when the i-Sb levels are empty.

One must also suppose that the acceptor level belonging to Sb-pairs is below the Fermi level in all experiments mentioned.

This viewpoint makes it possible to explain all the features of the recovery process observed in the present work. These features are:

I. There is no stage I on recovery curves for γ-irradiated InSb. Table I shows that at the carrier concentrations used, stage I can occur during irradiation. This fact enables us to estimate the upper limit of the activation energy for Sb-pair recovery as 0.28 eV, which corresponds to.[8]

* The level $E_c-0.03$ eV observed in[4] is not the effective acceptor in our experiments.

II. The temperature of stage II is rather different from that of the p-type in n-InSb. This can be seen in Fig. 5. While the conductivity of the p-type is preserved, annealing takes place at 95°K, but as soon as convertion to n-type occurs, the Fermi level rises up to the conductive band, and the recovery activation energy increases rapidly from 0.24 eV to 0.4 eV, which results in an increase of recovery temperature for the same type of defects. This effect causes splitting of stage II into two parts.

III. The fact that the hole concentration increases in p-InSb under X-irradiation leads to the conclusion that atom displacement in collisions with electrons is not the only process of defect creation. It was supposed that some ionizing mechanism (e.g.[9]) was responsible for defect creation in this case. The different conduct of n-InSb and p-InSb under X-irradiation can be accounted for because only p-type material has a non-equilibrium hole lifetime (i.e. the lifetime of the ionized atom) large enough for a defect to be created. It was possible to estimate the probability of defect formation due to this mechanism in p-InSb. It was found that one defect can arise if 10^9 atoms are ionized. The role of this mechanism in defect formation under γ ^{60}Co-irradiation was also estimated. It was shown that the rate of formation of defects by the displacement process is $\sim 10^2$ times greater than by the Varley mechnaism. The coincidence of the recovery stage for X-irradiated p-InSb with the recovery stage II for γ-irradiated material (see Table I) suggests that the creation of In-pairs is due to the ionizing mechanism.

Acknowledgement

The authors wish to express their gratitude to professor S.M. Ryvkin for his attention and interest in this work.

REFERENCES

1. F. H. Eisen, *Phys. Rev.*, **123**, 736 (1961).
2. F. H. Eisen, *Phys. Rev.*, **135**, A1394 (1964).
3. G. W. Arnold and F. Vook, *Phys. Rev.*, **137**, A1839 (1965).
4. B. Chelustka, R. Ju. Khansevarov, T. V. Mashovetz and I. R. Koslova, *Fisika Tverdogo Tela*, **9**, 338 (1967).
5. T. V. Mashovetz and R. Ju. Khansevarov, *FTT*, **7**, 2229 (1965).
6. T. V. Mashovetz, R. Ju. Khansevarov, *FTT*, **8**, 1960 (1966).
7. F. H. Eisen, in *Radiation Damage in Semiconductors*, p. 163, Paris, 1965.
8. F. H. Eisen, *Phys. Rev.*, **148**, 828 (1966).
9. J. H. O. Varley, *Phys. Chem. Solids*, **23**, 985 (1963).
10. L. W. Aukerman, *Phys. Rev.*, **115**, 1125 (1959).

11. K. D. Alexopoulos and R. B. Oswald, in *Radiation Damage in Solids*, **111**, p. 49, Venice, 1962.
12. C. Georgopulos, M. Verdone and D. Dautreppe, *Comp. Rend.*, **257**, 2640 (1963).

Neutron-induced Defects in Cadmium Telluride

NATKO B. URLI

Department of Solid State Physics,
Institute "R. Bošković", Zagreb, Yugoslavia

ABSTRACT

Effects of reactor-neutron bombardment on p-type CdTe are discussed. It has been shown that thermal neutrons play a significant role in the creation of the lattice defects in CdTe. As a result of cadmium recoils, subsequent to absorption, Frenkel-type defects are introduced showing a donor action. Photoconductivity, optical absorption, and TSC measurements on a series of various samples were performed and compared with the results of previous measurements of the Hall coefficient and resistivity. A quasi-continuous distribution of energy states in the lower half of the gap has been confirmed by TSC and optical absorption measurements. A high photosensitivity has been detected at low temperatures after fast-neutron bombardment.

I. INTRODUCTION

Radiation damage studies on compound semiconductors have become of greater interest in recent years. Since it was realized that even the elemental semiconductors are not especially convenient for investigation of the basic radiation damage processes, the main object of studies has shifted to learning about the changes in the physical properties of the semiconducting materials as a result of irradiation, and trying to relate these changes to particular lattice defects that have been formed.

In this work neutron-induced defects in cadmium telluride are studied. There are several reasons why CdTe single crystals have been chosen as an object of investigation. Beside the fact that as a semiconductor CdTe is very sensitive to irradiation, it also has desirable nuclear properties, as one of the components, cadmium, has an extremely high thermal-neutron absorption cross section. The relative abundance of the ^{113}Cd isotope is 12.26% with a thermal-neutron absorption cross section of about 20,000 barns, which is almost two orders of magnitude greater than the cross section of any other cadmium or tellurium isotope. Thus, the nuclear reaction ^{113}Cd (n, γ) ^{114}Cd plays a predominant role in the creation of lattice defects even in the presence of the reactor fast-neutron flux. The recoils of energetic ^{114}Cd nuclei introduce relatively simple Frenkel-type defects.

There are only a few published data on the effects of irradiation on CdTe. Lorenz *et al.*[1] have bombarded n-type CdTe with 1.5 MeV-electrons resulting in the introduction of acceptors. There is also an older report[2] indicating the change from p-type to n-type after reactor-neutron bombardment. Chester[3] has investigated the effects of gamma and thermal-neutron irradiation on n-type CdTe. Higher energy gammas as well as higher doses of thermal neutrons introduce acceptors, and even a change from n- to p-type was observed, while lower energy gammas and lower doses of thermal neutrons introduce donors.

To the author's knowledge there is also one more unpublished work[4] dealing with neutron irradiation of n- and p-type CdTe.

P-type CdTe has been studied to a lesser degree than n-type, probably due to the experimental difficulties in making good ohmic electrical contacts on the crystals.

In the present report the effects of thermal- and fast-neutron bombardments on p-type CdTe are presented. The results of photoconductivity, optical absorption, and thermally-stimulated-current (TSC) measurements are discussed and compared with those of Hall and resistivity measurements reported elsewhere.[5] The aim of these studies was to discuss the main features of neutron-introduced defects, and to comment on some contradictions in results previously reported.

II. EXPERIMENTAL

Measurements were performed on p-type samples of different origin and methods of preparation with the initial carrier concentrations in the range from 4×10^{13} to 4×10^{16} cm^{-3}.

The problem of obtaining good ohmic contacts for electrical measurements on p-type crystals was solved by immersion of masked samples in the solution of $AuCl_3$; gold deposits immediately at unmasked spots forming good ohmic contacts. The attempt to obtain satisfactory contacts by putting small drops of the solution directly on the etched crystals proved unsuccessful due to poor wetting of surfaces.

Gold wires as pressure contacts were applied in the temperature range from 77° to about 320°K in photoconductivity and TSC measurements. Filtered incandescent radiation was used as a light source.

TSC data were taken after photoexcitation at 77°K, and after excitation while cooling (decayed TSC).

Samples used for optical absorption measurements were carefully ground to a thickness of about 0.4 mm, then optically polished with aluminum oxide and diamond paste (1/4 μ).

A UNICAM double-beam spectrophotometer was used for absorption studies in the near infrared at room temperature.

Measurements were done before and after irradiation in a graphite-moderated reactor. The thermal neutron flux was for an order of magnitude greater than the fast-neutron flux at the position of the samples, and the time integrated doses varied from 4×10^{14} to 1×10^{16} n/cm^2. These were determined by cobalt activation detectors. The reactor was operated at a lower power level such that the temperature at the sample positions would not exceed 50°C.

Seven pairs of samples were used for optical measurements, and eight pairs for electrical. In each pair one of the samples was protected with a cadmium thermal-neutron shield. Thus it was possible to separate thermal-neutron effects from those of fast neutrons.

III. RESULTS AND ANALYSIS

The results of measurements after irradiation have shown that they depend upon several factors, such as the ratio of the initial carrier concentration to the neutron doses, the degree of compensation, i.e. the methods of sample preparation, and the state of the sample surfaces.[5]

Although under the experimental conditions the ratio of displaced atoms due to thermal and fast neutrons was at least 20, and the thermal-neutron defects were predominant, there are also some significant qualitative differences in their effects on some physical properties.

All samples show an increase in the Hall coefficient and resistivity after irradiation. Highly compensated p-type samples change to n-type after higher thermal-neutron doses ($\sim 10^{16}$ n/cm^2) indicating a 0.02 eV energy state below the conduction band.[5] The rate of increase in resistivity for fast-neutron irradiation is much smaller, and no change to n-type was observed.

Optical absorption spectra of some samples are shown in Fig. 1. Both thermal and fast neutron defects introduce absorption beyond the fundamental absorption edge at about 1.47 eV. However, as is shown, thermal neutrons introduce absorption at lower energies than fast neutrons indicating deeper energy states in the gap.

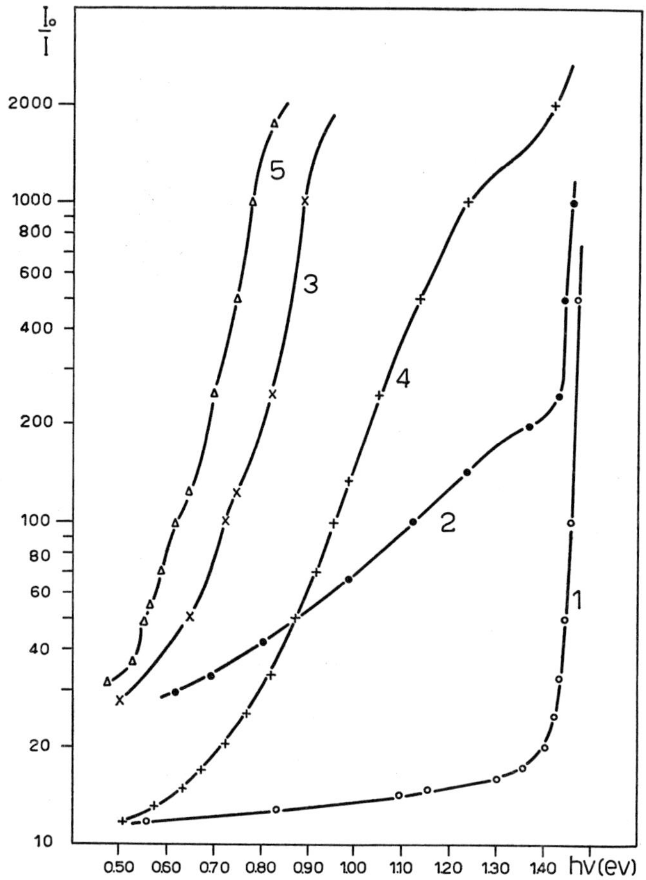

Fig. 1. Absorption spectra of the samples T-11, R-12, N-1, N-2 before and after irradiation. Curves: 1, before irradiation; 2, R-12 after 5.3×10^{14} n/cm^2; 3, T-11 after 3.4×10^{15} n/cm^2; 4, N-1 after 4.1×10^{14} n/cm^2; 5, N-2 after 2.2×10^{15} n/cm^2 (thermal) and 4.1×10^{14} n/cm^2 (fast). The curves 2, 3, 4 are after fast-neutron bambardment, and 5 after thermal and fast-neutron bombardment. T and R are multiple zone-refined samples, and N are grown by Bridgman-Stockbarger technique.

Besides some more or less pronounced peaks, TSC measurements (Fig. 2) indicate the existence of a quasi-continuous distribution of energy states. Assuming the traps remain effectively in thermal equilibrium with the valence band during the period of trap emptying their depths may be obtain using the relationship

$$E = kT \ln (N_V/p) \tag{1}$$

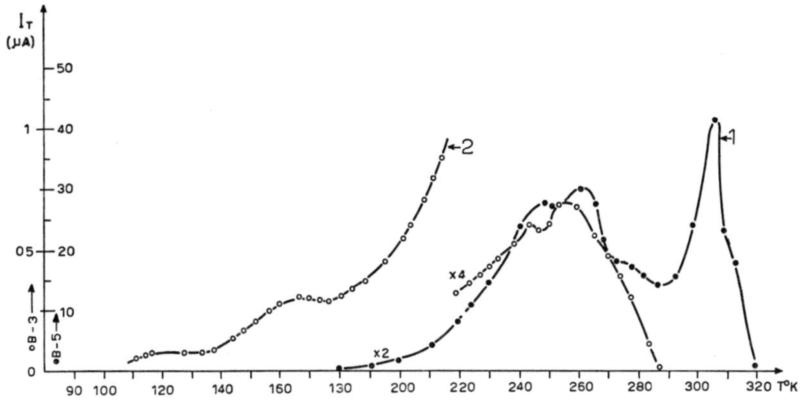

Fig. 2. TSC spectra of a pair of initially identical samples after irradiation. Curves: 1, B-5 after 7.6×10^{14} n/cm^2 (fast); 2, B-3 after 7.6×10^{14} n/cm^2 (fast) and 6.3×10^{15} n/cm^2 (thermal). Heating rate: 0.12°K/sec

where N_V is the density of states in the valence band, and p the concentration of holes at the temperature T of the correspondent TSC peak.

The main peaks of the curves give energy levels in the range from 0.30–0.33 eV above the valence band, in accordance with values obtained from Hall data of multiply zone-refined samples after thermal-neutron bombardment. Hall data also indicate the introduction of a level at about 0.20 eV above the valence band. After a two-month annealing at room temperature this level disappears but a new one is formed at 0.31 eV. Heating for half an hour at 300°C causes further changes, and an even deeper level (0.35 eV) appears with the increase in the resistivity of samples.

A gradual increase in resistivity with time was recorded for all samples that had received more than about 10^{15} n/cm^2.

Fast-neutron bombarded samples, like the one shown in Fig. 2, exhibit the main TSC peaks at 0.25 and 0.30 eV above the valence band superposed on the quasi-continuous distribution of energy levels.

There is also another useful method, independent of retrapping processes, for determining trap depths introduced by Garlick and Gibson[6] utilizing the initial exponential increase of TSC peaks. Shallower traps which could obscure the initial TSC increase may be removed by a previous heating to a temperature somewhat less than the desired TSC peak. Such curves are presented in Fig. 3, where low temperature slopes of the curve enable one to calculate the trap depths. In this case a value of 0.35 eV

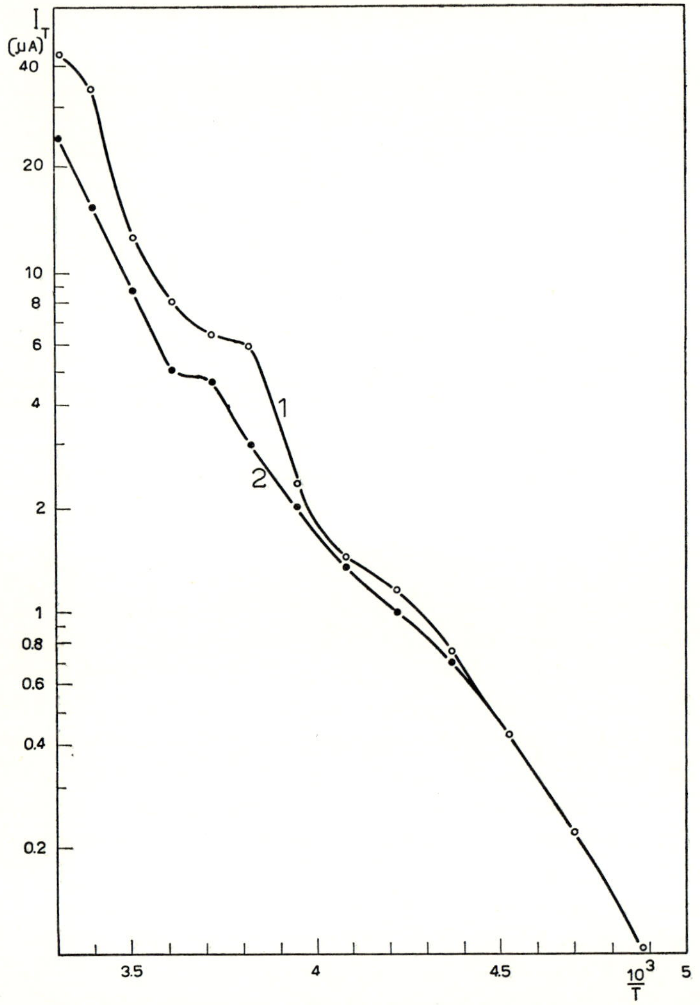

Fig. 3. Decayed TSC spectra of the sample I-5 after irradiation. Curves: 1, after decay to 221°K; 2, after decay to 265°K. Neutron doses: 5.9×10^{14} n/cm^2 (thermal) and 7.7×10^{13} n/cm^2 (fast). Heating rate: 0.27°K/sec.

was obtained; the same energy level was also observed in highly compensated samples before irradiation.[5] An indication of deeper levels may also be seen.

Photoconductivity measurements at different temperatures show a significant increase in photosensitivity especially after fast-neutron bom-

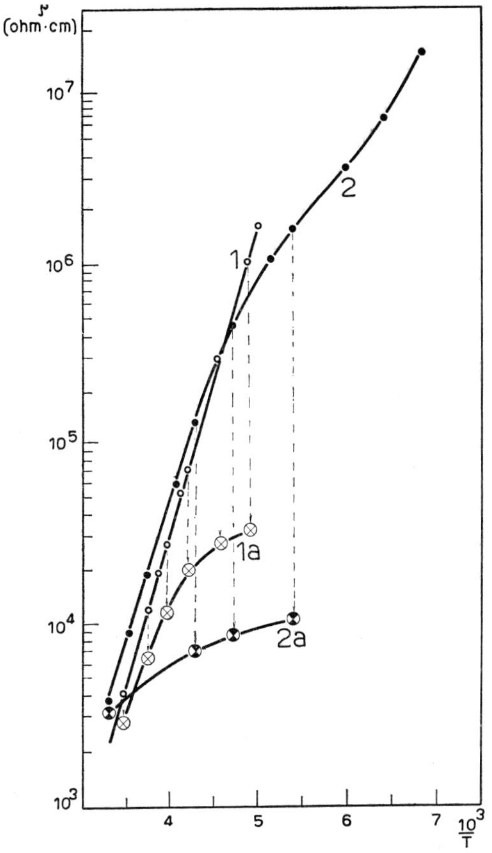

Fig. 4. The temperature dependence of resistivity in the dark and after illumination of two fast-neutron bombarded samples.

Curves: 1a, U-1 after 1.2×10^{15} n/cm^2 in the dark; 1a, the same sample after illumination; 2, I-3 after 7.7×10^{13} n/cm^2 in the dark; 2a, the same sample after illumination.

bombardment. A decrease in resistivity at lower temperatures amounts almost to two orders of magnitude (Fig. 4). Thermal neutrons introduce a much smaller change in photoconductivity.

Measuring the photoconductive decay after excitation at 77°K, an abrupt drop in photocurrent followed by a very slow decay with time constants on the order of hours, was observed.

IV. DISCUSSION

An inspection of electrical and optical measurements shows clearly that the overall effect of thermal-neutron defects is a donor action, and partial or total compensation of the initially present acceptor states results as the Fermi level rises to the upper half of the gap with the increase of neutron doses. Such an action is attributed to the cadmium interstitials resulting from recoils as is discussed elsewhere.[5]

As a fast-neutron flux was always present, although to a smaller extent, its effects could not be neglected. It is well known that fast neutrons tend to introduce disordered regions in semiconductors. It is quite probable that complexes involving impurities and Frenkel-type defects or more complicated structural defects are formed by irradiation and thermal action during and subsequent to neutron bombardment. There is also a tendency to increase the trap depths by thermal processes after irradiation.

TSC and optical absorption measurements have indicated the existence of a quasi-continuous distribution of acceptor levels in the lower half of the gap that could be attributed to the fast-neutron defects. Similar optical attenuation beyond the fundamental absorption edge was observed by Aukerman *et al.*[7] in the case of GaAs, CdS, and CdTe after fast-neutron irradiation.

The increase in photoconductivity and very long photo-decay times after fast-neutron bombardment show that those traps have a high capture cross section for minority carriers; centers surrounded by a Coulomb-repulsive barrier may be assumed.

Some discrepancies between the previous results on neutron irradiation of CdTe[2–4] are probably due to variation in purities of the samples and degree of compensation, or insufficiently high neutron doses.[4]

In order to obtain a deeper insight and understanding of the radiation damage effects observed in CdTe, further investigations are necessary. Annealing studies which are already in progress seem to be very promising.

Acknowledgements

The author wishes to thank Dr. Š. Mesarić for assistance with the optical absorption measurements.

REFERENCES

1. M. R. Lorenz, B. Segall and H. H. Woodbury, *Phys. Rev.*, **134**, A751 (1964).
2. L. W. Aukerman, *J. Appl. Phys.*, **30**, 1239 (1959).
3. R. O. Chester, Thesis, University of Tennessee, 1964.

4. C. E. Barnes, private communication.
5. N. B. Urli, *Proc. of Inter. Conf. on Physics of Semiconductors, Kyoto, 1966*, pp. 259–262.
6. G. F. J. Garlick and A. F. Gibson, *Proc. Phys. Soc.* (London), **60**, 574 (1948).
7. L. W. Aukerman, P. W. Davis, R. D. Graft and T. S. Shilliday, *J. Appl. Phys.*, **34** 3590 (1963).

Effects of Defects Produced by Irradiation on Exciton Spectra at Low Temperatures

M. GROSMANN AND S. NIKITINE

Laboratoire de Spectroscopie et d'Optique du Corps Solide (Laboratoire associé au C.N.R.S.), Institut de Physique, 5, rue de l'Université, Strasbourg, France

ABSTRACT

The effects of irradiations by fast neutrons in a reactor on the exciton spectra of Cu_2O have been studied by spectrophotography and spectrophotometry at 77° and 4°K.

The variations of the positions of the lines, the oscillator strengths and the lineshapes have been measured for a broad range of neutron fluxes.

The effects of neutrons are compared to those of other perturbations of the crystal: electric fields, impurities, uniaxial pressures and thermal quenching.

The results are particularly interesting in the case of the yellow series of Cu_2O where the results may be compared for a great number $(1<n<6)$ of lines of the same series.

The theory of the interaction of neutron defects with excitons allows us to deduce from spectroscopic measurements the number and dimensions of the defects introduced by the bombardment in the crystal.

The variation of the rate of creation of defects with the radiation flux is studied and analyzed.

I. INTRODUCTION

The study of the excitonic spectra of irradiated crystals has proved to be an interesting tool of investigation on radiation damages, especially in the case of pile irradiations of copper oxide and copper halides.[1] However a spectrophotographic method was used which did not allow the evaluation of the oscillator strengths. As a result the quantitative interpretation was rather difficult.

In order to obtain this information the absorption coefficient $K(\nu)$ of pile irradiated copper oxide has been measured at 4.2°K. The results obtained on the yellow series are analyzed. They are explained in the framework of the theory proposed by K. S. Song.

II. EXPERIMENTAL RESULTS

A. *Effects of fast neutrons on the yellow series of Cu_2O*

Poly and monocrystalline samples produced in our laboratories[2,3] having thicknesses from 10 to 200 μ have been used. The irradiation conditions and devices were described previously[4] as well as the apparatus[5] used for the measurements of the absorption coefficients. The oscillator strengths were evaluated by the Krawetz method. Figure 1 shows the variation of the absorption coefficients as a function of wave numbers when the number of neutrons received by the crystal is increased.

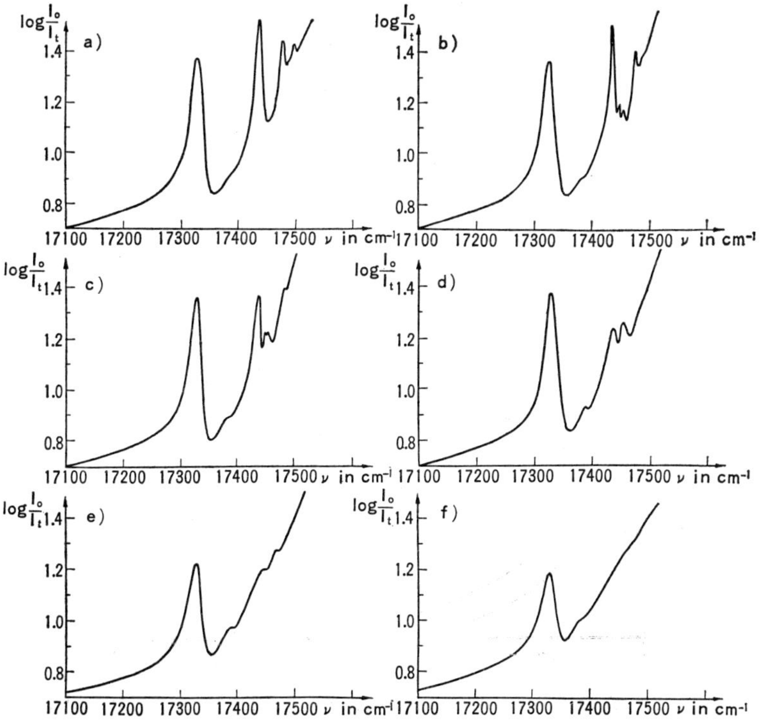

Fig. 1. Yellow series of the absorption spectrum of Cu_2O at 4°K. Integrated fluxes received by the samples (n/cm²). a) 0, b) 6.5×10^{15}, c) 1.3×10^{16}, d) 3.7×10^{16}, e) 9.4×10^{16}, f) 2.7×10^{17}.

Four different phenomena occur as the radiation doses increase:

a) The intensities of the principal lines (corresponding to p excitonic states) decrease. The intensity of the high quantum number lines decrease before the lines of lower quantum number are affected.

b) Between the lines of the initial series new lines (fine structure corresponding to s, d, . . . excitonic states) appear, first increase in intensities, and decrease later for strong irradiation rates.

c) The absorption edge shifts to longer wavelengths and, for increasing neutron fluxes, overlaps the lines.

d) The whole spectrum shifts very slightly (about 1 Å) to longer wavelengths.

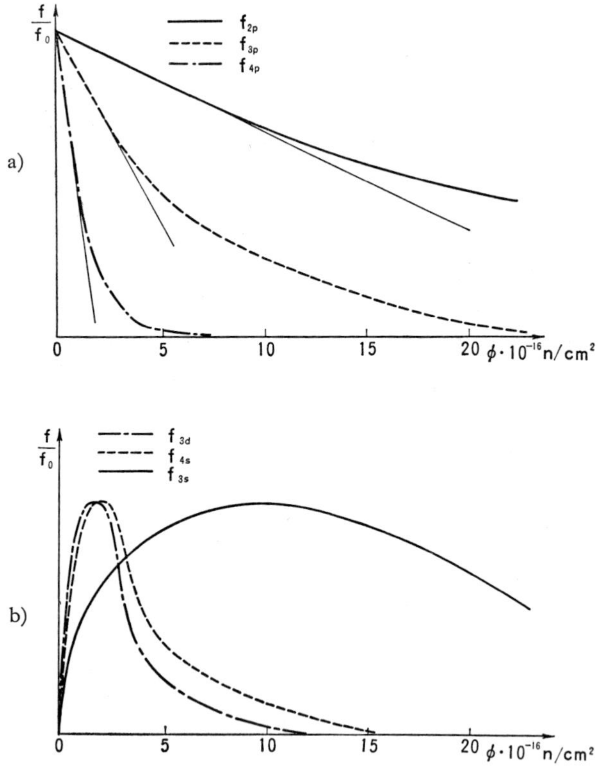

Fig. 2. a) Oscillator strength of the regular lines versus integrated neutron fluxes.

b) Oscillator strength of the fine structure lines versus integrated neutron fluxes. f_0 corresponds in each case to the maximum of f.

e) The whole spectrum reappears by annealing the crystal for two hours at 1000°C.

Effects a) and b) have been particularly analyzed theoretically and are hence the most interesting ones.

Figure 2a shows the variations of the oscillator strengths of lines 2p, 3p, 4p.

Figure 2b shows the variations of the oscillator strengths of lines 3s, 3d and 4s. Oscillator strengths of lines such as 4d or 5s could not be measured accurately because of a too strong overlapping with other lines. The oscillator strength of line 2s is not measurable with accuracy because this line is very weak and broad (not visible at the scale of the figures).

B. *Comparison with the effects of other perturbations*

1. *Electric fields.* The effects observed on the excitonic spectrum when a uniform electric field is applied to the crystal[6] present close analogies with the effects of neutrons. Some differences occur in the shape of the curves' oscillator strength versus perturbation and in the positions of the fine structure lines. The variations of the oscillator strengths of the lines as functions of the applied electric field are in good agreement with a quadratic law. Figure 3 shows the variation of the position of line 3s for increasing neutron fluxes compared with its position in an electric field.

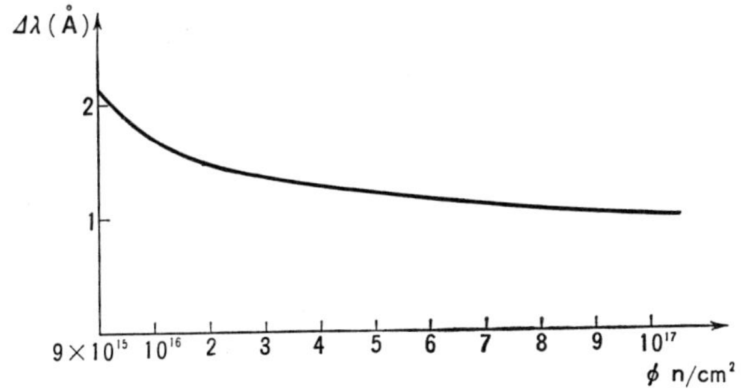

Fig. 3. Variation of position of line 3s with the radiation dose (with regard to its fixed position in an electric field).

2. *Thermal quenching.* The effects of thermal quenching are very similar to those of neutrons, but they have been less extensively studied. The quantitative reproductivity of the effects is not simple to obtain and the dimensions and shapes of the samples play an important role.

3. *Gamma-rays.* The effects of gamma-rays are very weak compared to those of neutrons[7] but are very interesting. We have irradiated samples

with a ^{60}Co source. After irradiation at 5×10^8 Rad. the spectrum showed:

a) two more lines of the hydrogenic series ($n=7$ and $n=8$);

b) a small increase of the oscillator strengths of lines of high quantum number ($n=5$ and $n=6$). This increase is difficult to measure because the lines are weak and in the vicinity of the absorption edge which limits the series;

c) the whole spectrum is slightly shifted (about 0.5 Å) to shorter wavelengths.

4. *Impurities.* The effects of point defects produced for instance by the presence of silver in Ag-doped copper oxide are quite different from those of neutrons. In this case[8] no fine structure occurs, but the whole series is shifted towards the longer wavelengths. This shift is proportional to the concentration of impurities.

III. DISCUSSION

A. *Theory of the effect of large defects in the excition on the exciton in the effective mass approximation*

The yellow series of Cu_2O corresponds to second class excitonic transitions. The excitons are formed in P states. Transitions towards S and D states are forbidden. No degenerescence occurs because the dielectric constant varies with the dimensions of the exciton.

When the crystal contains large defects, for instance dislocation loops, it is possible to calculate the deformation potentials produced by those defects[9] as a function of their dimensions.

Introducing the deformation potentials in the effective mass approximation Schrödinger equation allows us to calculate the variation of the oscillator strengths of the lines in the approximation of small perturbation

$$\left| -\frac{\hbar^2}{2M}\nabla_R{}^2 - \frac{\hbar^2}{2\mu}\nabla_r{}^2 - \frac{e^2}{\varepsilon|r|} + E_d\, r \operatorname{grad} \Delta(R) \right| \phi'(r, R) = E\phi'(r, R)$$

where

$M = m_e{}^x + m_h{}^x$, $\mu = m_e{}^x m_h{}^x / M$, E_d = deformation potential coefficient, $\Delta(R)$ = dilatation.

Perturbed wave functions ϕ' are formed with unperturbed wave functions ϕ, for instance, $\phi'_{2p,0}(r, R)$ perturbed wave function for $S=2p$, $K=0$ will be

$$\phi'_{2p,0}(r, R) = \frac{1}{\sqrt{NB}} \sum_{S,K} a_S(K)\, e^{iKR}\, \varphi_S(r)$$

where

$$a_{2p}(0) \gg a_S(K) \text{ for } S \neq 2p$$

and

$$\frac{a_{2p}(K)}{a_{2p}(0)} = 0, \qquad \frac{a_{2s}(K)}{a_{2p}(0)} = \frac{\langle K, 2s \mid E_d\, r \operatorname{grad} \Delta(R) \mid 0, 2p\rangle}{\varepsilon_{2p} - \varepsilon_{2s} - \dfrac{\hbar^2 K^2}{2M}}$$

$\phi'_{2s,K_0}(r, R)$ perturbed wave function for $S=2s$, $K=K_0$ will be

$$\phi'_{2s,K_0} = \frac{1}{\sqrt{NB}} \sum_{S,K} b_S(K)\, e^{iKR}\, \varphi_S(r)$$

where

$$b_{2s}(K_0) \gg b_S(K) \text{ for } S \neq 2s,\ K \neq K_0$$

and

$$\frac{b_{2s}(K)}{b_{2s}(K_0)} = 0, \qquad \frac{b_{2p}(K)}{b_{2s}(K_0)} = \frac{\langle K, 2p \mid E_d\, r \operatorname{grad} \Delta(R) \mid K_0 2s\rangle}{\varepsilon_{2s} - \varepsilon_{2p} + \dfrac{\hbar^2 (K_0^2 - K^2)}{2M}}$$

the corresponding oscillator strengths are (compared to the unperturbed oscillator strength f°_{2p})

$$\frac{f_{2p}}{f^{\circ}_{2p}} = 1 - \sum_K \left| \frac{\langle K, 2s \mid E_d\, r \operatorname{grad} \Delta(R) \mid 0, 2p\rangle}{\varepsilon_{2p} - \varepsilon_{2s} - \dfrac{\hbar^2 K^2}{2M}} \right|^2$$

$$\frac{f_{2p}}{f^{\circ}_{2p}} = 1 - \frac{\delta N}{R}$$

where N is the number of dislocation loops and R their dimension. And

$$\frac{f_{2s}}{f^{\circ}_{2p}} = \sum_{K_0} \left| \frac{\langle 0, 2p \mid E_d\, r \operatorname{grad} \Delta(R) \mid K_0, 2s\rangle}{\varepsilon_{2p} - \varepsilon_{2s} - \dfrac{\hbar^2 K_0^2}{2M}} \right|^2$$

$$\frac{f_{2s}}{f^{\circ}_{2p}} = \frac{\delta N}{R}$$

As the dilatation in the real space $\Delta(r)$ varies slowly, the Fourier transfrom $\Delta(K)$ is different from zero only in the vicinity δK of $K=0$ in K space.[10] The vicinity is defined by $\delta K \cdot \delta L = 2\pi$ where δL characterizes the dimension of the perturbation in real space.

The transitions occur for E:

$$E = E_{2s} + \frac{\hbar^2 K_0^2}{2M}$$

K_0 being a mean value of K in the interval δK. For a crystal perturbed by an electric field the transitions occur for $E = E_{2s}$.

There is a displacement

$$\delta\nu = \frac{1}{ch}\frac{\hbar^2 K_0^2}{2M}$$

therefore if $\delta K = aK_0$

$$\delta L = C \cdot (\delta\nu)^{-1/2}$$

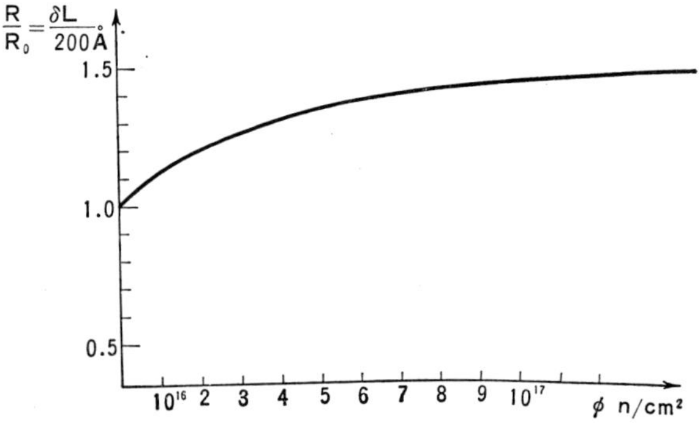

Fig. 4. Variation of the dimension of the defects with radiation dose.

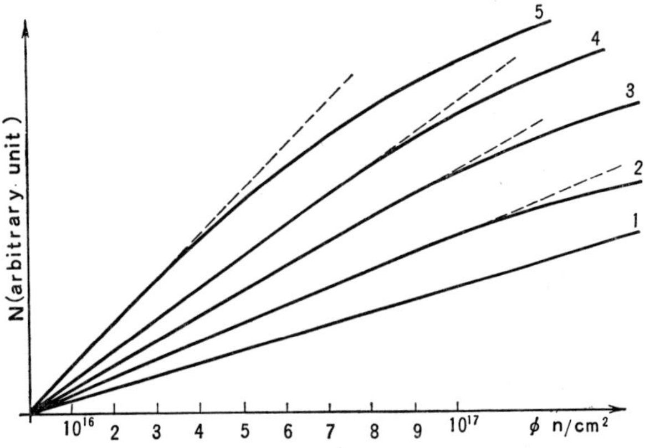

Fig. 5. Variations of N calculated from the date obtained on different lines (arbitrary units).

$$1: \frac{\mathfrak{N}}{\delta}\left(1-\frac{f_{2p}}{f^{o}_{2p}}\right), \qquad 2: \frac{\mathfrak{N}}{\delta} f_{3s}, \qquad 3: \frac{\mathfrak{N}}{\delta}\left(1-\frac{f_{3p}}{f^{o}_{3p}}\right),$$

$$4: \frac{\mathfrak{N}}{\delta} f_{3d}, \qquad 5: \frac{\mathfrak{N}}{\delta} f_{4s}.$$

the value of C is different for different models of perturbating potentials. With the model of dislocation loops it is possible to find $\delta L \sim 200$ Å.

Figure 4 shows the variation of $\Re = \delta L/200$ Å with the radiation dose in this case.

From this results it is possible to deduce with the curves of Fig. 2 the variation of N which is proportional to $(1-f_p/f_p^0)R$ or to $f_s x R$. Figure 5 shows these variations.

The linearity is better for the case of line 2p than for other cases. This is due to the fact that the theory is valid only for " weak " perturbations. Any given perturbation affects in a more important way lines of higher quantum number.

From the data on line 2p the number of large defects created by the bombardment is linear at least up to doses of 5×10^{17} n/cm^2. For higher doses the linearity can no longer be proved but it is difficult to tell from the available data whether the rate of creation of defects changes, or whether the theory itself is no longer valid.

IV. CONCLUSION

We have measured the oscillator strength of the lines of the yellow series of Cu_2O as a function of the neutron fluxes received by the crystal. Comparison of the results with the effects of electric fields allows to calculate in the framework of Song's theory the dimensions of the defects as well as their number. These quantities are determined with only slight precision because the shape of the defects are unknown. However the variations of these quantities during irradiation can be measured fairly accurately. It is shown that the rate of creation of defects remains constant during the irradiation, at least up to fluxes of 5×10^{17} n/cm^2.

Acknowledgements

This work was made possible by the financial support of the Commissariat à l'Energie Atomique (contrat n° 8249/h). The authors wish to thank Dr. Song for helpful discussions and communication of results.

REFERENCES

1. M. Grosmann and S. Nikitine, **1**, 277 (1963).
2. J. B. Grun, *Thèse, Revue d'Optique*, **41**, 439 (1962).
3. M. Grosmann, **62**, 1129 (1965).
4. M. Grosmann, *Thèse 3ème Cycle*, Strasbourg (1962).
5. S. Nikitine, J. B. Grun and M. Sieskind, *J. Phys. and Chem. Solids*, **17**, 292 (1961).

6. S. Nikitine, J. L. Deiss and K. S. Song, *C. R. 7ème Congress Intern. Phys. Semiconducteurs, Paris*, 909 (1964); J. L. Deiss, *Thèse 3ème Cycle*, Strasbourg (1963).
7. S. Nikitine, M. Grosmann and J. B. Grun, *C. R. Acad. Sci.*, **255**, 1087 (1962).
8. C. Schwab, F. Marvillet, J. P. Adloff and S. Nikitine, *J. Phys. Rad.*, **25**, 381 (1964); C. Hottier, *D. E. S.*, Strasbourg (1965).
9. S. Nikitine, M. Grosmann, C. Schwab, J. L. Deiss, P. Soxhlet and K. S. Song, *7ème Congrès Intern Semiconducteurs*, Paris-Royaumont, 245 (1964).
10. K. S. Song, S. Nikitine, J. B. Grun and M. Grosmann, *C. R. Acad. Sci.*, **259**, 1087 (1964).

Electron Paramagnetic Resonance of Unstable Centers in Photon-irradiated BN and ^{10}BN

G. RÖMELT

Institut für Angewandte Physik der Technischen Hochschule, Clausthal, Germany

ABSTRACT

Electron paramagnetic resonace (EPR) of BN with natural isotopic composition (81.17% ^{11}B, 18.83% ^{10}B) and of ^{10}BN with 91% ^{10}B, both irradiated with ultraviolet light, X- or gamma-rays, shows two different kinds of centers. In the first kind an electron interacts with *three* surrounding nuclei of boron and gives 10 lines for BN and should give 19 lines for ^{10}BN, but these 19 lines are not resolved. This center must lie in the center of B_3N_3-rings or on places of nitrogen atoms.

The second kind of centre gives 4 lines for BN and 7 lines for ^{10}BN, i.e. the unpaired electron interacts with *one* boron-nucleus only. These one-boron centres are not stable and disappear; the higher the temperature the quicker they disappear. EPR of these centers occurs between 150 and 450°K, and the 4-line spectrum was found at 1.5°K also.

I. INTRODUCTION

Study of the 10-line EPR-spectrum of BN, which was found by Geist,[1] has shown that those centers causing the 10-line spectrum can be produced *thermally* by heating BN to more than 1850°C or by *irradiation* with ultraviolet light, X- or gamma-rays.[2] The splitting results from Fermi contact hyperfine interaction[3] of an unpaired electron with *three* surrounding nuclei ^{11}B with spin $I=3/2$, magnetic moment $\mu=2.69$ and $\mu/I=1.8$. For natural boron there are 81.17% ^{11}B. The remaining 18.83% are ^{10}B with $I=3$ and $\mu/I=0.6$. The 10-line spectrum can bc synthesized by adding the EPR-spectra of centers with 3, 2, 1, and 0 ^{11}B (others are ^{10}B), each occuring with proper probability.

In ^{10}BN with 91% ^{10}B the expected 19 EPR-lines merge into one, as the distance between the 19 lines is too small. Moreover irradiation produces centers that interact with *one* boron nucleus only and give 4 EPR-lines for BN and 7 lines for ^{10}BN. These spectra and centers primarily will be discussed in this paper.

II. EXPERIMENTALS

EPR measurements were made with a Varian EPR-spectrometer in the X-band and for temperatures between 1.5 and about 500°K. Most measurements were made with *sintered* BN and ^{10}BN, as anisotropic effects could be observed. Furthermore large crystallites as normally found in sintered material yield large spin densities under irradiation. In fine powders the spin densities are about 10 times smaller. Irradiation was done at room temperature with a mercury lamp Hg 3 (Osram) as a source of ultraviolet light, and with X-rays (25–60 kV; Mo-, Cu-anode) and gamma-rays from a ^{60}Co-source.* Although the photon-energies of these three kinds of rays differ from 1 to 10^6 eV, the produced centers and resulting EPR-lines are the same.

Anisotropic hyperfine structure

Figure 1 shows the EPR spectra of X-irradiated BN and ^{10}BN. For comparison the middle part is shown separately in Fig. 2. From Fig. 1 the following facts are seen at once:

1) The number of lines is $2I+1$, $I=3/2$ and $I=3$ respectively.

2) The distance between the lines of the 4-line spectrum is just three times that of the 7-line spectrum of ^{10}BN, due to the relations $\mu/I=1.8$ for ^{11}B and 0.6 for ^{10}B. This proves that there are principally the same centres in BN and ^{10}BN.

3) The intensities of the lines are not equal, but their relation is 1:2:2:1 for BN and odd for ^{10}BN.

There are two reasons for this inequality:

a) As stated in the introduction, both materials used were mixtures of the two isotopes. Therefore all observed spectra are superpositions of " pure " spectra. Nearly coinciding lines can reinforce each other remarkably, e.g. $m=\pm1/2$ of ^{11}B and $m=2$ resp. $m=-1$ of ^{10}B, Fig. 1.

b) Anisotropy of the recorded spectra, as seen in Figs. 3–5. One can explain this assuming that a single crystal of ^{11}BN or ^{10}BN has an anistropic axial symmetric hyperfine splitting. In the sintered material crystallites of various orientations are mixed, but not fully irregular, as

* I thank Professor Dr. H. Luther, Institut für Chemische Technologie und Brennstofftechnik, Technische Hochschule Clausthal, who kindly made available his facilities for the gamma-irradiations.

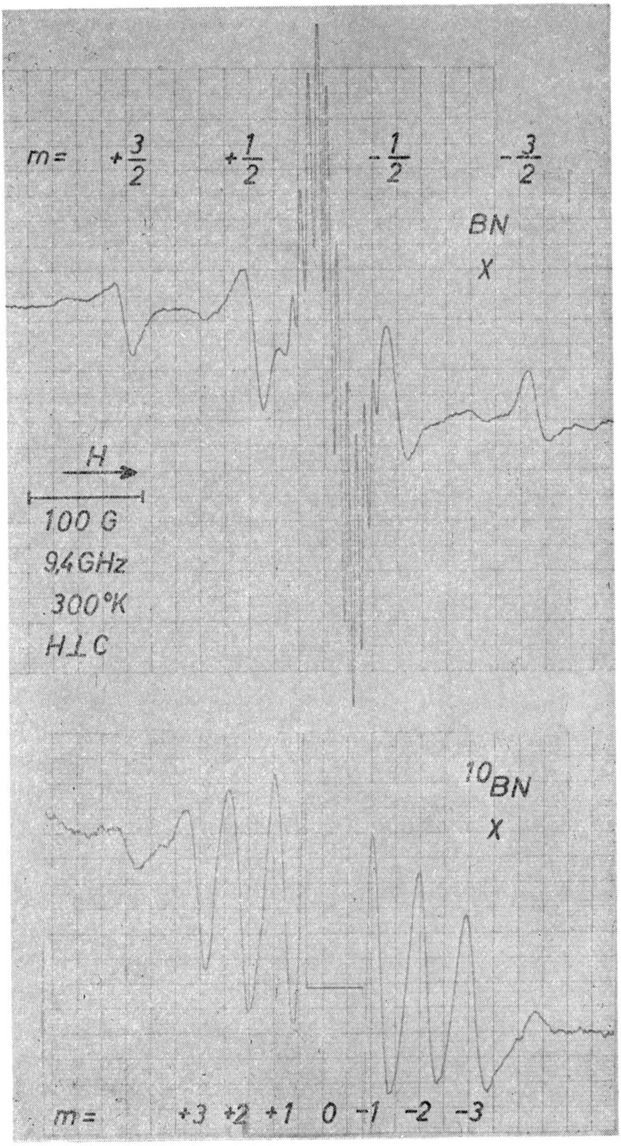

Fig. 1. EPR of X-irradiated boronnitride.

a) BN: 4 lines from a centre with *one* nucleus ^{11}B ($I=3/2$, $\mu/I=1.8$). (Sample Ca II).
10 lines from a center with *three* ^{11}B.

b) ^{10}BN: 7 lines from a center with *one* ^{10}B ($I=3$, $\mu/I=0.6$)
19 lines from a center with *three* ^{10}B, merged into *one* line, which was suppressed to avoid overloading of the recorder (see Fig. 2). (Sample W2).

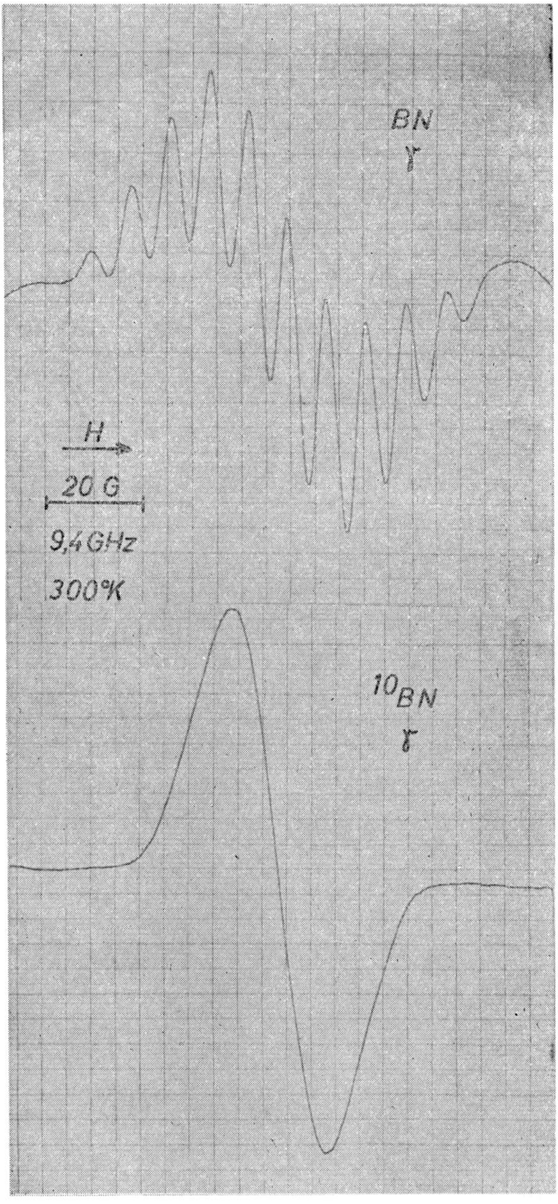

Fig. 2. a) 10-*line spectrum* of BN (three ^{11}B-nuclei). (Sample CaXIII).
b) *Unresolved* 19-*line spectrum* of ^{10}BN (three ^{10}B-nuclei). (Sample W).
Both γ-irradiated material.

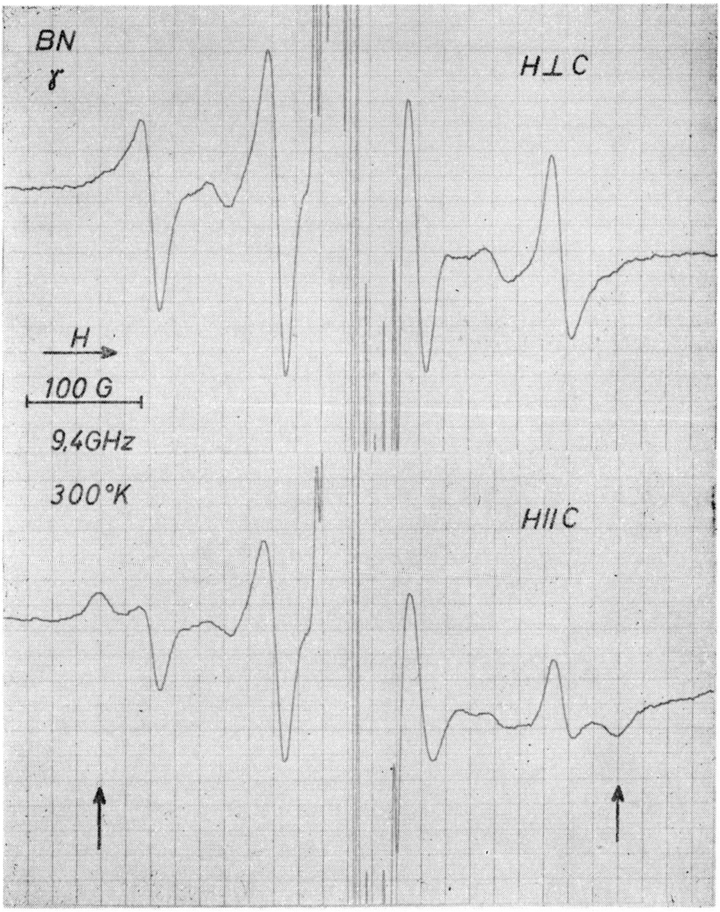

Fig. 3. Anisotropic 4-line EPR-spectrum of one-boron centers in sintered BN. C is the axis of the slab. (Sample U, γ-irradiated)

the slab axis and hexagonal c-axis tend to coincide† as proved by Figs. 3–5. Therefore the recorded spectrum is the superposition of the individual anisotropic spectra. Obviously the lines of BN for $m=\pm 3/2$ are more disturbed and have a smaller intensity than those for $m=\pm 1/2$. The lines for $m=\pm 3/2$ also show the larger anisotropic effect, as seen in Fig. 3. For H⊥C these lines are single but they split for H//C. For the 7-line spectrum of ^{10}BN anisotropy is similar (Fig. 4). Fig. 5 shows, for comparison, that for the 10 lines of three-boron centers in BN the anisotropy

† A modest orientation effect is found in the Debye-Scherrer diagrams also.

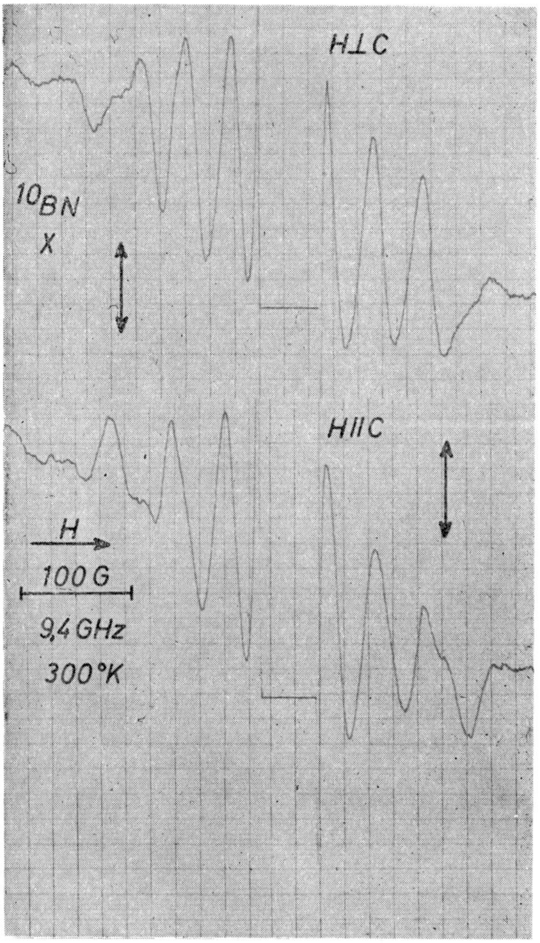

Fig. 4. Anisotropic 7-line EPR-spectrum of one-boron centers in sintered ^{10}BN. (Sample W2, X-irradiated)

results only in a better resolution for H$\perp$C than for H//C.

The constants for the energy expression for axial symmetry[4]

$$h\nu = g\beta H + Am + B\{I(I+1) - m^2\}$$

for EPR with $M=1/2 \longleftrightarrow M=-1/2$ and $\Delta m=0$, and for H$\perp$C are shown in Table I.

As the hyperfine splitting is not very anisotropic, it is mainly caused by Fermi contact-interaction. Therefore $A=(8\pi/3)(\mu/I)\,|\,\psi(0)\,|^2$. $|\,\psi(0)\,|^2$ is

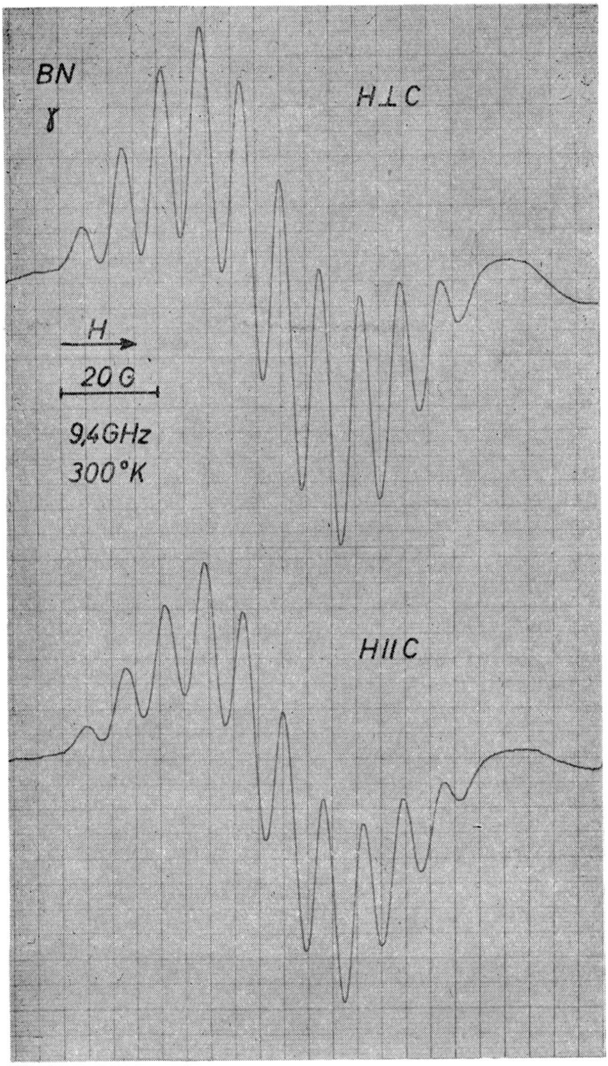

Fig. 5. Anisotropic 10-line EPR-spectrum of three-boron centers in BN. (Sample U, γ-irradiated)

TABLE I.

	Number of lines	$2I+1$	g	$A/g\beta$ (gauss)	$B/g\beta$ (gauss)
BN	10	—	2.0027 ± 0.0003	(7.8)	—
(^{11}B)	4	4	2.004 ± 0.002	122 ± 4	2.5 ± 0.5
^{10}BN	19(1)^2	—	2.0029 ± 0.0003	(2.6)	—
(^{10}B)	7	7	2.003 ± 0.001	40 ± 2	—

the probability of the electron to be on the nucleus. For one-boron centers it is 16 times larger than for three-boron centers.

Formation and decay of one-boron centers

By comparison of height and width of the BN-EPR-signals with the line of $CuSO_4$, spin densities of BN centres were measured as a function of the duration of irradiation. Figure 6 shows that after some hours of irradiation the spin densities cease to increase. The final spin density is reached when production and annealing rate for the centers become equal.

Figure 6 contains for comparison the behavior of the three-boron centres, which reach a final density of about 5×10^{17} spins/g BN, and which are essentially stable. The final spin density for the one-boron centres both in BN and ^{10}BN is about 1×10^{17} g^{-1}; this means not more

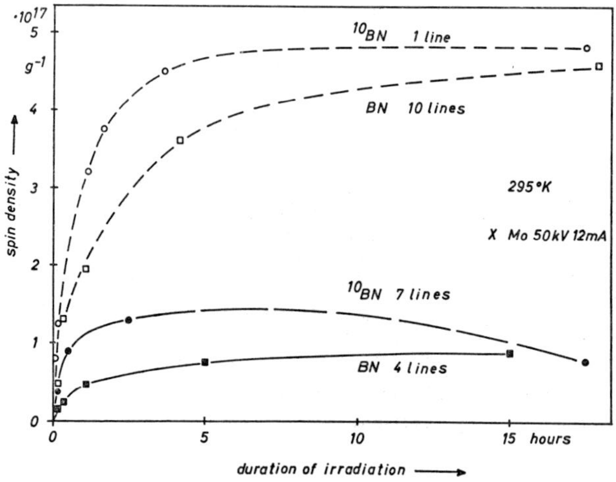

Fig. 6. Increase of spin density in BN and ^{10}BN with duration of X-irradiation.

than one spin on 2×10^5 boron atoms. As BN and ^{10}BN were made from very distinct raw materials, this same number for BN and ^{10}BN shows that the one- (and also the three-) boron centers perhaps do not need an impurity for their existence.

The one-boron centers are very unstable, and their EPR signals disappear at room temperature after some days for ^{10}BN and after a few weeks for BN. The description of decrease needs at least two time constants. Preliminary measurements give for the function

$$A = A_1 \exp(-t/\tau_1) + A_2 \exp(-t/\tau_2)$$

the values in Table II.

TABLE II.

Temperature (°C)	BN		^{10}BN	
	τ_1 (h)	τ_2 (h)	τ_1 (h)	τ_2 (h)
0	>200	—	40	200
22	11.4	507	10	60
40	2	62	—	—
53	—	—	1	1
$\frac{A_1}{A_1+A_2}$	0.67		0.4	

Because of the very different τ_2 of ^{11}B-centres in BN and ^{10}B-centers in ^{10}BN at room temperature, and as there are 9% ^{11}B in ^{10}BN, one may expect a small 4-line spectrum after disappearance of the 7 lines. But the experiments do not show these 4 lines. So ^{11}B-centers may have the same τ_2 as ^{10}B-centers. This fact may be a hint that impurities are responsible for recombination, but little is known about impurities in BN. After annealing of the one-boron centers, they can be produced again by irradiation. The small lifetime of one-boron centers at elavated temperatures may be the reason that these centers cannot be found after heating to more than 1850°C, in contrast to the three-boron centers.[2]

Temperature dependence of EPR intensity

The spectra of three-boron centers can be found between 1.5 and 500°K with nearly the same intensity. The EPR-spectra of one-boron centres, however, can be found only between 150 and 450°K (Fig. 7). The 4 lines appear again at 1.5°K, but the 7 lines do not. Line width seems to be nearly equal in the whole interval of temperature. So the reason

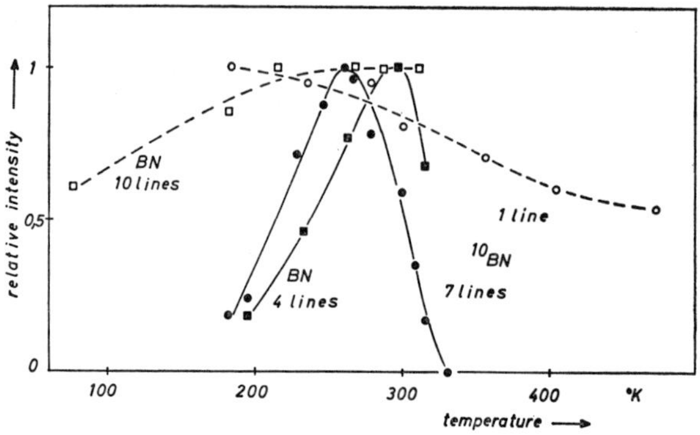

Fig. 7. Intensity of the EPR-spectra versus temperature of X-irradiated BN and ^{10}BN. The decrease of 19-[1] and 10-line spectrum is largely fainted by changes in spectrometer sensitivity.

for the temperature dependence may be a different probability, that a level is occupied by an electron (or a hole). These reversible changes are a true temperature dependence and mean no annealing.

III. DISCUSSION

The measurements of EPR on irradiated BN yield the existence of a so-called one-boron center, in which one unpaired electron interacts with *one* boron nucleus. This is proved by the different hyperfine splitting for ^{10}BN and ^{11}BN. The main part of the unpaired electron is s-like.

As there is a certain connection between the *production* of EPR centres (both one- and three-boron-),[2] the *degree of crystallization*, and the *content of* B_2O_3 *in BN*,[5] it may be guessed that the unpaired electron stems from oxygen, substituting nitrogen atoms, as oxygen has one electron more than nitrogen. Another possible source of unpaired electrons may be nitrogen vacanies with electrons trapped, somewhat similar to the F-centers, known from alkali halides.

The *resolved* spectra of the one-boron centers strongly support our proposal to explain the three-boron spectra, of which that for ^{10}BN had not been resolved.[1, 2]

Other spectra are found for neutron irradiated BN.[6] There are at least four systems of one or more lines, which may be understood after measurements on neutron irradiated ^{10}BN.

Acknowledgments

Thanks are due to Professor Dr. D. Geist, who made this work possible in his institute, especially with materials and measuring equipment kindly lent by the Deutsche Forschungsgemeinschaft.

SEFERENCES

1. D. Geist and G. Römelt, *Solid State Communications*, **2**, 149 (1964).
2. G. Römelt, *Z. Naturforschg.*, **21a**, 1970 (1966).
3. E. Fermi, *Z. Physik*, **60**, 320 (1930).
4. W. Low, in *Paramagnetic Resonance in Solids*, p. 61, Academic Press, New York, 1960.
5. A. H. Silver and B. J. Bray, *J. Chem. Phys.*, **32**, 288 (1960).
6. D. Geist, G. Römelt and W. Klein, unpublished communication at the symposium on lattic defects in semiconductors, Tokyo, 1966.

Spectral Analysis of Spatial Distribution of Radiation Damage in Silicon

A. V. M. FERRIS-PRABHU*

The Catholic University of America, Washington, D. C. 20017, U.S.A.

ABSTRACT

A model is presented whereby the spatial distribution of the radiation damage to a solar cell may be determined by a spectral analysis of the photovoltaic current. The theory of how to solve the resultant integral equation is discussed.

I. INTRODUCTION

In an attempt to analyze the spatial distribution of radiation damage in silicon semiconductor devices, an interesting method has been suggested[1] whereby a spectral analysis of the measured photovoltaic current before and after irradiation may be used to provide information about the damage to a solar cell as a function of distance from the irradiated surface. The method as presented depends on the introduction of a so-called conversion efficiency parameter $F(\lambda)$ defined through the equation

$$i_\lambda = \int_0^t d\xi \, e^{-\alpha(\lambda)\xi} F(\lambda) \, I(\xi, \lambda) \tag{1}$$

where i_λ is the measured photovoltaic current from a solar cell of thickness t due to absorption of photons of energy hc/λ, and the intensity of light of wavelength λ at a depth ξ from the surface is for a homogeneous medium

$$I(\xi, \lambda) = I(0, \lambda) e^{-\alpha(\lambda)\xi} \tag{2}$$

where $\alpha(\lambda)$ is the optical absorption coefficient.

From Eqs. (1) and (2)

$$F(\lambda) = \frac{\alpha(\lambda)}{1-e^{-\alpha(\lambda)t}} \cdot \frac{i_\lambda}{I(0, \lambda)} \tag{3}$$

---

* Presently at Goddard Space Flight Center, Greenbelt, Maryland 20771

and can be interpreted as the ratio of the number of current carriers generated to the number of photons absorbed. It is not always unity[2] because next to the fundamental absorption band there exist others, from which absorption of light produces excitons whose motion through the crystal does not produce electric current.[3] Thermal excitations may cause some excitons to dissociate into electrons and holes which contribute to the current, but not all excitons are thus annihilated. It is thus possible as indicated in Eq. (3) for $F(\lambda)$ to differ from unity for at least some values of λ. However, it is evident that not all the current carriers generated by photon absorption at some depth ξ will arrive at the electrodes at which the current i_λ is actually measured. There will be some attenuation due to, for example[1], local recombinations, the effect of the junction field, trapping due to structural defects and other such mechanisms whose spatial distribution is sought. Since the model of Eq. (1) does not account for the spatial distribution of such mechanisms, we propose in this note an alternative model, which leads to an equation the theory of the solution of which will be discussed.

II. THE MODEL

Since the detailed description of each of the attenuating mechanisms poses a rather severe problem, it is proposed to represent all of them by a " collective " space dependent function $G(\xi)$. Differences in the value of $G(\xi)$ after successive irradiation dosages will then provide some information about the collective effect of the radiation damage as a function of distance from the irradiated surface. In addition to requiring $G(\xi)$ to be continuous and single valued, no assumption is made as to its form other than that it be zero for $\xi<0$ and $\xi>t$. The last requirement expresses the fact that there can be no contribution to the photovoltaic current from outside the cell. Equation (1) may then be replaced by

$$i_\lambda = F(\lambda) \int_0^t d\xi\, G(\xi) \left[-\frac{d}{d\xi} \{(1-R(\lambda))I(\xi, \lambda)\} \right] \tag{4}$$

where the quntity in brackets is the number of photons absorbed at a depth ξ corrected for reflection of the incident light by introduction of the reflection coefficient $R(\lambda)$. Rearrangement gives

$$\tilde{i}_\lambda = \int_0^t d\xi\, G(\xi) e^{-\alpha(\lambda)\xi} \tag{5}$$

where

$$\tilde{i}_\lambda = i_\lambda/[F(\lambda)\alpha(\lambda)I(0, \lambda)\{1-R(\lambda)\}] \tag{6}$$

We have tacitly made the assumption that $F(\lambda)$ and $G(\xi)$ are separable functions. Without this assumption the problem seems intractable. A further assumption is necessary if Eq. (5) is to be solved, since it contains two unknowns, whereas only i_λ is experimentally known. As a first approximation let us assume that in the range of λ of interest, which is 0.35 μ to 1.1 μ for silicon, $F(\lambda)$ is unity. It is also convenient to replace $\alpha(\lambda)$ by α, which is justifiable since in the range of λ under discussion the absorption coefficient is a single valued isomorphic function of the wavelength. Equation (5) then becomes

$$I_\alpha = \int_0^t d\xi\, G(\xi)e^{-\alpha\xi} \tag{7}$$

This is a Fredholm homogeneous linear integral equation of the first kind, for which solutions do not always exist.[4] One might expect to solve it by replacing the integral by a sum and then inverting the resultant matrix equation. This turns out to be impracticable because the range of α from 1 cm^{-1} to 10^5 cm^{-1} is such that the coefficient matrix is ill-conditioned, that is, its inverse is essentially singular.[5] However, $\exp(-\alpha t)$ is practically zero for almost the entire range of α, so negligible error results from replacing t by ∞. We then obtain

$$I_\alpha = \int_0^\infty d\xi\, G(\xi)e^{-\alpha\xi} \tag{8}$$

The right hand side is merely the Laplace transform of $G(\xi)$ so formally

$$G(\xi) = \frac{1}{2\pi i}\int_{\alpha_M - i\infty}^{\alpha_M + i\infty} d\alpha\, I_\alpha e^{\alpha\xi} \tag{9}$$

where all the poles of I_α lie in the complex plane to the left of some real positive number α_M.

An immediate difficulty in evaluating Eq. (9) is that it requires a knowledge of the behavior of I_α in the complex plane, whereas I_α or rather i_λ is known only as a set of experimentally observed points along the real axis. The information sought can however be obtained by application of the Padé approximant method which has been used in a wide variety of physical problems.[6–9]

III. THE PADÉ METHOD

A complete discussion of the Padé method has been given by Baker[6] and will not be repeated here. We shall merely indicate how it may be used to evaluate the inversion integral, reserving the details of its application for a subsequent paper.*

Since the left hand side of Eq. (8) is known as a set of values for various α, it may be approximated in the least squares sense by a suitable set of orthogonal polynomials. It is necessary to use orthogonal polynomials to eliminate matrix inversion which leads to computational difficulties because of the range of α. Thus

$$I_\alpha = \sum_{m=0}^{M} \phi_m \varphi_m(\alpha) \tag{10}$$

where ϕ_m are the coefficients of $\varphi_m(\alpha)$ and

$$\varphi_m(\alpha) = \sum_{j=0}^{m} f_j^{(m)} \alpha^j \tag{11}$$

Then

$$I_\alpha = \sum_{n=0}^{N} k_n \alpha^n \tag{12}$$

where the coefficients k_n are obtained by introducing Eq. (11) into Eq. (10) and rearranging.

The right hand side of Eq. (12) may now be replaced by its Padé approximant

$$P_{n-1,\,n}(\alpha) = A_{n-1}(\alpha)/B_n(\alpha), \qquad 2n-1 \leqslant N \tag{13}$$

where $A_{n-1}(\alpha)$, $B_n(\alpha)$ are polynomials in α of degree $n-1$, n respectively and are defined through the relationship

$$B_n(\alpha)\cdot I_\alpha - A_{n-1}(\alpha) = 0 + O(\alpha^{2n} + \ldots\ldots) \tag{14}$$

Baker[6] has proved that an infinite sequence of approximants $\{P_{mn}(\alpha)\}$ with $m+n\to\infty$ to a function $f(z)$, converges to $f(z)$ over certain regions of the complex z-plane in which the sequence is uniformly bounded. The

* Application of Padé Approximants to Laplace Transform Theory, to be published elsewhere.

stringency of this requirement has been relaxed by Chisholm who has shown[9] that in certain regions of meromorphy it is not necessary to assume uniform boundedness, even though one must assume boundedness of the number of poles in the region. Since I_α from physical considerations can be assumed to be meromorphic in the finite complex α-plane, and to have a bounded number of poles, its zeroes and poles can be approximated with increasing accuracy by a sequence of Padé approximants. If all the poles then lie to the left of some real positive number α_M, the shift theorem of the Laplace transform theory[10] may be invoked to rewrite Eq. (9) as

$$G(\xi) = \exp(-\alpha_M \xi)\frac{1}{2\pi i}\int_{-i\infty}^{i\infty} d\alpha' P_{n-1,n}(\alpha')e^{\alpha'\xi} \tag{15}$$

where $\alpha'=\alpha+\alpha_M$ and all the poles of the shifted function $P_{n-1,n}(\alpha)$ lie in the left half plane. For such cases, Schmittroth[11] has shown that Eq. (15) may be replaced by either

$$G(\xi) = \exp(-\alpha_M \xi)\frac{2}{\pi}\int_0^\infty dy \,\mathrm{Re}\, P_{n-1,n}(y)\cos \xi y$$

or

$$G(\xi) = \exp(-\alpha_M \xi)\frac{2}{\pi}\int_0^\infty dy \,\mathrm{Im}\, P_{n-1,n}(y)\sin \xi y \tag{16}$$

where α' has been rewritten as y. Either of these two integrals can be evaluated by Gaussian quadrature.

Acknowledgements

I am indebted to Dr P.H. Fang for encouraging this analysis. I also wish to thank Professor George Pratt for introducing me to the Padé method and Professor Arnold Stokes and Dr John Marburger for many illuminating discussions.

REFERENCES

1. P. H. Fang, in *Proc. of the 7th International Congress on the Physics of Semiconductors*, p. 303, Royaumont, Dunod 1964,
2. V. S. Vavilov, in *Effects of Radiation on Semiconductors*, Consultants Bureau, 1963.
3. Ibid.
4. R. Courant and D. Hilbert, in *Methods of Mathematical Physics*, Interscience, 1953.
5. A. V. M. Ferris-Prabhu and J. H. Marburger, in *Report of the Radiation Physics Group*, Goddard Space Flight Center, Summer 1964 (unpublished).

6. G. A. Baker, *Advan. Theoret. Phys.*, **1**, 5 (1965).
7. S. Tani, *Phys. Rev.* **139**, B1011 (1965).
8. A. V. M. Ferris-Prabhu, *Phys. Stat. Solidi*, **15**, K23 (1966).
9. J. S. R. Chisholm, *J. Math. Phys.*, **7**, 39 (1966).
10. R. Churchill, in *Operational Mathematics*, 2nd Ed., New York, 1958.
11. L. A. Schmittroth, *Communications of the ACM*, pp. 171–173, March 1960.

Part V

Surface

Electrical Properties of the Defects near the Surface of Thermally Oxidized Silicon

SHUICHI SATO AND SHINJI KAWAJI

Department of Physics, Faculty of Science, Gakushuin University, Mejiro, Tokyo, Japan

ABSTRACT

The energy position, the effective capture cross section for holes or electrons and the concentration of the trapping centers distributed in the upper half of the band gap near the surface of thermally oxidized silicon were measured by the pulsed field effect technique. The Fermi level at the surface of thermally oxidized silicon crystals was controlled from the bottom of the conduction band to 0.6 eV below the conduction band by means of the bias temperature treatments. Two midgap states usually observed on etched surfaces (CP-4) were not observed on thermally oxidized silicon surfaces. Trapping levels whose electric and annealing characters are similar to those of the Si-A center, the Si-E center and other defects studied so far in the silicon bulk were observed.

I. INTRODUCTION

Electrical properties of electron or hole trapping centers in etched or oxide covered silicon surfaces have been studied by many workers.[1–12] However, no definite atomistic model has been established for the surface traps in the real silicon surfaces. On the other hand, the electronic and annealing properties of radiation defects introduced by bending crystals have been studied in detail for the volume of silicon crystals by many workers[13–30] and some definite models like the Si-A center[22, 23] or Si-E center[20] have been proposed.

In the present investigation, it is intended to clarify the models of the surface traps in silicon that are distributed in the upper half of the band gap. The electronic and annealing properties of levels observed by pulsed field effect technique are compared to those of levels studied so far in the volume of silicon crystals.

Whether a trap level can be observed or not depends on the position of the Fermi level relative to that particular trap level. In the usual experiments for the investigation of surface traps, the position of the Fermi level at the surface is controlled by adsorption of gases. In such cases it is always questionable whether the adsorbed gases introduce additional measurable trap levels or whether the gases affect the electrical properties of the trap levels. In the present study on thermally oxidized silicon surfaces, the surface potential was controlled by bias-temperature treatment, i.e. heat treatment under the application of an electrostatic field between the silicon substrate and the field effect electrode. This treatment enables us to control the position of the Fermi level at the silicon surface from the bottom of the conduction band to the middle of the energy gap. Consequently, the trapping centers near the surface whose energy levels are located in the upper half of the band gap were studied by the pulsed field effect eliminating a possible effect of the adsorbed gases on the surface traps.

II. EXPERIMENTAL PROCEDURES

Silicon samples used in the present experiments were p-type 50–70 $\Omega\cdot$cm and n-type 50–100 $\Omega\cdot$cm floating zone single crystals. The surfaces investigated were parallel to $\{111\}$ plane. These samples were oxidized at 1,200°C for 5 hours in an atomosphere of mixed oxygen gas and water vapor. The thickness of the oxide layer was about 0.57 μ. Samples for field effect measurements were made by sandwiching 0.02 mm thick mica spacers between silicon slabs of about $10\times2\times0.5$ mm^3 and tantalum field effect electrodes. The samples were mounted in a cryostat and experiments were carried out in a vacuum of 10^{-4} mmHg.

At first, the surface conductivity type was measured by the large signal AC field effect at room temperature. Then the polarity of the step voltage in the pulsed field effect was chosen to deplete the majority carriers in the surface space charge region. The temperature dependence of the relaxation time in the pulsed field effects was measured in the temperature range between 250°K and 400°K. The energy position below the conduction band edge (E_C-E_T) or above the valence band edge (E_T-E_V) and the effective capture cross section for electrons ($\gamma\sigma_n$) or holes (σ_p/γ) were determined from Rupprecht's theory.[2] Here, γ is the statistical weight factor of the trapping centers. The surface concentration of the traps, N_S, was estimated from the increment of the conductance between the initial and the steady state in the pulsed field effect at the step voltage of 600 V. If the traps are distributed uniformly in the surface space charge layer, the volume concentration of the traps N_V can be estimated from the energy position of the levels and the shape of the space charge layer.

After measurement of several runs of AC and pulsed field effects, bias-temperature treatments were carried out with a DC field of about 10^6 V/cm and a temperature of 145–160°C. After proper bias-temperature treatments, samples were immediately cooled down to room temperature by pouring liquid nitrogen into the cryostat and AC field effect and pulsed field effect measurements were repeated.

III. EXPERIMENTAL RESULTS

Typical results of AC field effect patterns and pulsed field effects for p-type samples in the case of the first or second run are shown in Fig. 1. The surface conductivity type is n-type and the bending of the energy band at the surface is 0.59 eV downward, assuming that the lowest surface conductance value is the minimum of the conductance. Two discrete levels of 0.38 eV and 0.05 eV below the conduction band with effective capture cross section of 3×10^{-16} cm^2 and 3×10^{-23} cm^2 for electrons, respectively, were observed. The surface concentration and the volume concentration for 0.38 eV levels are 5×10^{11} cm^{-2} and 1×10^{18} cm^{-3} and those for 0.05 eV levels are 5×10^{11} cm^{-2} and 4×10^{17} cm^{-3} respectively. From the effective capture cross section for electrons, 0.38 eV levels are acceptor- or neutral-like and 0.05 eV levels are acceptor-like. Repeating the pulsed

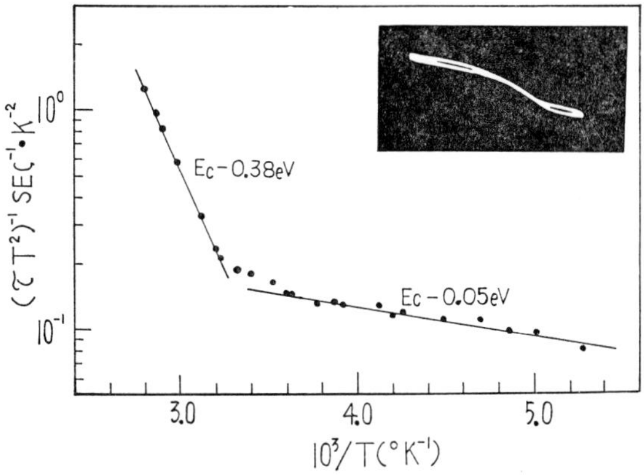

Fig. 1. AC field effect pattern and pulsed field effect results in the first or second run for p-type silicon.

field effect around 100°C, these two levels disappeared and different levels were observed as shown in Fig. 2. The effective capture cross section, the surface concentration and the volume concentration for 0.46 eV levels below the conduction band are 2×10^{-12} cm^2 for electrons, 1×10^{11} cm^{-2} and 4×10^{16} cm^{-3}, and those for 0.64 eV levels above the valence band are 9×10^{-13} cm^2 for holes, 2×10^{11} cm^{-2} and 4×10^{15} cm^{-3} respectively. The former level is donor-like and the latter acceptor-like. These levels also disappeared after a few runs of pulsed field effect or bias-temperature treatments.

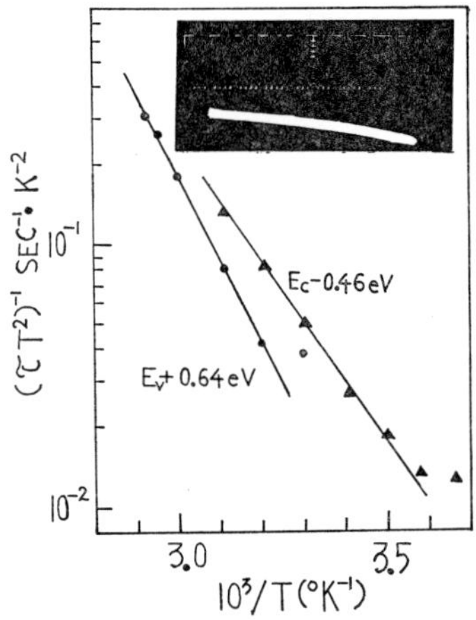

Fig. 2. AC field effect pattern and pulsed field effect results after several runs of the pulsed field effect measurements for p-type silicon.

In the course of the experiments shown in Figs. 1 and 2, step voltage with opposite polarity was applied. Relaxation phenomena with a long time constant were then observed. The temperature dependence of the relaxation time gives the activation energy of 0.99 eV as shown in Fig. 3. Considering the position of the Fermi level at the surface of the samples, the levels are believed to be located at 0.99 eV above the top of the valence band. This level was always observed when a high voltage step of about 1,500 V was applied to the field effect electrode even after the bias-temperature treatments were carried out and the energy band at the surface became nearly flat. The effective capture cross section for holes, the surface concentration and the volume concentration of the levels were 2×10^{-12} cm^2, 8×10^{12} cm^{-2} and 2×10^{21} cm^{-3} respectively. This level is acceptor-like.

With bias-temperature treatments, the AC field effect patterns changed

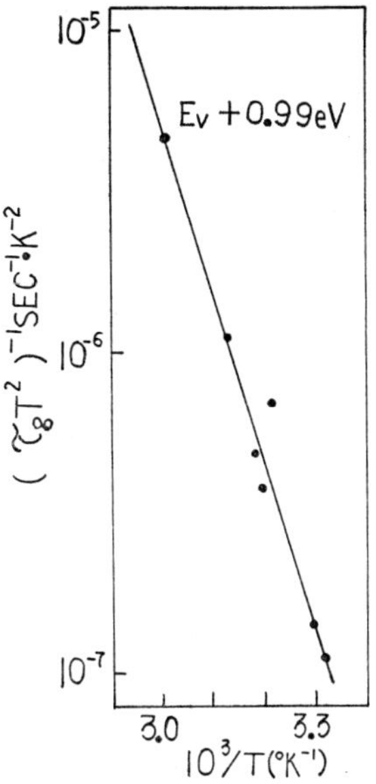

Fig. 3. Results of pulsed field effect with a long time constant for p-type samples.

as shown in Fig. 4. Here the sign Si(+) shows that the bias voltage applied to the silicon is positive. The bending of the energy band at the surface for each field effect pattern is also shown in Fig. 4. By reversing the polarity of the bias voltage or bias-temperature treatments in Si(−), the surface potential of the silicon sample recovered from Fig. 4(d) to Fig. 4(a). The AC field effect patterns in Fig. 4 were measured after cooling the samples from heat treatment temperature to room temperature in one or two minutes. When the sample was left after cutting off the heater current and the bias voltage, and was cooled down to room temperature in several hours by natural radiation cooling, the p-type field effect pattern seen in Fig. 4(d) was not observed. These results show that the migration of charged particles in the oxide layer takes place in an electric field at elevated temperatures. The concentration of charged particles in the oxide layer that can explain the change in the surface potential in the present experiments is about 2×10^{16} cm^{-3}.

In the course of the bias-temperature treatments, pulsed field effect measurements were performed. No signal of the transient phenomena due to the trap levels was observed for the surface of Fig. 4(b), (c) and (d) when negative voltage was applied to the field effect electrode. The

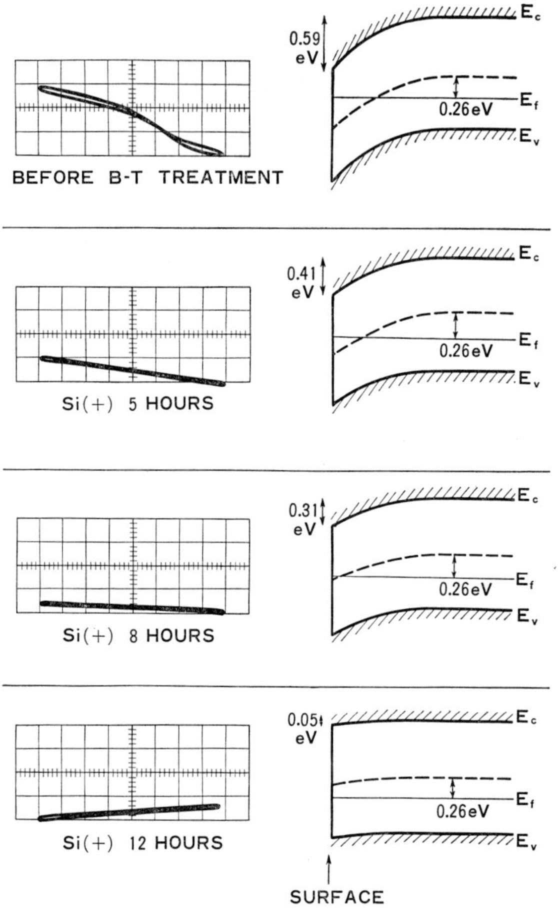

Fig. 4. Changes in AC field effect pattern and the bending of the energy band at the surface in p-type silicon by bias-temperature treatments.

0.99 eV level above the valence band was always observed when positive voltage was applied to the field effect electrode. In n-type samples without bias-temperature treatments, a discrete level of 0.38 eV below the conduction band was observed in the first or second run of the temperature cycle of the pulsed field effect as shown in Fig. 5. The effective capture cross section for holes, the surface concentration and the volume concentration of the level were 2×10^{-17} cm^2, 1×10^{11} cm^{-2} and 2×10^{15} cm^{-3} respectively. This level is acceptor-like. After a few runs of the pulsed field effect, this level disappeared and another level of 0.51 eV below the conduction band

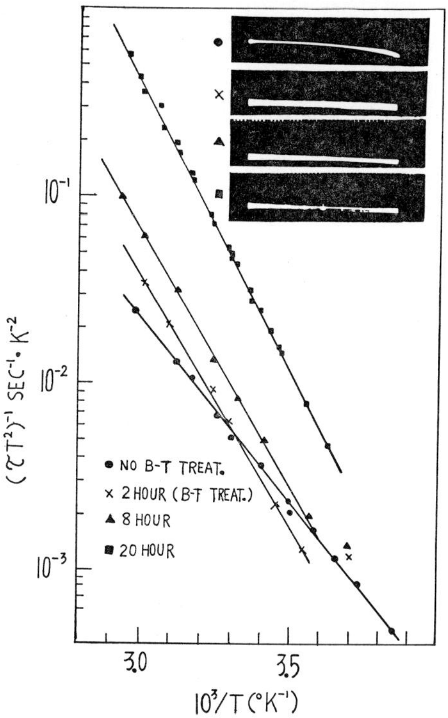

Fig. 5. AC field effect pattern and pulsed field effect results for n-type silicon.

was observed. The effective capture cross section for electrons, the surface concentration and the volume concentration of the 0.51 eV level are 4×10^{-15} cm^2, 4×10^{11} cm^{-2} and 2×10^{16} cm^{-3} respectively. The energy position and the annealing character of the 0.38 eV level in n-type samples are similar to those of the 0.38 eV level in p-type samples. But the capture cross section of the levels for electrons in n-type samples is about one tenth that in p-type samples. The donor-like deep level of 0.51 eV did not disappear after a few runs of the temperature cycle or bias-temperature treatment. The surface potential of n-type samples changed similarly to that of p-type samples with bias-temperature treatments.

After resolving the oxide layer in hydrofluoric acid, the levels of 0.53 eV above the valence band were measured in p-type samples and the levels of 0.67 eV below the conduction band were measured in n-type samples. They could not be observed in p- and n-type oxidized samples, however. The levels of 0.54 eV above the valence band and the levels of 0.53 eV below the conduction band are usually observed in etched surfaces.

IV. DISCUSSION

Energy levels of the trapping centers observed on thermally oxidized silicon surfaces are shown in Fig. 6. The effective capture cross sections and concentration of these levels are summarized in Table I. In Fig. 6, the levels observed in thermally oxidized surfaces irradiated by γ-ray, and in the etched surface[6] and levels reported so far for silicon bulk[13–30] are shown for comparison. The following conclusions can be derived from our experimental results:

As shown in Fig. 6, the two midgap states usually observed on etched surfaces since Rupprecht could not be observed on thermally oxidized silicon surfaces. These levels could be observed after removing the oxide layer from the samples. Accordingly, it will be concluded that the two midgap levels are surface states on etched surfaces.

Both levels 0.38 eV and 0.05 eV below the conduction band were observed in p-type samples, and the 0.38 eV level was observed in n-type samples. These levels are acceptor-like and disappeared after a few runs of the temperature cycle of the pulsed field effect or bias-temperature treatments. The energy position and the charge character of the annealing process for the two levels 0.38 eV and 0.05 eV below the conduction band

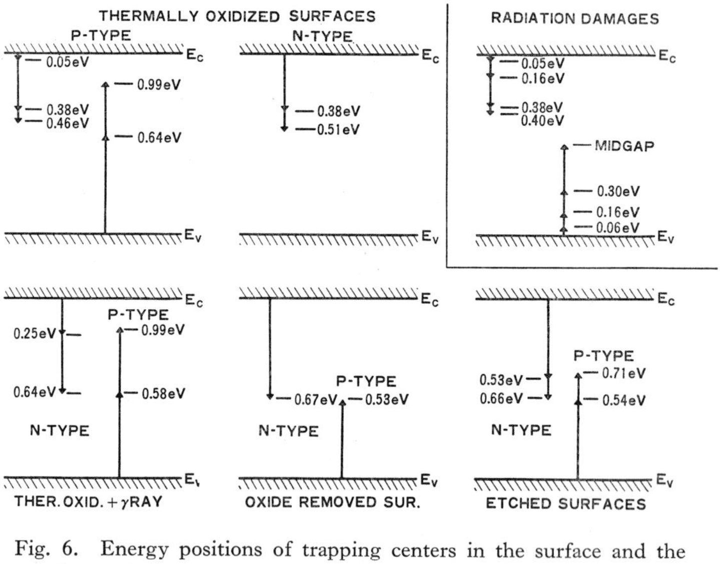

Fig. 6. Energy positions of trapping centers in the surface and the volume of silicon. See ref. 6 for the etched surfaces and the γ-ray irradiated thermally oxidized surfaces.

TABLE I. Electrical properties of trapping centers on thermally oxidized silicon surfaces.

Surfaces	Sample type	Energy position (eV)	Effective capture cross section (cm^2)	Surface concentration of traps (cm^{-2})	Volume concentration of traps (cm^{-3})
Oxidized	p	$E_c-0.05$	$\gamma\sigma_n=3\times10^{-23}$	5×10^{11}	4×10^{17}
		$E_c-0.38$	$\gamma\sigma_n=3\times10^{-16}$	5×10^{11}	1×10^{18}
		$E_c-0.46$	$\gamma\sigma_n=2\times10^{-12}$	1×10^{11}	4×10^{16}
		$E_v+0.64$	$\sigma_p/\gamma=9\times10^{-13}$	2×10^{11}	4×10^{15}
		$E_v+0.99$	$\sigma_p/\gamma=2\times10^{-12}$	8×10^{12}	2×10^{21}
	n	$E_c-0.38$	$\gamma\sigma_n=2\times10^{-17}$	1×10^{11}	2×10^{15}
		$E_c-0.51$	$\gamma\sigma_n=4\times10^{-15}$	4×10^{11}	2×10^{16}
Oxide removed	p	$E_v+0.53$	$\sigma_p/\gamma=1\times10^{-14}$	7×10^{10}	1×10^{14}
	n	$E_c-0.67$	$\gamma\sigma_n=6\times10^{-12}$	4×10^{11}	2×10^{16}
γ-ray irradiated oxidized	p	$E_v+0.58$	$\sigma_p/\gamma=8\times10^{-13}$	4×10^{11}	2×10^{17}
	n	$E_c-0.64$	$\gamma\sigma_n=4\times10^{-13}$	3×10^{11}	1×10^{15}
		$E_c-0.25$	$\gamma\sigma_n=1\times10^{-19}$	3×10^{11}	3×10^{15}

in p-type samples are in good agreement with the levels of dislocation introduced by bending n-type samples of floating zone crystals reported by Matsuura and Inuishi.[11] It is expected that there are a greater number of dislocations or lattice defects near the surface region, more than those of bulk due to the distortion in the interface between silicon and silicon dioxide. Therefore, it may be considered that these levels are associated with the lattice defects near the surface introduced by oxidation. The energy position, the charge character and the annealing character of the 0.38 eV level are also in good agreement with those of the Si-E center. Experiments reported so far by many authors were performed in n-type silicon bulk. In the present experiments, a 0.38 eV level was observed in n-type and p-type samples without irradiation. Accordingly, it may be considered that 0.38 eV levels below the conduction band are not Si-E centers, which are associations of lattice imperfections and phosphorus atoms. However, the levels observed in the present experiments are considered to be not the aurface states but the bulk levels in the surface space charge layer. The energy position and the annealing character of 0.38 eV levels in p-type samples are in very good agreement with those in n-type samples. But the effective capture cross section for electrons of the former levels is almost ten times that of the latter levels. This discrepancy is not yet understood.

Energy levels of 0.99 eV above the valence band are acceptor-like and did not disappear in the temperature cycle of the pulsed field effect and bias-temperature treatments. The energy positions and the charge character for these levels are in good agreement with those of the Si-A center 0.16 eV below the conduction band which is an association of radiation damages and oxygen atoms. The Si-A center was observed in n-type pulled silicon crystals having an oxygen concentration of more than 10^{18} cm^{-3} and was not observed in floating zone single crystals. But the samples used in the present experiments are p-type floating zone crystals and the oxygen concentration is believed to be less than 10^{13} cm^{-3}. Accordingly, if the levels observed are the Si-A center in the surface region, it will be considered that oxygen atoms diffused into the surface region of the silicon crystals during the oxidation process and the lattice imperfections associated with these oxygen atoms produced the Si-A centers.

Acknowledgements

The authors would like to thank Mr. H. Arata of the Electrical Communication Laboratory for the oxidation of silicon single crystals. Part of this work was supported by Grants-in-Aid from the Scientific Research Funds of the Ministry of Education.

REFERENCES

1. H. Statz, G. A. deMars, L. Davis, Jr. and A. Adams, Jr., *Phys. Rev.*, **101**, 1272 (1955); *ibid.*, **106**, 455 (1957).
2. G. Rupprecht, *J. Phys. Chem. Solids*, **14**, 208 (1960); *Ann. N. Y. Acad. Sci.*, **101**, 960 (1963).
3. J. A. Champion, *Proc. Phys. Soc.*, **79**, 662 (1962).
4. G. C. Alexzndrakis and G. C. Dousnanis, *J. Appl. Phys.*, **34**, 3077 (1963).
5. G. Gerlich, *J. Phys. Chem. Solids*, **23**, 837 (1963).
6. S. Sato, S. Kawaji and A. Kobayashi, *Surface Sci.*, **3**, 98 (1965); *ibid.*, **4**, 299 1966).
7. M. M. Atalla, E. Tannenbaum and E. J. Sheibner, *Bell Sys. Tech. J.*, **38**, 749 (1959).
8. H. Edagawa, Y. Morita, H. Ishikawa, S. Maekawa and Y. Inuishi, *Japanese J. Appl. Phys.*, **3**, 644 (1964).
9. S. Horiuchi, J. Yamaguchi and K. Shibata, *Japanese J. Appl. Phys.*, **4**, 394 (1965).
10. S. Horiuchi and J. Yamaguchi, *Japanese J. Appl. Phys.*, **4**, 613 (1965).
11. Y. Miura, *Japanese J. Appl. Phys.*, **4**, 958 (1965).
12. K. Sawamoto, M. Takahashi, H. Tomishima and H. Toyoda, *Japanese J. Appl. Phys.*, **5**, 191 (1966).
13. J. A. Hornbeck and J. R. Haynes, *Phys. Rev.*, **97**, 311 (1955).
14. J. R. Haynes and J. A. Hornbeck, *Phys. Rev.*, **100**, 606 (1955).

15. G. W. Wertheim, *Phys. Rev.*, **105**, 1730 (1957); *ibid.*, **110**, 1272 (1958); *ibid.*, **111**, 1500 (1958); *J. Appl. Phys.*, **30**, 1166 (1959).
16. H. Y. Fan and A. K. Ramdas, *J. Appl. Phys.*, **30**, 1127 (1959).
17. G. K. Wertheim and D. N. E. Bechanam, *J. Appl. Phys.*, **30**, 1232 (1959).
18. E. Sonder, *J. Appl. Phys.*, **30**, 1186 (1959).
19. G. Bemski, *J. Appl. Phys.*, **30**, 1195 (1959).
20. G. D. Watkins, J. W. Corbett and R. H. Walter, *J. Appl. Phys.*, **30**, 1198 (1959).
21. C. A. Klein, *J. Appl. Phys.*, **30**, 1222 (1959).
22. G. D. Watkins and J. W. Corbett, *Phys. Rev.*, **121**, 1001 (1961).
23. J. W. Corbett, G. D. Watkins, R. M. Crenko and R. S. McDonald, *Phys. Rev.*, **121**, 1015 (1961).
24. G. D. Watkins, *J. Phys. Soc. Japan*, **18**, Suppl. II, 22 (1963).
25. K. Matsuura and Y. Inuishi, *J. Phys. Soc. Japan*, **17**, 874 (1962).
26. V. S. Vavilov and A. F. Plotnikov, *J. Phys. Soc. Japan*, **18**, Suppl. III, 230 (1963).
27. Y. Inuishi and K. Matsuura, *J. Phys. Soc. Japan*, **18**, Suppl. III, 240 (1963).
28. H. Saito, M. Hirata and T. Horiuchi, *J. Phys. Soc. Japan*, **18**, Suppl. III, 246 (1963).
29. T. Nakano, K. Nakasima and Y. Inuishi, *J. Phys. Soc. Japan*, **20**, 2140 (1965).
30. H. Saito, M. Hirata and T. Horiuchi, *J. Appl. Phys.*, **37**, 1867 (1966).

Paramagnetic Defects Introduced near a Silicon Surface

TAKAO WADA, TERUYOSHI MIZUTANI, MASATAKA HIROSE AND TETSUYA ARIZUMI

Department of Electronics, Faculty of Engineering, Nagoya University, Nagoya, Japan

ABSTRACT

There are two kinds of paramagnetic defect centers in mechanically crushed Si powder. One is a broad line, and its line width is ~6 Oe (at $g=2.0055$). From the experiments on the properties of this line, carried out in ultra-high vacuum and by exposure to various gases, Chung and Haneman[4] proposed that the surface structure for the cleaved Si surface involves dangling bonds, the concentration being on the order of 20%. The experimental results of low energy electron diffraction by Lander *et al*[5]., however, suggests a surface structure with double bonds, differing from Chung and Haneman's proposal. We studied the strength of the resonance line through isochronal and isothermal annealing. At lower annealing temperatures there are close similarities in the EPR properties between this center and the Si-A center of an irradiated silicon, i.e. the g-value, the line width, the activation energy (~1.4 eV) for annealing and isochronal annealing behaviors, and possible mechanisms of formation of the Si-A center were postulated. At higher temperatures the resonance center behaves like an O_3-vacancy complex in an irradiated Si. The complex is surely formed by the diffusion of oxygen adsorbed on the surface.

Besides the above-mentioned broad line there are two kinds of narrow lines: a line width of ~0.8 Oe (at $g=2.0024$), and ~3.2 Oe (at $g=2.0029$). Both lines are due to a certain oxygen content near the surface. These line widths broaden logarithmically with pressure above 1 Torr upon admission of oxygen. Müller *et al.*[7] concluded that the broadening of the lines is due to the dipole interaction of the oxygen adsorbed at the surface with the paramagnetic centers produced near the surface.

I. INTRODUCTION

There have been several works[1-3] published on the paramagnetic resonance of defects mechanically introduced near the silicon surface. For an n- or p-type sample crushed in air, the broad resonance line appears with $g=2.0061\pm0.0002$, with a relaxation time T_1 of 10^{-5} sec, and a width of 6 Oe.[2] Walters and Estle[3] concluded that the paramagnetic centers introduced by mechanical damage do not originate from the Si surface alone but are distributed within a layer $\sim 10^{-4}$ cm thick. They pointed out the similarities in paramagnetic resonance behaviors between a neutron irradiated and a mechanically damaged Si.

Recently, Chung and Haneman[4] found that an EPR signal appears when Si is crushed in an ultra-high vacuum ($\sim 10^{-9}$ Torr). The signal was studied relating to the surface area, heat treatment, and exposure to various gases, and they proposed that the surface structure for a cleaved Si surface involves dangling bonds, the concentration being on the order of 20%. Later, however, a surface structure with double bonds was suggested from the results of low energy electron diffraction by Lander and Morrison.[5]

Following these works, the paramagnetic resonance properties of the defects in mechanically crushed Si powder were studied by annealing the samples isochronally and isothermally at a temperature range between 200°C and 900°C. The centers were annealed in two stages: at lower temperatures they behaved like an oxygen-vacancy pair of an irradiated Si, while at higher temperatures they behaved like O_3-vacancy complexes.

In addition to the above broad line, two kinds of narrow lines were detected by heating Si powder in a sealed quartz tube at $\sim 10^{-5}$ Torr at $g=2.0024$ (width ~ 0.8 Oe)[6] and at $g=2.0029$ (width ~ 3.2 Oe).[7] Müller *et al.*[7] observed that the line width of both lines did not change until about 1 Torr, but it broadened logarithmically with pressure above 1 Torr. The broadening can be interpreted by the dipole interaction.

II. BROAD LINE

1) *General properties*

Fletcher *et al.*[1] observed at first an EPR line associated with the surface of mechanically damaged silicon samples. Feher[2] studied some properties of this resonance line; it appeared when silicon was sandblasted or crushed in air but disappeared when a surface layer $\sim 10^{-4}$ cm thick was removed by the etchant, $HF+HNO_3$, but the line could not be eliminated by treating the samples for a few minutes in concentrated HC1, HNO_3, or HF. When the silicon sample was sandblasted with 600 mesh of SiC, the density of the centers was of the order of 3×10^{14}. The resonance line was homogeneously broadened ($T=1.25°K$, $\nu_e=9000$ Mc) with a relaxation time of $\sim10^{-5}$ sec and a line width of $\sim$6 Oe. The electron g-value was 2.0061 ± 0.0002. The same resonance line was observed both in n- and p-type silicon.

Further studies of this signal were undertaken by Walters and Estle.[3] The resonance line had a g-value of 2.0055 ± 0.0002 and the line was easily saturated at a low microwave power level at 1.3°K. The density of centers of the sample powdered to about 1 μ average particle size was between

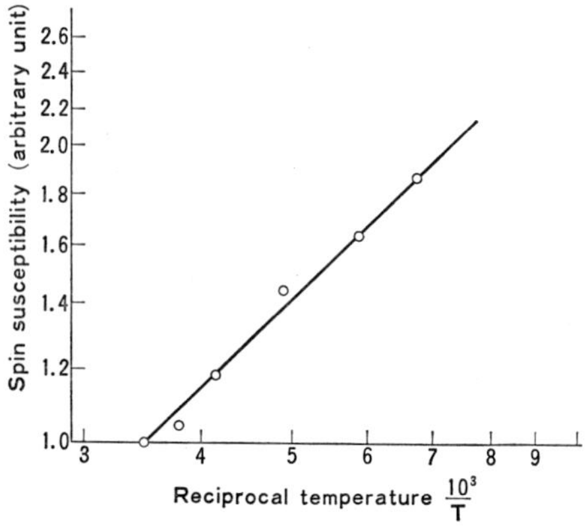

Fig. 1. Paramagnetic susceptibility of crushed silicon powder as a function of temperature.

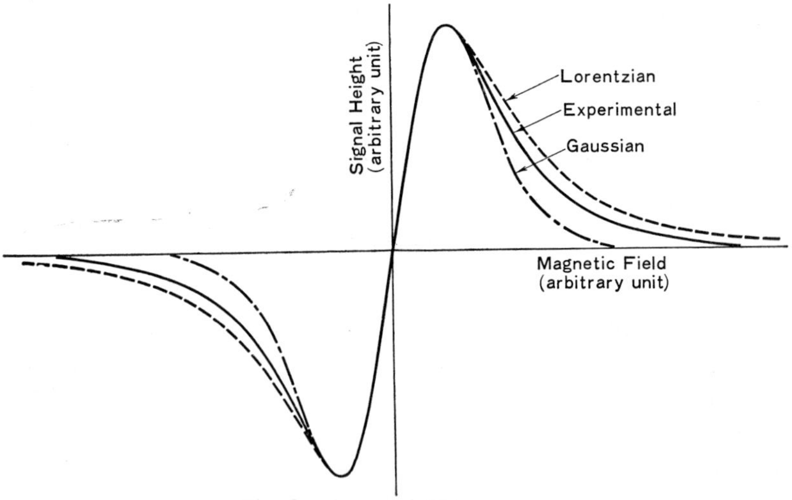

Fig. 2. A typical EPR spectrum.

1×10^{18} and 5×10^{18} per gram and its resonance characteristics were not influenced by the interaction with ambient atmospheres. Heating silicon powder in vacuum at 650°C for 6 hours had no effect on the density of paramagnetic damage centers and on the properties of the resonance line, but heating in air for 10 min at 400°C resulted in the destruction of 80 to 90% of centers. For removing the final 10 to 20% of centers the sample had to be heated at a temperature above 800°C. The resonance line of a mechanically damaged Si had the same *g*-value as a neutron irradiated Si.

It was observed[8] that the *g*-value and the line width of the centers were nearly independent of the temperature between −170°C and 120°C, and the paramagnetic susceptibility was inversely proportional to the temperature, obeying Curie's law as shown in Fig. 1. The line shape was intermediate between Gaussian and Lorentzian (Fig. 2).[3,8]

2) *Experiments in ultra-high vacuum and effects of exposure to various gases*

2. 1) *Experimental results*

Müller *et al.*[7] studied samples crushed in an ultrahigh vacuum by a magnetically operated hammer. The vacuum was from 0.25 to 1.5×10^{-9} Torr during the whole process. By resonance measurements made within 5 min after crushing with higher precaution, no EPR signal, broad or narrow, was found within the sensitivity of $10^{12} \times \Delta H$ (spins × Gauss) of the spectrometer. Thus it was suggested that the origin of these signals is associated with the surface adsorption of gases.

However, Chung and Haneman[4] reported recently that silicon powder crushed in an ultrahigh vacuum ($\sim 10^{-9}$ Torr) displayed an EPR signal close to $g = 2.0055$ with a width of 7.0 Oe. The signal was strongly

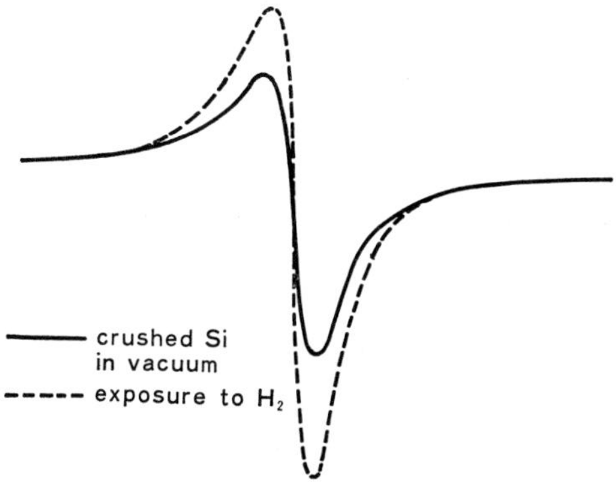

Fig. 3. Trace of EPR signal from vacuum-crushed silicon before (solid line) and after (broken line) exposure to molecular hydrogen at 10^{-2} Torr.[4]

TABLE I. Effect of exposure to H_2 and air, % change in signal height.

Sample No.	Upon exposure to H_2 at 10^{-2} Torr	Upon exposure to air after H_2
1	Sample at liquid-N_2 temperature +60%	Liquid-N_2 temperature −50%
2	Sample at liquid-N temperature +11%	
3	Sample at room temperature +53%	
4		Powdered in H_2 −24%
5		Powdered in H_2 −17%
6		Powdered in H_2 −35%
7	+65%	−20%
8	+50%	−35%
9	+63%	−31%
10	+80%	
11	+42%	
	On pumping out H_2 afterwards −8%	−20%

influenced by the exposure to molecular hydrogen, oxygen, and water vapor. When vacuum-crushed samples were exposed to 10^{-2} Torr of molecular hydrogen, the signal height increased about 60% as shown in Fig. 3, but the line width hardly changed. The experimental results for the effect of exposure to H_2 and air are shown in Table I. The increase of signal height scattered perhaps due to the difference of the surface area. From BET experiments the hydrogen coverage on silicon crushed in high vacuum was found to be less than 3% of the mono-layer within a factor of 2. Chung and Haneman[4] could not satisfactorily explain the apparently large (60%) increase in EPR signal, which was observed when only a few percent of the cleaved surface was covered by molecular hydrogen. Table I shows the effects of subsequently exposing the sample to air. In all cases the signal was reduced but not to its original value. The signal was reduced similarly on samples which had been powdered in hydrogen at room or liquid-nitrogen temperature. The decrease of signal height for the exposure to air after H_2 was probably due to water vapor.[4] When vacuum-crushed samples were exposed to oxygen of 10^{-2} Torr the signal height increased, as shown in Fig. 4, accompanied with a slight decrease in width. Results for five samples are shown in Table II, indicating that the increase

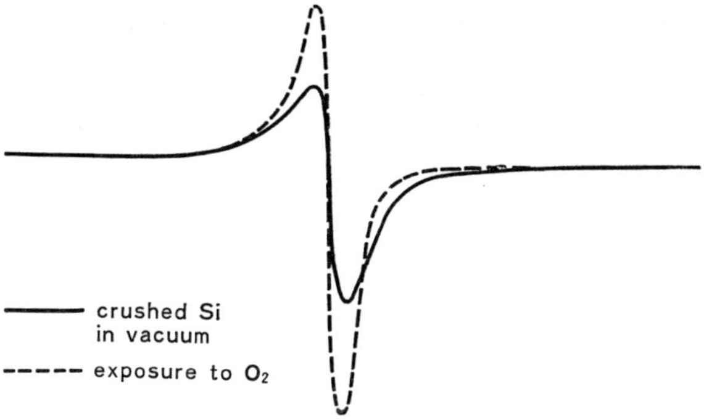

Fig. 4. Trace of EPR signal from vacuum-crushed Si before (solid line) and after (broken line) exposure to oxygen at 10^{-2} Torr.[4]

TABLE II. Effect of exposure to O_2, % change in signal height.

Sample No.	Upon exposure to O_2 at 10^{-2} Torr	Upon exposure to air after O_2
12	Liquid-N_2 temperature	
	+47%	
	Room temperature	
13	+75%	−13%
14	+108%	
15	+75%	−14%
16	+148%	−13%

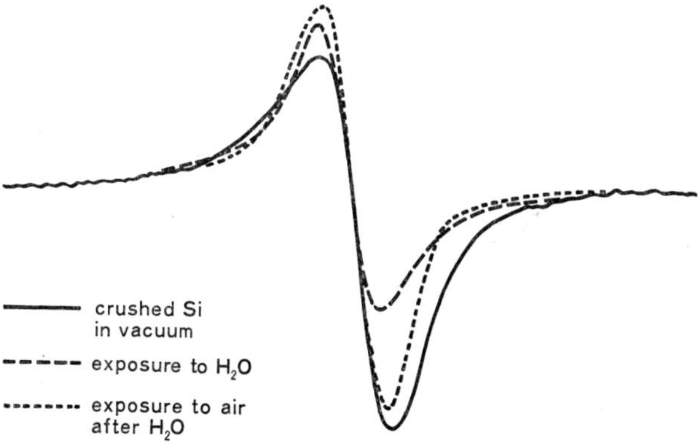

Fig. 5. Trace of EPR signal from vacuum-crushed Si before (solid line) exposure to gases, after (broken line) exposure to water vapor and after a subsequent (dotted line) exposure to air.[4]

of signal height is larger than those for the exposure to hydrogen. Exposure of these samples to air caused a small decrease of signal. Exposure of fresh sample surfaces to the pressure of 10^{-1} Torr of inert gases Ar and Kr, to nitrogen and to mixtures of CO and CO_2 had no detectable effect on the EPR signal. However, the exposure to room air following the above gases caused signal increases of 20% to 60%. When the crushed samples were exposed to water vapor, the signal height decreased, but subsequent exposure of the surface previously covered by water-vapor to air caused increases, as shown in Fig. 5.

2. 2) *Dangling bonds*

Chung and Haneman[4] proposed from the experimental results in section 2.1 that there exist unpaired electrons at the surface, which may be the dangling bonds predicted on a simple bond rupture process for separating a tetrahedrally bonded substance along (111) planes. The ratio of dangling bonds to surface atoms was determined to be approximately 1 to 5 from the measurements of surface area.[4]

However, the results of low energy electron diffraction (LEED) by Lander *et al.*[5] indicate that the cleaved silicon surface has a different structure. The substrate atom arrangement, which would be the surface arrangement if no atom displacements occurred with cleavage, is shown in Fig. 6 (a), and the structure of the surface inferred from the diffraction results is shown in Fig. 6 (b). The silicon (111) surface has a surface structure with a rectangular unit mesh formed by large displacements within the mesh of the first two atom layers, involving no dangling bond, but double bonds. Even if the cleaved surface has dangling bonds as proposed by

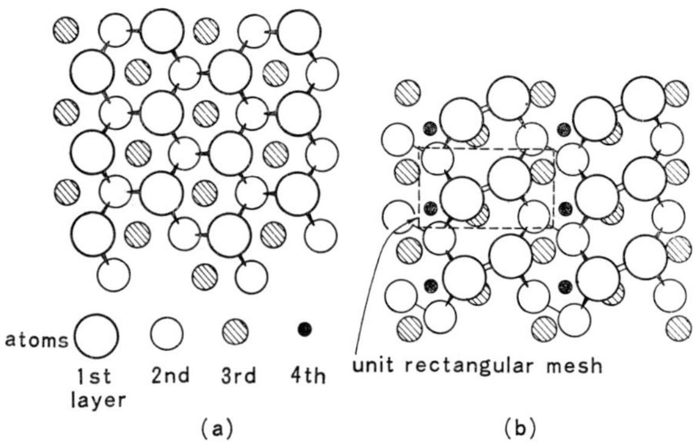

Fig. 6. a) Idealized cleavage structure before rearrangement
b) Si (111)–$1\times\sqrt{3}$ structure determined by analysis of the diffraction data.[5]

Chung and Haneman, the dangling bonds may have to be eliminated by removing the surface layer, i.e. by treating the sample for a few minutes in concentrated HNO_3. But Feher[2] reported that the signal did not change with this treatment. Even in the experiments of Chung *et al.*, oxygen must cover the new cleaved surface in the time interval between the crushing process and the resonance measurement, destroying a number of dangling bonds on the surface. In other words the origin of the center is not associated with the dangling bonds.

3) *Annealing effects*[8]

3. 1) *Experimental results*

The isochronal (15 min) annealing behavior of paramagnetic defects is shown in Fig. 7 by a solid line. The results of infrared absorption measurement in an irradiated bulk silicon[9] are also shown by the dotted lines. In this figure the fraction f of centers which were not annealed at a temperature of T°C is represented by $N(T)/N_0$, where N_0 is the signal strength before annealing and $N(T)$ the strength at T°C. The density of centers steadily decreases with rise in temperature between 100°C and 400°C (region I), but it increases again with rising temperature between 400°C and 600°C and after passing a maximum decreases above 600°C (region II). The results indicate the existence of at least two annealing stages for paramagnetic centers.

The activation energy for annealing was determined to be 1.4 eV from isothermal annealing curves of centers in region I (Fig. 8). Isochronal

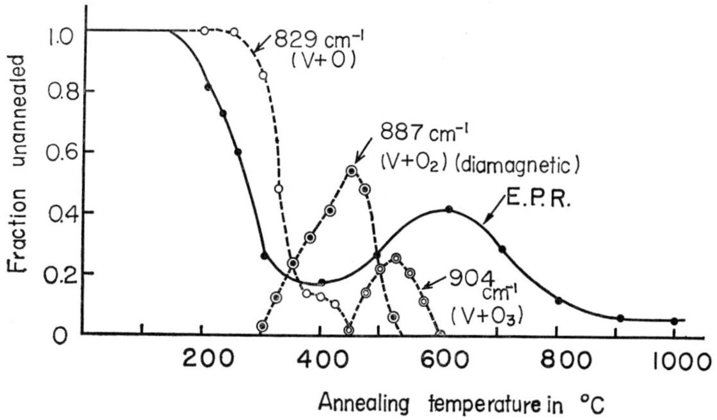

Fig. 7. Isochronal (15 min) annealing curves of the centers together with that of infrared absorption measurement in an irradiated Si.[9]

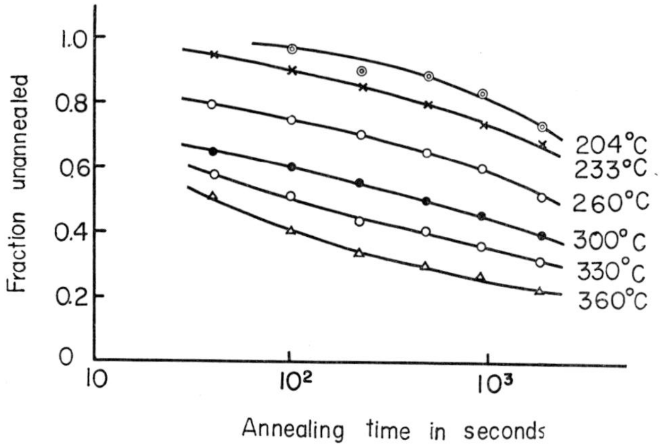

Fig. 8. Isothermal annealing curves at different temperatures in region I.

annealing curves of centers are shown in Fig. 9 at different partial pressures of oxygen. At a partial pressure of about 10^{-7} Torr the paramagnetic signal did not vanish up to a higher temperature, and the g-value and the line width of EPR signal did not change up to a higher temperature. These results suggest that the paramagnetic centers are mainly destroyed by oxygen migrating into silicon during the heat treatment.

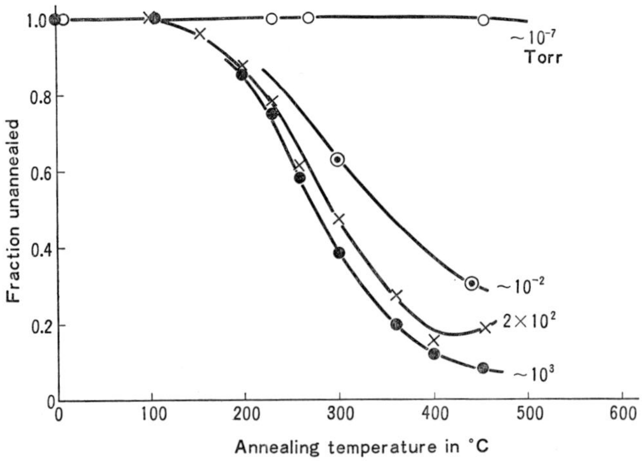

Fig. 9. Isochronal (10 min) annealing curves at different partial pressure of oxygen.

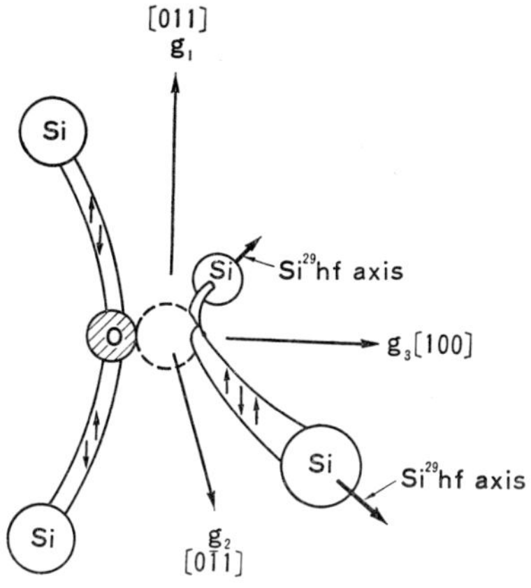

Fig. 10. Model of the Si-A center as a substitutional oxygen atom.

3. 2) *Si-A center and O_3-vacancy complex*

a) *Radiation-damaged center.*[10,11] The Si-A center is a major radiation-damaged defect produced in pulled Si by high energy electron irradiation. It was first observed in spin resonance by Bemski, Feher, and Gere.[10] In a subsequent study Watkins and Corbett reported that oxygen, present in pulled Si of a concentration of $\sim 10^{18}$ cm^{-3}, is necessary for the formation of this defect. Figure 10[11] presents an approximate description of the model of the defect. The values of the *g*-tensor along the principal axes were measured to be[11]

$$g_1 = 2.0093 \pm 0.0003$$
$$g_2 = 2.0025 \pm 0.0003$$
$$g_3 = 2.0031 \pm 0.0003$$
(the mean value $g = 2.0050 \pm 0.0003$).

This defect is formed in the following way: there are four broken bonds around a vacancy, one for each of the four silicon atoms surrounding it. The role of the oxygen is to bridge two of these broken bonds, forming an Si-O-Si " molecule." The remaining two Si atoms pull together to form a Si-Si molecular bond. Figure 11 shows the Si-Si bond considered as a simple diatomic molecule. In the neutral state, two electrons are paired off in the bonding orbital. An additional trapped electron goes into the anti-bonding orbital at $E_c - 0.17$ eV and spin resonance arises from it. The activation energy for annealing of the Si-A center was found to be about 1.3 eV by Bemski and Augustyniak.[12] Isochronal annealing curves for the oxygen-vacancy pair in an irradiated Si crystal are shown in Fig. 12.

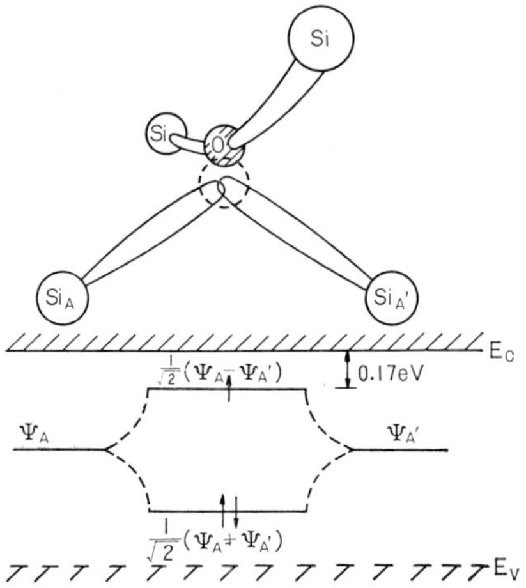

Fig. 11. The Si-Si bond considered as a simple diatomic molecule. In the neutral state, two electrons (one contributed by each atom) are paired off in the bonding orbital. The additional trapped electron goes into the anti-bonding orbital at $(E_c-0.17)$ eV.[11]

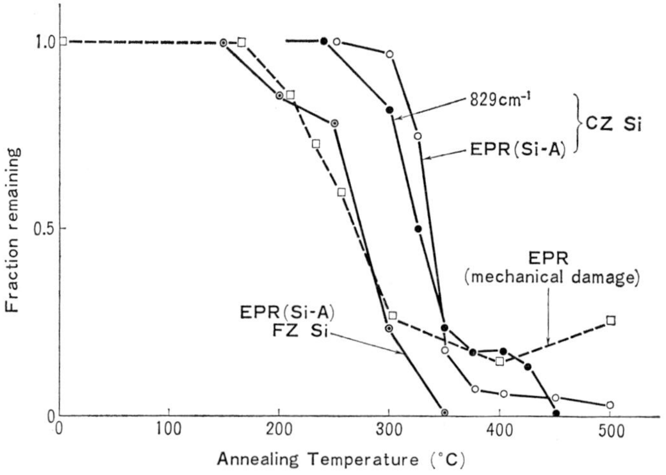

Fig. 12. Comparison of the fractional recovery observed in an isochronal (15 min) anneal for the 829 cm^{-1} IR band (sample with an irradiation of 8×10^{18} e/cm^2), EPR (sample-6.3×10^{16} e/cm^2) of Si-A centers[9] and EPR of mechanical damage centers.[9]

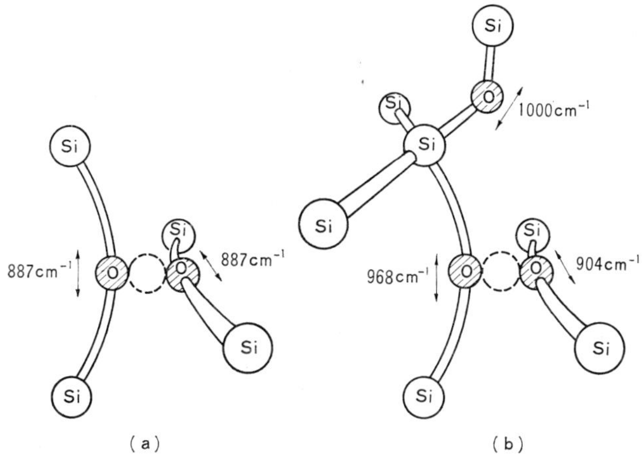

Fig. 13. Tentative models for (a) O_2-vacancy complex and (b) O_3-vacancy complex. The dashed circles indicate the position of the lattice vacancy.[9]

Corbett *et al.*[9] reported that the absorption coefficients of the various infrared bands (Fig. 7) was observed in an isochronal (15 min) annealing of n-type (phosphorus $\sim 10^{16}$ cm^{-3}) pulled Si irradiated by 8×10^8 e/cm^2. The annealing stages are explained by successive variation of the defect configuration, i.e. oxygen-vacancy pair (829 cm^{-1} infrared band), O_2-vacancy complex (887 cm^{-1} infrared band), and O_3-vacancy complex (the 904, 968, and 1000 cm^{-1} bands). Tentative models for O_2-vacancy and O_3-vacancy complexes are shown in Fig. 13.

b) *Mechanical damaged center in region I.* As described previously in section II-1, the line shape is intermediate between Gaussian and Lorentzian, and the line width is nearly independent of the temperature. These results suggest that the line width ΔH is not determined by the electronic relaxation, but by the anisotropy of the *g*-value. The value of ΔH agrees with that (<10 Oe) estimated from the anisotropy of the *g*-value of Si-A center in an irradiated Si crystal.[11]

We see many close similarities in paramagnetic resonance properties between the centers induced by the mechanical damage and the Si-A centers in an irradiated Si, i.e. the *g*-value, the line width, the activation energy for annealing and isochronal annealing behavior.

Since oxygen adsorbed on the silicon surface cannot migrate into the bulk so deeply at temperatures below 200°C, the centers may be mainly present near the silicon surface. The density of the defects is of the order

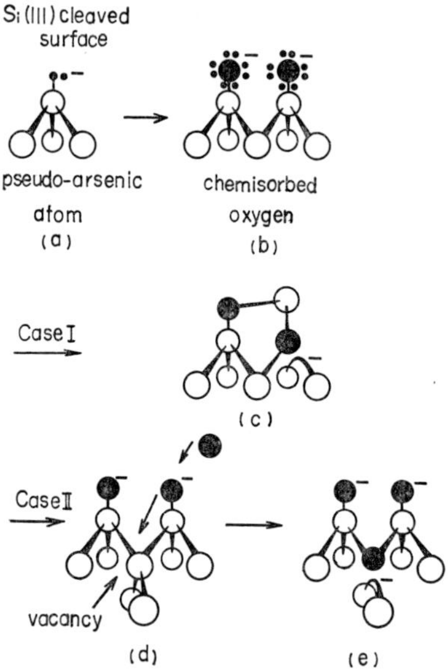

Fig. 14. Possible mechanisms for the formation of the Si-A center.

of $10^{18}/cm^3$, being independent of resistivity of FZ or CZ samples, temperature below 200°C, and type of conduction. This defect density is much larger than the oxygen concentration ($10^{16}/cm^3$) of the FZ Si. Thus these centers are possibly associated with oxygens adsorbed on the surface, or vacancies.

As discussed thus far, many experimental results suggest strongly that the paramagnetic defects are Si-A centers induced near the silicon surface. Possible mechanisms for the formation of Si-A centers in the crushing process are referred to in Fig. 14 a, b, c, d, and e. A surface atom of cleaved Si (111) traps a valence electron (a pseudo-arsenic atom[13]) (Fig. 14a) and an ambient oxygen is adsorbed chemically by the process

$$1/2\ O_2\ (\text{gas}) + e \longrightarrow O^-(\text{ads.}) \qquad \text{(Fig. 14b)}$$

Si-A centers may be formed by the following two processes:

Case I: A chemisorbed oxygen alternates its position with a surface Si atom by the repulsion between neighboring oxygens.[14] As the valencies of the three underlying silicon atoms previously bonded to the silicon atom which now exists on the surface must be satisfied, one of these silicon atoms bonds to the oxygen atom now in the second layer, while the other two silicon atoms bond to one another. Then the extra electron of O^- is trapped by the antibonding orbital of the Si-Si molecular bond (Fig. 14c).

Case II: It may be formed in a similar process mentioned in section

3.2 *a*. A large amount of vacancies introduced during the damaging process migrate to the surface (Fig. 14d), and oxygen adsorbed on the surface may make a jump to a vacancy and occupy a substitutional site.[11] Also, in this case, an Si-Si molecular bond can trap an extra electron of O^- (Fig. 14e).

When the sample was heated in an atmosphere containing oxygen at higher temperatures, Si-A centers near the silicon surface are perhaps annealed by the trapping of an excess oxygen atom by the center, which forms a diamagnetic O_2-vacancy pair.

c) *Mechanical. damaged center in region II.* The isochronal annealing behavior in region II resembles that of O_3-vacancy complexes observed by infrared absorption in an irradiated Si (Fig. 7). The negative annealing of the paramagnetic centers in the temperature range between 400°C and 600°C may be interpreted in the manner mentioned in section 3.2 *a*. At still higher temperatures the O_3-vacancy complex becomes diamagnetic by capturing an additional oxygen atom.

III. NARROW LINE

1) *Experimental results*

Kusumoto and Shoji[6] found a narrow line of 0.8 G width with a *g*-value of 2.0024 (named line A) upon heating between 400°C to 700°C various

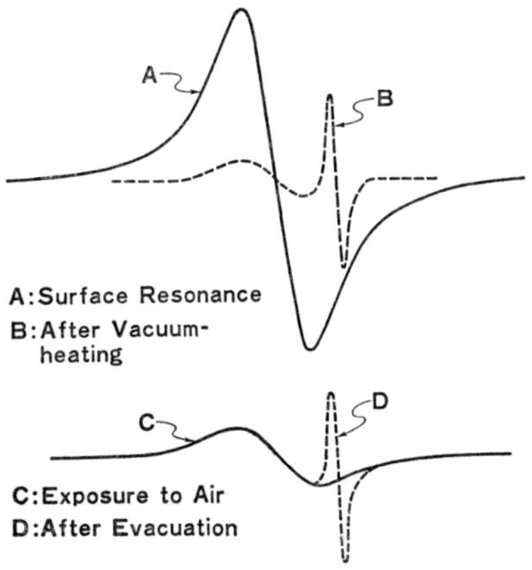

Fig. 15. The effect of vacuum heating upon surface resonance and the creation of a sharp and broad resonance. The process was observed on a powder sample of a floating-zoned single crystal vacuum-heated at 650°C for 30 min.[6]

silicon powders crushed in air and sealed in a quartz tube at 10^{-5} Torr. The line A appeared together with the broad line under these conditions. The line A disappeared immediately and the broad line did not change upon exposure to air, but the line A reappeared after evacuating again to 10^{-3} Torr, as shown in Fig. 15. This was ascribed to broken Si-Si bonds at the free silicon surface.

As described in the previous section, Müller *et al.*[7] found no EPR signal, broad or narrow, when silicon was crushed in an ultra-high vacuum ($\sim 10^{-9}$ Torr) by a magnetically operated hammer. However, heating in an ambient containing oxygen caused a narrow line (named line B) of 1.8 to 3.2 Oe width at $g = 2.0029$ to appear together with the broad line. The line B of 3 to 4.8 Oe width had apparently been created by outgassing the sample at a pressure of 10^{-3} to 10^{-5} Torr at a temperature between 350°C and 400°C. By subsequent heating of the powder at a temperature between 450°C and 470°C in a quartz tube for 12 hours, the line B was considerably enhanced and the line width was reduced to 1.8∼2.6 Oe. During this procedure the broad line weakened or disappeared.

When Si powder having the line B was brought into contact with air, by opening the ultra-high vacuum system, the line broadened to 7∼10 Oe

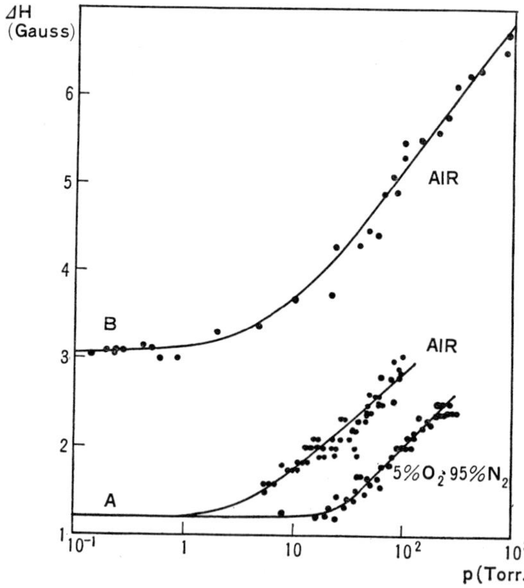

Fig. 16. The width of lines A and B due to paramagnetic centers near the surface of silicon as a function of logarithm of pressure at 300°K for air (20% O_2-80% N_2) and 5% O_2-95% N_2.[7]

but the g-value did not change at all. The widths of line A and B are shown in Fig. 16 as a function of the logarithm of pressure P.[7] If the system is not exposed too long to atmospheric pressure, the line width could be reproduced by evacuation. The line width did not change until about 1 Torr, but it broadened logarithmically with pressure above 1 Torr. The slope of the curve and the pressure P_0 of line A at which the knee occurs are nearly the same as those obtained with line B. Using 80% N_2-20% O_2, a mixture of 90% N_2-10% O_2 or 95% N_2-5% O_2 (see Fig. 16), nearly the same slope was obtained but the broadening started at a higher pressure P_0. Introducing pure diamagnetic gases such as argon, nitrogen, CO_2, or helium did not affect the line at all, up to the pressure of 400 Torr.

2) *Line width due to dipole interaction*

From the above results, Müller *et al.*[7] concluded that the broadening of the line is due to the dipolar interaction of the paramagnetic oxygen physically adsorbed onto the surface with the paramagnetic centers produced near the surface. This interaction changes the relaxation times T_1 and T_2 of the centers. For not too thick layers the times will be inversely proportional to the number Γ_{O_2} of adsorbed oxygen molecules,

$$\Delta H - \Delta H_0 = K\Gamma_{O_2} \qquad (1)$$

where K is a proportionality constant and ΔH_0 is the line width at the pressure of 1 Torr. The line width has been found to broaden logarithmically with pressure within the experimental error at the pressures above 1 Torr; it follows from Eq. (1) that the number of oxygen molecules at the Si surface is a logarithmic function of the pressure.

The van der Waals equation of adsorbed molecules at a surface is

$$(\pi + a\Gamma^2) = \Gamma kT$$

neglecting the term $(1-b\Gamma)$, where Γ is the number of adsorbed molecules of O_2 per unit surface and π is the reduced pressure. Using the Gibbs adsorption equation $d\pi = \Gamma d\mu^s$, the following expression is obtained for the surface chemical potential

$$\mu^s = \mu_0{}^s - 2a\Gamma + kT \ln \Gamma$$

Equating μ^s to the potential μ^g of the oxygen molecules in the gas, which due to the low pressure obey the ideal gas equation, i.e.

$$\mu^g = \mu_0{}^s + kT \ln P_{O_2}$$

as Γ increases with P, a must be negative for a large a; i.e. the van der Waals forces between the adsorbed molecules are repulsive. Then it is justified that the term $(1-b\Gamma)$ was neglected. For a large a the logarithmic dependence of Γ for P_{O_2} is obtained.

REFERENCES

1. R. C. Fletcher, W. A. Yager, G. L. Pearson, A. N. Holden, W. T. Read and F. R. Merritt, *Phys. Rev.*, **94**, 1392 (1954).
2. G. Feher, *Phys. Rev.*, **114**, 1219 (1959).
3. G. K. Walters and T. L. Estle, *J. Appl. Phys.*, **32**, 1854 (1961).
4. M. F. Chung and D. Haneman, *J. Appl. Phys.*, **37**, 1879 (1966).
5. J. J. Lander and J. Morrison, *J. Appl. Phys.*, **34**, 1403 (1963); J. J. Lander, G. W. Gobelli and J. Morrison, *ibid.*, **34**, 2298 (1963).
6. H. Kusumoto and M. Shoji, *J. Phys. Soc. Japan*, **17**, 1678 (1962).
7. K. A. Müller, P. Chan, R. Kleiner, D. W. Ovenall and M. J. Sparnaay, *J. Appl. Phys.*, **35**, 2254 (1964).
8. T. Wada, T. Mizutani, M. Hirose and T. Arizumi, to be published.
9. J. W. Corbett, G. D. Watkins and R. S. McDonald, *Phys. Rev.*, **135**, A1381 (1964).
10. G. Bemski, G. Feher and E. Gere, *Bull. Am. Phys. Soc.*, **3**, 135 (1958).
11. G. D. Watkins and J. W. Corbett, *Phys. Rev.*, **121**, 1001 (1961).
12. G. Bemski and W. M. Augustyniak, *Phys. Rev.*, **108**, 645 (1957).
13. P. Handler, *in Semiconductor Surface Physics* (K. H. Kingston, ed.), p. 36, Univ. of Pennsylvania Press, Philadelphia, 1953.
14. M. Green and K. H. Maxwell, *J. Phys. Chem. Solids*, **13**, 145 (1960).

Low-energy Electron Diffraction Study of the Superstructure of a Clean Si (111) Surface

YOSHIYUKI TAKEISHI AND KANJI HIRABAYASHI

Central Research Laboratory, Toshiba Electric Co.,
Kawasaki, Japan

ABSTRACT

Experimental and theoretical studies are reported of the diffraction of low-energy electrons by a cleaned and annealed Si(111) surface whose plane periodicity is known to be seven times as large as the bulk one. Intensities of diffracted beams were extensively measured as functions of incident beam energy in the range 30 to 400 eV at several fixed incident angles. It was found in the intensity profiles that the several characteristic peaks have fine structures which are sensitive to whether the surface is free or adsorbed by foreign atoms and the intensities of many fractional-order beams are higher than one hundredth of those of normal-order beams. Concerning the analysis of the data, a pseudo-kinematical theory is discussed using a screened Coulomb potential (with screening length R) with a constant imaginary part, $-iW_0$, for the interaction of incident electrons with ion cores and electron gas, under the assumption that the diffracted waves are much weaker than the incident wave. Numerical calculations of the intensities were carried out along this theory by using a trial model of the superstructure. The model is made up of hexagonal dislocation networks that have been shown to be elastically stable and yield the 7×7 plane periodicity. The results reproduce to some extent the general features of the observed intensity characteristics by choosing $R=0.21$ Å and $W_0=4$ eV. Discussion of the effect of multiple scattering is presented.

I. INTRODUCTION

It has been revealed from the diffraction patterns of low-energy electrons that the structures of the cleaned and annealed (111) surfaces of diamond-type crystals are markedly different from a simple termination of the bulk structure, and that the (111) surfaces of diamond,[1,2] silicon,[3–6] and germanium[3–8] have a 2×2, 7×7,* and 8×8* plane periodicity, respectively, provided the bulk plane periodicity is taken to be unit. The diffraction pattern itself, however, does not allow us to deduce any definite information as to the surface structure. For the structural analysis, the intensity characteristics of diffracted beams against incident beam energy and or incident angle have to be analyzed with a theory relevant to the involved problem, under adequate consideration of possible atomic arrangement (a speculative model) on the reconstructed surface layers. No acceptable method of unequivocal structural determination by low-energy electron diffraction has been established, at least in cases as intricate as these surfaces. The effect of possible multiple scattering[9,10] in the diffraction would complicate the problem.

Lander and Morrison[3] proposed, with a view to eliminating dangling bonds of surface atoms, a model made up of warped benzene-like rings (or a " vacancy " model) as a basic unit for the superstructures of the (111) surfaces of the diamond-type crystals. For germanium surface structures, Seiwatz[11] suggested a " conjugated chain " model, also eliminating dangling bonds, and Haneman suggested a " rumpled top-layer " model, raising and depressing topmost atoms. These investigators calculated the intensities of diffracted beams as functions of incident beam energy for the Ge(111)–12×12 (see footnote) and compared with the observed ones by Lander and Morrison. Their method of calculation is common and follows a kinematical theory of electron diffraction, provided the atoms within few atomic layers down to the third or at most the fourth beneath the surface contribute the diffraction in the sense of very strong damping of incident beams. The results are seemingly in good agreement with the observed ones.

In this paper we report experimental and theoretical studies of diffraction by the Si(111)–7×7 surface. The intensity characteristics of the diffracted beams have been extensively measured as functions of incident beam energy in the range 30 to 400 eV at several incident angles, and theoretical interpretations have been attempted using a pseudo-kinematical theory reported elsewhere[13] and adapting another possible model, a " dislocation networks " model, suggested by one of the authors.[14] The intensity

* Lander *et al.*[4,5] found a 5×5 structure in lieu of the 7×7 on the Si (111) surface and also a 12×12 structure on the Ge(111) surface, after annealing the samples at particular temperatures, but Jona[6] could not observe either Si(111)-5×5 or Ge(111)-12×12 structures, following their treatment. Schlier and Farnsworth[3] reported some different results on these surfaces. These structures are now thought to be presumably extrinsic.

characteristics calculated in this way have to a certain extent reproduced general features of the observed ones. The effects of multiple scattering have also been pointed out through comparison of the calculated results with the observed ones.

In this paper, experimental procedures and some introductory results are described in section II, and the method of calculating the intensity of diffracted beams is discussed in section III. A trial model on the superstructure is presented in section IV. Observed intensity characteristics, calculated results, and related discussions are given in section V.

II. EXPERIMENTAL METHOD

Several U-shaped silicon samples were cut from 20–40 $\Omega\cdot$cm, p-type material and attached to tantalum electrodes for heating by electrical conduction. The front face was polished parallel to a (111) face to a few minutes. Prior to installation into a diffraction equipment the target was chemically etched in a mixture of HNO_3 and HF. The diffraction tube used is of a display type provided with spherical grids and a fluorescent screen.[15] Background pressures during the measurements were always less than 5×10^{-10} Torr, equivalent N_2 pressure.

The target surfaces were cleaned in three ways: high temperature heating up to about 1,300°C, H_2 gas etching, and ion bombardment and annealing. Resultant diffraction patterns were always of a well-defined 7×7 structure, independent of these cleaning techniques. Lesser background patterns, however, were more readily obtained on the ion-bombarded and annealed surface, which was used for extensive measurements of the intensities of diffracted beams.

A typical diffraction pattern is reproduced in Fig. 1. The appearance of the pattern, derived from the relative intensity of the diffraction spots, is different at different incident beam energies. The patterns obtained in the present experiments agree very well with the patterns published by Lander and Morrison[5, 9] and Jona.[6] This means that the intensity characteristics of diffracted beams against the beam energy are essentially the same for Si(111) surfaces cleaned in different ways. We have not succeeded, as Jona did, in finding a 5×5 structure, in spite of long annealing of the surface at temperatures in the range 550 to 700°C, under which conditions Lander and Morrison found it. Thus this structure is ruled out in this paper, and it is assumed that the 7×7 pattern does reflect the intrinsic surface structure of the cleaned and annealed Si(111) surface, and does not depend on the topographical structure of the actual surfaces, minor-impurity distribution, and history of processing of the material.

A nomenclature for the reciprocal lattice, which will be used throughout this paper, is indicated in Fig. 1. The fractional-order beam is denoted by $\left(\frac{n}{7}\frac{m}{7}\right)$; n and m are integers. The intensities of all normal-order beams and most of the observable fractional-order beams under experimental conditions have been measured. Almost all fractional-order beams are clearly seen in the patterns at the incident beams around 75 eV. All intensity data reported in section V were taken at room temperature after flashing the target at temperatures of 800–1,000°C.

Intensity measurements were made with a spot-photometer. The intensity indicated is proportional to the output current of a photomultiplier which is linear with diffracted beam intensity, and measured at a constant target current (1 μA). Correction to the constant incident beam current has not been made. For the yield of secondary electron emission (total elastic and inelastic current), Lander and Morrison's work[9] should be referred to. Measurement of very weak intensity is limited by finite background in the patterns as well as the noise level in the photomultiplier. Thus, it is not clear whether the very weak intensity is as measured, or is actually higher than the true relative intensity, and whether other weaker maxima and minima are concealed in the background. The intensity of

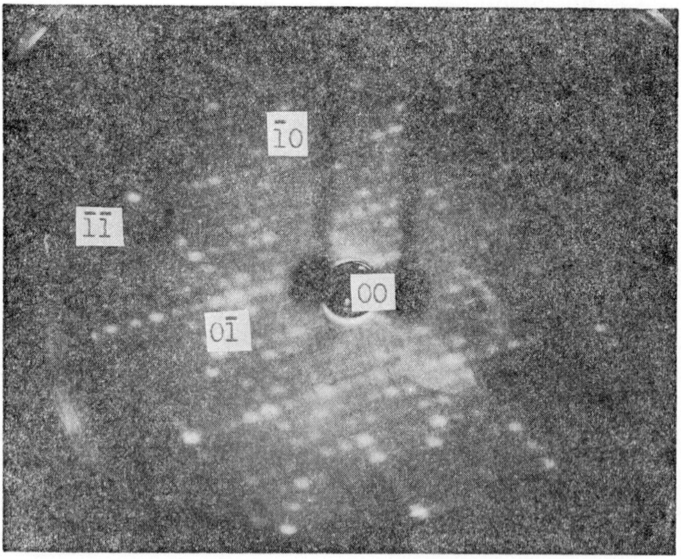

Fig. 1. A diffraction pattern of the Si(111)-7×7 at 76 eV beam.

the diffracted beams was measured as a function of incident beam energy at several fixed incident angles (the angles between surface normal and the direction of incident). The incident angle is varied by rotating the target around the vertical axis. Actually the angle indicated is half the angle between the incident beam direction and the (00) spot (specular reflection). The azimuthal angle of the crystal is fixed, as seen in the diffraction pattern of Fig. 1.

The energy of the incident electron beam indicated in the intensity profiles is the voltage difference between the target and the cathode (a ThO_2-coated Ta cathode), and is not corrected by the contact potential difference of about 2 eV.

III. THEORETICAL

Particular problems in the dynamical theory of the diffraction are how to approximate the effective potential that describes the interaction of incident electrons with ion cores and electron gas and how to calculate the wave function of the electrons multiple-scattered in this field.

The characteristic peaks in the observed intensity profiles against the incident beam energy are always of a finite width. This may be caused not only by inelastic scattering and multiple scattering, but also by the turbulence in the atomic arrangement, especially near the surface. We have, however, poor information in this respect.

The effect of inelastic scattering is taken up phenomenologically by the introduction of complex potential, which is an analogue of the complex dielectric constant in optics. If a beam with energy $(\hbar^2/2m)k_0{}^2$ moves in a complex potential $V(\mathbf{r})$, we have

$$\nabla^2\Psi - \left(\frac{2m}{\hbar^2}V - k_0{}^2\right)\Psi = 0 \qquad (1)$$

The density ρ and the current vector $\mathbf{j}$ are given by

$$\rho = \Psi^*\Psi, \qquad \mathbf{j} = \frac{\hbar}{i2m}(\Psi^*\nabla\Psi - \Psi\nabla\Psi^*)$$

Then

$$\nabla\mathbf{j} = \frac{\hbar}{i2m}(\Psi^*\nabla^2\Psi - \Psi\nabla^2\Psi^*) = \frac{1}{i\hbar}(V - V^*)\Psi^*\Psi = \frac{1}{\hbar}(\text{Im } V)\rho$$

Thus, sinks are distributed in the region $\text{Im } V(\mathbf{r}) < 0$. Though the spatial dispersion of Im V may be important, we neglect it here and assume that the potential has the following form:

$$V(\mathbf{r}) = -Ze^2\sum_m \frac{\exp(-R^{-1}|\mathbf{r}-\mathbf{r}_m|)}{|\mathbf{r}-\mathbf{r}_m|} - iW_0 \tag{2}$$

where Z is the atomic number, R the screening length, $\mathbf{r}_m$ the coordinates of the ion cores, and W_0 a positive constant.

If we choose z axis as normal to the surface of the crystal, which spreads over the negative half-space, $z \leq 0$, we may expand the potential as follows:

$$V(\mathbf{r}) = \sum_j V_j(z) \exp[i2\pi\,\mathbf{A}_j(\mathbf{x}+\mathbf{y})] \tag{3}$$

Here $\mathbf{A}_j = h_j\mathbf{a}^* + k_j\mathbf{b}^*$, and $\mathbf{a}^*$ and $\mathbf{b}^*$ are the unit reciprocal lattice vectors for a diperiodic lattice in an xy plane, and $h_j = k_j = 0$ when $j=0$. The solution of Eq. (1) has the form

$$\Psi(\mathbf{r}) = \sum_j \Psi_j(z) \exp[i(\mathbf{K}_0 + 2\pi\,\mathbf{A}_j)(\mathbf{x}+\mathbf{y})] \tag{4}$$

where

$$k_0^2 - \mathbf{K}_0^2 = k_0^2 \cos^2\theta$$

and θ is the angle between the direction of incidence and the surface, normal. The introduction of Eqs. (3) and (4) into Eq. (1) leads to

$$\frac{d^2\Psi_j}{dz^2} + [k_0^2 - (\mathbf{K}_0 + 2\pi\,\mathbf{A}_j)^2]\Psi_j - \frac{2m}{\hbar^2}\sum_l \Psi_l V_{j-l}(z) = 0 \tag{5}$$

We solve Eq. (5) assuming that $|\Psi_j| \ll |\Psi_0|$, and thus

$$\frac{d^2\Psi_0}{dz^2} + \left(k_0^2\cos^2\theta - \frac{2m}{\hbar^2}V_0(z)\right)\Psi_0 = 0,$$

$$\frac{d^2\Psi_j}{dz^2} + \left[k_0^2 - (\mathbf{K}_0 + 2\pi\,\mathbf{A}_j)^2 - \frac{2m}{h^2}V_0(z)\right]\Psi_j = \frac{2m}{\hbar^2}\Psi_0 V_j(z)$$

$$\text{for } j \neq 0$$

which requires that the diffracted beams are much weaker than the incident beam and may not be true if the effect of multiple scattering is large. This will be discussed below in section V. After matching the wave function thus obtained to the wave function in the region outside the surface, we have the following expressions for the intensity of the diffracted beams:[13]

$$I_0 = \left| \frac{k - k_0\cos\theta}{k + k_0\cos\theta} + i\frac{2m}{\hbar^2}\cdot\frac{2k_0\cos\theta}{(k + k_0\cos\theta)^2}\int_{-\infty}^{0} V_0(z)\exp(-i2kz)dz \right|^2$$

$$\text{for (00) beam,} \tag{6}$$

$$I_j = \left| \frac{2m}{\hbar^2} \frac{2k_0 \cos\theta}{(\kappa+\kappa_j)(k+k_0\cos\theta)} \int_{-\infty}^{0} V_j(z) \exp\left[-i(\kappa+k)z\right] dz \right|^2$$

$$\text{for } (h_j k_j) \text{ beam,} \tag{7}$$

where

$$\left.\begin{aligned}
k &= k'+ik'', \qquad \kappa = \kappa'+i\kappa'' \\
k' &= \left\{\frac{1}{2}\left(k_0^2\cos^2\theta+\frac{2m}{\hbar^2}U_0\right)+\frac{1}{2}\left[\left(k_0^2\cos^2\theta+\frac{2m}{\hbar^2}U_0\right)^2\right.\right. \\
&\quad \left.\left.+\left(\frac{2m}{\hbar^2}W_0\right)^2\right]^{1/2}\right\}^{1/2} \\
k'' &= \frac{2m}{\hbar^2}\frac{W_0}{2k'} \\
\kappa' &= \left\{\frac{1}{2}\left(\kappa_j^2+\frac{2m}{\hbar^2}U_0\right)+\frac{1}{2}\left[\left(\kappa_j^2+\frac{2m}{\hbar^2}U_0\right)^2+\left(\frac{2m}{\hbar^2}W_0\right)^2\right]^{1/2}\right\}^{1/2} \\
\kappa'' &= \frac{2m}{\hbar^2}\frac{W_0}{2\kappa'},
\end{aligned}\right\} \tag{8}$$

and

$$U_0+iW_0 = \lim_{L\to\infty}\left[-\frac{1}{L}\int_{-L}^{0} V_0(z)dz\right]$$

$$\kappa_j = [k_0^2-(\mathbf{K}_0+2\pi\,\mathbf{A}_j)^2]^{1/2}$$

Let the x, y and z components of the coordinate vector $\mathbf{r}_m$ be x_m, y_m, and z_m, respectively, and the unit direct lattice vectors for a diperiodic lattice in an xy plane be $\mathbf{a}$ and $\mathbf{b}$. Then we may write

$$\mathbf{x}_m+\mathbf{y}_m = \xi_m\mathbf{a}+\eta_m\mathbf{b} \tag{9}$$

We can readily obtain the following relations for the (111) surface of a diamond-type crystal:

$$\mathbf{A}_j^2 = \frac{4}{3a^2}(h_j^2+h_jk_j+k_j^2), \qquad a^2 = \mathbf{a}^2 = \mathbf{b}^2,$$

$$\mathbf{A}_j(\mathbf{x}_m+\mathbf{y}_m) = h_j\xi_m+k_j\eta_m$$

$V_0(z)$, $V_j(z)$, U_0 and κ_j in Eqs. (6), (7) and (8) are rewritten as follows in this special case:

$$V_0(z) = -\frac{4\pi Ze^2R}{\sqrt{3}\,a^2}\sum_m \exp(-R^{-1}\,|\,z-z_m\,|)-iW_0 \tag{10}$$

$$V_j(z) = -\frac{4\pi Ze^2}{\sqrt{3}\,a^2}\left[R^{-2}+\frac{16\pi^2}{3a^2}(h_j^2+h_jk_j+k_j^2)\right]^{1/2}$$

$$\times \sum_m \exp\left[-\left\{R^{-2}+\frac{16\pi^2}{3a^2}(h_j^2+h_jk_j+k_j^2)\right\}^{1/2}|z-z_m|\right.$$
$$\left.-i2\pi(h_j\xi_m+k_j\eta_m)\right] \tag{11}$$

$$U_0=\frac{8\sqrt{2}\,\pi Ze^2}{a^3}R^2 \tag{12}$$

$$\kappa_j=\left[k_0^2\cos^2\theta-\frac{8\pi}{\sqrt{3}\,a}k_0(h_j^2+h_jk_j+k_j^2)^{1/2}\sin\theta\cos\varphi_j\right.$$
$$\left.-\frac{16\pi^2}{3a^2}(h_j^2+h_jk_j+k_j^2)\right]^{1/2} \tag{13}$$

where φ_j is the angle between $\mathbf{K}_0$ and $\mathbf{A}_j$. The length a is related to the interatomic distance in bond, d, through

$$a=\frac{2\sqrt{6}}{3}d.$$

The intensity of diffracted beams can be numerically calculated using Eqs. (6)–(13), as functions of the incident beam energy $(\hbar^2/2m)k_0^2$ and the incident angles, θ and φ_j, if R, W_0, ξ_m, η_m and z_m are fixed. To fix ξ_m, η_m and z_m means to choose an appropriate model for the superstructure observed on the (111) surface of the crystal. Parameter R is related to the peak position in the intensity profile against incident beam energy, since k' and κ' depend on U_0 (Eq. [8]) which is directly connected with R (Eq. [12]). Parameter W_0 concerns the half-width of the peak, since k'' and κ'' are determined by W_0 (Eq. [8]).

IV. A TRIAL MODEL OF THE SUPERSTRUCTURE

As is mentioned in the introduction, the superstructures on the (111) surfaces of diamond-type crystals are observed if and only if the surfaces are annealed after being cleaned. It is plausible to conceive that these reconstructions are caused so that the surface free energy at annealing temperatures is lowered. The lowest temperatures necessary for annealing are 600°C of Si(111) and 200°C of Ge(111),[5] while the temperatures at which plastic deformation (understood to be caused by slip between (111) planes in [1$\bar{1}$0] direction) starts are 600°C and 350°C, respectively.[16] We note that the (111) surfaces of diamond-type crystals have lower surface energy per unit area than the surface of any other orientation, if surface reconstruction is not considered.

In diamond-type crystals, the surface stress is compressive. We assume that the stress may be reduced by introducing a distribution of edge dislocations having Burgers vectors parallel to the surface, with two layers of atoms less on the side toward the surface than on the side toward the interior.[17] Thus, the surface free energy is lowered. The increment of bulk free energy is assumed to be less than the lowering.

In a previous paper,[14] the forces between two parallel dislocations on a common slip plane which is parallel to the surface and located at a distance l beneath it were calculated by neglecting the elastic anisotropy of the crystal. It has been shown that the two dislocations attract each other and are unstable if the dislocation is formed by pure slip, but they are stable at a definite distance $2\sqrt{3}\,l$ if the dislocation is formed by removal of a slice of uniform thickness or two layers of atoms. Of the dislocations of the latter type we can construct a stable network plane if we neglect the interactions between crossed dislocations. We may expect the existence of regular hexagonal networks on a (111) plane located at a distance l beneath the (111) surface of a diamond-type crystal (see Fig. 2). The side length of each hexagon is $2l$ and the unit length of a superstructure formed by these hexagons is $2\sqrt{3}\,l$ as is illustrated in Fig. 2.

Though the limit of validity of the theory of elasticity for this problem is not obvious, we try to explain the superstructure observed on the (111) surface of silicon along this line. When we set $2\sqrt{3}\,l = 7a$, the network plane should lie between the sixth and the seventh atomic layer beneath the (111) surface. The unit mesh of the proposed superstructure is illustrated in Fig. 3. The number of atoms in each atomic layer down to

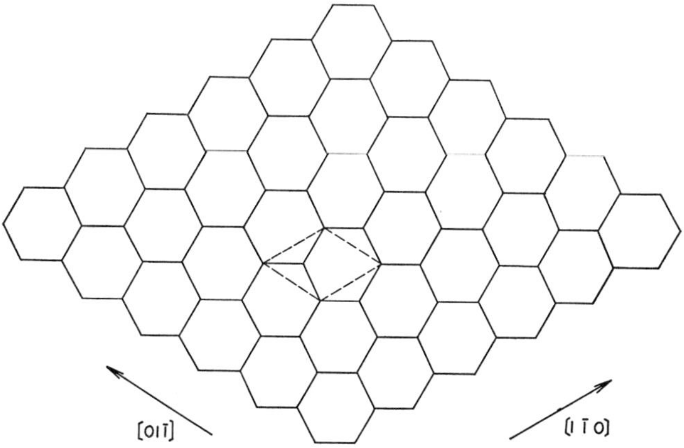

Fig. 2. A superstructure formed by regular hexagonal networks of dislocations. The unit mesh is indicated by dotted lines.

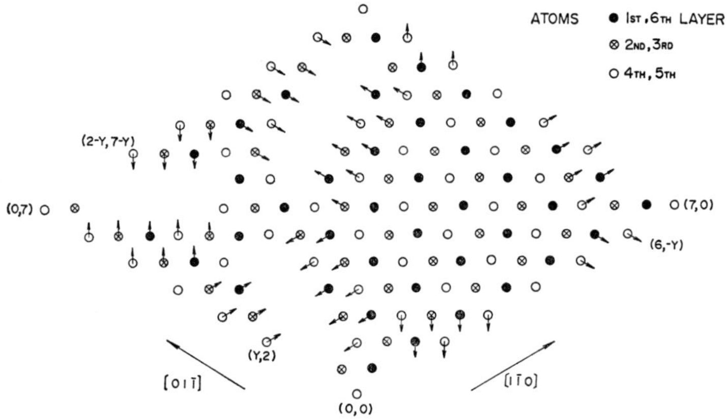

Fig. 3. Atomic arrangement in the unit mesh of the proposed superstructure. Displacements of atoms are indicated by arrows.

the sixth is 40 per unit mesh and below the seventh the bulk structure holds, that is, 49 atoms per unit mesh are in each layer.

In the numerical calculations of the intensity of the diffracted beams, using this model, two parameters, X and Y, are introduced; X adjusts the displacement of atoms along [111] direction and Y and along $\langle 1\bar{1}0\rangle$ directions as follows: inter-layer spacing is adjusted such that

$1/3\,Xd$ for 1st-2nd, 3rd-4th, and 5th-6th,
Xd for 2nd-3rd, 4th-5th, and 6th-7th,
$1/3\,d$(normal) for 7th-8th, etc.
d(normal) for 8th-9th, etc.

The displacements of atoms in the (111) planes are indicated by arrows in Fig. 3. The amount, Y, of the displacement is taken to be same for all the displaced atoms. Coordinates ξ and η of some of the displaced atoms are shown in the figure.

V. EXPERIMENTAL AND CALCULATED RESULTS AND DISCUSSIONS

The intensity profiles observed as functions of the energy of incident beam in the range 30 to 200 eV at $\theta=2.5^\circ$ are shown in Figs. 4(a), 5(a), 6(a), 7(a), and 8(a). The diffracted beams in each figure are arranged in order on the line connecting two adjacent normal-order beams (see Fig. 1). The intensity profiles of the (00) and (0$\bar{1}$) beams against the incident beam energy over 200 eV are shown in Figs. 9 and 10, respectively. Intensity

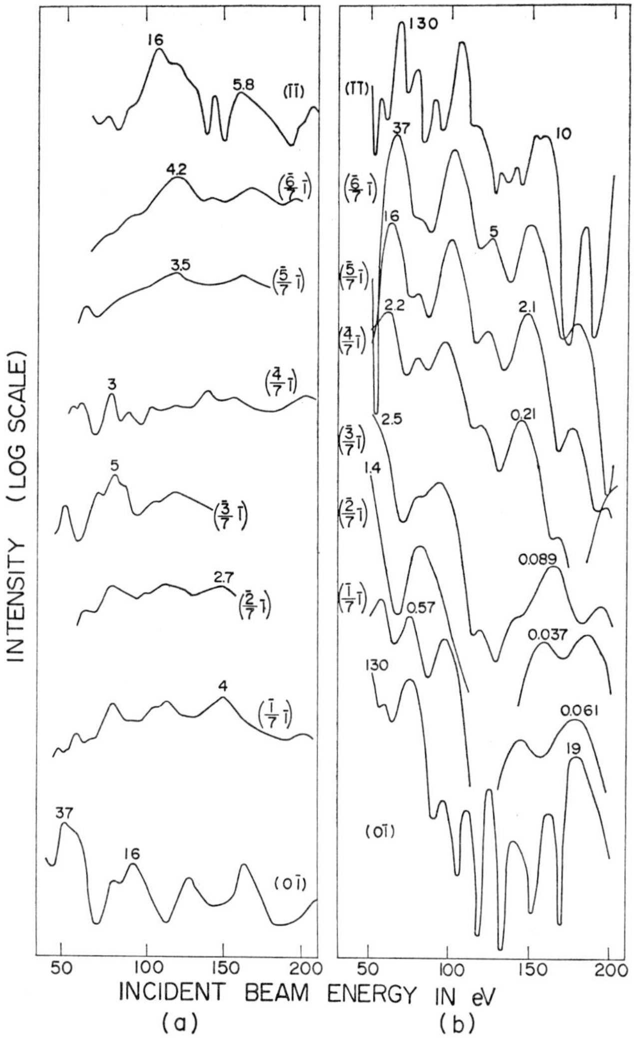

Fig. 4. The observed (a) and calculated (b) results of the intensity profiles of the diffracted beams which are $\left(\frac{\bar{n}}{7}\bar{1}\right)$ with $n=7, 6, 5, \ldots, 0$.

measurement of the fractional-order beams close to the (00) beam was hindered under experimental conditions. The numbers indicated on the profiles are relative intensity in logarithmic scale of arbitrary unit in the order of 10^{-4}, measured at constant target current. True intensities, which should be measured at constant current of incident beam, in a lower energy region, in particular below 100 eV, would be higher than those

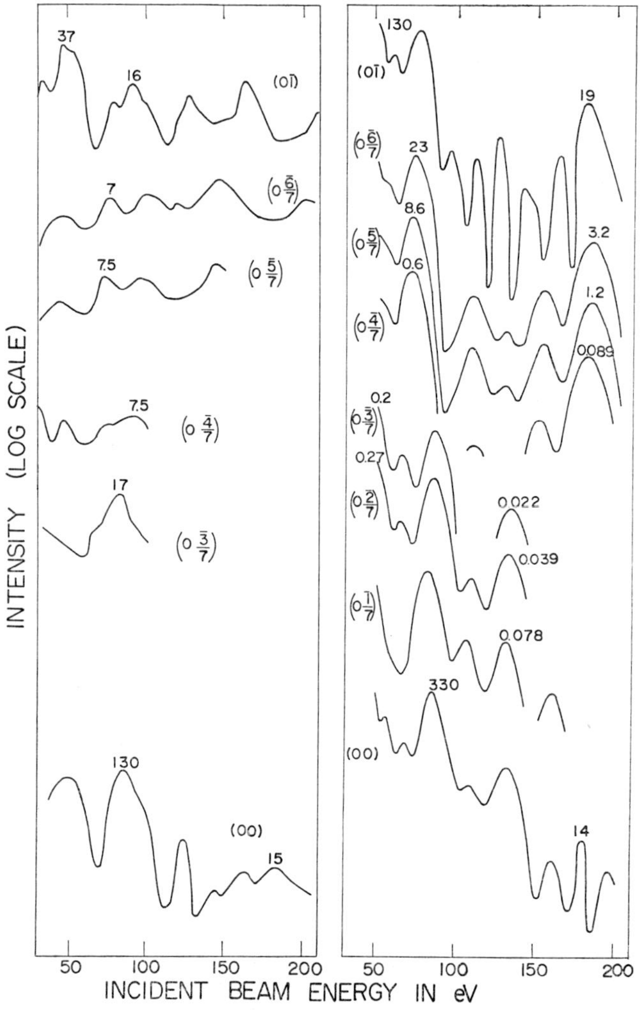

Fig. 5. The observed (a) and calculated (b) results of the intensity profiles of the diffracted beams which are $\left(0\frac{\overline{m}}{7}\right)$ with $m=7, 6, 5, \ldots, 0$.

reported (see section II). The intensities can be compared relative to one another for all diffracted beams, however.

The intensity of the diffracted beams has been calculated along the theory described above, based on the trial model. Parameter R and W_0 are determined by comparison of the calculated intensity profiles of the (00) beam with the observed ones. R is chosen to be 0.21 Å, which is

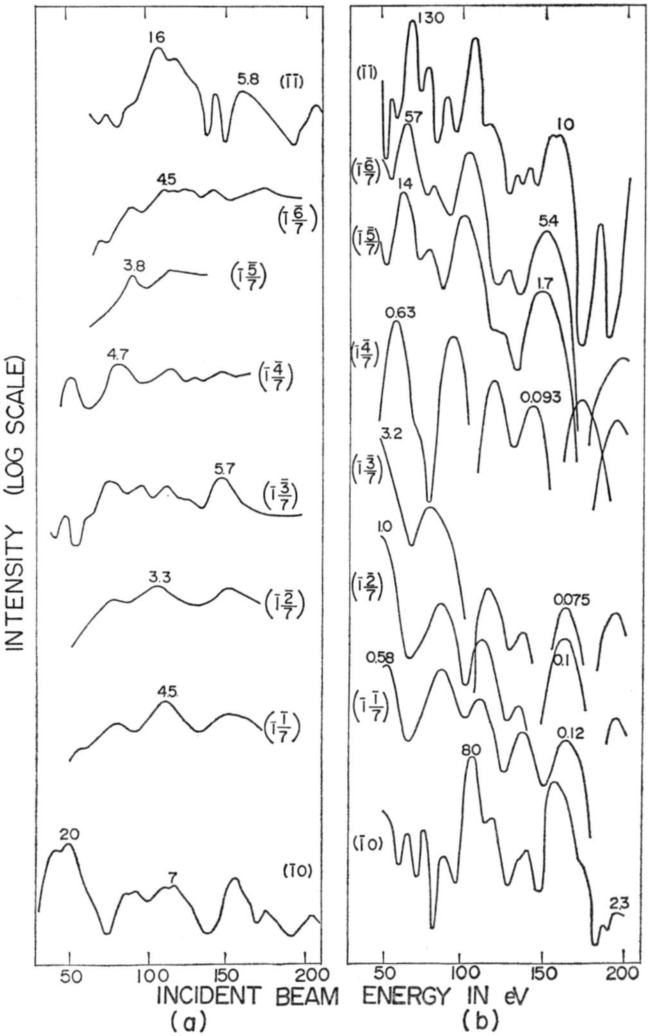

Fig. 6. The observed (a) and calculated (b) results of the intensity profiles of the diffracted beams which are $\left(\bar{1}\frac{\bar{m}}{7}\right)$ with $m=7, 6, 5, \ldots, 0$.

very close to the value derived from the Thomas-Fermi approximation.[13] W_0 is 4.0 eV. If the surface were not reconstructed, the calculated profile of the (00) beam at $\theta=2.6°$ would show peaks only at about 60, 90, 180, 240, and 300 eV in the range from 50 to 350 eV. The observed profile, however, shows in addition peaks at 130 and around 150 eV. The peak at 130 eV may be taken as the (666) reflection for which the structure

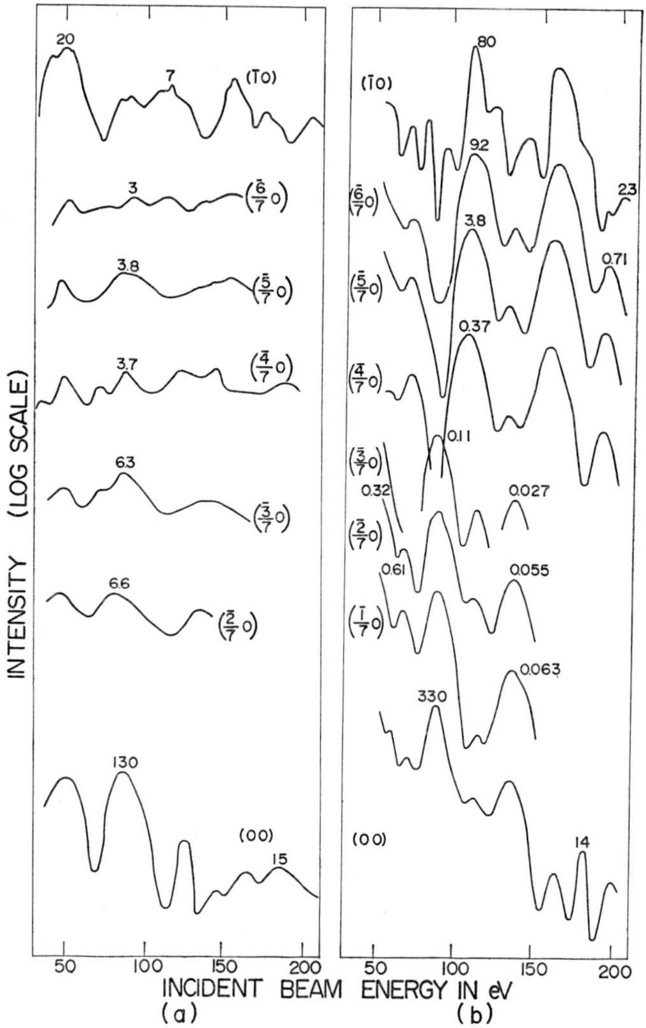

Fig. 7. The observed (a) and calculated (b) results of the intensity profiles of the diffracted beams which are $\left(\frac{\bar{n}}{7}0\right)$ with $n=7, 6, 5, \ldots, 0$.

factor of the diamond lattice vanishes. It is very interesting to compare the observed profile at incident angles around 5° (see Fig. 9) with those of an indium-adsorbed Si(111)[18] and an iodine-adsorbed Si(111),[4] reported by Lander and Morrison, and of a tantalum-adsorbed Si(111).[19] In the latter cases the peaks are only at about 50, 90, 130, 180, and 240 eV. When oxygen is adsorbed on the Si(111) at room temperature, the profile in the

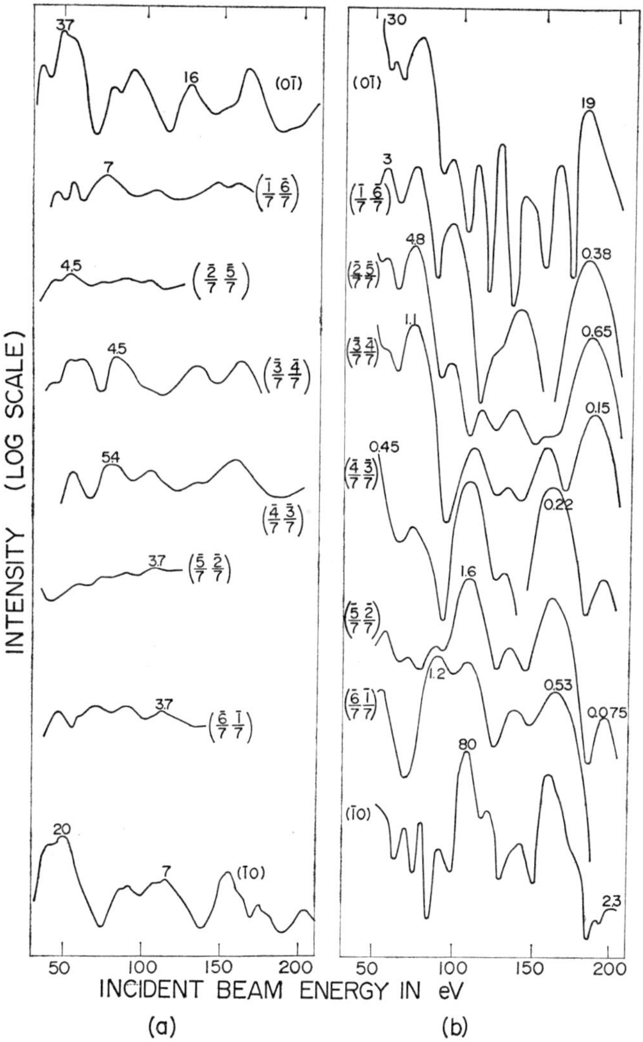

Fig. 8. The observed (a) and calculated (b) results of the intensity profiles of diffracted beams which are $\left(\frac{\bar{n}}{7}\,\frac{\bar{m}}{7}\right)$ with $(nm)=(07), (16), (25), \ldots, (70)$.

range from 100 to 200 eV is considerably changed, except for the peaks at 130 and 180 eV.[20] Thus, we shall proceed with the assumption that the peaks other than 60, 90, 180, 240, and 300 eV are caused by reconstruction in the surface layers.

Preliminary calculations yield these extra peaks provided that inter-layer

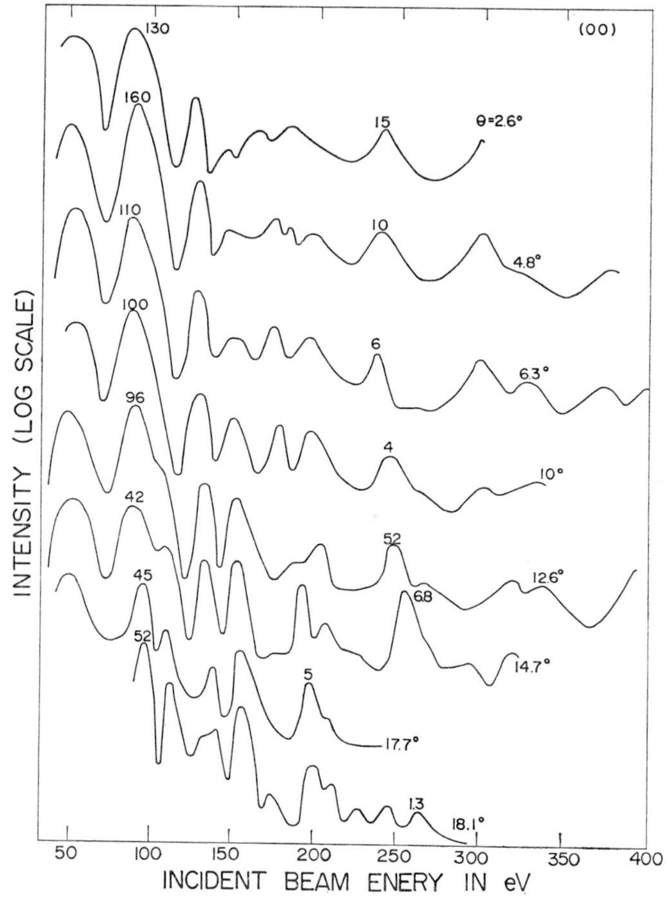

Fig. 9. The intensity profiles of the (00) beam at several incident angles.

spacings are decreased appropriately in the surface layers. In the present model we fix parameter X at 0.82. When we displace the surface atoms in a reasonable range from their normal positions in the (111) plane, the fractional-order beams appear with appreciable intensity, as expected. The larger the parameter Y in our model, the larger the intensity of the fractional-order beams. Y is actually chosen to be 0.22. The calculated results using these values of the parameters are reproduced in Figs. 4(b), 5(b), 6(b), 7(b), and 8(b). Here the incident angle is 2.5° for the (00) beam and zero for all other beams. The intensity of the diffracted beams is relative to that of the incident beam. (The numbers in the figures should be multiplied by 10^{-4}.)

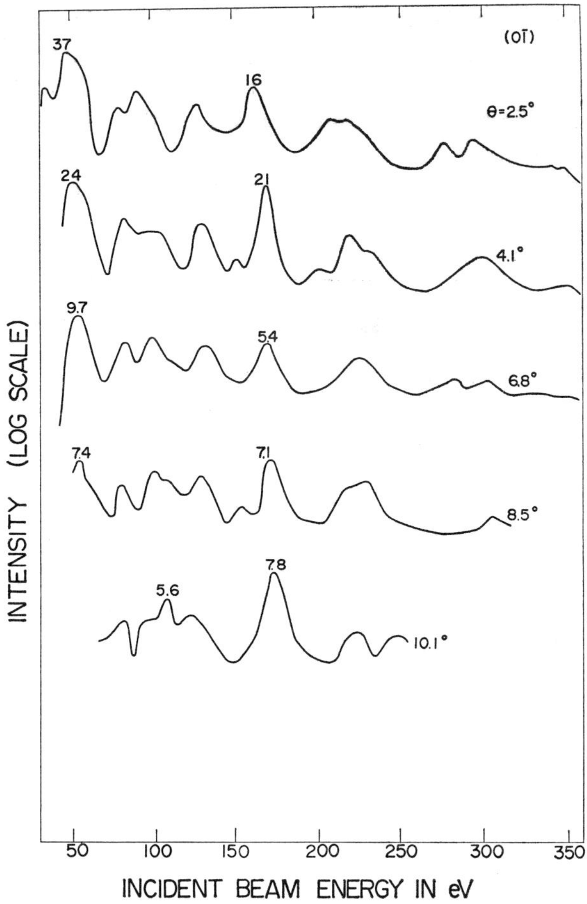

Fig. 10. The intensity profiles of the (0$\bar{1}$) beam at several incident angles.

The calculated intensity profiles of the normal-order beams show fine structures that are absent in the observed ones. These fine structures or splitting of peaks are thought to be caused by the abrupt change in the inter-layer spacing at the seventh atomic layer in our model. The observed intensity of the fractional-order beams is generally some fraction of that of the normal-order beams, while the calculated intensity of some fractionalorder beams is much smaller. This might be caused by the fact that only the atoms along the " dislocation " lines are displaced in the (111) plane, other atoms remaining in normal positions. Another remarkable feature of the calculated results is that the intensity profile of the fractional-order beam in each sequence changes gradually from that of one normal-order beam to that of the other. Such a tendency is seen in some sequences of the observed results. Overall, the use of only two parameters X and Y for describing the atomic arrangement near the " dislocations " may be an over-simplification. An improvement of this treatment has not been attempted, however.

Next we reproduce the observed intensity profiles of (00), $(0\bar{1})$, and

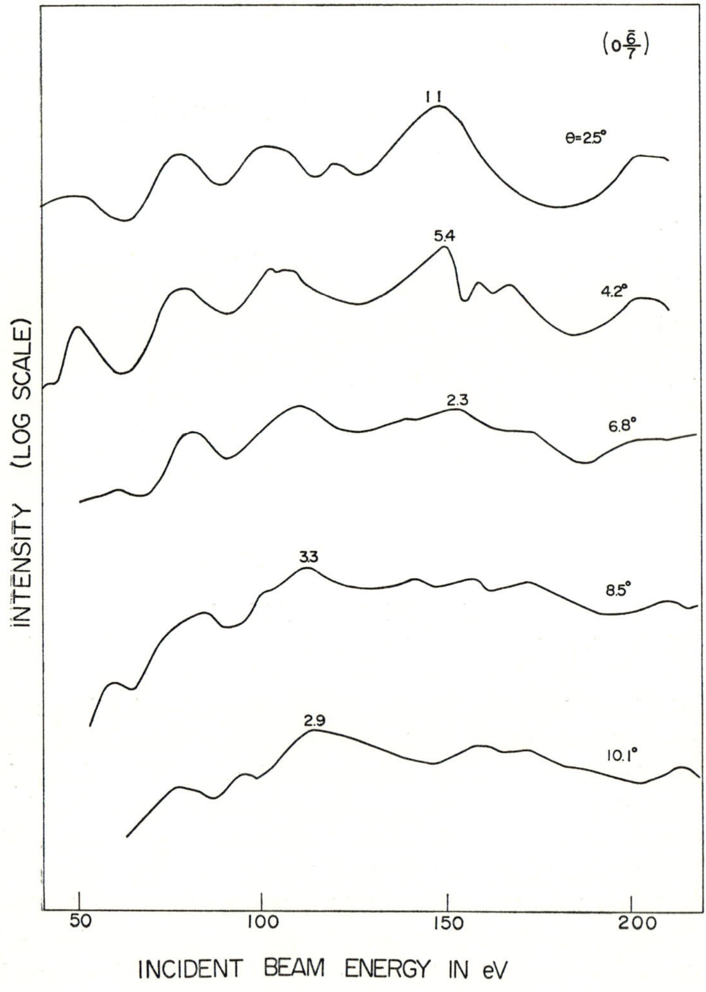

Fig. 11. The intensity profiles of the $\left(0\frac{\bar{6}}{7}\right)$ beam at several incident angles.

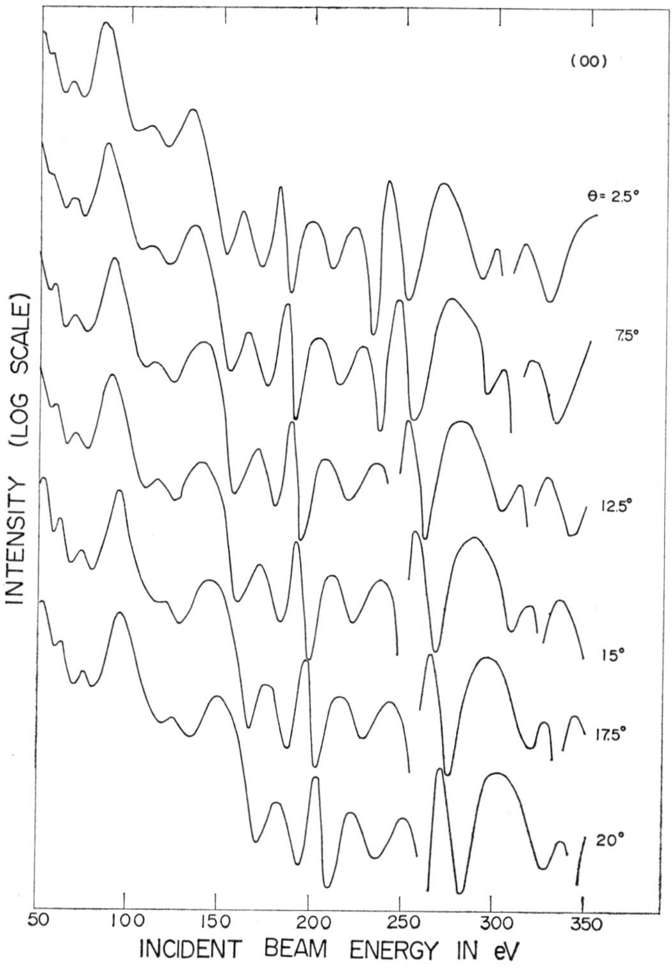

Fig. 12. The calculated intensity profiles of the (00) beam at several incident angles.

$\left(0\frac{\bar{6}}{7}\right)$ beams at several incident angles in Figs. 9, 10, and 11, respectively. The results of the cases of $(0\bar{1})$ and $\left(0\frac{\bar{6}}{7}\right)$ are representative of those of the other normal-order and fractional-order beams, respectively. It appears in the data on the (00) beam that (1) the positions of the main peaks shift to higher energies with increasing θ; (2) the intensity of some peaks is remarkably changed with varying θ; and (3) the splitting of some peaks depend on θ, for example, the peak at 90 eV for $\theta=2.5°$ begins to split with increasing θ over 10°. Similar features are seen in the data on the $(0\bar{1})$ and the $\left(0\frac{\bar{6}}{7}\right)$ beams.

The calculated results of the (00) beam with varying θ are plotted in Fig. 12. The results indicate that all peaks shift to higher energies with increasing θ as is observed, but the intensity and the splitting of the peaks are little changed, in contrast with the observed results. This discrepancy is not due to the model. As was mentioned earlier, different models give different intensity profiles with respect to the intensity and the splitting of peaks. The peaks, however, shift monotonically to higher energies with little change in their shapes when θ is increased. Thus the method of calculating the intensity is thought to be responsible for the discrepancy.

We must note that the theory of diffraction described in section III is based on the assumption that $|\Psi_j| \ll |\Psi_0|$, which requires that the diffracted beams are much weaker than the incident beam and the effect of multiple scattering is not so large. If this is not the case and $|\Psi_j|$ is comparable to $|\Psi_0|$, the comparable terms in the sum of Eq. (5) must not be neglected. Then strong coupling between different Ψ_j will occur and the calculated intensity profiles may be modulated; for instance, the shift, broadening, and splitting of peaks as well as the increase in the intensity of the fractional-order beams relative to that of the normal-order beams. These may partly elucidate the discrepancy between the observed and the calculated results. If these effects are truly not negligible, it is inevitable, though very laborious, to solve Eq. (5) with a more exact method.

In view of these effects the extra peaks or the fine structures as well as the intensity of the fractional-order beams with which we are concerned for the present structural analysis might not directly reflect the reconstruction, although the extra peaks are certainly sensitive to the structural change by adsorption of foreign atoms. If this is the case, the method of calculating the intensity must be refined as indicated above before proceeding with a detailed examination of the superstructure.

Acknowledgments

The authors wish to acknowledge with thanks the able technical assistance of I. Sasaki, the temporary help of Y. Ioki and A. Nakamura, and the glass blowing services of T. Oikawa. Acknowledgment is also made to the computation center in the laboratory for numerical calculations.

REFERENCES

1. J. B. Marsh and H. E. Farnsworth, *Surface Sci.*, **1**, 3 (1964).
2. J. J. Lander and J. Morrison, *Surface Sci.*, **4**, 241 (1966).
3. R. E. Schlier and H. E. Farnsworth, *J. Chem. Phys.*, **30**, 917 (1959).
4. J. J. Lander and J. Morrison, J. *Appl. Phys.*, **34**, 1403 (1963); *J. Chem. Phys.*, **37**, 729 (1962).
5. J. J. Lander, G. W. Gobeli and J. Morrison, *J. Appl. Phys.*, **34**, 2298 (1963).
6. F. Jona, *IBM Res. Develop.*, **9**, 375 (1965).
7. R. L. Jacobson and G. K. Wehner, *J. Appl. Phys.*, **36**, 2674 (1965).
8. W. Peria, cited in ref. 7.
9. J. J. Lander and J. Morrison, *J. Appl. Phys.*, **36**, 3517 (1963); J. J. Lander, in *Progress in Solid State Chemistry*, (H. Reiss, ed.), Vol. 2, p. 26, Pergamon Press, London, 1965.
10. S. Miyake and K. Hayakawa, *J. Phys. Soc. Japan*, **21**, 363 (1966); E. Bauer, *Phys. Rev.*, **123**, 1206 (1961).
11. R. Seiwatz, *Surface Sci.*, **2**, 473 (1964).
12. D. Haneman, *Phys. Rev.*, **121**, 1093 (1961); N. R. Hansen and D. Haneman, *Surface Sci.*, **2**, 566 (1964).
13. K. Hirabayashi and Y. Takeishi, *Surface Sci.*, **4**, 150 (1966).
14. K. Hirabayashi, *J. Phys. Soc. Japan* (to be published).
15. J. J. Lander, J. Morrison and F. Underwald, *Rev. Sci. Instr.*, **33**, 784 (1962).
16. G. L. Pearson, W. E. Read, and W. I. Feldmann, *Acta Met.*, **5**, 181 (1957).
17. C. Herring, in *The Physics of Powder Metallurgy* (W. Kingston, ed.), McGraw-Hill, New York, 1951.
19. Y. Takeishi, I. Sasaki and Y. Ioki, *Surface Sci.*, **4**, 317 (1966).
20. Y. Takeishi and I. Sasaki (to be published).

Part VI

Dislocations

A Light Effect on the Bending Creep of Germanium*

A. Fortini and A. De Bouard

Faculté des Sciences, Université de Caen, France

ABSTRACT

Bending creep experiments were performed on parallelepipedic germanium samples, at 380°C in the presence or absence of the light produced by a well focused 250 W white source. The plastic deformation is observed to be faster in the presence than in the absence of light. The difference is equivalent to an increase of about 32° of the sample temperature which is much larger than the experimental imprecision.

The effect of light on the plastic behavior may be understood as an electronic one. From a fundamental point of view two kinds of effect may be expected: a variation of the shear modulus due to the excess electron-hole density; a decreased of the dislocation energy due to the occupation of the associated acceptor levels.

I. INTRODUCTION

Various photo-mechanical effects have been recently reported in semiconductors. Kuczynski and Hochman[1] found important variations of micro-hardness in illuminated crystals of Ge, Si, InSb, and related[2] this later to the generation of a high density of dislocations near the fracture of a germanium slab when illuminated by a 150 W flood lamp. Photo reduction of the hardness was also reported in antimony by P. P. Kuzmenko *et al.*[3] and in germanium and silicon by V. M. Beilin and Yu. Kh. Vekilov.[4]

On the other hand effects probably related to the same phenomena have been reported in doped germanium. Ablova and Regel[5] observed a reducing of the yield point; Patel *et al.*[6] measured an increment of dislocation velocities in the case of n-doped germanium and a reduction in the p-doped case.

From these experiments it may be thought that in semi-conductors

* The work presented here was supported by the " Direction des Recherches et Moyens d'Essais " (Convention N° 65, 34, 188).

plastic properties depend to a great extent upon the free carrier density. The exact mechanism of such an electronic effect is not as yet known. Present ideas will be briefly discussed below.

In the present paper an experimental set designed to investigate the effect of light on the bending creep curve of a germanium slab at 380°C is described. Quantitative results are given and roughly interpreted in terms of the present phenomenological theories of plastic creep in semiconductors.

II. EXPERIMENTAL METHOD

a) Apparatus

All samples tested were cut as $18 \times 3 \times 0.240$ mm^3 parallelpipeds out of a high purity germanium single crystal ingot the resistivity of which, at room temperature, was about 50 Ω·cm. The largest face of the slab was (111) oriented and the longest edge parallel to a [$\bar{1}$10] axis (Fig. 1).

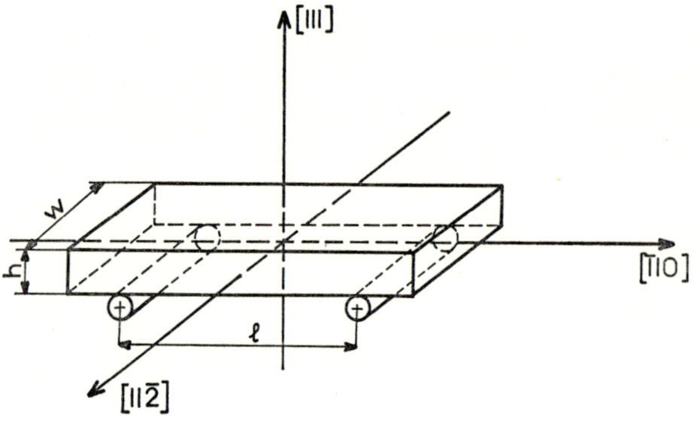

Fig. 1. Orientation of the germanium specimens used in bending creep tests.

The samples which were identical in length and width were given the same thickness by the following metallurgical treatments: grinding of the two large faces with 303 and 305 silica, followed by a 1h-alumina, then a 12h-alumina polishing; chemical polishing, a few seconds, in the usual CP4 solution (without Br_2).

Bending creep tests were performed by using the classical three-points bending apparatus described elsewhere.[7–9] The sample was supported by two self-aligning invar rolls, $l = 15$ mm apart, and the bending moment was impressed to it by a loading force F applied in the center through a third roll (Fig. 2). In order to avoid oxidation the sample was immersed

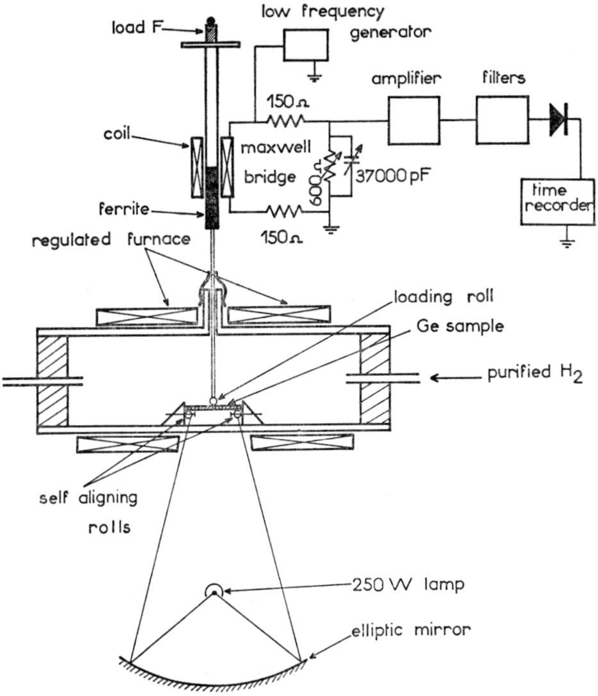

Fig. 2. Schematic representation of the experimental arrangement designed for creep tests on germanium under illumination.

in a well purified hydrogen atmosphere. An external tube furnace allowed the temperature to be raised up to definite value for experiments conducted in the dark, and the temperature to be adjusted to the same value for experiments with light.

Bending tests on illuminated samples were achieved by using a 250 W lamp, the filament of which was well focused on one large face with an elliptic mirror (Fig. 2).

Plastic deformations were measured through the vertical displacement z of the loaded central region of the sample. A cylindrical ferrite magnet followed the translation z and induced self inductance variations in a little coil, inserted in a Maxwell bridge. After selective amplification the displacement z was directly plotted versus time by recording (Fig. 2). During the very first moments of the deformation the latter may be connected with the stress tensor by the elasticity theory. The maximum displacement z is related to F, the sample dimensions W, h (Fig. 1), and the distance l of the rolls

$$z = \frac{3}{2} \frac{Fl}{Wh^2}$$

The maximum macroscopic strain ε, at the surfaces, is proportionnal to z

$$\varepsilon = \frac{6h}{l^2} z$$

The question arises as to the exact relation between ε and z farther in the plastic domain. The simple proportionality with the same coefficient is rather doubtful. Nevertheless, to make matters simpler we shall assume this proportionality below.

b) Results

Typical results obtained with samples tested in the dark at 380°C, and with illuminated samples, in the same conditions, are presented in Fig. 3. Deformations were performed near the limiting stress, the experimental determination of which leads to the adopted value: $F = 218$ g. As shown, the reproducibility of the tests is fairly good. Because of the strong dependence of the shear stress tensor intensity upon the sample thickness h, it may be thought that the residual scatter is due to the error in h which was controlled within about 2%.

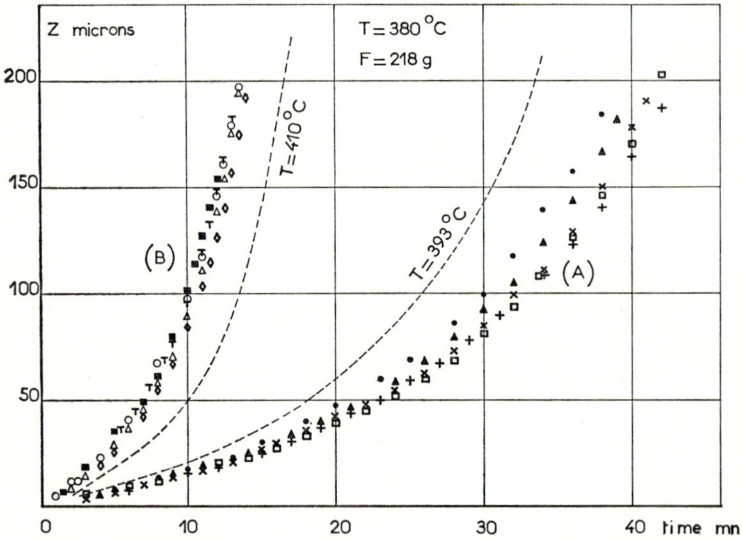

Fig. 3. Creep curves of several germanium samples in the first stage: (A) In obscurity; (B) With illumination on one face. Different symbols relate to different samples. Dashed curves correspond to obscurity tests at higher temperatures.

The general feature of the creep curves is the same as in previous experiments.[7–9] Only the first stage of the usual behavior is present here, for the fracture of the sample occurred generally a little later. Experiments conducted with smaller loads revealed the three classic stages: stage I in which the creep rate increases gradually up to a saturation value; stage II in which this maximum rate is maintained; stage III, in which the creep rate decreases as a result of work hardening.

Observation of dislocations generated by bending was achieved by the usual CP4 etching process. Typical results are shown in the microphotographs in Figs 4, 5 and 6. By bending, a high etch pits density appears on the (111) faces and on the $(11\bar{1})$ planes, which are the equivalent (111) planes nearest the $(11\bar{2})$ faces. The observed directions of etch pits patterns are consistent with dislocation half-loops lying in part along the $[\bar{1}01]$ and $[0\bar{1}1]$ directions and emerging at the (111) face along the [011], [101] directions, as discussed earlier by Kabler[10]

At first sight, illumination has the same effect as an increase of about 32° of the sample temperature, as was checked by the intermediate obscurity creep curves obtained at 393°C and 410°C (Fig. 3). As first shown below, this observation agrees with the simple phenomenological theory. Because of this equivalent thermal effect great care was taken in controlling the sample temperature before and during the tests. As its transverse dimensions were much smaller than the dimension of the furnace and a little smaller than the transverse dimensions of the light spot no transverse

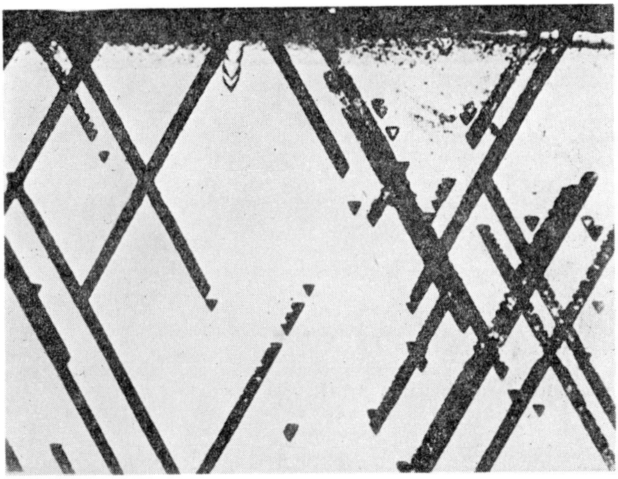

Fig. 4. Etch pits patterns on the (111) face. Magnification ×228.

Fig. 5. Etch pits patterns on the $(11\bar{1})$ plane. Magnification $\times 918$.

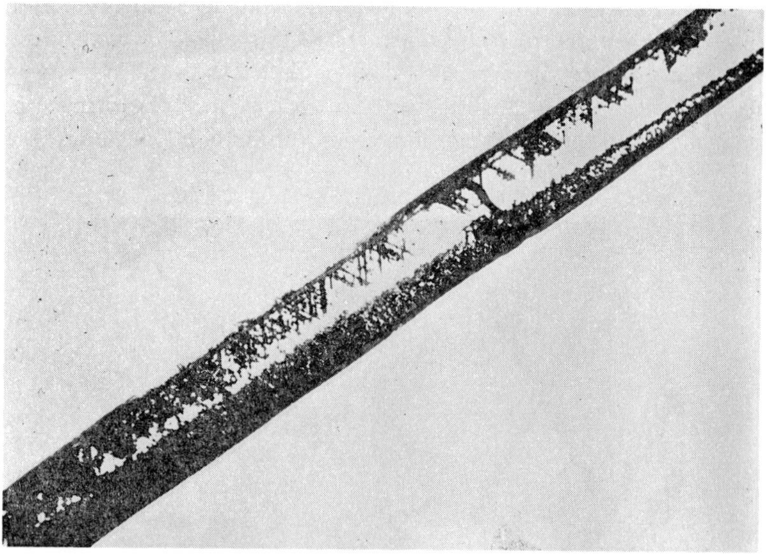

Fig. 6. Distribution of etch pits in the $(11\bar{1})$ plan along the $(\bar{1}10)$ direction. Magnification $\times 147$.

temperature gradient might be expected. Moreover direct differential measurements of the temperatures of the large faces with thermocouples soldered to the germanium revealed temperature differences of at most 1–2°, probably greater than the real ones, owing to direct heating by the light beam of the couple which was soldered to the illuminated face.

III. PHENOMENOLOGICAL INTERPRETATION

The creep of germanium by bending, traction or compression tests has been extensively studied before[11–16]; in particular, it has been attempted[11–13, 15] to determine the analytic form $\varepsilon(t)$ of the creep curve which agrees with experimental results. These investigations, more or less based on qualitative physical arguments, showed that the theoretical interpretation of the observed creep curves depends mainly upon the exact dislocation multiplication mechanism underlying the macroscopic deformation, rather than the breaking-away from the impurities. According to the predominant mechanism which is theorized two different forms are obtained for the law of ε versus time, in stage I: a cubic law proposed by Van Bueren[12] who assumes that the number of sources increases proportionnally with time; an exponential law proposed by Johnston and Gilman[11] who showed that in LiF the number of loop-sources increases exponentially with time, according to the equation

$$dN = 2\delta\, vNdt \qquad (1)$$

where δ is a stress dependent parameter and v the dislocation velocity.

Combining this with the basic equation for the macroscopic strain-rate

$$\frac{d\varepsilon}{dt} = Nbv \qquad (2)$$

(b=Burger's vector) one can deduce that

$$dN = \frac{2\delta}{b} d\varepsilon \qquad \text{or} \qquad N = \frac{2\delta}{b}\varepsilon$$

and

$$\frac{d\varepsilon}{dt} = 2\delta v\varepsilon \qquad (3)$$

In fact, Haasen[16, 17] pointed out that the applied shear stress σ must be replaced by the " effective shear stress " σ_e which takes into account the internal shear stress resulting from the elastic interaction between dislocations. The latter can be shown to be proportionnal to $N^{1/2}$ so that the effective shear stress can be written

$$\sigma_e = \sigma - AN^{1/2}$$

With this improvement a good agreement between theory and experiment was found in germanium, provided that reasonable stress dependences[13, 16] of δ and v were adopted.

In stage I, to which we shall restrict ourselves hereafter, the dislocation interaction may be neglected, so that in Eqs. (1) and (2) δ and v are not N dependant and Eq. (3) is valid.

The stress and temperature dependence of the dislocation velocity may be described by relation[16]

$$v = v_0 \exp\left(-\frac{U}{kT}\right)$$

with

$$v_0 \sim \left(\frac{\sigma}{\sigma_0}\right)^m$$

U is an activation energy associated vith the dislocation movement. Exact knowledge of its microscopic meaning is not as yet available in spite of the extensive work which has been devoted to that question.[18–27] Nevertheless in highly pure semi-conductor materials such as intrinsic germanium, in which the density of point defects is low, it is extremely probable that the dislocation movement is limited by the crystal matrix, i.e. by the breaking of the covalent bonds. In such materials the consideration of the Peierls stress, which has a rather high value, suffices to explain many features of the plastic behavior. Recent measurements of dislocation mobility in semi-conductors[10, 28] are in good agreement with a model of dislocation movement based on kinks nucleation and migration. The experimental value of U is near the theoretical estimate of the kinks pair formation energy.[27, 28] The latter turns out to be a function of the Peierls stress and the line tension. In germanium

$$U \sim 1.6\ \text{eV}$$

Eqs. (3) and (4) now yield the creep equation in stage I

$$\varepsilon = \varepsilon_0 \exp\left(2\delta\, v_0\, e^{-U/kT} t\right) \tag{5}$$

In Figs. 6 and 7 a log-log and semi-log plot of the measured creep curves (Fig. 3) are represented. It can be observed that beyond the first 10 or 20 μ the exponential law is in much better agreement with the experimental data than the t^3 law. Since the temperature and applied stress have the same value in both the darkness and the illumination case the effect of light may be interpreted, in the simplest way, as a variation ΔU of the activation energy U. As far as definite slopes can be associated with the

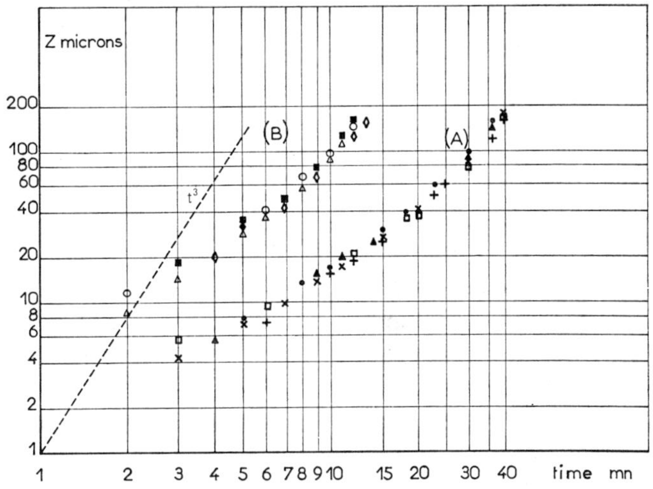

Fig. 7. Log-log plot of the measured creep curves of Fig. 3 and comparison with the t^3 law.

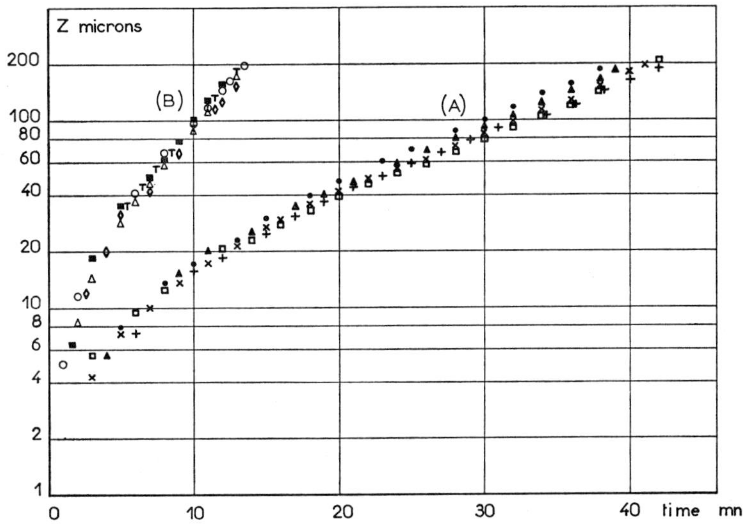

Fig. 8. Semi-log plot of the measured creep curves of Fig. 3.

straight parts of the bundles (A) and (B), Fig. 7, a value of ΔU can be roughly deduced by

$$\Delta U = -kT \log \alpha$$

where α is the ratio of the illumination (B) over the darkness (A) slope

$$\alpha = \frac{\Delta \log \varepsilon_B / \Delta t}{\Delta \log \varepsilon_A / \Delta t} = 2.6 \pm 0.7$$

which yields

$$\Delta U = (0.055 \pm 0.016) \text{ eV}$$

The error to be taken into account here results from the scatter of the creep curves.

It should be noticed further, in Eq. (5), that a decreased value of U is equivalent to an increased temperature T, as mentioned above from an experimental point of view. For a relatively small variation ΔU, the equivalent temperature effect is given by:

$$\frac{\Delta T}{T} = -\frac{\Delta U}{U} = (3.5 \pm 1) \times 10^{-2}$$

or

$$\Delta T \simeq (23 \pm 6.5)\ ^\circ\text{C}$$

in the present experimental conditions. This data agrees approximately with the measured value of 32° because of the roughness of the adopted value for U.

IV. FUNDAMENTAL BACKGROUND AND DISCUSSION

Present results should be compared with previous micro-hardness tests on illuminated semi-conductors and plastic deformations performed on doped germanium. As suggested before[4,5,16] one may expect a direct effect of free electrons on the plastic behavior. The free carrier density influences the Peierls energy (or Peierls stress) on which directly depends the activation energy U of a kinks pair nucleation. A theoretical approach to such a situation appears extremely difficult, for an electronic effect on the dislocation energy would be located mainly in the core. Existing estimates of the core energy are rather rough and the validity of the elastic approximation which is more or less used in that region is doubtful.

Although we do not have detailed knowledge of the electronic effect on dislocation energy, it may be very helpful to get a qualitative insight into the possible mechanisms. The effect of free carriers, mainly the electrons, on the shear modulus c_{44} of germanium has been theoretically dealt with. The Keyes[31] and Bir and Tursunov[32] approach remains valid in the case of optically injected electron-hole pairs because, owing to the very long electron-hole recombination time compared with the energy relaxation time, between the carriers and the lattice, electrons and holes may be separately considered in thermodynamical equilibrium with the lattice. Recently a modulation of the shear stress of germanium by light has been observed.[33] As pointed out by Haasen[16] the core energy of a dislocation might be appreciably influenced, through the Peierls potential and the line energy, by a free-carrier induced c_{44} variation.

Another, more probable explanation of the light effect may be found in the occupation by electrons of the acceptor levels[34] associated with the dangling bonds of the edge-dislocations. Kuczinski[1] underlined the strong dependence of the energy of a dislocation in germanium upon the electronic contribution, as first analyzed by Read[35] who carried out a calculation of the electrostatic contribution to the energy of a dislocation in the case of impurities. This point of view was reconsidered by Haasen[16] as a possible interpretation of the effect of impurities on dislocation mobility. An analogous work[36] is possible in the case of optically generated electron-hole pairs and one may expect, owing to the electrostatic compensation of electrons and holes, that the occupation rate of the acceptor levels becomes important, at 650°C, if the pair density is near the value of a few 10^{17} cm^{-3} for which the quasi-Fermi-level of electrons reaches the acceptor site energy. It would be very interesting to know if such an excess density exists very near the illuminated surface in our experiments. By conductivity modulation a mean density of about 2×10^{15} cm^{-3} was measured, but the superficial value is probably much higher. Superficial density measurements could be achieved through radio-frequency reflectivity. It would also be desirable to determine the relative efficiency of the different wavelengths of the light spectrum.

It should be added that the present type of experiment is more convenient than deformation tests on doped materials, in which the effect of free carriers is never well separated, because of the simultaneous hardening action of the impurity atoms.

Acknowledgments

The authors express their thanks to Prof. P. Aigrain and Prof. J. Friedel for their suggestions, to Prof. Thureau for his technical advices and to the Research and Development Department of the Radiotechnique, Caen, whose technical assistance was much appreciated. The single crystal ingot of germanium was supplied by this firm, which also carried out the orientation and first cutting out of the samples.

REFERENCES

1. G. C. Kuczynski and R. F. Hochman, *Phys. Rev.*, **108**, 946 (1957).
2. G. C. Kuczynski and R. F. Hochman, *J. Appl. Phys.*, **30**, 267 (1959).
3. P. P. Kuz'menko, N. N. Novikov and N. Ya. Gorid'ko, *Sov. Phys. Sol. State*, **4**, 1950 (1963).
4. V. M. Beilin and Yu. Kh. Vekilov, *Sov. Phys. Sol. State*, **5**, 1727 (1964).
5. M. S. Ablova and A. R. Regel, *Sov. Phys. Sol. State*, **4**, 775 (1962).
6. J. R. Patel and A. R. Chauduri, *J. Appl. Phys.*, to be published.
7. C. J. Gallagher, *Phys. Rev.*, **88**, 721 (1952).
8. H. G. Van Bueren, *Physica*, **24**, 831 (1958).
9. P. Penning and G. De Wind, *Physica*, **25**, 756 (1959).
10. M. N. Kabler, *Phys. Rev.*, **131**, 54 (1963).
11. W. G. Johnston and J. J. Gilman, *J. Appl. Phys.*, **30**, 129 (1959).
12. H. G. Van Bueren, *Physica*, **25**, 775 (1959).
13. E. Peissker, P. Haasen and H. Alexander, *Phil. Mag.*, **7**, 1279 (1962).
14. V. G. Govorkov, V. L. Indenbom, V. S. Papkov and V. R. Regel, *Sov. Phys. Sol. State*, **6**, 802 (1964).
15. J. J. Gilman, *J. Appl. Phys.*, **36**, 2772 (1965).
16. P. Haasen, " Colloque International sur les défauts dans les solides " Toulouse, Mars 1966, *J. Phys. Suppl.*, to n° 7–8, **27**, 30 (1966).
17. P. Haasen, *Z. Phy.*, **167**, 461 (1962).
18. J. Lothe and J. P. Hirth, *Phys. Rev.*, **115**, 543 (1959).
19. A. D. Brailsford, *Phys, Rev.*, **122**, 778 (1961).
20. P. D. Southgate and A. E. Attard, *J. Appl. Phys.*, **34**, 855 (1963).
21. V. Celli, M. Kabler, T. Ninomiya and R. Thomson, *Phys. Rev.*, **131**, 58 (1963).
22. T. Suzuki and Ch, Flbaum, *J. Appl. Phys.*, **35**, 1539 (1964).
23. T. Ninomiya, R. Thomson and F. Garcia-Moliner, *J. Appl. Phys.*, **35**, 3607 (1964).
24. A. D. Brailsford, *Phys. Rev.*, **137**, A 1562 (1964).
25. W. T. Sanders, *J, Appl. Phys.*, **36**, 2822 (1965).
26. J. J. Gilman, *J. Appl. Phys.*, **36**, 3195 (1965).
27. G. Alefeld, *J, Appl. Phys.*, **36**, 2642 (1965).
28. A. R. Chauduri, J. R. Patel and L. G. Rubin, *J, Appl. Phys.*, **33**, 2736 (1962).
29. H. Suzuki, *J. Phys. Soc. Jap.*, **18**, suppl. I, 182 (1963).
30. R. Labusch, *Phys. St. Sol.*, **10**, 645 (1965).
31. R. W. Keyes, *I.B.M. Journal*, 266 (Oct. 1961).
32. G. L. Bir and An Tursunov, *Sov. Phys. Sol. State*, **4**, 1925 (1963).
33. A. Fortini and J. Hober, *J. Phys, Chem. Sol.*, **28**, 1099 (1967).
34. G. L. Pearson, W. T. Read Jr., F. J. Morin, *Phys. Rev.*, **93**, 666 (1953).
35. W. T. Read, *Phil. Mag.*, **45**, 775 (1954).
36. A. Fortini, to be published.

Yield Point and Dislocation Mobility in Silicon

H. SIETHOFF AND P. HAASEN

Institut für Metallphysik, University of Göttingen, Germany

ABSTRACT

The dynamic yield of pure and doped silicon single crystals is measured and compared with a dislocation theory. Dislocations move viscuously in silicon with an activation energy of 2.3 eV. In doped silicon deformation proceeds by Lüders bands as is shown by etching. This is to be taken into account if yield behavior is analyzed.

I. INTRODUCTION

The plastic deformation of semiconductors with the diamond cubic structure (DS) has recently been the object of intense study. For a summary of results see refs. 1 and 2. While results on Ge are relatively complete data on silicon are still scarce. Goworkow[3] has deformed some Si single crystals of special orientations in compression at a constant strain rate between 700° and 950°C. The data have not been evaluated with respect to activation energy etc. Reppich *et al.*[4] investigated the creep behavior of silicon single crystals of two different grown-in dislocation densities between 750° and 1,025°C. These authors compared their results with a dislocation theory of the initial stage of deformation of the DS.[5,6] This theory covers both deformation at constant strain rate (dynamic experiment) and under constant stress (static experiment). In the dynamic case it explains the characteristics of a pronounced yield point and leads to the following expression for the lower yield stress τ_{ly} as a function of temperature T and strain rate $\dot{\varepsilon}$

$$\tau_{ly} = C_{ly}\dot{\varepsilon}^{1/(2+m)} \exp\left[\frac{U}{(2+m)kT}\right] \tag{1}$$

Here U is the activation energy of the dislocation velocity which has been observed refs. 7 and 8 to depend on stress and temperature as

$$v = B\,\tau^{m} \exp\left(-\frac{U}{kT}\right) \tag{2}$$

The upper yield stress shows a dependence on T and $\dot{\varepsilon}$ similar to Eq. (1) except for the factor C_{uy} being different from C_{ly}. While C_{ly} contains besides the parameters B and m [Eq. (2)] only known constants, C_{uy} depends also on grown-in dislocation density and somewhat on τ_{uy}, see ref. 1.

In this work we attempt to test critically the theory leading to Eq. (1) in dynamic experiments on pure and doped Si. It is already known in Ge from direct velocity measurements as well as from dynamic tests[9] that n-doping Ge in the extrinsic range decreased τ_{uy} and U while m was not changed. p-doping in the extrinsic range had the opposite effect. Patel *et al.*[9] did a few experiments also on Si doped with 10^{19} B or As/cm^3 and found no effect on τ_{uy}. Direct velocity measurements on doped Si were not done previously because of etching difficulties. Recently work in Russia[10, 11] has become known, in which doped Si single crystals of ⟨111⟩ orientation were deformed dynamically. These authors plot τ_{uy} as a function of T, $\dot{\varepsilon}$, and concentration of doping (B, Al, Sb, P, As) between 10^{14} and 10^{20}/cm^3. In this work a strong effect of doping on τ_{uy} was found in the extrinsic range of Si contrary to Patel *et al.*[9] In the following experimental reasons for such inconsistencies will be demonstrated on phosphorus doped silicon.

II. EXPERIMENTAL RESULTS AND DISCUSSION

The crystals to be deformed in compression between 750° and 1,100°C all had the same orientation ⟨123⟩ favorable for single slip. Dislocation

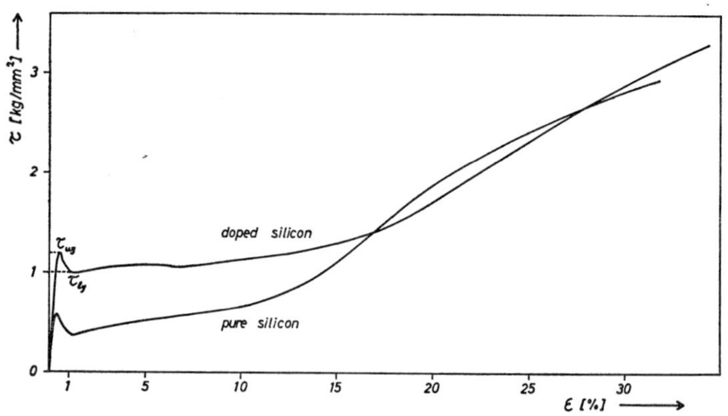

Fig. 1. Shear stress (τ) vs. shear strain (ε) curves of pure and doped ($n=10^{20}$ phosphorus/cm^3) silicon deformed at $T=1{,}000$°C at a strain rate $\dot{\varepsilon}=2.39\times10^{-4}$ sec^{-1}.

etch pits were observed on a {111} side surface of the crystals. As has been pointed out to us by Dr. Theiss (AEG Frankfurt) dislocation etch pits can be produced also in heavily doped silicon by prolonged action of the Sirtl reagent.[12] The "pure" crystals had resistivities (p-type) of 1,200 Ω·cm, 200 Ω·cm, and 0.04 Ω·cm (4×10^{17} phosphorus/cm³), respectively. They all showed the same mechanical behavior. The doped crystals contained 10^{20} atoms of phosphorus per cm³ (0.001 Ω·cm).

Typical shear-stress shear-strain curves are shown in Fig. 1. Pure and doped silicon differ in their work hardening behavior in a similar fashion as do fcc metals and solid solutions. In the following only the upper and lower yield stresses are discussed further, as a function of T and $\dot{\varepsilon}$ and in relation to Eq. (1).

a) Yield stresses of "pure" silicon

The pure Si crystals contained about 10^4 grown-in dislocations/cm². Figures 2 and 3 show τ_{uy} and τ_{ly} as a function of T and $\dot{\varepsilon}$. The results

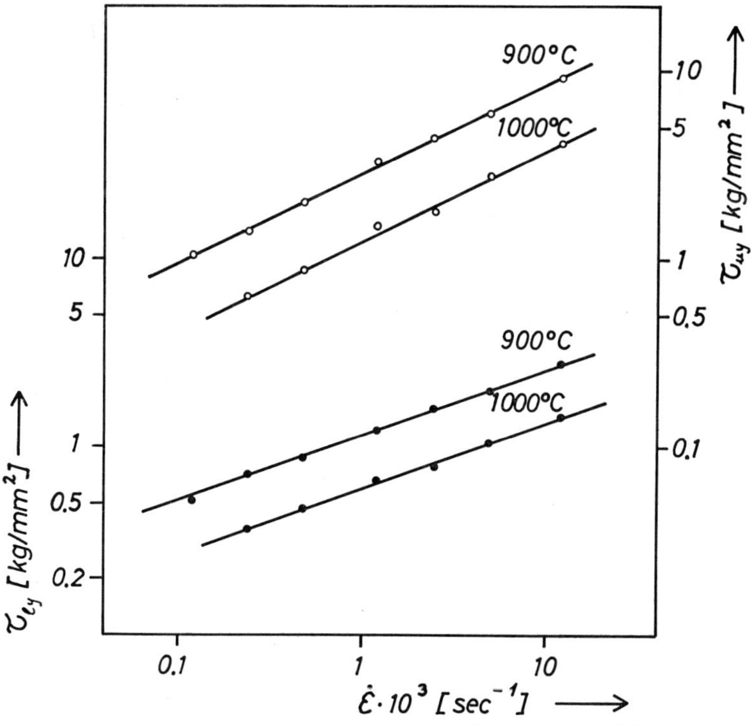

Fig. 2. Pure silicon. Lower yield stress τ_{ly} (●) and upper yield stress τ_{uy} (○) versus strain rate $\dot{\varepsilon}$ at two different temperatures.

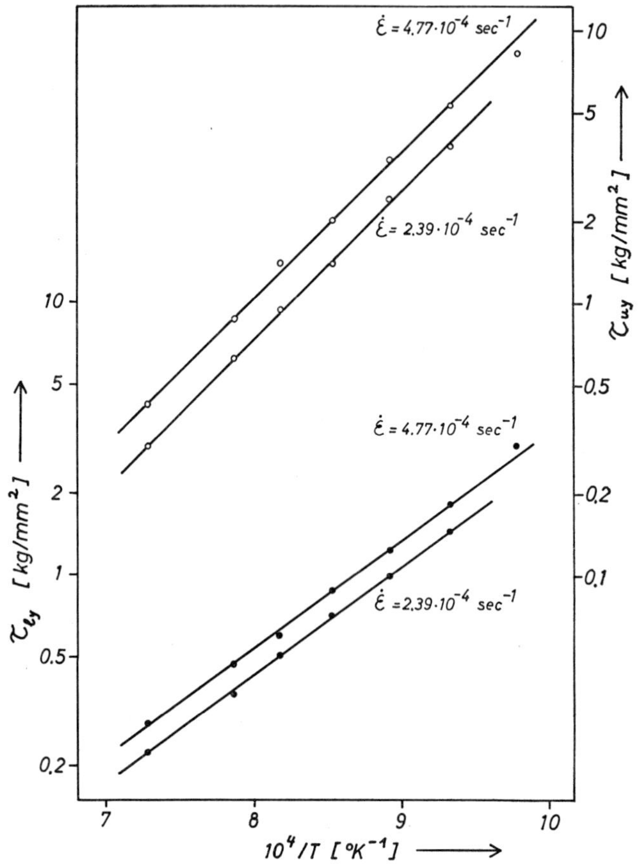

Fig. 3. Pure silicon. Lower yield stress τ_{ly} (●) and upper yield stress τ_{uy} (○) versus inverse temperature at different strain rates.

fit Eq. (1) very well. The values of $(m+2)$ and U obtained from these curves using Eq. (1) are shown in Table I. Also given in this table are the results of Reppich *et al.*[4] and Mildwidskii *et al.*[11] The agreement among different investigators is satisfactory. By means of the theory [Eqs (1) and (2)] U in Table I can be identified with the measured activation energy of the dislocation velocity. Such measurements on silicon have given the following results: $U = 2.17$ eV (ref. 7) and $U = (2.4 \pm 0.2)$ eV (ref. 8) in excellent agreement with our data. The thermally activated process underlying dislocation motion in Si most probably is the formation of kinks in the Peierls potential. The energy of formation of a double kink has been calculated by Suzuki[13] to 2.30 eV and by Labusch[14] to 2.14 eV

TABLE I.

	$m_{ly}+2$	$m_{uy}+2$	U_{ly}(eV)	U_{uy}(eV)	C_{ly}(kg/mm²)*
This work	2.9±0.1	2.1±0.2	2.3±0.1	2.3±0.2	$(4.5\pm0.5)\times10^{-3}$
Reppich *et al.*[4]	3.0±0.01	—	2.4±0.05	—	$(2.8\pm0.5)\times10^{-3}$
Mildwidskii *et al.*[11]	—	1.82±0.05	—	2.1±0.06	—

* If $\dot{\varepsilon}$ is measured in units of sec^{-1}.

for silicon. These values agree again with experiments. Table I shows that m at the lower yield point, m_{ly}, and in stationary creep[4] is about one. The direct velocity measurements of Chaudhuri *et al.*[7] give $m=1.4$, while Suzuki *et al.*[8] find $m=1.5$ at 800°C. There are experimental reasons for suspecting that the latter values are too high.[15] Because of the stress dependence of C_{uy} it is difficult to discuss m_{uy}. U_{uy} should not be affected by this difficulty and is found indeed to agree with U_{ly}. Nearly the same value of C_{ly} is measured in static and dynamic experiments. This again supports our theory.[1, 2]

b) *Yield stresses of heavily doped silicon*

The crystals doped with 10^{20} phosphorus/cm³ were dislocation-free. Figures 4 and 5 show τ_{uy} and τ_{ly} as a function of T and $\dot{\varepsilon}$. The results on pure Si described above are also indicated in these plots. The most striking feature of Fig. 4 is that the yield stresses, in particular τ_{ly}, become independent of $\dot{\varepsilon}$ at small strain rates while there is still a finite temperature dependence. This discrepancy leads one to suspect that $\dot{\varepsilon}$ is not well defined in this range because deformation is not homogeneous any more. Etch pit studies along the stress strain curve show indeed that the beginning of deformation in this range proceeds by the spreading of Lüders bands. Figure 6 shows etched crystals at τ_{uy} (a), and and at τ_{ly} (b). Lüders band propagation has also been found in deformation of pure Ge which had a small $N_0<5{,}000$ cm^{-2}.[16] The Lüders bands are, however, much sharper in doped Si, and contrary to pure Ge no deformation occurs outside the band. The deformation is still inhomogeneous at τ_{ly} in doped Si, while it is already homogeneous in low N_0 pure Ge. Therefore in the following we will not further discuss data taken at $\dot{\varepsilon}<10^{-3}$ sec^{-1} in connection with our theory, Eq. (1), which assumes homogeneous strain. In the range of higher $\dot{\varepsilon}$ the charateristic kink bands (Fig. 6) formed in the wake of the sharp Lüders band front have not been observed. We expect that deformation is more homogeneous at high $\dot{\varepsilon}$ in the doped crystals and certainly over the whole $\dot{\varepsilon}$ range in the

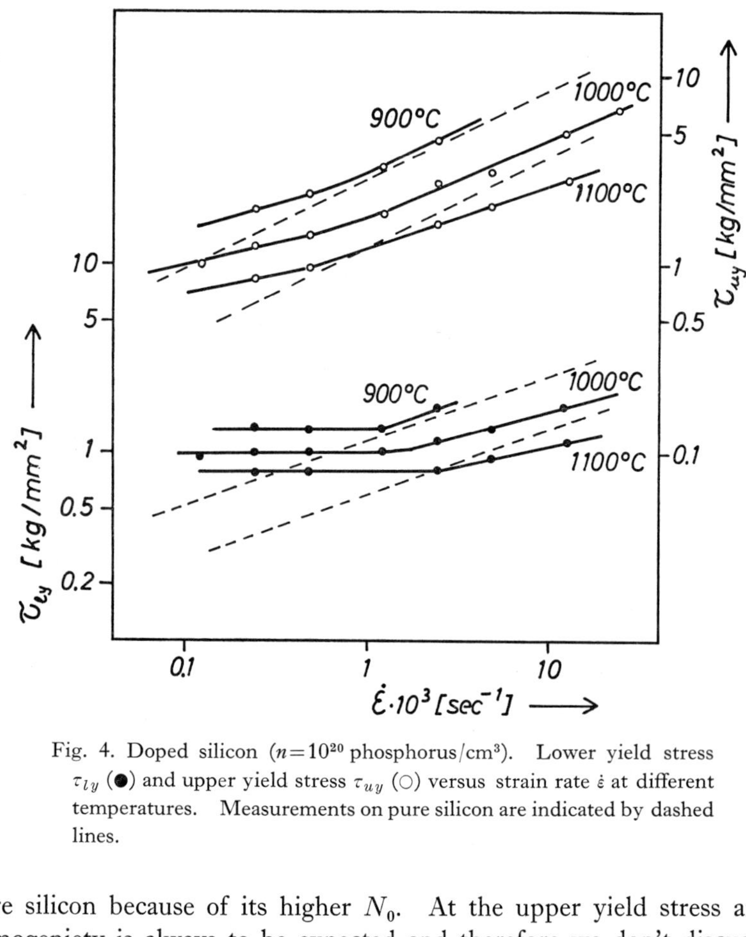

Fig. 4. Doped silicon ($n=10^{20}$ phosphorus/cm³). Lower yield stress τ_{ly} (●) and upper yield stress τ_{uy} (○) versus strain rate $\dot{\varepsilon}$ at different temperatures. Measurements on pure silicon are indicated by dashed lines.

pure silicon because of its higher N_0. At the upper yield stress an inhomogeniety is always to be expected and therefore we don't discuss τ_{uy} further even at $\dot{\varepsilon}>10^{-3}$ see^{-1} for the doped crystals. Patel *et al.*[9] and Mildwidskii[10, 11] do, however, describe the mechanical effect of doping using the upper yield stress. This might explain some inconsistencies pointed out above.

The high $\dot{\varepsilon}$ data on τ_{ly} in Fig. 4 of the doped crystals seem to approach the curves for pure crystals from above. We therefore assume that the n values of doped crystals approach those of the pure material for the case of homogeneous deformation. This is in agreement with direct velocity measurements on Ge[9] which show m to be unchanged by doping. Using $(2+m)=2.9$ also for doped Si one obtains from τ_{ly} (T) for the highest $\dot{\varepsilon}$ (Fig. 5) an activation energy $U=1.74$ eV. If this value has any meaning

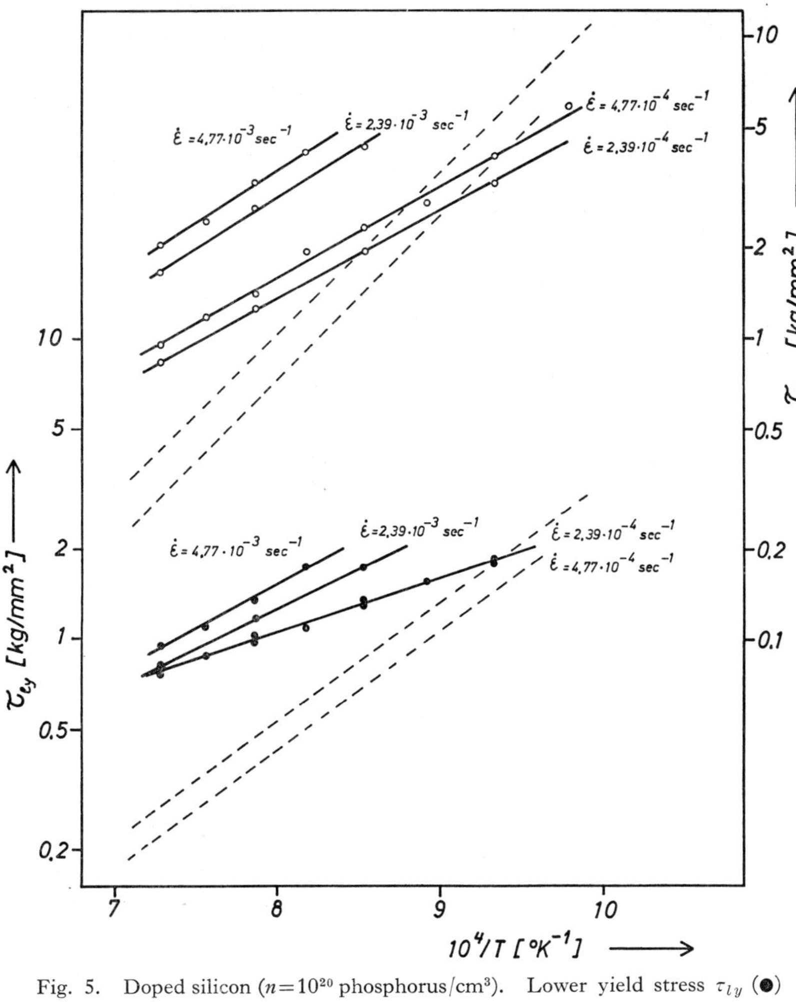

Fig. 5. Doped silicon ($n=10^{20}$ phosphorus/cm^3). Lower yield stress τ_{ly} (●) and upper yield stress τ_{uy} (○) versus inverse temperature at different strain rates. Measurements on pure silicon are indicated by dashed lines.

at all because of the possibly still not homogeneous deformation, it is lower than the activation energy for pure silicon. A decrease in U with n-doping is in agreement with the published data on dislocation velocities in Ge[9] and, with all the reservations discussed above, with measurements of τ_{uy} (Si)[10, 11] and τ_{uy} (Ge).[9] It would be extremely useful to measure dislocation velocities directly on doped silicon using the etching method described above.

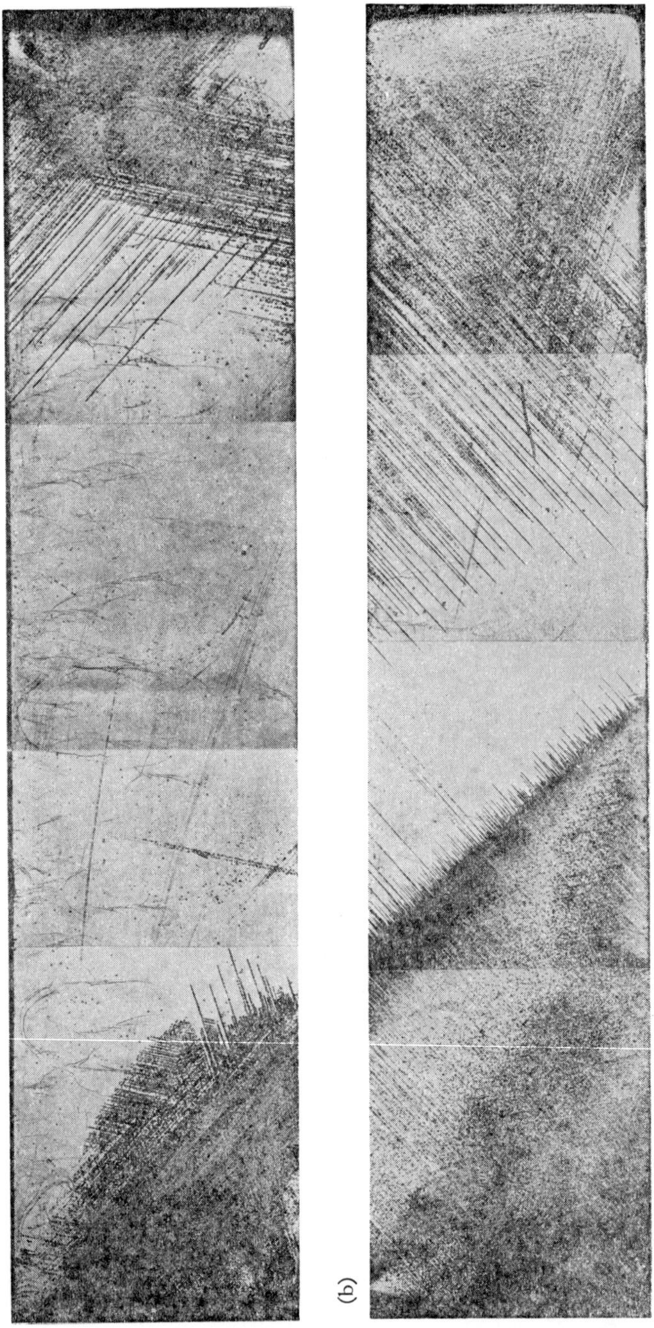

(a) (b)

Fig. 6. Dislocation distribution by etch pits on two crystals of doped silicon ($n=10^{20}$ phosphorus/cm^3) deformed at $T=1{,}000°$C and at $\dot{\varepsilon}=4.77\times10^{-4}$ sec^{-1}
(a) deformed to the upper yield stress; (b) deformed to the lower yield stress.

Acknowledgements

We wish to thank the AEG-Forschungsinstitut Frankfurt for some of the materials and Deutsche Forschungsgemeinschaft for financial support.

REFERENCES

1. P. Haasen, *Festkörperprobleme*, **3**, p. 137, Vieweg, Braunschweig, 1964.
2. P. Haasen, *Discussion of the Faraday Society* No. 38, held at the University of Göttingen, Sept. 1964, p. 191.
3. W. G. Goworkow, *Sov. Phys. Cryst.*, **6**, 634 (1961/62).
4. B. Reppich, P. Haasen and B. Ilschner, *Acta Met.*, **12**, 1283 (1964).
5. E. Peissker, P. Haasen and H. Alexander, *Phil. Mag.*, **7**, 1279 (1962).
6. P. Haasen, *Z. Physik*, **167**, 461 (1962).
7. A. R. Chaudhuri, J. R. Patel and L. G. Rubin, *J. Appl. Phys.*, **33**, 2736 (1962).
8. T. Suzuki and H. Kojima, *Acta Met.*, **14**, 913 (1966).
9. J. R. Patel and A. R. Chaudhuri, *Phys. Rev.*, **143**, 601 (1966).
10. M. G. Mildwidskii, O. G. Stolarow and A. V. Berkova, *Sov. Phys. Sol. State*, **6** 2531 (1965).
11. M. G. Mildwidskii, W. B. Oswenski and O. G. Stolarow, *Neorgamtscheski Materials*, Vol. 1, No. 9, 1449 (1965).
12. E. Sirtl and A. Adler, *Z. Metallkunde*, **52**, 529 (1961).
13. H. Suzuki, *J. Phys. Soc. Japan*, **18**, Suppl. 1, 182 (1963).
14. R. Labusch, *Phys. Stat. Sol.*, **10**, 645 (1965).
15. S. Schäfer, *Thesis*, Göttingen, 1966, to be published in *Phys. Stat. Sol.*
16. W. Schröter, H. Alexander and P. Haasen, *Phys. Stat. Sol.*, **7**, 983 (1964).

X-ray Observation of Gold-induced Dislocation Loops in Silicon Crystals*

TAKASHI IIZUKA AND MAKOTO KIKUCHI

Electrotechnical Laboratory, Nagata-cho, Tokyo

ABSTRACT

The nature of dislocation loops (up to 600 μ in diameter) formed by the diffusion of gold into crucible-grown and dislocation-free silicon crystals has been studied in some detail by X-ray diffraction topography and etching techniques. It has been found that faulted loops with stacking faults are most frequently formed. On the basis of the diffusion mechanism of gold into silicon, the faulted loop is considered to be a Frank sessile interstitial loop. Generation of the loop is found to be closely related to the distribution of pre-existing oxygen atoms, the concentration of the doped impurity and the grown-in dislocation density in the sample. It is concluded that the loop source is associated with oxygen precipitate, and the loop is believed to be formed by a process analogous to the Bardeen-Herring mechanism.

I. INTRODUCTION

Gold has a large diffusion coefficient[1–4] and introduces deep levels[5–9] in silicon. Therefore, gold is used in the control of lifetime and resistivity in silicon. Gold-doped silicon is employed to make high-speed, high-voltage semiconductor devices. Recently, an electrical oscillation phenomenon has been found in gold-diffused silicon by Kikuchi.[10]

Because of such applications and the interesting characteristics of gold-doped silicon crystals, the diffusion of gold into silicon has been investigated by many authors.[1–4, 11–16] It has become apparent from these publications that gold diffuses into silicon by an interstitial-substitutional mechanism proposed for the diffusion of copper into germanium,[17] and the diffusion is sensitive to the perfection of silicon crystals.

It was observed by Dash[11] and Milevskii[18] that gold diffusion resulted

* A full account will be published in JJAP.

in the climb of dislocations. Dash[11] suggested that the climb could be explained by the interstitial-substitutional mechanism.

In addition to the climb of straight dislocations, dislocation loops on {111} planes have been found in gold-diffused silicon crystals.[11,19] Dash,[11,19] and Phillips and Dash[20] reported briefly the generation and the nature of the dislocation loop. They reported that such loops were produced only in oxygen-free crystals by the introduction of other impurities, probably copper and nickel, prior to the in-diffusion of gold. The loops decorated with copper were observed by the use of a microscope and infrared image tube,[11,19] and the undecorated loops were observed by transmission electron microscopy.[20] The detailed nature of the loop, however, has not been reported.

It is the purpose of this paper to report the nature of the gold-induced dislocation loops in some detail. The loops investigated here are formed in silicon crystals containing oxygen without any deliberately introduced impurities before the in-diffusion of gold.

X-ray observation of the bulk and the optical microscopic observation of the etched surface were performed. X-ray observation showed the details of the loop structure which were not clarified by copper decoration and etching techniques.

II. EXPERIMENTAL METHODS

The silicon crystals used in this study were pulled from quartz-crucibles with the growth axis of [111]. The specimens were sectioned along either the (111) plane perpendicular to the growth axis or the $(1\bar{1}0)$ plane including the growth direction. The doped impurity was phosphorus, arsenic, boron or gallium, and the resistivities ranged from 0.001 to 300 ohm-cm. The grown-in dislocation densities were between zero and $5\times10^3/\text{cm}^2$.

The samples were lapped and etched to the thickness of about 1.0 to 1.5 mm and rinsed with deionized water. Droplets of a water solution of gold chloride were applied to the specimen surface. The sample was then heat-treated at a temperature between 1,000°C and 1,200°C for a period of time ranging from a few minutes to 5 hours in argon atmosphere, and quenched to room temperature within a few seconds. The samples were then lapped and etched to reveal gold-induced defects on the specimen surface and to remove surface damages for X-ray observation.

For the X-ray observation, Lang's technique[21] was used. $AgK\alpha_1$ and $MoK\alpha_1$ radiations were used under the operating condition of 50 kV and

800 μA. An Ilford nuclear plate G5 (emulsion thickness 100 μ) was used as the photographic plate.

An etchant consisting of 5 parts HNO_3, 3 parts HF and 2 parts CH_3COOH was used for chemical polish. The Sirtl etchant[22] was used to reveal the defect on the specimen surface after the chemical polishing.

III. RESULTS

1. *Generation of faulted loops*

X-ray and optical observations showed that various kinds of loops are generated in gold-diffused silicon crystals. Among these loops, faulted loops are most frequently observed. In the following, the nature of the faulted loops is mainly described.

The observations showed the presence of a plane defect within the loop.

Figure 1 is an X-ray topograph of two dislocation loops lying on the inclined {111} plane and shows the characteristic fringe patterns inside the loop. From this contrast it is concluded that the defect must be either a stacking or twinning fault.

Straight etch grooves in Fig. 2 indicate lines of intersection of loops

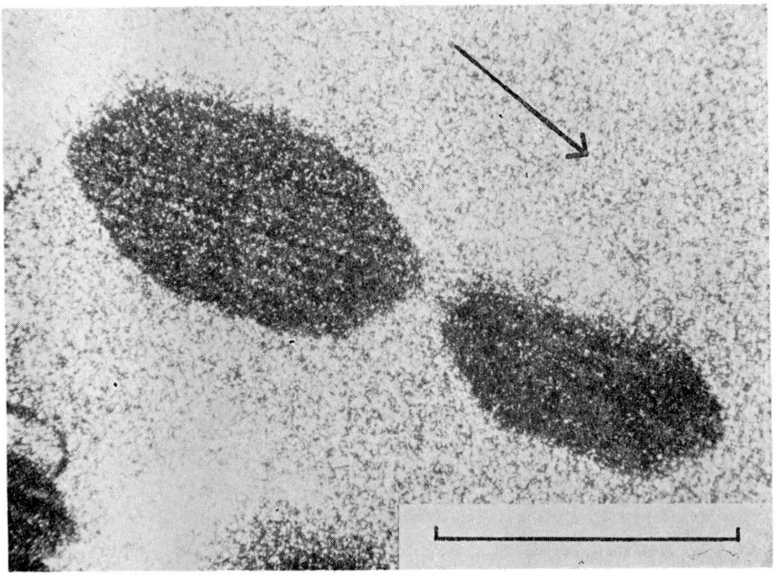

Fig. 1. X-ray topograph of hexagonal dislocation loops showing fringe pattern. $(1\bar{1}1)$ topograph, Mo$K\alpha_1$ radiation. Arrow shows the direction of reflection. Scale mark 300 μ.

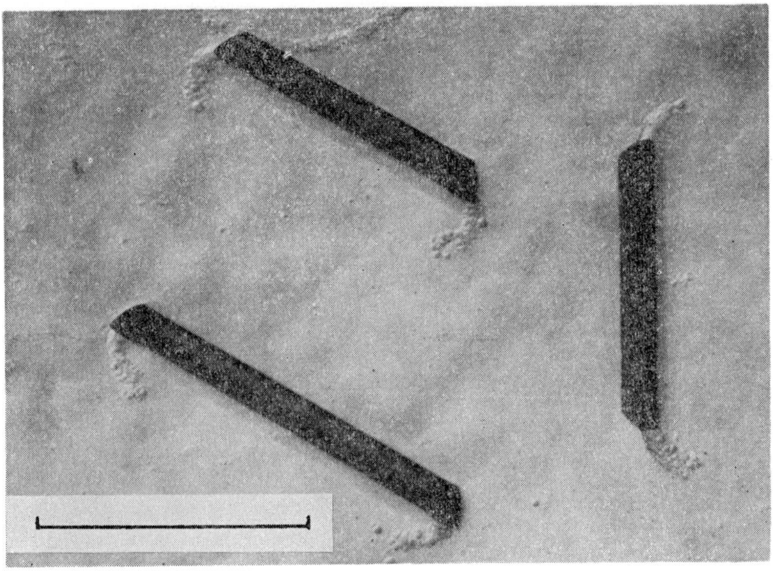

Fig. 2. Etch grooves revealed by the Sirtl etch on a (111) surface. Scale mark 200 μ.

lying on the inclined $\{111\}$ planes and the specimen surface. The grooves lie along $\langle 110 \rangle$ directions on the (111) surface. Small etch pits at the ends of the groove represent the outcrops of a surrounding dislocation. The appearance of such grooves also confirms the existence of a plane defect inside the loop.

On the other hand, Phillips and Dash[20] verified by electron microscope observations that the gold-induced dislocation loops contain stacking faults. The loop observed by them without copper decoration is probably the same in nature as the loop described here, except for the loop source and size as will be discussed later.

From these results the plane defect in this study is considered to be a stacking fault.

Figures 3(a) and 3(b) are X-ray topographs of the $(1\bar{1}1)$ and $(20\bar{2})$ reflections, respectively, showing the loops lying parallel and inclined to the surface. The disappearance of the stacking fault contrast and the appearance of double-arc images of the loop are clearly seen in the $(20\bar{2})$ topograph whose reflection plane is perpendicular to the fault plane.

The determination of the Burgers vector of the surrounding dislocation loop is difficult because of the precipitation of impurity, probably gold, and detailed studies will be reported elsewhere.

(a)

(b)

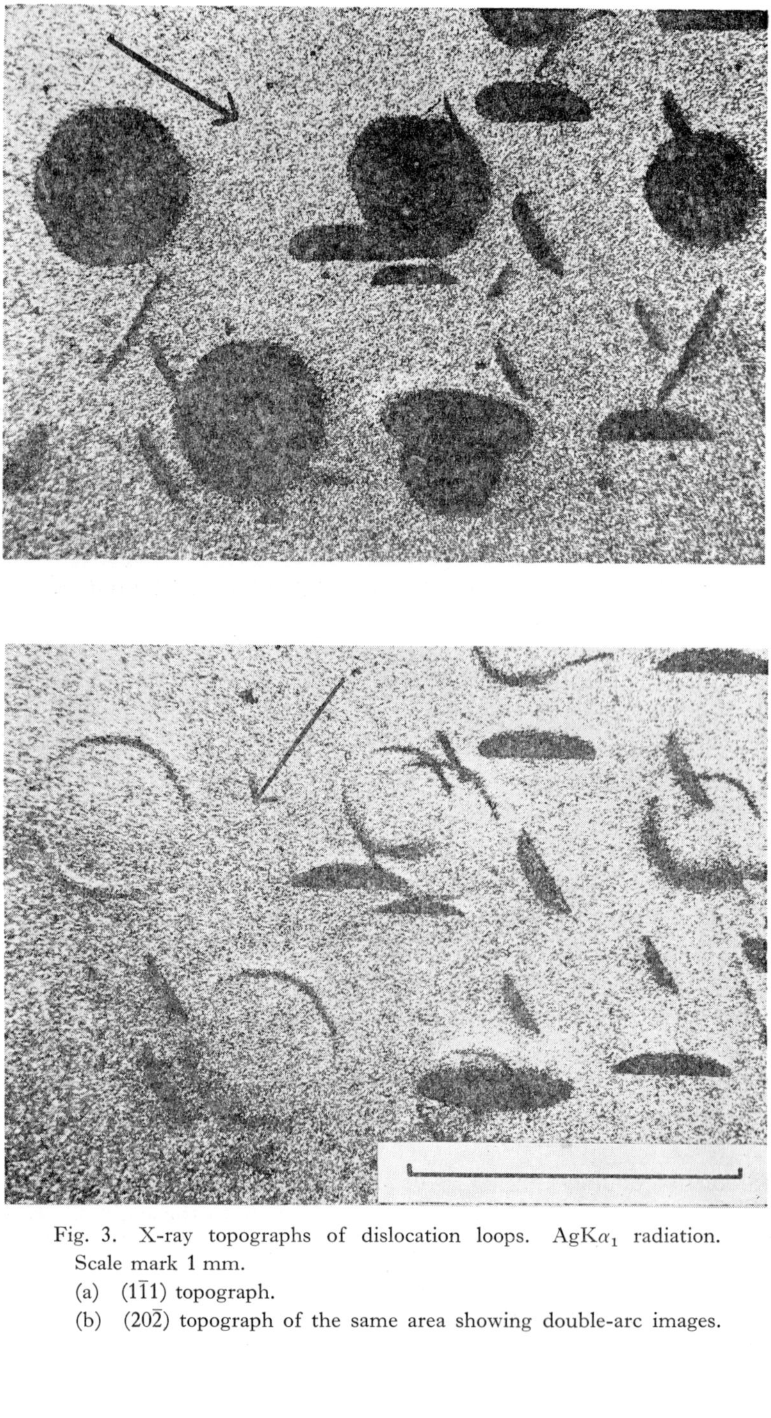

Fig. 3. X-ray topographs of dislocation loops. AgKα_1 radiation. Scale mark 1 mm.

(a) $(1\bar{1}1)$ topograph.

(b) $(20\bar{2})$ topograph of the same area showing double-arc images.

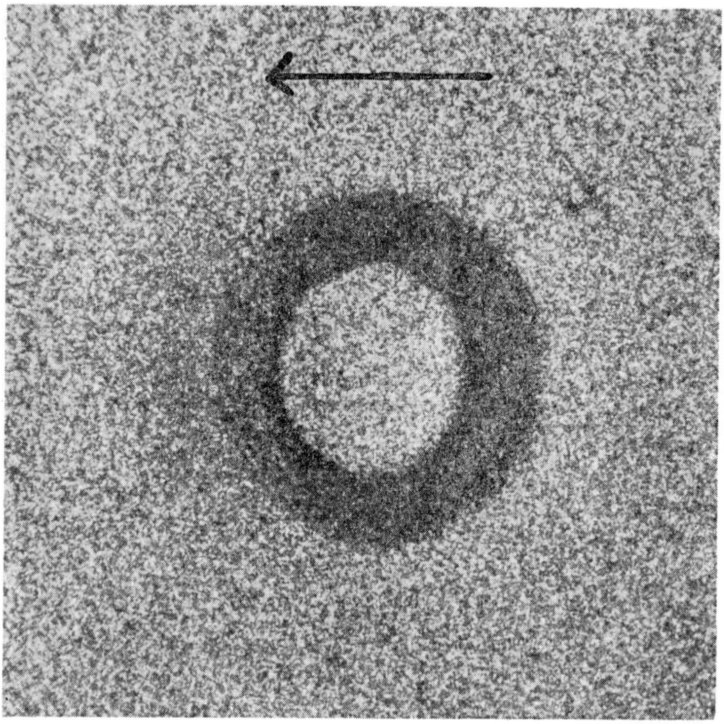

Fig. 4. X-ray topographs of concentric loops. Mo$k\alpha_1$ radiation.
(a) $(1\bar{1}1)$ topograph showing doughnut-shaped stacking fault contrast.

The loop is, however, presumed to be a Frank sessile dislocation loop with the extrinsic stacking fault[23] as will be described later.

In addition to the above-mentioned single faulted loops, concentric loops have occasionally been observed. They have inner loops within a loop. The inner loop is usually located at the center of a loop and there is no definite size relation between them, as shown in Figs. 4 and 5. Figure 4(a) is an X-ray $(1\bar{1}1)$ topograph, showing the doughnut-shaped stacking fault contrast. This contrast, however, disappears in Fig. 4(b), a $(20\bar{2})$ topograph of the same area. Similar loops revealed by the Sirtl etch on a (111) surface are given in Fig. 5. In the case of such a surface observation, no information about the existence of stacking fault is obtained. The arrangement of these concentric dislocation loops is similar to that expected from the operation of a source of the Bardeen-Herring type.[24]

By optical microscope observation, a small *dot* has been frequently observed just at the center of the concentric loops as shown in Fig. 5,

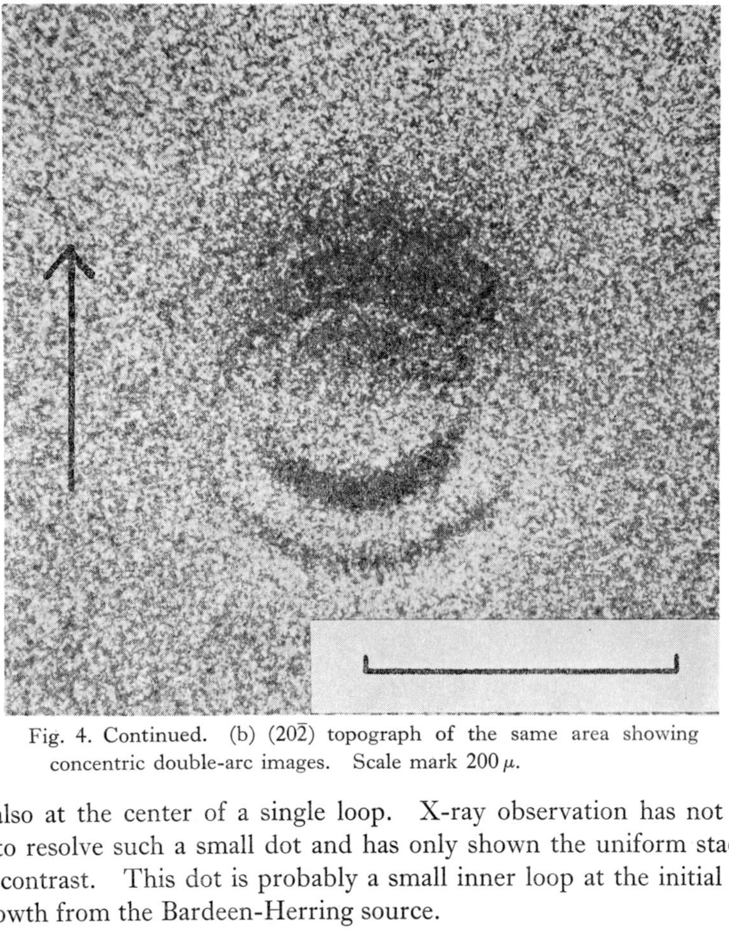

Fig. 4. Continued. (b) $(20\bar{2})$ topograph of the same area showing concentric double-arc images. Scale mark 200 μ.

and also at the center of a single loop. X-ray observation has not been able to resolve such a small dot and has only shown the uniform stacking fault contrast. This dot is probably a small inner loop at the initial stage of growth from the Bardeen-Herring source.

2. *Density and distribution of loops*

The density and distrubution of loops have been found to depend on nature of crystals. The average density of loops, in general, increases with the decrease in either the grown-in dislocation density or the doped impurity concentration. It is found that many loops are preferentially formed in specimens with the dislocation densities less than about $10^2/cm^2$ and the impurity concentration less than about 10^{17} atoms/cm^3. X-ray observation showed that the loop is rarely formed in impure crystals containing more than 10^{18} impurity atoms/cm^3. In specimens of relatively higher densities of the grown-in dislocation, i.e. more than $10^3/cm^2$, only a small number of loops have been observed. A large number of loops,

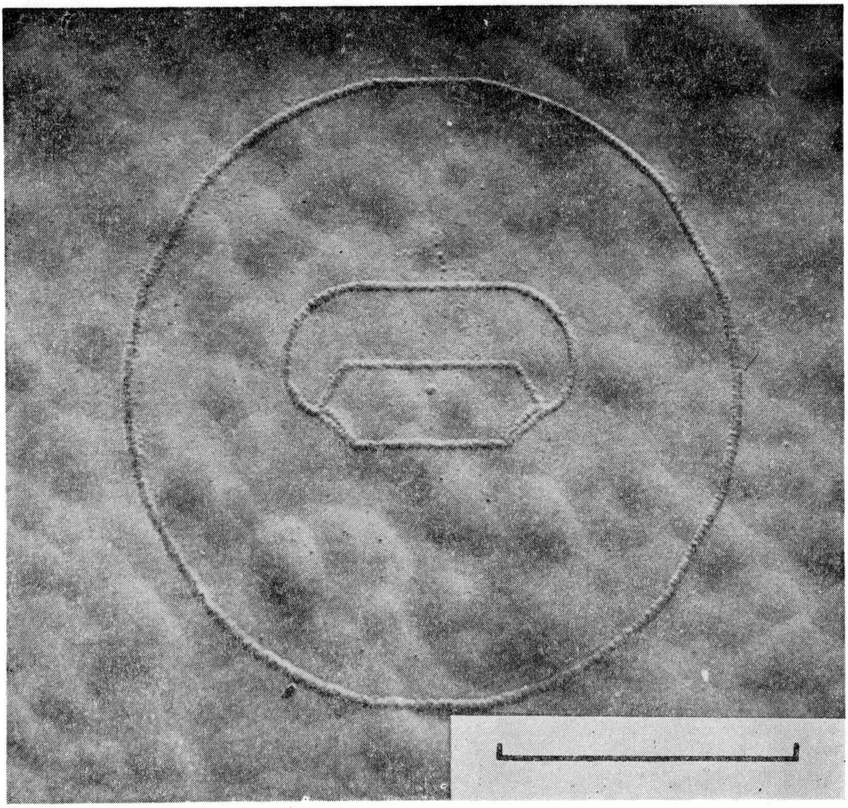

Fig. 5. Concentric loops revealed by the Sirtl etch on a (111) surface. A small dot at the center should be noticed. Scale mark 200 μ.

however, have been observed in relatively dislocation-free regions even in such specimens. An example of such distribution is shown in Fig. 6. Figure 6(a) is an X-ray topograph of a wafer cut from an as-grown ingot. It was confirmed from etch-pit counting and X-ray observation that the central region is nearly dislocation-free. Figure 6(b) shows the same region with the same reflection plane as Fig. 6(a) after the in-diffusion of gold at 1200°C for one hour. Many dislocation loops are observed in cluster at the central region. It is clear from these photographs that the loops are preferentially generated in a dislocation-free region.

The average density of loops varies from one sample to another under otherwise indentical experimental conditions. Furthermore, the density varies with the location in a sample. Generally, the density is larger (10^5–10^6 loops/cm^3) at the center region, and smaller (10 loops/cm^3) at the

(a)

(b)

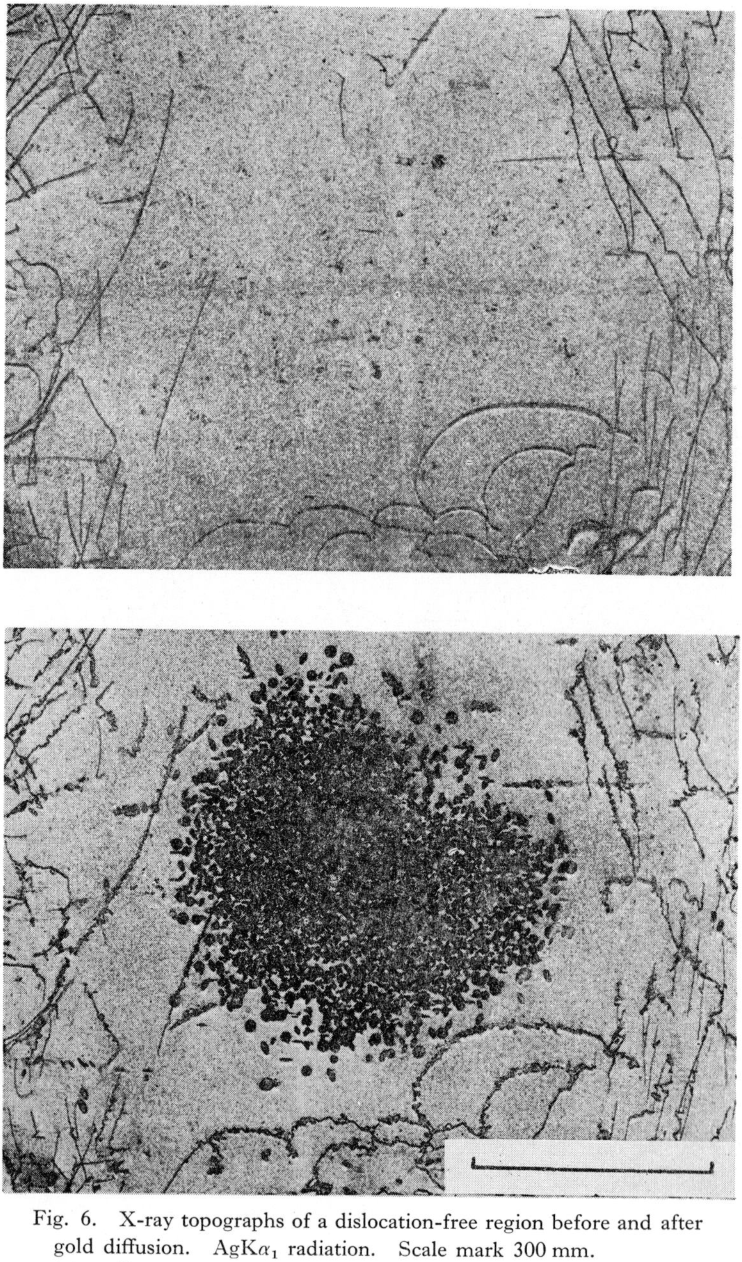

Fig. 6. X-ray topographs of a dislocation-free region before and after gold diffusion. AgKα_1 radiation. Scale mark 300 mm.

(a) (1$\bar{1}$1) topograph, before gold diffusion.

(b) (1$\bar{1}$1) topograph of the same area, after gold diffusion.

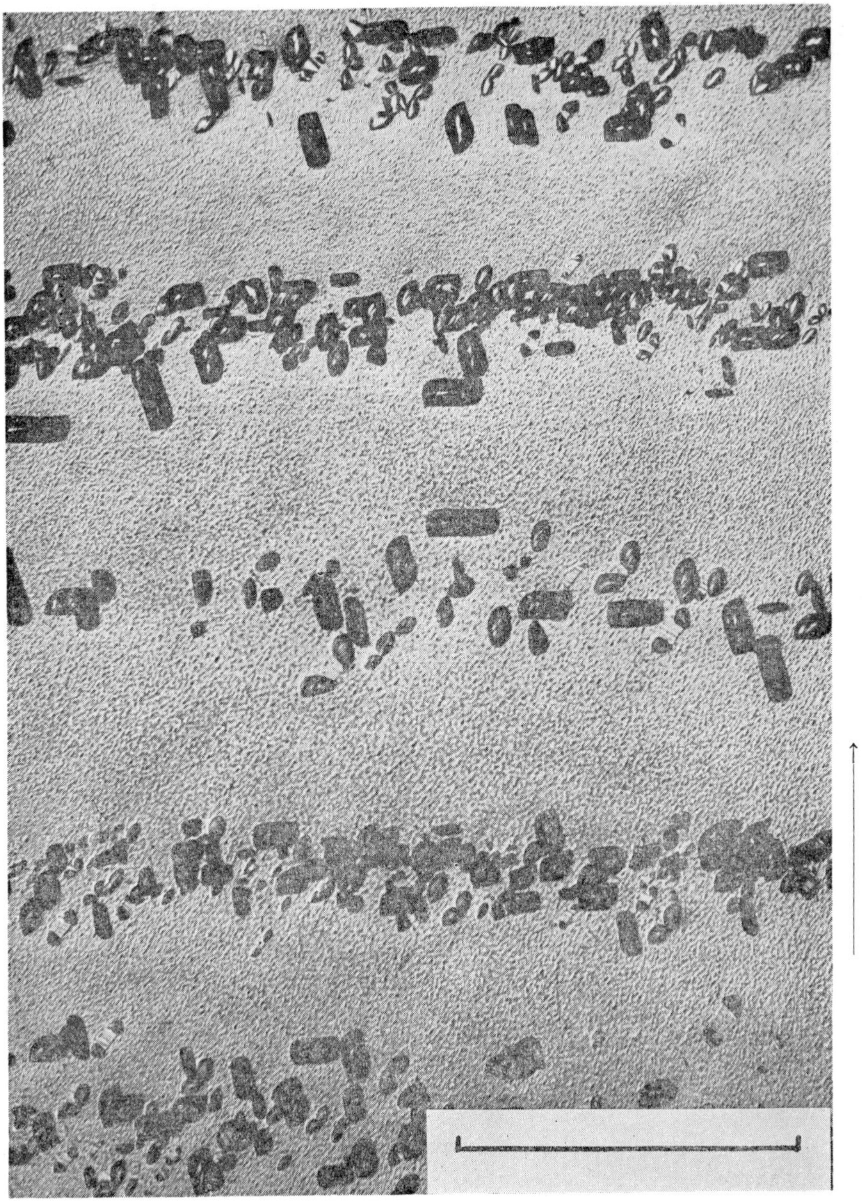

Fig. 7. Etch grooves striations revealed by the Sirtl etch. The surface shown is a $(1\bar{1}0)$ plane parallel to the growth axis. Arrow indicates pulled direction. Scale mark 300 μ.

periphery in slices cut perpendicular to the growth axis. In some cases, it has been observed that the loops are distributed as stripes in the $(1\bar{1}0)$ slices cut parallel to the growth axis (Fig. 7). The black and short rods are the same kind of etch grooves as shown in Fig. 2. The striations were also observed in the $(1\bar{1}0)$ slice before the in-diffusion of gold by X-ray topograph as black and white streaks which were similar to those reported by many authors[25–27] as oxygen striations.

These distributions may indicate that the loops are heterogeneously nucleated at pre-existing imperfections, such as oxygen precipitates, as will be described later.

3. *Size and shape of loops*

At a given temperature the diameter of the loop increases rapidly at first with diffusion time, and approaches a fixed value after about 30 minutes of heat-treatment. It also increases with the diffusion temperature during a given time of heat-treatment. The size as well as the density varies from one sample to another. There is also wide variation in size from area to area in a sample. In general, the diameter is smaller (of the order of 10 μ) at the region of higher loop density and larger (of the order of 100 μ) at the region of lower density. The largest loop observed in this study is about 600 μ in diameter.

Almost all the loops are circular or hexagonal in shape and lie on $\{111\}$ planes. The sides of hexagonal loops lie along $\langle 110 \rangle$ directions on a $\{111\}$ plane.

IV. DISCUSSION

1. *Source of loops*

It is well known that Czochralski-grown silicon crystals have inhomogeneous impurity distribution. It is occasionally observed that the impurity is distributed spirally in a slice cut perpendicular to the growth axis, or it is distributed as stripes in a plane containing the growth direction. In general, the associated impurity is the III- or V-group element, or oxygen atoms.

In some specimen with high purity, the loop is distributed inhomogeneously as mentioned in section III-2. Such an inhomogeneous distribution of the loops is considered to be associated with the above-mentioned pre-existing inhomogeneous impurity distribution in the crystal. This fact may indicate that the loop is heterogeneously nucleated at the precipitated impurity.

On the other hand, the loop is rarely observed in highly doped silicon crystals as mentioned in section III-2. Therefore, it is concluded that the nucleus of the loop is not associated with the III- or V-group doping elements, but another kind of impuritiy. Although there is no direct evidence, this impurity is considered to be oxygen from the well-known fact that oxygen is distributed as striations in pulled silicon crystals.

In a specimen containing grown-in dislocations, oxygen mainly precipitates along dislocation lines,[11, 18, 28] but in a dislocation-free sample or region, oxygen precipitates as an isolated spot.[19, 29] Such an isolated oxygen precipitate may act as a dislocation climb source and nucleate the dislocation loop in silicon crystals into which gold is diffused.

Therefore, in a sample with dislocation densities higher than $10^3/cm^2$, there are less isolated oxygen precipitates and only a few dislocation loops are generated as mentioned in section III-2. The effect of grown-in dislocation on the formation can be explained by considering the site of oxygen precipitation.

In Dash's experiment,[11] the introduction of some impurities, probably copper and nickel, into silicon during short pre-heating prior to the in-diffusion of gold is necessary for the loop formation. These facts may indicate that if some kind of impurities, i.e., copper, nickel or oxygen, precipitate as a cluster, they may act as a dislocation source and may generate dislocation loops by a process of climb as will be discussed in section IV-2.

The true nature of the source has not yet been understood, but the source may be an edge dislocation at the interface of the oxygen precipitate and the matrix, because a screw dislocation may climb into a helix.[11, 19, 30]

2. *Growth mechanism of loops*

It has become apparent that gold diffuses into silicon almost entirely as an interstitial and combines with vacancy to form a substitutional. Dash[11] deduced from his experiment that this conversion causes the depletion of vacancies, which is the driving force for their further generation at the sources such as dislocations. The experiment was performed as follows:

Left-handed screw dislocations were introduced into previously dislocation-free crystals by indentation and deformation. Upon in-diffusion of gold it was found that the screw dislocations were transformed into right-handed helices. Geometrical considerations of such a transformation indicate that gold causes a vacancy deficiency in the neighborhood of the screw dislocations and the subsequent generation of vacancies at the dislocation core as the climb processes.

On the other hand, Milevskii[18] observed the gold-induced climb of edge dislocations by an optical polarization method. The observations showed that edge dislocations increased in the size of the extra-half plane in gold-diffused silicon crystals. This result also indicates that gold causes a vacancy depletion, and then the edge dislocations operate as vacancy sources to maintain the equilibrium concentration of vacancies as well as Dash's experiment in the screw dislocations.

Therefore it is generally established that if there are screw or edge straight dislocations in gold-diffused silicon crystals, these will operate as vacancy sources and result in climb.

However, in dislocation-free samples, such a large source does not exist and another kind of microscopic defect as mentioned in section IV-1 will operate as the vacancy source. Therefore, in the case of dislocation-free samples containing oxygen, isolated oxygen precipitates can nucleate the loops and their subsequent growth will be controlled by a similar mechanism to that controlling the climb of straight edge dislocations.[18] Upon in-diffusion of gold, it is considered that the climb source, probably an edge dislocation at the interface of the oxygen precipitate and the matrix, creates vacancies and absorbs interstitial silicon atoms with the subsequent growth of an additional plane of atoms as a negative climb process.

By such a mechanism, a Frank sessile loop with the extrinsic stacking fault[23] can be generated in gold-diffused silicon crystals. Therefore, the loop formation mechanism is considered to be analogous to the Bardeen-Herring type.[24] An example of the concentric loops in Fig. 5 is considered to be an advanced stage of the Bardeen-Herring source whose shape is rather distorted.

Acknowledgements

The authors wish to express their thanks to Prof. K. Kohra for his valuable discussions and suggestions.

REFERENCES

1. J. D. Struthers, *J. Appl. Phys.*, **27**, 1560 (1956); **28**, 516 (1957).
2. B. I. Boltaks, G. S. Kulikov and R. S. Malkovich, *Soviet Physics-Solid State*, **2**, 2134 (1960/61).
3. W. R. Wilcox and T. J. LaChapelle, *J. Appl. Phys.*, **35**, 240 (1964).
4. D. R. Collins, D. K. Schroder and C. T. Sah, *Appl. Phys. Letters*, **8**, 323 (1966).
5. G. Bemski, *Phys. Rev.*, **111**, 1515 (1958).
6. R. O. Carlson, *J. Appl. Phys.*, **29**, 1001 (1958).

7. C. B. Collins, R. O. Carlson and C. J. Gallagher, *Phys. Rev.*, **105**, 1168 (1957).
8. W. D. Davis, *Phys. Rev.*, **114**, 1006 (1959).
9. A. E. Bakanowski and J. H. Forster, *Bell Syst. Tech. J.*, **39**, 87 (1960).
10. M. Kikuchi, to be published in *J. Appl. Phys.* (Sept, 1966).
11. W. C. Dash, *J. Appl. Phys.*, **31**, 2275 (1960).
12. W. R. Wilcox, T. J. LaChapelle and D. H. Forbes, *J. Electrochem. Soc.*, **111**, 1377 (1964).
13. G. J. Sprokel and J. M. Fairchild, *J. Electrochem. Soc.*, **112**, 200 (1965).
14. G. J. Sprokel, *J. Electrochem. Soc.*, **112**, 807 (1965),
15. J. Martin, E. Haas and K. Raithel, *Solid-State Electronics*, **9**, 83 (1966).
16. W. H. Bullis, *Solid-State Electronics*, **9**, 143 (1966).
17. F. C. Frank and D. Turnbull, *Phys. Rev.*, **104**, 617 (1956).
18. L. S. Milevskii, *Soviet Physics-Solid State*, **4**, 1792 (1963).
19. W. C. Dash, *J. Appl. Phys.*, **30**, 459 (1959); *Growth and Perfection of Crystals*, p. 361, John Wiley & Sons, New York, 1958.
20. V. A. Phillips and W. C. Dash, *J. Appl. Phys.*, **33**, 568 (1962).
21. A. R. Lang, *Acta cryst.*, **12**, 249 (1959); *J. Appl. Phys.*, **30**, 1748 (1959).
22. E. Sirtl and A. Adler, *Z. für Metallk.*, **52**, 529 (1961).
23. F. C. Frank, *Phil. Mag.*, **42**, 809 (1951).
24. J. Bardeen and C. Herring, *Imperfections in Nearly Perfect Crystals*, p. 261, John Wiley & Sons, New York, 1952.
25. Z. Ishii, T. Furuoya, Y. Sasaki and K. Kohra, *J. Phys. Soc. Japan*, **15**, 267 (1960).
26. G. H. Schwuttke, *Direct Observation of Imperfections in Crystals*, p. 497, Interscience, New York, 1962; *J. Electrochem. Soc.*, **109**, 27 (1962); *J. Appl. Phys.*, **33**, 2760 (1962).
27. K. Kohra, S. Nakano and M. Yoshimatsu, *J. Phys. Soc. Japan*, **18**, Suppl. II, 341 (1963).
28. W. C. Dash, *J. Appl. Phys.*, **29**, 705 (1958).
29. G. H. Schwuttke, *J. Electrochem. Soc.*, **108**, 163 (1961).
30. W. C. Dash, *Phys. Rev. Letters*, **1**, 400 (1958).